Chemical Oceanography

VOLUME 7

2ND EDITION

Chemical Oceanography

Edited by

J. P. RILEY

and

R. CHESTER

Department of Oceanography,
The University of Liverpool, England

VOLUME 7
2ND EDITION

1978

ACADEMIC PRESS

LONDON   NEW YORK   SAN FRANCISCO

A Subsidiary of Harcourt Brace Jovanovich, Publishers

ACADEMIC PRESS INC. (LONDON) LTD.
24/28 Oval Road,
London NW1

United States Edition published by
ACADEMIC PRESS INC.
111 Fifth Avenue
New York, New York 10003

Library of Congress Catalog Card Number: 74–5679
ISBN: 0–12–588607–1

Printed in Great Britain by

PAGE BROS (NORWICH) LTD
NORWICH

Contributors to Volume 7

S. R. ASTON, *Department of Environmental Sciences, University of Lancaster, Lancaster LA1 4YQ, England*

WALTER W. BERG, JR., *National Center for Atmospheric Research, Boulder, Colorado 80307, U.S.A.*

J. KIRK COCHRAN, *Department of Geology and Geophysics, Yale University, New Haven, Connecticut 06520, U.S.A.*

G. ROSS HEATH, *Graduate School of Oceanography, University of Rhode Island, Kingston, Rhode Island 02881, U.S.A.*

E. J. W. JONES, *Department of Geology, University College London, Gower Street, London WC1E 6BT, England*

L. D. MEE, *Centro des Cienas del Mar y Limnología Universidad Nacional Autónoma de México, México 20, D.F.*

TED C. MOORE, JR., *Graduate School of Oecanography, University of Rhode Island, Kingston, Rhode Island 02881, U.S.A.*

WILLIAM M. SACKETT, *Department of Oceanography, Texas A and M University, College Station, Texas 77843, U.S.A.*

BERND R. T. SIMONEIT, *Institute of Geophysics and Planetary Physics, University of California, Los Angeles, California 90024, U.S.A.*

KARL K. TUREKIAN, *Department of Geology and Geophysics, Yale University, New Haven, Connecticut 06520, U.S.A.*

JOHN W. WINCHESTER, *Department of Oceanography, Florida State University, Tallahassee, Florida 32306, U.S.A.*

Preface to the Second Edition

Rapid progress has occurred in all branches of Chemical Oceanography since the publication of the first edition of this book a decade ago, and in no field have the advances been more dramatic than in that of marine geochemistry. Topics which have received increasing attention include the metalliferous sediments found at some active centres of sea-floor spreading, the chemistry of interstitial waters, the formation of deep-sea carbonates and the chemistry and mineralogy of both atmospheric and sea-water particulates. Many of the most important developments in the study of deep-sea sediments themselves have been linked to the Deep-Sea Drilling Project. This was initiated in 1968, and for the first time the entire length of the marine sedimentary column was sampled. The extent of these many advances has made it necessary to devote three volumes, i.e. the fifth, sixth and seventh, to various topics in marine geochemistry.

The present volume is devoted to a number of fields of geochemical research, some of which were little more than embryonic a decade ago. These fields include those of marine atmospheric chemistry, sea-water particulates and sedimentary organic geochemical cycles, all of which are covered within the present volume. Other topics discussed are sea-floor spreading, sampling of deep-sea sediments, isotopic dating and the chemistries of lagoonal and estuarine environments.

This series is not intended to be a practical handbook and if such details are required the original references given in the text should be consulted. In passing, it should be mentioned that, although those practical aspects of sea-water chemistry which are of interest to biologists are reasonably adequately covered in the "Manual of Sea Water Analysis" by Strickland and Parsons, there is an urgent need for a more general laboratory manual.

The editors are most grateful to the various authors for their helpful co-operation which has greatly facilitated the preparation of this book. They would particularly like to thank Miss Helen Rowson, Dr Joanna Sharples and Mr M. Preston for their willing assistance with the arduous task of proof reading; without their aid many errors would have escaped detection. They would also like to acknowledge the courtesy of the various copyright holders, both authors and publishers, for permission to use tables, figures and photographs. In conclusion they wish to thank Academic Press, and in particular Mr T. Lincoln, for their efficiency and ready co-operation which has much lightened the task of preparing this book for publication.

Liverpool   J. P. RILEY
July, 1978   R. CHESTER

CONTENTS

Chapter 35 by E. J. W. JONES

Sea-floor Spreading
and the
Evolution of the Ocean Basins

Chapter 36 by TED C. MOORE, JR. and G. ROSS HEATH

Sea-floor Sampling Techniques

Chapter 37 *by* WILLIAM M. SACKETT

Suspended Matter in Sea-water

Chapter 38 *by* WALTER W. BERG, JR. and JOHN W. WINCHESTER

Aerosol Chemistry of the Marine Atmosphere

Chapter 39 *by* BERND R. T. SIMONEIT

The Organic Chemistry of Marine Sediments

Chapter 40 *by* KARL K. TUREKIAN and J. KIRK COCHRAN

Determination of Marine Chronologies Using Natural Radionuclides

Chapter 41 *by* S. R. ASTON

Estuarine Chemistry

Chapter 42 *by* L. D. MEE

Coastal Lagoons

Contents of Volume 1

Contents of Volume 2

Contents of Volume 3

Contents of Volume 4

Contents of Volume 5

Contents of Volume 6

Symbols and units used in the text

Concentration. There are several systems in common use for expressing concentration. The more important of these are the molarity scale (g molecules 1^{-1} of solution = mol 1^{-1}) usually designated by c_i, the molality scale (g molecules kg^{-1} of solvent* = mol kg^{-1}) designated by m_i and the mole fraction scale usually denoted by x_i, which is of more fundamental significance in physical chemistry. In each instance, the subscript i indicates the solute species; when i is an ion the charge is not included in the subscript unless confusion is likely to arise. Some other means of indicating the concentration are also to be found in the text, these include: g or mg kg^{-1} of solution (for major components), μg or ng 1^{-1} or kg^{-1} of solution (for trace elements and nutrients) and μg-at 1^{-1} of solution (for nutrients).

UNITS

Where practicable, SI units (and the associated notations) have been adopted in the text except where their usage goes contrary to established oceanographic practice.

LENGTH

Å	= Ångstrom	= 10^{-10} m
nm	= nanometre	= 10^{-9} m
μm	= micrometre	= 10^{-6} m
mm	= millimetre	= 10^{-3} m
cm	= centimetre	= 10^{-2} m
m	= metre	
km	= kilometre	= 10^3 m
mi	= nautical mile (6080 ft)	= 1·85 km

WEIGHT

pg	= picogram	= 10^{-12} g
ng	= nanogram	= 10^{-9} g
μg	= microgram	= 10^{-6} g
mg	= milligram	= 10^{-3} g
g	= gram	
kg	= kilogram	= 10^3 g
ton	= metric ton	= 10^6 g

*A common practice is to regard sea-water as the solvent for minor elements.

VOLUME

μl	= microlitre	$= 10^{-6}$ l
ml	= millilitre	$= 10^{-3}$ l
l	= litre	
dm^3	= litre	

CONCENTRATION

ppm	= parts per million ($\mu g\, g^{-1}$ or mg l^{-1})
ppb	= parts per billion (ng g^{-1} or μg l^{-1})
μg-at l^{-1}	= μg atoms l^{-1} = (μg/atomic weight) l^{-1}

TIME

s	= second	h	= hour
ms	= millisecond	d	= day
min	= minute		

ENERGY AND FORCE

J	= Joule	= 0·2390 cal
N	= Newton	$= 10^5$ dynes
W	= Watt	

RADIOACTIVITY

dpm	= disintegrations per minute
pCi	= picocurie

GENERAL SYMBOLS

K	= equilibrium constant
M_x	= molarity of component x
T	= temperature in K
t	= temperature in °C

CHAPTER 35

Sea-floor Spreading
and the
Evolution of the Ocean Basins

E. J. W. JONES

Department of Geology, University College London, London, England

SECTION 35.1. INTRODUCTION 1

35.2. SEA-FLOOR SPREADING 5

35.3. MAGNETIC ANOMALIES 7
    35.3.1. The Vine–Matthews hypothesis 7
    35.3.2. Dating the oceanic crust 8

35.4. SEA-FLOOR SPREADING AND GLOBAL TECTONICS............................... 17
    35.4.1. Lithospheric plates and their interactions............................. 17
    35.4.2. Plate motions in the major ocean basins 19

35.5. THE CONSTITUTION OF THE OCEANIC BASEMENT 28
    35.5.1. Evidence from seismic refraction studies 28
    35.5.2. Results of sampling 30
    35.5.3. The magnetization of the oceanic crust.............................. 35

35.6. IMPLICATIONS OF RECENT GEOPHYSICAL OBSERVATIONS...................... 37
    35.6.1. Bathymetry and heat flow 37
    35.6.2. Seismicity 42
    35.6.3. Gravity 46
    35.6.4. Magnetic anomalies and the zone of crustal accretion 49
    35.6.5. Asymmetrical spreading 55

35.7. DEEP-SEA SEDIMENTS AND PLATE MOTIONS: PALAEOCEANOGRAPHY 57

35.8. CAUSES OF SEA-FLOOR SPREADING... 64

35.9. CONCLUDING REMARKS 66

ACKNOWLEDGEMENTS 68

REFERENCES 68

35.1. INTRODUCTION

Our reconstruction of the history of the ocean basins is largely based on developments which have taken place in the Earth Sciences over the past two decades. Before the early 1950s, the question of the origin of the oceans

1

fell almost entirely within the realm of speculation because little was known about the constitution of the Earth's crust beneath the deep-sea floor. While it is true that sea-going expeditions mounted during the nineteenth and the early parts of this century collected many bottom samples and soundings of great value to geologists, the observations did not, however, provide the vital stratigraphical and structural information needed to determine how the ocean basins were formed. Even forty years after the publication of Alfred Wegener's compelling, but controversial, arguments in support of continental drift during the last 200 million years (Wegener, 1912) the un-certainties surrounding the age of the oceans were not resolved.

Since 1950, the floors of the deep oceans have been intensively studied, mainly for academic reasons, although military and economic motives are by no means unimportant. The limitations of the traditional oceanographic methods, practised since the days of the H.M.S. *Challenger* expedition, have stimulated the development of new ways of investigating the form of the seabed and the nature of the crust below it. It has been the application of these new exploration techniques which has provided definitive evidence of the age and origin of the ocean basins.

A major instrumental advance was made in the early 1950s with the construction of the precision depth recorder (PDR). Continuous bathy-metric profiles from the PDR have made possible detailed mapping of submarine topography lying beyond the limits of the continental shelves. Many features, such as the Mid-Atlantic Ridge and the Pacific trenches, were already well known when the PDR came into routine use, but these had been delineated only in broad outline, thus making assessments of their geological significance uncertain. Over the past few years, many hundreds of bathy-metric maps based on PDR soundings have been published to illustrate the principal morphological features of the ocean basins and their margins—the continental shelf, slope and rise, the trenches, the abyssal plains and the large oceanic ridges, with their often rugged topography and transverse fracture zones.

The major oceanic ridges, so vividly portrayed on the maps by Heezen and Tharp (1967, 1968, 1969, 1971), are the dominant topographical elements of the ocean basins. They form an almost world-encircling system, extending across the Pacific into the Indian, Atlantic and Arctic Oceans (Fig. 35.1). For much of its length the ridge crest is defined by a narrow belt of earth-quakes. Indeed, Heezen and Ewing (1961) used the occurrence of earth-quakes to infer continuity of the ridge system in areas with little or no sounding data. Those sections of ridge existing in the Atlantic and Indian Oceans occupy median positions when related to the continental margins. In each ocean, the deep basins lie roughly 2 km below the level of the ridge crests, which are approximately 3 km deep. In the Atlantic, Indian and

Arctic Oceans, the basins are generally occupied by exceedingly flat abyssal plains (bottom gradients less than 1/1000) and gently sloping sections of the continental rises (bottom gradients about 1/300). Abyssal plains are rare in the Pacific but, in marked contrast to other oceans, there are many marginal trenches.

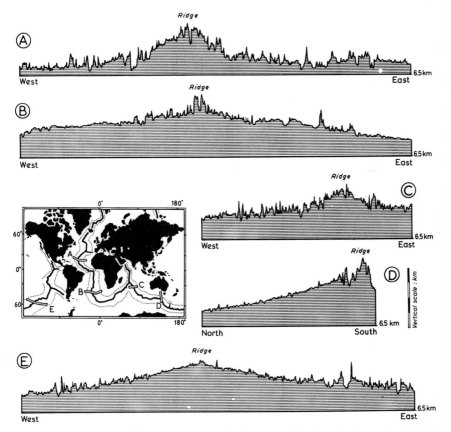

FIG. 35.1. Bathymetric profiles across the oceanic ridge system. The base-line for each section is 6·5 km. The vertical scale represents a depth of 4 km. (After Heezen, 1962; profiles reproduced by kind permission of the author.)

It was during the early 1950s that the first determinations of the thickness and structure of the oceanic crust were made by seismic refraction shooting. The apparently uncomplicated sections derived from the travel times of refracted waves proved to be of fundamental significance because they demonstrated that the Earth's crust under the oceans is quite different from the continental crust (Ewing and Press, 1955; Hill, 1957). Beneath the continents, the base of the crust (or the Moho), where compressional wave velocities of about $8 \cdot 0 \, \text{km s}^{-1}$ are first encountered, lies at an average depth of about 35 km. In the deep basins of both the Atlantic and Pacific, however, the Moho was found to lie at an average depth of only 12 km below sea level. It is interesting to note that long before the seismic work was carried out, Hecker (1903, 1908), Duffield (1924) and Vening Meinesz (1948) had shown, from barometric and pendulum gravity measurements, that the ocean basins are in approximate isostatic equilibrium. The refraction data indicate that the mass deficiency caused by the water layer is largely compensated by a shallow Moho, a structure which is also compatible with seismic surface wave observations.

The refraction shooting also clearly revealed that the seismic layering beneath the floor of the Atlantic and Pacific (Table 35.1) is quite different

TABLE 35.1

Structure of the oceanic crust (after Hill, 1957)

Layer	Average thickness (km)	Average P-wave velocity (km s^{-1})
Sea-water	4·5	1·5
1	0·45	2·0
2	1·75	4·0–6·0
3	4·7	6·71
——————————— MOHO ———————————		
4	—	8·09

from that of the continental crust. The crust in continental regions is variable in structure, but in many areas the top 10–20 km has been shown to consist of rocks with granitic seismic velocities ($\sim 6 \cdot 2 \, \text{km s}^{-1}$), with intermediate or basic crystalline basement below (velocity $\sim 7 \cdot 2 \, \text{km s}^{-1}$). The thinner oceanic crust appears to be made up of three layers. *Layer 1* is characterized by a low seismic velocity ($\sim 2 \, \text{km s}^{-1}$), which leaves little doubt that it is composed of unconsolidated or semi-consolidated sediments. Velocities in *Layer 2* fall over a fairly broad range (4–6 km s^{-1}), which can be attributed to basaltic lavas, certain types of metamorphic rocks and, also, to sedimentary rocks such as limestone. *Layer 3* ($6 \cdot 7 \, \text{km s}^{-1}$) was believed by Hill (1957) and others to be made up of gabbroic rocks.

There were some who interpreted these structural contrasts between continents and oceans in terms of continental drift. The thinner oceanic crust had been created as a result of rifting processes during the slow drifting apart of the continental blocks (Heezen, 1962). The "anti-drifters", notably Sir Harold Jeffreys, argued that the differences had arisen as the outer shell of the earth solidified in the early part of the Precambrian (Jeffreys, 1970). Continental drift could not have taken place at a later time because of the great strength of the lithosphere.

Many were swayed towards the former explanation. Powerful geological arguments, particularly those based on palaeoclimatic data, were reinforced by the quantitative evidence for large horizontal movements of the continents which had emerged from the palaeomagnetic work of the 1950s. Systematic variations had been found between the magnetization directions of lavas and sediments of the same age from different continents, a feature which could be interpreted in terms of continental drift during Palaeozoic and later time (Runcorn, 1962). This view was strengthened by the fact that palaeogeographic reconstructions based on palaeomagnetism are in broad agreement with geological field evidence. The existence of Permo-Carboniferous glacial deposits on portions of the continents now located near the Equator, and the presence of Triassic evaporites at present high latitudes, for example, can be convincingly explained by moving the continental blocks to positions indicated by the palaeomagnetic results (Opdyke, 1962). Although there were strong reasons for believing that the ocean basins are relatively young features, the processes involved in the creation of the oceanic crust were obscure.

35.2. SEA-FLOOR SPREADING

In the early 1960s there was a growing body of opinion that continental drift was caused by convective motions in the mantle. The concept of a convecting mantle was not new. Holmes, Vening Meinesz and others had long considered deep convection to be responsible for major surface deformation (Vening Meinesz, 1952), and it seems to be able to provide the forces necessary to move the continents. Some authors have argued that the present configuration of the continental masses was reached in several stages. Runcorn (1962) noted that there have been four major thermal events on the continents during the past 4000 million years which, he suggested, represent times when continental blocks broke up and new ocean basins were formed. He related each period of fragmentation to a change in the pattern of mantle-wide convection cells, brought about by a slow growth of the Earth's core.

The ephemeral nature of the ocean basins was also emphasized by H. H.

Hess in a paper in 1962. Written to provide "a useful framework for testing various and sundry groups of hypotheses relating to the oceans" his contribution has had an immense influence on geological thinking. He put forward the view that the sea floor itself takes part in convective motions (Fig. 35.2). According to Hess (1962), the crests of major oceanic ridges represent the traces of the rising limbs of convection cells, whereas trenches and island

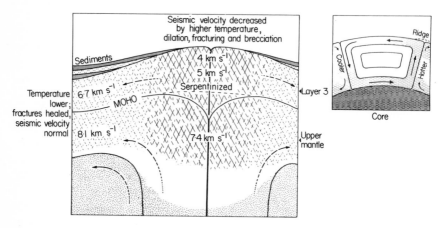

FIG. 35.2. The evolution of the ocean floor according to Hess (1962). (Diagram reproduced with the permission of the Geological Society of America.)

arcs are situated over descending portions. Oceanic crust is formed beneath the ridge crests by a process of hydration of olivine-rich mantle material (Fig. 35.2). As new crust is produced so the older ocean floor moves laterally from its site of generation at rates of the order of 1 cm year^{-1}. The sea floor thus increases in age with distance from the ridge crests. Dietz (1962) coined the term sea-floor spreading for this process.

Hess interpreted the 6·7 km s^{-1} seismic layer of the oceans as being hydrated mantle formed above the 500°C isotherm. The anomalously low seismic velocities beneath the ridge crests, reported by Ewing and Ewing (1959) were attributed by him to fracturing and high thermal gradients. The newly generated crust only assumes the structure typical of the deep ocean basins after cooling and annealing have taken place on the ridge flanks. Hess pointed out that such a widening mechanism for the oceans avoids the need for considering that the continents have to plough through oceanic crust as they drift; they simply ride passively on a convecting mantle.

35.3. MAGNETIC ANOMALIES

35.3.1. THE VINE–MATTHEWS HYPOTHESIS

Hess (1962) suggested an important test of his convective model; "the crest of the ridge should only have recent sediments on it and Recent and Tertiary sediments on its flanks; the whole Atlantic Ocean and possibly all the ocean should have little sediment older than Mesozoic". At the time, relatively little was known about the stratigraphy of oceanic sediments. However, Wilson (1963a) noted that the age of oceanic islands tends to increase with distance from the crests of the seismically active ridges. Such evidence suggests that Hess' concept is essentially correct, although it could be argued that islands are not at all typical of the ocean floor around them.

Strong support, now considered proof, of sea-floor spreading came in 1963 from a totally unexpected source—the magnetic measurements made from surface ships. In the 1950s, a pattern of remarkable linear magnetic anomalies was discovered in the NE. Pacific (Mason, 1958; Vacquier *et al.*, 1961; Mason and Raff, 1961; see Fig. 35.3). The explanations which were

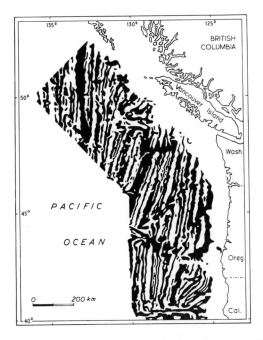

FIG. 35.3. Pattern of magnetic anomalies in the NE. Pacific. Positive anomalies are indicated in black. (After Raff and Mason, 1961; reprinted with the permission of the authors and the Geological Society of America.)

then put forward for its origin were not entirely satisfactory. Mason and Raff (1961) suggested that the anomalies resulted from compositional variations within the basement, but it was difficult to visualize how contrasts could arise in such a striking linear fashion. Vine and Matthews (1963) proposed a radically different explanation, which is illustrated in Fig. 35.4.

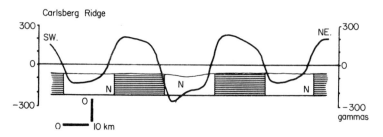

FIG. 35.4. Magnetic anomalies computed for the Carlsberg Ridge. N represents a normally magnetized block; reversely magnetized blocks are shaded. The normally magnetized material under the median valley is assumed to have an effective susceptibility of 0·0053 emu. Other blocks have susceptibilities of 0·0027 emu. Total field value = 37 600 gammas; magnetic inclination = −6°; magnetic declination = 44°. (After Vine and Matthews, 1963.)

In their model, the oceanic crust possesses a high remanent magnetization and is formed by sea-floor spreading during periods when intermittent reversals of the Earth's magnetic field occur. New oceanic crust is magnetized in the contemporary direction of the Earth's field and, because of high remanence, retains its magnetization during its slow movement away from the ridge crest. Depending on the polarity of the palaeofield, the magnetization is either normal (same direction as present-day field) or reversed (opposite sense). Thus, strips of ocean floor having alternate magnetic polarity are produced as sea-floor spreading takes place. Assuming reasonable values for the magnetization of the crust, Vine and Matthews calculated the form of the magnetic anomalies which such a model would predict over the Carlsberg Ridge and clearly showed that there was a close agreement with the anomalies which are actually observed.

35.3.2. DATING THE OCEANIC CRUST

When the Vine–Matthews hypothesis first appeared in the literature little was known about the magnetization of the oceanic crust and the time-scale of magnetic reversals. It is, therefore, not surprising that many workers failed to make immediate use of the idea (see e.g. Heirtzler and Le Pichon, 1965). However, the situation soon changed. Firstly, it became clear, from examination of dredge samples, that the assumption of remanent magnetization dominating the induced component in the oceanic crust is correct, at least at shallow levels in the basement (Ade-Hall, 1964; Vogt and Ostenso, 1966). Secondly, Cox *et al.* (1964) were able to establish a time-scale of magnetic reversals back to about four million years B.P. from studies of lava sequences on land.

The time-scale was a vital piece of new information. It enabled the widths of the magnetized blocks in the Vine–Matthews model (Fig. 35.4), and hence the form of the anomaly pattern, to be computed for any given rate of spreading (Vine and Wilson, 1965). It was Vine (1966) and Pitman and Heirtzler (1966) who demonstrated that the observed anomalies over the crest of the ridge system could be matched remarkably closely using a spreading model and the reversal time-scale. The comparison for the Juan de Fuca Ridge in the NE. Pacific is illustrated in Fig. 35.5. To obtain optimum fits it is necessary to vary the spreading rate along the length of the ridge system. Computed velocities range from about 1 cm year^{-1} per limb (Reykjanes Ridge) to 4·6 cm year^{-1} per limb (East Pacific Rise) (Table 35.2), values which are of the same order as the rates of continental separation deduced from terrestrial palaeomagnetism (Runcorn, 1962). The validity of the spreading model was also greatly strengthened by the symmetry of the magnetic anomalies about the ridge crest. The symmetry is particularly well displayed on the aeromagnetic map of part of the Reykjanes Ridge (Fig. 35.6c), which was published in the same year that the anomaly simulations were carried out (Heirtzler *et al.*, 1966). Moreover, the pattern of reversals in time that was used to simulate a magnetic profile across the Pacific–Antarctic Ridge proved also to be capable of reproducing the profile across the Reykjanes Ridge (Pitman and Heirtzler, 1966). By the end of 1966, it was difficult to resist the attractions of the sea-floor spreading hypothesis.

Surveys completed over the past ten years have revealed that linear magnetic anomalies characterize enormous areas of the ocean basins and there now seems little doubt that they do, indeed, reflect the pattern of sea-floor spreading. Oceanic crust also appears to be formed by spreading in the marginal basins behind island arcs. Linear magnetic anomalies have been found in the Scotia Sea (Barker, 1972), west of the Tonga Trench (Sclater *et al.*, 1972) and in the Philippine Basin (Ben-Avraham *et al.*, 1972), supporting earlier suggestions that these areas develop by crustal extension (Karig, 1971). A recent compilation of magnetic trends in the oceans by Pitman *et al.* (1974a) is reproduced in Fig. 35.7. Examples of detailed surveys from the Atlantic, Indian and Arctic Oceans are illustrated in Fig. 35.6. Figure 35.6a shows a well-defined set of lineations in the Eurasian Basin of the Arctic. In this region the anomalies are associated with spreading on the Gakkel Ridge, an extension of the Mid-Atlantic Ridge passing close to the North Pole. Linear anomalies, displaced by a series of NE.–SW. faults related to

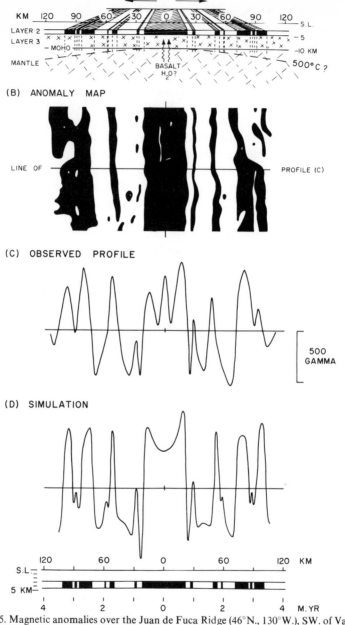

FIG. 35.5. Magnetic anomalies over the Juan de Fuca Ridge (46°N., 130°W.), SW. of Vancouver Island. (A) Crustal model. Shaded material in *Layer 2* is normally magnetized; unshaded is reversely magnetized. (B) Part of magnetic map over Juan de Fuca Ridge. (After Raff and Mason, 1961.) Areas of positive anomalies, black; white areas are regions of negative anomalies. (C) Magnetic profile along line indicated in (B). (D) Computed profile using reversal time-scale and spreading model. (From F. J. Vine, "Magnetic Anomalies Associated with Mid-Ocean Ridges" *in "The History of the Earth's Crust"* (Robert A. Phinney, ed.) (copyright © 1968 by Princeton University Press) Figure 1, p. 75. Reprinted by permission of the author and Princeton University Press.)

TABLE 35.2

Rates of sea-floor spreading during the past four million years (after Vine, 1968)

Location (see also Fig. 35.7)	Spreading rate (cm year^{-1} per limb.)
Reykjanes Ridge	1·0
Mid-Atlantic Ridge (32°N.)	1·5
Mid-Atlantic Ridge (31°S.)	2·3
Carlsberg Ridge (NW. Indian Ocean, 5°N.)	1·5
East Pacific Rise (51°S.)	4·6
Juan de Fuca Ridge (NE. Pacific, 46°N.)	3·0

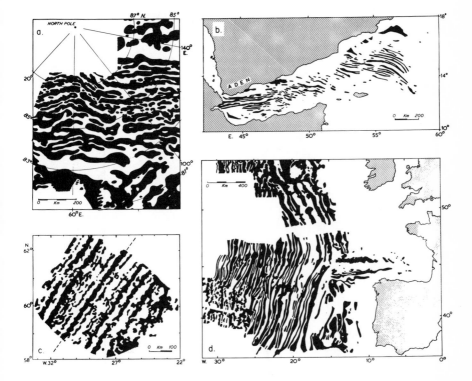

FIG. 35.6. Magnetic surveys (positive anomalies indicated in black). (a) Gakkel Ridge in the Eurasian Basin of the Arctic. (After Karasik, 1968.) (b) Gulf of Aden. (After Whitmarsh, *in* Laughton *et al.*, 1970.) (c) Crestal province of Reykjanes Ridge south of Iceland. (After Heirtzler *et al.*, 1966.) The position of the ridge axis is indicated by the dashed line. (d) Bay of Biscay and environs. (After Williams, 1975.)

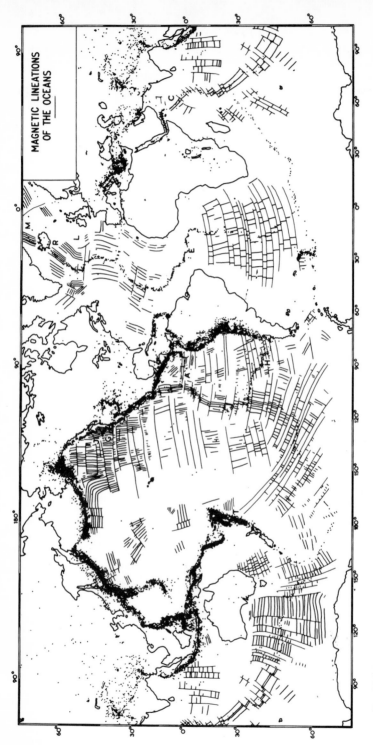

FIG. 35.7. Trends of linear magnetic anomalies in the ocean basins (after Pitman et al., 1974a), with recent additions from Hayes and Rabinowitz (1975), Larson (1975) and Williams (1975). Earthquake epicentres are plotted after Barazangi and Dorman (1969). Features referred to in the text are labelled as follows: M, Jan Mayen Ridge; R, Reykjanes Ridge; L, Rockall Plateau; E, Romanche fracture zone; J, Juan de Fuca Ridge; G, Gorda Ridge; A, Galapagos Ridge; H, Chile Rise; T, Tonga Trench; C, Carlsberg Ridge. (Reproduced with the permission of the authors and the Geological Society of America.)

the opening of the Gulf of Aden, are illustrated in Fig. 35.6b. Figure 35.6c illustrates the striking symmetry of the linear anomalies about the crest of the Reykjanes Ridge, just south of Iceland. Figure 35.6d shows the pattern of anomalies in the region of the Bay of Biscay. In the western half of the map the anomalies trending NNW. and roughly N.–S. were formed by sea-floor spreading on the Mid-Atlantic Ridge. Within the Bay of Biscay the lineations run approximately E.–W., reflecting a spreading axis associated with the separation of the Iberian block from France. The easternmost anomalies of the Atlantic set can be seen to join the anomalies running closest to the continental margins of France and Iberia, indicating that sea-floor spreading was taking place simultaneously on the Mid-Atlantic Ridge and in the Bay of Biscay about a "triple point" near 14°W. Triple points have been found elsewhere in the Atlantic, Pacific and Indian Oceans. Their geometrical properties have been discussed by McKenzie and Morgan (1969) and Le Pichon et al. (1973).

The work of Vine (1966) and Pitman and Heirtzler (1966) clearly demonstrates that it is possible to determine crustal ages from these widespread linear anomalies if the time-scale of geomagnetic reversals is known. A detailed time-scale from terrestrial measurements is, however, only available for the past 4·5 million years (Cox, 1969). Observations of reversals in long deep-sea piston cores, taken from areas where deposition appears to be continuous, make a tentative extension back to about nine million years B.P. possible (Foster and Opdyke, 1970). In order to date crust which is older than uppermost Tertiary, it has been necessary to derive the reversal sequence from the magnetic anomalies observed at sea (Vine, 1968; Heirtzler et al., 1968). Heirtzler et al. chose the anomalies recorded across the Mid-Atlantic Ridge in the South Atlantic to compile a "standard" profile. In this region the anomalies can be correlated over large distances and are symmetrical about the ridge crest. By assuming that continuous spreading has taken place at 1·9 cm year^{-1} per limb (the spreading rate obtained for the past four million years at the ridge crest) they obtained the sequence of magnetic reversals for the past 80 million years. On applying this time-scale to simulate anomalies on other sections of the ridge system, excellent agreements between observed and computed profiles were found. Support for the assumption of a constant spreading rate was later provided by drilling data from the Ridge at about 30°S. The ages of the sediments resting on the basement surface (generally assumed to give a close approximation to the crustal age) increase uniformly with distance from the ridge axis at a rate remarkably close to that predicted from a 1·9 cm year^{-1} spreading velocity (Fig. 35.8).

An extension of the reversal time-scale from 80 million years. B.P. (Upper Cretaceous) to 160 million years. B.P. (Upper Jurassic) has been made by Larson and Pitman (1972) (Fig. 35.9) who pieced together three sets of lineations in the Pacific west of Hawaii (Fig. 35.7), and derived a time-scale which they then used to reproduce the so-called Keathley sequence in the western Atlantic. Such a successful application to other groups of anomalies suggests that the time-scale is essentially correct although, at present, it is only calibrated at a few points by drilling. Modifications will undoubtedly be made as more basal sediments are recovered from areas of Mesozoic crust (Larson and Hilde, 1975).

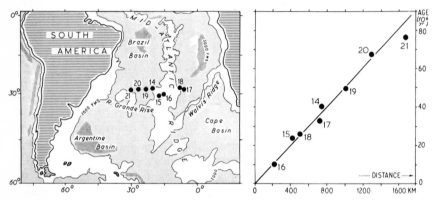

FIG. 35.8. Age of oldest sediment at DSDP Sites 14–21, drilled by the D.V. *Glomar Challenger* in the South Atlantic, plotted against distance from the axis of the Mid-Atlantic Ridge. The sediment column was not fully penetrated at Site 21. (After Maxwell *et al.*, 1970. Reproduced by permission. Copyright © 1970 by the American Association for the Advancement of Science.)

The use of the time-scale in Fig. 35.9 has enabled basement ages to be inferred in most regions where magnetic information is available. Figure 35.10 summarizes the patterns of crustal ages. The ocean floor appears to be no older than Mesozoic, as Hess (1962) believed. Tertiary crust occupies about 50% of the ocean basins—a region created in less than 2% of geological time (Vine, 1970). Ambiguities in crustal dating occur, however, where frequent jumps of the ridge axis have taken place or where magnetic anomalies are absent. Dating the latter, magnetically quiet, regions (Fig. 35.11) poses a major problem because there is some question over whether they result from spreading during long intervals of constant polarity (Heirtzler and Hayes, 1967), or from processes that demagnetize the crust (Poehls *et al.*, 1973). The quiet zones adjacent to the continental margins in the Atlantic and Indian Oceans are especially difficult to date since they are flanked by identifiable anomalies on one side only. Furthermore, in many places it is not clear where the earliest crust formed by spreading is situated.

Consequently, the initial rifting of the continents has not been related in any detailed way to the start of sea-floor spreading in the major oceans. Most workers adopt the view that quiet zones result from spreading during long periods of constant polarity. Two principal periods can be seen in Fig. 35.9, one in the Middle Cretaceous and the other in the Jurassic. Earlier ones are known, such as the 45 million year Kiaman interval in the late Palaeozoic (Irving, 1966). Pitman and Talwani (1972) have suggested that the marginal, quiet zones in the North Atlantic were formed during the period of normal polarity (the Graham interval) in the Jurassic. The ages of other magnetically quiet zones have been discussed by Poehls et al. (1973).

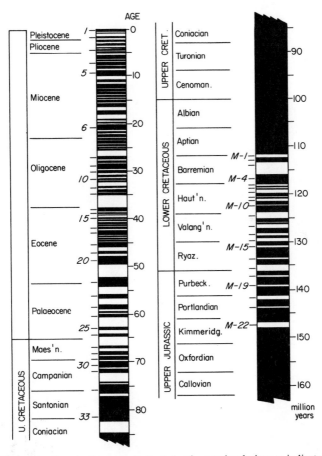

FIG. 35.9. Time-scale of magnetic reversals. Periods of normal polarity are indicated in black. (Diagram reproduced from Larson and Pitman (1972), with the permission of the authors and the Geological Society of America.) Larson and Hilde (1975) have recently revised the late Jurassic and early Cretaceous section of the time-scale.

FIG. 35.10. Crustal ages in the ocean basins determined from magnetic anomalies. (After Pitman *et al.*, 1974b; reproduced with the permission of the authors and the Geological Society of America). A small area of Jurassic crust south of the Java Trench, described by Larson (1975), is also included.

AGE OF THE OCEAN BASINS

million years

Pliocene to Holocene
5
Miocene-Oligocene
38
Eocene-Palaeocene
65
Cretaceous
135
Jurassic
190

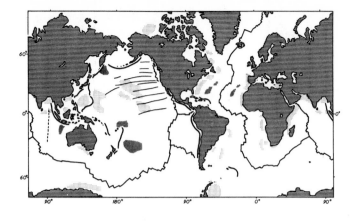

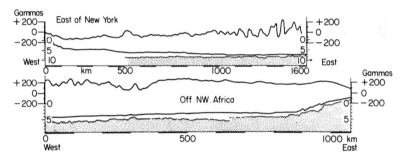

FIG. 35.11. Magnetically quiet zones in the oceans. (After Poehls *et al.*, 1973.) Sea floor apparently formed during the long period of normal polarity in the Middle Cretaceous (Fig. 35.9) has dark shading. Other quiet zones have light shading. (Reproduced by permission of the authors. Copyright © American Geophysical Union.)

35.4. SEA-FLOOR SPREADING AND GLOBAL TECTONICS

35.4.1. LITHOSPHERIC PLATES AND THEIR INTERACTIONS

The pattern of magnetic anomalies indicates that the present production of oceanic crust at ridge crests is $\sim 3 \text{ km}^2 \text{ year}^{-1}$ (Chase, 1972). Since the rate of expansion of the Earth required to accommodate the new material is unacceptably large, oceanic crust must be resorbed in the mantle. Recent seismological studies clearly demonstrate that the sites of crustal consumption are marked by the deep-sea trenches and their associated island arcs and mountain systems, as Hess and Maxwell (1953) suggested over twenty years ago. Beneath these areas lies a seismically active zone, which dips into the mantle at an angle of about 45° from the trench axis, reaching a maximum depth of 700 km (Fig. 35.12). Investigations of focal mechanisms and seismic propagation (Section 35.6.2) show that each seismic zone identifies a 70–100 km thick slab of relatively cold oceanic crust and upper mantle material (the lithosphere) sinking into softer mantle material (the asthenosphere). A simple model of crustal evolution, involving oceanic ridges as the sources and the trenches as the sinks of crustal material, can be applied to the Pacific without much modification since trenches occur along a large part

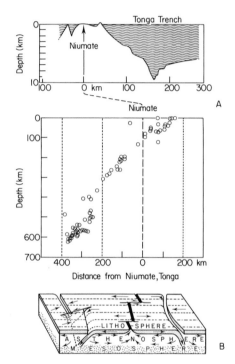

FIG. 35.12. (A) Depth distribution of earthquakes recorded during 1965 over a 300 km segment of the Tonga arc. Earthquakes as deep as 700 km have been recorded in this region. (B) Schematic representation of the interaction of rigid plates of lithosphere. Arrows on the lithosphere indicate relative motions. Arrows in the asthenosphere represent possible compensating flow in response to the sinking of the lithosphere. (After Isacks *et al.*, 1968. Reproduced by permission of the authors. Copyright © American Geophysical Union.)

of its margin. In the Atlantic, however, active trenches are absent except along the edges of the Caribbean and Scotia areas, raising the question of the ultimate fate of crust created along the axis of the Mid-Atlantic Ridge.

An explanation, now widely accepted, is provided by the plate theory of tectonics, which considers the Earth's surface to be composed of a number of rigid lithospheric plates in relative motion, sea-floor spreading being part of a global pattern of plate interaction. The seismological basis for plate tectonics is well established. The map of world seismicity (Fig. 35.7) shows that earthquakes are not randomly distributed, but generally occur in well-defined zones. One important seismic belt is associated with the spreading ridges and another with the trench–island arc systems of the Pacific. A zone stretching through the Mediterranean, Middle East and Himalayas is also very active seismically, but it is one in which the epicentres are not so spatially confined.

The regions between the earthquake belts are seismically quiet to such an extent that they can be thought of as rigid plates of lithosphere. Earthquakes can then be considered to be produced by interactions at their margins (McKenzie and Parker, 1967; Morgan, 1968; Isacks et al. 1968). Three kinds of plate boundary are recognized.

(i) Conservative. Plates slide past one another without creation or destruction of crust.

(ii) Constructive. Boundaries of this type are marked by the crests of actively spreading ridges. As two plates move apart at a ridge crest fresh material is added to their margins.

(iii) Destructive. Consumption of lithosphere occurs by one plate moving beneath another. Trenches with island arcs or young continental mountain belts are present.

These types of boundaries are illustrated schematically in Fig. 35.12.

Six major plates can be identified (Fig. 35.13) at present (the Eurasian, American, Indian, Pacific, African and Antarctic) together with a number of minor ones, some of which are named in Fig. 35.13. The relative movement of any two plates over the asthenosphere can be described by a rigid rotation about an axis through the centre of the Earth, the simple geometry being a consequence of Euler's theorem applied to motion on a sphere (Bullard et al., 1965). Figure 35.13 gives the direction of present movements derived from magnetic and seismological data (Le Pichon, 1968). It should be noted that many plate boundaries do not fall at continental margins. With the exception of the Pacific, the main plates in Fig. 35.13 possess both oceanic and continental portions. Because of their buoyancy, the latter regions do not undergo subduction beneath the island arcs (McKenzie et al., 1974).

35.4.2. PLATE MOTIONS IN THE MAJOR OCEAN BASINS

The relative motion between plates at an accretionary plate boundary can be determined from the magnetic anomaly pattern. If subduction does not occur, then the anomalies give the time-varying geometry of an ocean basin if allowances are made for vertical motions of the continental crust and deposition of sediments at continental margins. One example, taken from Pitman and Talwani (1972), is presented in Fig. 35.14 to show how the shape of the

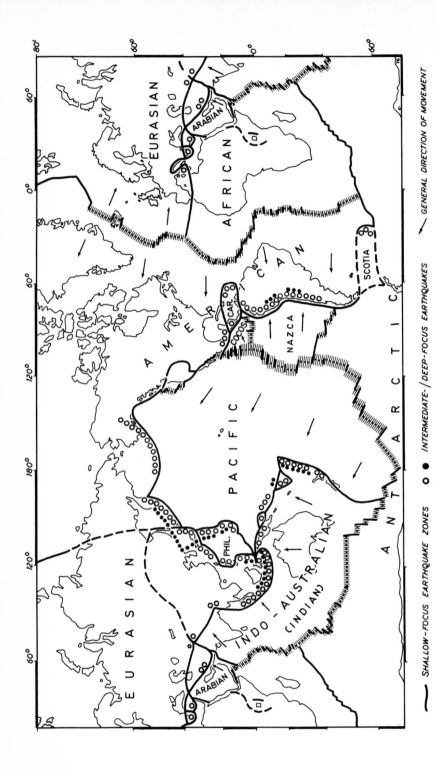

Fig. 35.13. Extent of lithospheric plates. Many minor plates, such as those in the Mediterranean, are not named. The broken lines indicate zones of diffuse seismic activity. (After Vine and Hess, 1970. Reprinted by permission.)

— SHALLOW-FOCUS EARTHQUAKE ZONES   o INTERMEDIATE- / DEEP-FOCUS EARTHQUAKES   → GENERAL DIRECTION OF MOVEMENT

ⅢⅢⅢⅢⅢ ACTIVELY SPREADING RIDGE CRESTS

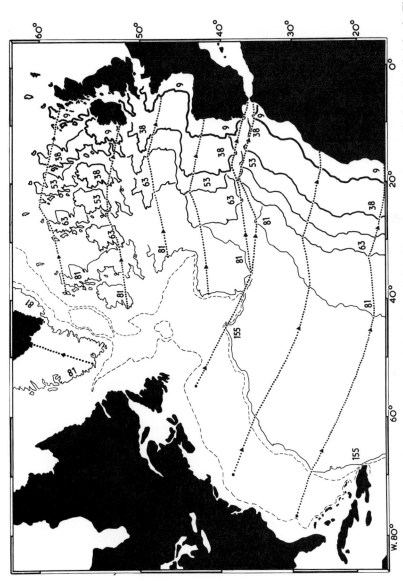

Fig. 35.14. The drift of Europe and Africa with respect to North America, as deduced from the magnetic anomalies in the North Atlantic. Times of reconstructions are given in millions of years. The arrows show the drift paths of specific parts of the coastline. Note the complex movements in the Azores–Gibraltar region, a feature consistent with the observed geological structure (Purdy, 1975). (The diagram is reproduced from Pitman and Talwani (1972) with the permission of the authors and the Geological Society of America.)

North Atlantic has changed during the past 155 million years. The relative positions of Europe, North Africa and North America have been computed by fitting together magnetic anomalies of the same age from opposite sides of the Mid-Atlantic Ridge. Similar kinds of analysis have been carried out in other areas of the oceans and a brief summary of the results is given below. Details of the plate motions and their relationship to the geology of the continental margins can be found in the references quoted.

35.4.2.1. North Atlantic

The relative positions of the continents prior to drift in the Atlantic have been most convincingly determined by Bullard *et al.* (1965). Some of the subsequent locations are illustrated in Fig. 35.15, which is taken from Phillips and Forsyth (1972). Magnetic anomalies clearly show that the North Atlantic did not open as one unit. The earliest stage of sea-floor spreading took place south of the Azores (Pitman and Talwani, 1972). Unfortunately, the onset of spreading cannot be dated precisely because of the absence of magnetic anomalies close to the continental margins of eastern North America and NW. Africa (Section 35.3.2). The quiet zone appears to identify crust formed during the Graham period of normal polarity, which ended 153 million years ago (Fig. 35.9; see Pitman and Talwani, 1972; Hayes and Rabinowitz, 1975). Sea-floor spreading must have certainly started by the end of the Middle Jurassic as sediments of Oxfordian to possible Callovian age have been found resting on oceanic basement at DSDP Site 100, drilled by D.V. *Glomar Challenger* east of the Bahamas (Hollister *et al.*, 1972).

As the proto-North Atlantic widened during the Upper Jurassic and Lower Cretaceous, a narrow wedge of ocean floor extending northwards as far as the continental margin of Ireland began to develop by the rotation of Europe from North America (Fig. 35.15). During the Cretaceous the Bay of Biscay was created by the rotation of Iberia away from France, a process involving the formation of a triple junction between 80 and 73 million years B.P. (Williams, 1975). The Labrador Sea and Baffin Bay began to open in the Upper Cretaceous (82 million years B.P.) and spreading ceased in the Eocene (47 million years B.P.) (Laughton, 1971). The Rockall Trough appears to be floored by oceanic crust (Jones *et al.*, 1970; Scrutton, 1972) but its age is uncertain since it is magnetically quiet. Formation of this trough during the long Cretaceous period of normal polarity is favoured by several workers (Roberts, 1975), although an early Mesozoic origin is possible (Jones *et al.*, 1970).

The basin between Greenland and the continental block of the Rockall Plateau opened during the Lower Tertiary, the oldest magnetic anomaly running close to their opposing margins being of Palaeocene (60 million year) age (Vogt *et al.*, 1969; Laughton, 1971). From 60 million years B.P. until

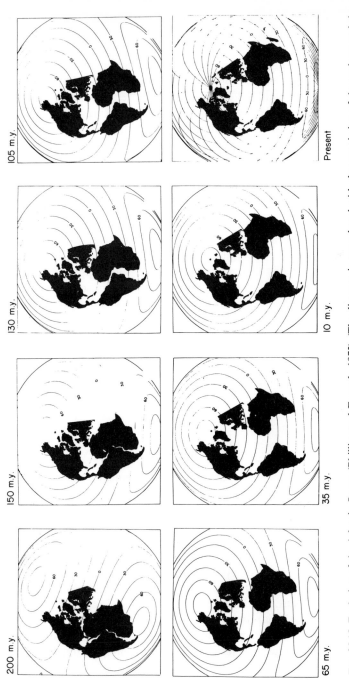

105 m.y.

130 m.y.

150 m.y.

200 m.y.

Present

10 m.y.

35 m.y.

65 m.y.

Fig. 35.15. Evolution of the Atlantic Ocean (Phillips and Forsyth, 1972). (The diagram is reproduced with the permission of the authors and the Geological Society of America.)

spreading ceased in the Labrador Sea (47 million years B.P.), a triple junction existed from the join of the Mid-Atlantic Ridge and the Reykjanes and Labrador axes (Laughton, 1971). The Norwegian Sea also began to open in the Palaeocene. In this region, however, the spreading centre jumped from a position mid-way between Norway and the Jan Mayen Ridge to one between the latter and Greenland, where spreading is presently taking place. (Fig. 35.7; see also Avery et al., 1968). The Jan Mayen Ridge, like the Rockall Plateau, is believed to be a continental fragment. The broad, transverse ridge running between Greenland and the Faeroes appears to have been formed during the Tertiary by excessive igneous activity associated with crustal accretion over the Iceland "hot-spot" (Bott et al., 1971).

35.4.2.2. South Atlantic

Dating of magnetic anomalies in the Cape Basin off SW. Africa indicates that oceanic crust first began to form between Africa and South America during the Lower Cretaceous (Valanginian, approximately 127 million years B.P.) (Larson and Ladd, 1973), the earliest period of sea-floor spreading coinciding with an important Valanginian marine transgression in southern Argentina and South Africa (Reyment and Tait, 1972). At this time the spreading rate was $1 \cdot 6 \, \text{cm year}^{-1}$ per limb, but later in the Cretaceous it increased to $2 \, \text{cm year}^{-1}$ per limb. It should be noted from the reconstructions in Fig. 35.15 that until late Cretaceous–early Tertiary times the connection between the North and the South Atlantic was narrow. This arose from the strike-slip character of the motion between the southern side of West Africa and northeast South America.

35.4.2.3. Arctic Ocean

The Arctic Ocean appears to have developed in two main stages. The important structural boundary during the opening process was the Lomonosov Ridge, a feature probably composed of continental rocks (Wilson, 1963b) which runs directly over the Pole from Greenland to the Siberian Shelf. The Ameriasian Basin, lying between the Lomonosov Ridge and North America, is magnetically quiet and seems to have been formed, like the oldest part of the North Atlantic, during the Jurassic period of normal polarity (Herron et al., 1974). The Eurasian Basin, situated between the Barents Shelf and the Lomonosov Ridge, began to open in the Lower Tertiary (63 million years B.P.) at the same time as spreading was initiated in the Greenland–Rockall and Norwegian Basins. The linear anomalies on the Gakkel Ridge (Karasik, 1968; see also Fig. 35.6a) are related to this stage of opening, which has continued to the present time. The Alpha-Mendeleyev Ridge (Vogt and Ostenso, 1970), previously believed to be a spreading centre, is now interpreted as being a subduction zone which accommodated the opening of the

North Atlantic about a pole in Northern Greenland 81–63 million years ago (Herron *et al.*, 1974).

35.4.2.4. *Indian Ocean*

The evolution of the Indian Ocean is complex as it involves the motions of seven large plates. At present, analysis of their movements from the magnetic anomalies (McKenzie and Sclater, 1971) can only begin at 75 million years B.P. (late Cretaceous), since older anomalies have not yet been dated. It is evident from the 75 million years B.P. reconstruction in Fig. 35.16, however,

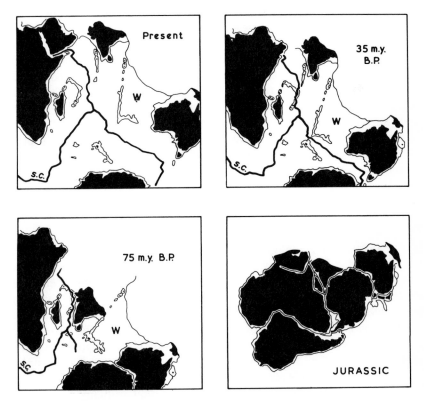

FIG. 35.16. Evolution of the Indian Ocean. (After McKenzie and Sclater, 1971.) W, Wharton Basin, S.C., spreading centre.

that an appreciable amount of older crust must occur. The area of the Wharton Basin (W) was particularly large at this time. If the continental assembly suggested by Smith and Hallam (1970) for the Jurassic is accepted (Fig. 35.16), the oldest oceanic crust in the Indian Ocean is of late Jurassic or early Cretaceous age. Dietz and Holden (1970), however, believe that fragmentation of the continents started in the Triassic. The results from deep-sea drilling favour the former interpretation because, in the Wharton Basin, the oldest sediments appear to be of late Oxfordian age. These have been found lying directly on basaltic basement rocks beneath the Argo Abyssal Plain (Heirtzler *et al.*, 1974). North-east-trending magnetic lineations of Kimmeridgian–Oxfordian age (M-22–M-25) have recently been reported in this area (Larson, 1975; see Fig. 35.7).

By the late Cretaceous, Africa, India and Australia were well separated, although Australia remained fixed to Antarctica. Madagascar had also parted from Africa (Fig. 35.16). Three subsequent phases of evolution have been distinguished by McKenzie and Sclater (1971). During the first of these from 75 to 45 million years B.P., two spreading centres existed, one south of India and the other north of the Seychelles (Fig. 35.16). At this time the Indian plate, with the Ninetyeast Ridge as its eastern boundary, moved rapidly northwards, at a rate of as much as 17 cm year^{-1} between 75 and 70 million years B.P. Motion stopped at 56 million years B.P. when India reached the Tethyan subduction zone to the north. The second period occurred between 45 and 35 million years B.P. when Australia, as part of the Indian plate, moved away from Antarctica at a rate of 5 cm year^{-1}. In the third period, 35 million years B.P. to the present, the African, Indian and Antarctic plates continued to separate by spreading along the inverted-Y system of ridges which is currently active (Fig. 35.16).

35.4.2.5. *Pacific Ocean*

It is clear from the geology of the Palaeozoic and early Mesozoic fold belts surrounding the Pacific (Miyashiro, 1972; Grindley, 1974; Matsumoto and Kimura, 1974; Rutland, 1974; Wheeler *et al.*, 1974) that extensive marginal subduction zones have existed over a period considerably longer than that represented by the entire Pacific crust. Thus, the geometrical inferences from the magnetic anomalies concern plate motions within the basin rather than the movements of the continental margins. The configuration of the anomalies in Fig. 35.7 show that these movements must have been complex.

One of the remarkable results of the application of the time-scale proposed by Heirtzler *et al.* (1968) is the demonstration that in many areas of the Pacific the crust becomes younger towards the edges of the basin. With the exception of small areas east of the Juan de Fuca and Gorda Ridges, the magnetic anomalies in the NE. Pacific decrease in age towards the western margin of North America (Figs 35.3, 35.7). The E.–W. anomalies south of the Aleutians become younger towards the trench. In the latter region the spreading axis which generated the present anomalies and which produced a corresponding plate to the north (the Kula plate of Grow and Atwater, 1970) has entirely disappeared by a northward migration beneath the Aleutian arc. Its subduction appears to have been responsible for a major mid-Tertiary thermal event on the Aleutian Islands (Grow and Atwater, 1970). In the Gulf of Alaska it is necessary to postulate a subduction zone along the

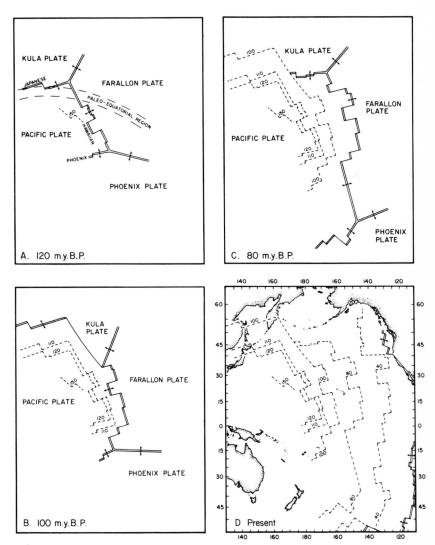

FIG. 35.17. Evolution of spreading centres in the Pacific, relative to a fixed Pacific plate. Double lines on plate boundaries represent spreading axes, with direction of spreading indicated; single lines are transform faults. (From Larson and Chase (1972). Reprinted with the permission of the authors and the Geological Society of America.)

western margin of North America (Atwater, 1970) to explain the consumption of the spreading axis which gave rise to the N.–S. lineations and the disappearance of the corresponding plate to the east (the Farallon plate of McKenzie and Morgan, 1969). Pitman and Hayes (1968) have suggested the presence of a triple junction in the area during the late Cretaceous, involving the Kula, Farallon and Pacific plates, to account for the sharp bend in the anomalies south of the Alaskan peninsula (Fig. 35.7).

Triple junctions have also been important in the evolution of the western Pacific. Three sets of magnetic anomalies can be distinguished in this region— an ENE.-trending "Japanese" group east of Hokkaido, a NW.–SE. "Hawaiian" set west of the Hawaiian Islands and an ENE.-trending "Phoenix" pattern in the equatorial region north of the Tonga Trench (Fig. 35.7). Larson and Chase (1972) have identified the Lower Cretaceous anomalies M-1 to M-10 in each of these sequences and have shown that the Japanese anomalies become younger towards the Japan and Kurile trenches, the Hawaiian lineations older westwards and the Phoenix anomalies older northwards. The differences in orientations of these synchronous patterns implies the existence of five Mesozoic spreading centres, which were joined at two triple points. The evolution of the triple junctions proposed by Larson and Chase (1972) is illustrated in Fig. 35.17. The history of the basins behind the subduction zones in the western Pacific has been reviewed by Moberly (1972).

The development of the eastern equatorial and south-eastern Pacific has also involved the interaction of several plates (Herron, 1972). Three spreading centres are currently active—the East Pacific Rise, the Galapagos Rift and the Chile Rise. This pattern is, however, not a long established one. Herron (1972) has identified a relict NNW.-trending Tertiary spreading centre which lies to the east of the East Pacific Rise crest between 10°S. and 30°S. and to the west of the ridge axis, between the Equator and 20°N. The NNW. trend of the spreading centre in the region north of the Equator has clearly been maintained since the late Cretaceous but, according to Herron's analysis, a jump of the ridge crest to the east occurred five to ten million years ago. Before ten million years B.P., spreading in the region to the south took place on the NNW.-trending ridge in the centre of the Nasca plate. The Chile Rise appears to be an active segment of this old ridge system.

35.5. THE CONSTITUTION OF THE OCEANIC BASEMENT

35.5.1. EVIDENCE FROM SEISMIC REFRACTION STUDIES

The early seismic refraction measurements indicated that the oceanic crust consists of three principal layers (Table 35.1), which were discussed briefly in Section 35.1. More recent observations, made with sound sources with high repetition rates, have shown that the crustal structure is much more complicated than this. *Layer 2* in many regions is divisible into two layers, designated *2A* and *2B*, which are characterized by average seismic velocities of about 3·6 km s^{-1} and 5·2 km s^{-1}, respectively (Houtz and Ewing, 1976). From model studies and Fourier inversion of magnetic data, Talwani *et al.* (1971) and Klitgord *et al.* (1973) have suggested that *Layer 2A* is highly magnetized and is largely responsible for the oceanic magnetic anomalies, an interpretation which is not, however, always supported by the results of drilling from D.V. *Glomar Challenger* (Section 35.5.3). The crust below

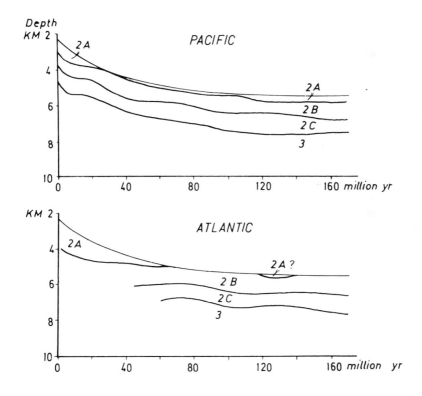

FIG. 35.18. The variation of upper crustal structure with basement age in the Atlantic and Pacific, based on the seismic observations of Houtz and Ewing (1976). (Diagram reprinted with the authors' permission. Copyright, American Geophysical Union.) The average P-wave velocities in *Layers 2A* and *2B* are approximately 3·6 km s^{-1} and 5·2 km s^{-1} respectively. *Layers 2C* and *3* are characterized by seismic velocities of 6·1 km s^{-1} and 6·7 km s^{-1} respectively. The generalized topographical profile is from Sclater and Francheteau (1970) (see Fig. 35.22).

Layer 2 also appears to be more complex than is indicated by the seismic profiles shot in the 1950s. In both the Atlantic and Pacific, Houtz and Ewing (1976) have found a layer, 0·6–1·2 km thick having a velocity of 6·0–6·2 km s^{-1}, which lies between *Layers 2B* and *3*. This they call *2C*. (Fig. 35.18) In the Pacific, Maynard (1970), Sutton *et al.* (1971) and Houtz and Ewing (1976) have detected an additional seismic layer, roughly 3 km thick with a velocity of about 7·4 km s^{-1}, between *Layer 3* (6·7 km s^{-1}) and the top of the mantle. The typical *Layer 3* velocities thus appear to occur in a more restricted region of the crust than was previously thought.

The present evidence suggests that there is a broad correlation between structure and crustal age. Figure 35.18 illustrates sections for the Pacific and Atlantic, compiled by Houtz and Ewing (1976) from airgun-sonobuoy data. A notable feature is the marked decrease in thickness of *Layer 2A* with increasing basement age, although, unexpectedly, the layer becomes thicker again in areas of Lower Cretaceous and Jurassic crust. The difference in the thickness of *Layer 2A* in the Atlantic and Pacific appears to be related to spreading rate. On the slow-spreading Mid-Atlantic Ridge, *Layer 2A* is about 1·5 km thick at the ridge crest, and thins to about 0·1 km on 60 million-year-old crust, whereas on the fast-spreading ridges of the Pacific it is consistently thinner than 0·8 km.

Sonic arrivals from *Layers 2C* and *3* have not been recorded by Houtz and Ewing from crust less than 50 million years old in the Atlantic, but other authors have shown that these layers are developed locally. Keen and Tramontini (1970) detected *Layer 3* velocities near the ridge crest at 45°N., and Whitmarsh (1973) and Poehls (1974) found velocities close to those of *Layer 2C* in the median valley near 37°N. In the Pacific, *Layers 2C* and *3* are persistent features of the structure. Both in the Pacific and on the flanks of the Mid-Atlantic Ridge *Layer 3* increases in thickness with crustal age. The rate at which it thickens bears a close relationship to spreading velocities, being greatest on the fastest spreading ridges.

35.5.2. RESULTS OF SAMPLING

The question of the origin of the crustal layers and the cause of the variation in structure with time will now be considered. Direct evidence of the composition of the oceanic basement has been obtained by dredging and drilling. Dredging in the fracture zones of the Atlantic and Indian Oceans, where deep crustal material appears to be exposed, has been particularly productive and has yielded a wide range of rocks, from serpentinites to fresh basalts. In the first five years of the Deep-Sea Drilling Project, D.V. *Glomar Challenger* recovered basement material at many sites but the core samples were only short (< 100 m) and often highly altered. In 1974, however, a 583 m penetration into the basement was achieved on the Mid-Atlantic Ridge near 37°N., and provided many fresh samples. The International Phase of Ocean Drilling, which is now in progress, is strongly orientated towards the recovery of deep basement cores.

The detailed composition of the oceanic crust is the subject of vigorous debate at the present time as controversies have arisen because of the many ambiguities inherent in the geophysical data, the shortness of drill cores from the oceanic basement and the vagueness of the relationship between dredge samples and the seismic layers. It is unfortunate that the places where dredging operations are most successful (the scarps of fracture zones) are the regions where the seismic structure has been least well determined. For this reason, the results of deep drilling at sites well away from structurally anomalous areas are awaited with considerable interest.

The bulk of the material recovered from the oceanic basement consists of basalts and their intrusive equivalents, and fresh to totally serpentinized ultra-basic rocks. Many of the basic rocks have suffered some degree of metamorphism; samples of zeolite-smectite, greenschist and amphibolite grade facies have been reported in many dredge hauls. From the observed distribution of rock types several models of the oceanic crust have been suggested. That proposed by Cann (1970, 1974) will be used as a basis of the present discussion. The basement is envisaged as consisting of basaltic lavas and their feeder dykes in the uppermost part, with a thick gabbroic layer below which has the Moho as its lower boundary. (Fig. 35.19). Serpentinite is considered to be only local in its distribution and is not included in Fig. 35.19. The close resemblance of this model to the structure of ophiolites (Gass and Masson-Smith, 1963) has been emphasized by Vine and Moores (1972), Gass and Smewing (1973) and many other authors during the past five years. (Fig. 35.20).

There seems little doubt that *Layer 2A* is composed of basaltic lavas. These have been dredged, drilled and photographed at many locations (Matthews, 1971). The seismic velocity in *Layer 2A* ($2 \cdot 3$–$3 \cdot 8$ km s^{-1}), observed by refraction shooting, is generally lower than the velocities measured in ocean-floor basalts (Fox *et al.*, 1973), but this is probably because the layer is very porous, as is indicated by the strong evidence for extensive hydro-thermal circulation (Section 35.6.1). Unaltered samples from *Layer 2A* generally have a high remanent magnetization, a feature consistent with interpretations of surface and near-bottom magnetic profiles (Talwani *et al.*, 1971; Klitgord and Mudie, 1974).

The base of *Layer 2A* appears to be expressed by both a magnetization and a marked velocity contrast (Talwani *et al.*, 1971). One explanation for this is that the underlying *Layer 2B* is largely composed of dykes, the slower cooling rates giving rise to higher seismic velocities and lower magnetiza-

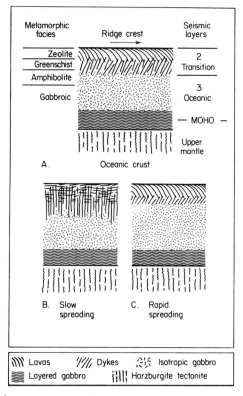

FIG. 35.19. Schematic representation of the structure of the oceanic crust (after Cann, 1974). (Diagram is reproduced by permission of the author and the Royal Astronomical Society.)

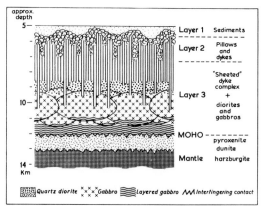

FIG. 35.20. Schematic section of oceanic crust, as inferred from a study of the Troodos Massif, Cyprus. (After Vine and Moores, 1972; reproduced by permission.)

tions; this is implicit in the interpretation given by Cann (1974) (Fig. 35.19). However, from their measurements of sound velocities in dredged samples, Christensen and Salisbury (1975) maintain that the seismic velocity in the dyke complex must exceed $6 \cdot 0 \, km \, s^{-1}$. If this literal interpretation of laboratory observations is accepted, then both *Layers 2A* and *2B* must largely consists of extrusives and only a small proportion of dykes, with *Layer 2C* containing a large part of the dyke complex. An explanation for the velocity contrast between *Layers 2A* and *2B* must, then, be sought in metamorphic reactions or in changes in the degree of fracturing, or both. The seismic boundary may represent a rapid transition from zeolite to greenschist facies (Fig. 35.19) or, perhaps, a sudden reduction in the porosity of the crust. It is likely that both factors would act in combination as the pervasiveness of greenschist metamorphism will be strongly related to the availability of circulating fluids.

The high amplitude of refracted waves from the base of *Layer 2B* indicates a major change in the physical properties of the crust. The interpretation by Hess (1962), that the $6 \cdot 7 \, km \, s^{-1}$ material consists of serpentinized peridotite, is no longer favoured by most workers, since the seismic velocities of dredge samples of this composition are too low (Barrett and Aumento, 1970; Fox *et al.*, 1973; Christensen and Salisbury, 1975). Moreover, Christensen (1972) has found that laboratory determinations of Poisson's ratio for serpentinites are much higher than values obtained from seismic shooting for the lower part of the oceanic crust. Nevertheless, serpentinite has often been recovered from the lower slopes of fracture zones (Bonatti *et al.*, 1971) and even from high levels of the crust (Cann and Funnell, 1967). However, it seems to be only local in its distribution, emplacement having taken place as a result of movements along deeply penetrating faults.

The crust below *Layer 2B* is generally thought to be of gabbroic composition, as originally was suggested by Hill (1957) and others. Gabbroic rocks have frequently been dredged from the lower parts of fracture zones and they have compressional-wave velocities in the range of those observed in refraction shooting. However, it is evident from the more recent seismic data that the structure of this region of the crust is far from uniform in character. Two factors, metamorphism and gravity differentiation in the cooling magma, may be responsible for this. Metamorphic reactions clearly take place below *Layer 2B* because metagabbros are by no means infrequent components of dredge collections. Their effect is to lower the seismic velocity of the parent gabbroic material, but the extent of the metamorphism is not known at present. A possible correlation of metamorphic facies with seismic layering, suggested by Cann (1974), is given in Fig. 35.19. Compositional and textural evidence for gravity differentiation has been presented in several papers. A study of fresh gabbros from the Atlantis Fracture Zone by Miyashiro *et al.*

(1970), for example, has revealed that the decrease in the anorthite content of modal plagioclase is closely related to increases in the FeO/MgO ratio, a feature of layered basic complexes. Furthermore, an aplite, a product of late-stage differentiation, was found associated with the basic rocks. Acidic or intermediate derivatives are by no means uncommon components of deep-ocean dredge hauls (Bonatti et al., 1970; Thompson, 1973). At the basic end of the differentiation sequence olivine-rich gabbroic cumulates have been recovered from the lower parts of fracture zones (Melson and Thompson, 1971; Bonatti et al., 1971).

The formation of olivine-rich cumulates offers a plausible explanation for the origin of the 7·4 km s^{-1} layer detected by Maynard (1970) and Sutton et al. (1971), but it is not yet clear whether the 6·7 km s^{-1} layer above consists largely of unaltered gabbro or gabbroic rock which has been extensively metamorphosed in the high temperature environment of the ridge crest. Fox et al. (1973) have argued that little alteration has occurred, since their measurements on dredge samples show that the metamorphic changes lower the sound velocity of gabbros to values well below 6·7 km s^{-1}. As they have pointed out, relatively few specimens of amphibolite have been recovered from the ocean floor, suggesting that it is not an important component of the crust. The velocities characteristic of *Layer 2C* are, however, compatible with those which they quote for greenschist and amphibolite facies metagabbros. Recently, their conclusions have been challenged by Christensen and Salisbury (1975) who believe, on the basis of a comprehensive compilation of seismic velocities in rocks, that a large part of the 6·7 km s^{-1} layer is made up of amphibolites. They have noted that Poisson's ratio for the 6·7 km s^{-1} layer is much closer to that determined for metagabbros than to that of unaltered gabbro. Controversies such as this, in which all sides place great emphasis on a literal extrapolation of laboratory measurements to what is probably a complex structural situation, are unlikely to be resolved until the results of drilling into *Layer 3* are known.

It was shown earlier in this section that the thicknesses of some of the seismic layers change with spreading rate and with time, a feature that provides an additional constraint on possible models of crustal composition. Both Menard (1967) and Shor et al. (1970) have observed that the thickness of the material having velocities between 3·5 and 6·0 km s^{-1} (*Layers 2A and 2B*) is inversely proportional to the spreading rate. This variation is to be expected if the upper parts of the basement consist of lavas and their feeder dykes (Cann, 1974). On fast-spreading ridges the rate of dyke injection is higher than in slow-spreading regions. Where the spreading rate is high each dyke will cool less before the subsequent one is intruded, so that the vertical extent of the dyke swarm is smaller, although the rate of extrusion of the lavas will remain relatively constant. With the decrease in thickness of the dyke–lava zone there is a concomitant increase in the thickness of the slower cooling gabbroic layer (Fig. 35.19).

The change in thickness of *Layer 2A* with time (Fig. 35.18) may be brought about by a decrease in its porosity as a result of the collapse of cavities and the filling of voids by hydrothermal deposits and infiltrating sediments. It is difficult to envisage, however, how such a process could explain a reduction in thickness by an order of magnitude in 60 million years. Since *Layer 2A* appears to thicken again in areas of Lower Cretaceous and Jurassic crust, these changes may be partly caused by variations in the rates of lava production at the ridge crest.

An increase in the crustal thickness below the level of *Layer 2B* with increasing basement age has been firmly established by Shor *et al.* (1970). Since *Layer 2C* does not appear to change significantly across the ridge flanks (Houtz and Ewing, 1976), this raises the possibility that the thickening is due to partial serpentinization of a zone of cumulates which is represented by the $7 \cdot 4$ km s^{-1} layer (Maynard, 1970) at the base of the crust (Le Pichon, 1969; Cann, 1974). However, the distribution of the basal high-velocity layer and changes in its seismic velocity are not yet known in sufficient detail to make a convincing test of this hypothesis, although such a test could be made with present seismic techniques. Another explanation is that the thickening described by Shor and others is due to intermittent intrusive activity in the lower crust off the ridge axis. Christensen and Salisbury (1975) picture this process as occurring for roughly 30 million years after the initial formation of the crust. The model is attractive in that it explains the occurrence of anomalous mantle velocities frequently observed on the upper flanks of slow-spreading ridges.

35.5.3. THE MAGNETIZATION OF THE OCEANIC CRUST

It was shown above that the magnetization of the oceanic basement has provided crucial evidence in favour of sea-floor spreading. However, it is by no means clear how the magnetization is distributed with respect to depth in the crust. Vine and Matthews (1963) chose the Curie point isotherm as the lower boundary of each magnetic block (Fig. 35.4), but other models can be devised to reproduce the observed magnetic anomalies. Recently, constraints on possible configurations of the source region have come from measurements on dredged samples and drill cores (Fox and Opdyke, 1973; Lowrie, 1974; Hall, 1976) and also from further studies of magnetic anomalies (Talwani *et al.*, 1971; Klitgord *et al.*, 1975).

Talwani *et al.* (1971) computed the remanent magnetization of the basement of the Reykjanes Ridge using magnetic anomalies which are associated with small-scale relief on profiles running parallel to the lineations. From these determinations, they demonstrated that the magnetic anomalies across the Ridge can be simulated with a layer only a few hundred metres in thickness, suggesting that the magnetization of the oceanic crust largely resides near the top of *Layer 2*. The magnetic layer is correlated with seismic *Layer 2A* (Section 35.5.1). Their model of an upper, 400 m-thick zone of strongly magnetized lavas, with a magnetization of 3×10^{-2} emu cm^{-3} at the ridge axis and $1 \cdot 2 \times 10^{-2}$ emu cm^{-3} on the ridge flanks, has gained much support from the Fourier inversion of near-bottom magnetic profiles (Klitgord *et al.*, 1975). However, measurements on dredged samples and drill cores have produced results which are in serious conflict with these interpretations. Fox and Opdyke (1973) have examined 72 basalt specimens from a wide variety of locations in the NW Atlantic and Caribbean and found a geometric average intensity of $0 \cdot 11 \times 10^{-2}$ emu cm^{-3}, about an order of magnitude lower than the value quoted for the flanks of the Reykjanes Ridge. Lowrie (1974) obtained a slightly higher geometric mean of $0 \cdot 21 \times 10^{-2}$ emu cm^{-3} from measurements on samples from 26 sites drilled by the D.V. *Glomar Challenger*, but this is still a figure requiring a magnetic source layer more than twice the thickness estimated by anomaly inversion. The structural and magnetic complexity of the upper oceanic basement, evident from the D.V. *Glomar Challenger's* recent drilling on the Mid-Atlantic Ridge (Hall and Ryall, 1976; Dick *et al.*, 1976), also make it necessary to consider a source region appreciably thicker than 400 m. At two drilling sites on the Mid-Atlantic Ridge near 37°N. (DSDP 332 and 333, with maximum basement penetrations of 583 m and 312 m, respectively) several inclination reversals in addition to well-scattered NRM inclinations were found by Hall and Ryall (1976), so it is clear that the basement is far from uniformly magnetized. It is also significant that nearly half of the basaltic samples at these locations possess a soft NRM component. Since soft magnetization in the crust is likely to be in the direction of the ambient field, its influence on the thickness of the magnetic blocks in the Vine–Matthews model may be appreciable, especially in reversely magnetized areas.

From the observed magnetizations of the drill cores at DSDP Sites 332 and 333, Hall and Ryall (1976) have estimated that the source layer of the magnetic anomalies is at least 2·5 km thick. The anomalies must therefore arise, in part, from the crust below seismic *Layer 2B* (Section 35.5.1). This result has important implications since, in this region of slower cooling, viscous remanent magnetization (VRM) may well be significant, causing the magnetization of the crust to change with time. Hall (1976) has pointed out that VRM may be dominant in some regions. Thus, the observed magnetic anomalies may not always relate to a period of accretion, but to some later time. Clearly, more deep drilling is required to investigate whether the results from this part of the Atlantic are typical of broad regions of the ocean floor.

35.6. IMPLICATIONS OF RECENT GEOPHYSICAL OBSERVATIONS

35.6.1. BATHYMETRY AND HEAT FLOW

The measurement of heat flow through the ocean floor has provided important indications of the origin of the regional changes in depth associated with
sea-floor spreading. In spite of the large scatter in heat flow values it is clear
that ridge crests are characterized by high heat flow, well in excess of the
oceanic average of $1 \cdot 2 \, \mu\mathrm{cal} \, \mathrm{cm}^{-2} \, \mathrm{s}^{-1}$ (Bullard, 1954; Bullard *et al.*, 1956;
Langseth and Von Herzen, 1970). It is also evident that there is a gradual
reduction in flux with increasing age (Fig. 35.21). In the trenches, heat flow

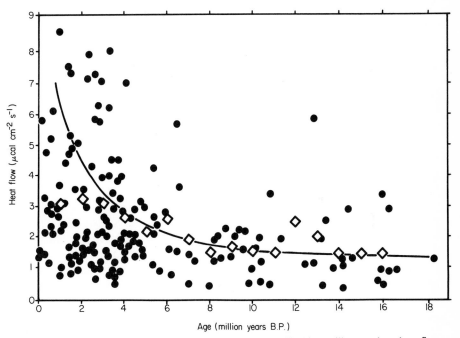

FIG. 35.21. Heat flow values (black dots) over the East Pacific Rise to illustrate how heat flow
varies with crustal age on a spreading ridge. Arithmetic means for each million year interval
are shown by diamonds. The solid curve indicates the heat flow pattern expected from a
spreading lithosphere (McKenzie and Sclater, 1969). (After Von Herzen and Anderson, 1972;
diagram reprinted by permission.)

values well below the oceanic average have frequently been measured. The
zone where these occur is narrow and appears to be restricted to the trench
floor and parts of the landward wall. In marked contrast, the heat flow on
the island arcs generally exceeds $2 \, \mu\mathrm{cal} \, \mathrm{cm}^{-2} \, \mathrm{s}^{-1}$ and is also high in the

back-arc basins, such as the Japan Sea. Langseth and Von Herzen (1970) have pointed out that, when the heat flow is integrated over the trench–island arc system, the total flux is appreciably greater than the oceanic average. Any thermal anomaly, due to the cool descending plate penetrating the asthenosphere, is obscured by active magmatic and hydrothermal processes on the island arc. Thus, both the generation and the consumption of the oceanic lithosphere are expressed by high heat flow.

With the exception of the trench–island arc areas, the pattern of heat flow is broadly consistent with the production of lithosphere at a ridge axis and its subsequent cooling on the ridge flanks. The contraction of the lithosphere as it cools appears to account for the major regional changes in depth of the ocean floor (Vogt and Ostenso, 1967; Menard, 1969). When related to crustal age the changes in depth appear to be remarkably predictable. From observations in the Pacific, Sclater et al. (1971) have derived a curve, relating depth to crustal age, which is of general application to other parts of the oceans (Fig. 35.22). Exceptional areas with "elevation anomalies" are known, notably in the Atlantic north of 40°N., an area much shallower than would be expected from its crustal age. However, the observed and the predicted depths are generally very close and, for this reason, several workers have used regional depths to infer crustal ages when magnetic data are lacking or have failed to provide an unambiguous answer. Recently, Sclater et al. (1975) have extended the age–depth curve back to 160 million years B.P., using the crustal ages in the western Pacific determined by Larson and Pitman (1972).

At present, there is no general agreement on the form of the cooling model which best explains the topographic variations. McKenzie (1967) and Sclater and Francheteau (1970) have shown that the conductive cooling of a 75 km-thick lithospheric slab, having a temperature of 1300°C at its base, can account for the elevation of the East Pacific Rise and the distribution of the heat flow values over it. Quantitative discussions have also been presented by Parker and Oldenburg (1973) and Davis and Lister (1974) who have demonstrated that the topography of Tertiary crust can be accounted for by a simple one-dimensional cooling model, without the need for assuming a finite plate thickness at large distances. However, their arguments have been questioned by Sclater et al. (1975) who maintain that heat flow in regions of older crust is uniform enough to suggest that the cooling plate has indeed a finite thickness.

Although a simple contraction model can explain the main features of topography and heat flow, it does not account for the "elevation anomalies". These anomalies, having amplitudes of ± 300 m and wavelengths of 500–2000 km, are by no means insignificant features of the ocean basins (Menard, 1973), and their close correlation with gravity variations suggest that they

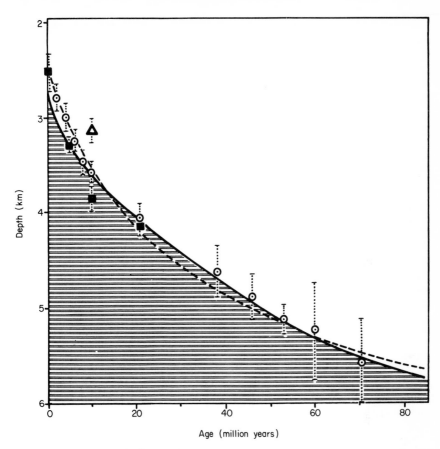

FIG. 35.22. Average depth of the sea floor versus crustal age for the North Pacific and Atlantic. (After Sclater *et al.*, 1971. Diagram reproduced with the permission of the authors. Copyright © American Geophysical Union.) – – – – Theoretical steady-state profile (Sclater and Francheteau, 1970). ——— Empirical elevation, N. Pacific. ⊙ S. Atlantic ($\sim 2 \cdot 0$ cm year^{-1}). ■ N. Atlantic, below 45°N. ($\sim 1 \cdot 0$ cm year^{-1}). Δ N. Atlantic, at 45°N. ($\sim 1 \cdot 4$ cm year^{-1}). Spreading rates are shown in parentheses.

are related to flow in the asthenosphere (see Section 35.6.3). Further, the great scatter in the heat flow data for spreading ridges cannot be explained by a contraction model. An illustration of the pronounced variability in these data is presented in Fig. 35.23, which is taken from a discussion of recent observations on the Galapagos spreading centre (Williams *et al.*, 1974). It should be noted that surprisingly low values (less than $0 \cdot 5 \,\mu\text{cal cm}^{-2}\,\text{s}^{-1}$) are found in locations less than 10 km from those in which values are an order of magnitude higher. Such rapid changes show that the lithosphere must be cooling by some process other than conduction. Convecting fluids within the upper parts of the lithosphere appear to be the only plausible means of heat transfer. In the Galapagos region there is direct evidence, from anomalously high bottom-water temperatures, for the existence of open, active hydrothermal vents at the sea floor. Such evidence is not unexpected as strong indications of the activity of hot circulating fluids in the oceanic crust have been provided by the hydrated mineral assemblage

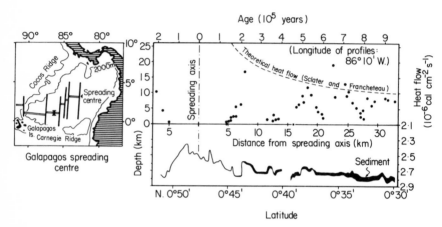

FIG. 35.23. Heat flow variations, bathymetry and sediment thickness across the Galapagos spreading centre. (Simplified, after Williams *et al.*, 1974.) The dashed curve is the heat flow predicted for an 85-km thick lithospheric plate, with a temperature of 1250°C on its lower boundary, an internal heat generation of $2 \cdot 0 \times 10^{-14}\,\text{cal cm}^{-3}\,\text{s}^{-1}$ and an average thermal conductivity of $6 \cdot 9 \times 10^{-3}\,\text{cal cm}^{-1}\,\text{s}^{-1}\,(°\text{C})^{-1}$ (Sclater and Francheteau, 1970).

and isotopic composition of metabasalts dredged and drilled from the oceanic basement (Miyashiro *et al.*, 1971; Spooner and Fyfe, 1973) and also by the discovery of basal metalliferous sediments (Corliss, 1971). Widespread hydrothermal activity is, of course, well known in Iceland, an exposed part of the Mid-Atlantic Ridge.

The effect of fluid circulation on the isotherms is illustrated in Fig. 35.24. If there is an impermeable sedimentary cap, the thermal gradients will be high over the rising parts of the flow. However, if the fluids are in direct contact with seawater small thermal gradients will occur over both the ascending and the descending portions of the convecting system. The observed heat flow pattern will depend greatly on the mode of circulation but, in general, heat flow maxima will be located close to minimum values. In the Galapagos region, the average horizontal extent of a convection "cell" appears to be about 6 km, with a depth of penetration of about 10 km. Ribando *et al.* (1976) have suggested that the mean residence time of the circulating fluid in the crust is about four years. During this time an appreciable amount of ion exchange takes place (Spooner and Fyfe, 1973), a process which is of vital importance in any consideration of the chemical balance of the oceans through time.

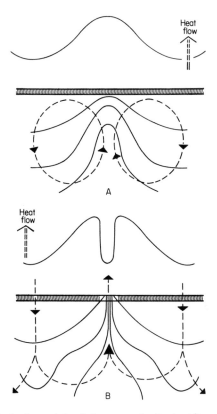

FIG. 35.24. Effects of hydrothermal circulation on conductive heat flow and geothermal gradients near the sea floor. Dashed lines indicate general convective pattern in the oceanic crust. Solid lines represent isotherms. In (A) convection takes place below an impermeable cover of sediments. In (B) the convection system is in direct contact with seawater. (After Williams *et al.*, 1974.)

The largest heat loss and the greatest variability of thermal gradients occur near to the spreading centre and decrease with increasing crustal age. Using the conductive-cooling model of Sclater and Francheteau (1970), Williams *et al.* (1974) have shown that hydrothermal circulation accounts for about 80% of the thermal losses from the Galapagos rift zone. Until these major shallow hydrothermal effects are fully evaluated it is clear that regional deviations from the heat flow pattern predicted by simple cooling models cannot be used to study deep-seated processes. The importance of taking measurements in areas where the sediment cover is thick enough to prevent hydrothermal fluids from reaching the sea floor has recently been emphasized by Sclater *et al.* (1976) and Anderson and Hobart (1976) who found that the variability in heat flow decreases markedly where a thick and uniform sedimentary blanket is present, with the mean values falling close to those predicted from thermal models. In these regions the permeability of the crust has probably become much reduced by hydrothermal deposits and infiltrating sediments (Palmason, 1967; Lister, 1970, 1972).

35.6.2. SEISMICITY

Rapid advances in our understanding of the tectonics of ocean basins have been made through the study of earthquakes associated with the oceanic ridge system and deep-sea trenches. Since the early 1960s, seismic activity has been recorded by a world-wide network of standardized seismographs and this has enabled focal positions to be determined much more precisely than previously; it has also allowed source mechanisms to be investigated in some detail. The map of epicentres in Fig. 35.7 shows the activity observed during the period 1961–67. Oceanic earthquakes have also been recorded at close quarters by bottom seismometers (Francis and Porter, 1972; Nowroozi, 1973; Prothero, 1974) and by free-floating sonobuoys (Brune *et al.*, 1972).

On actively spreading ridges, earthquakes occur in a confined zone over the ridge crest and along narrow belts between its displaced portions. The slower-spreading ridges (e.g. the Mid-Atlantic and Carlsberg Ridges; Table 35.2) have much higher teleseismic activity along them than do the fast-spreading ones, such as the East Pacific Rise. The former are also characterized by a rugged topography and have a central rift valley, whereas the latter lack a central rift and possess much smoother topography with regional gradients appreciably lower than those of rifted ridges (Fig. 35.1) (Menard, 1967). Van Andel and Bowen (1968) have suggested that the difference is related to the thickness of the surface layer in which brittle fracture occurs. On the fast-spreading ridges this layer is probably much thinner because of higher geothermal gradients.

The earthquakes in the narrow belts between displaced segments of the ridge crest are associated with linear fracture zones, such as the Romanche in the equatorial Atlantic. When fracture zones were first mapped they were thought to be transcurrent faults. From a study of earthquake focal mechanisms, however, Sykes (1967) showed that the sense of motion along the fault planes is opposite to that predicted by the transcurrent hypothesis. To explain this he adopted Tuzo Wilson's concept of a "transform" fault

(Wilson, 1965), which is illustrated in Fig. 35.25. In this model transcurrent faulting implies a left-lateral movement. However, if it is assumed that sea-floor spreading takes place then the blocks on either side of the fault move in the direction of the arrows, giving rise to a right-lateral motion along the fault plane. It is important to note that earthquakes are confined to the portions of the fracture zone between the ridge crests, a feature of the

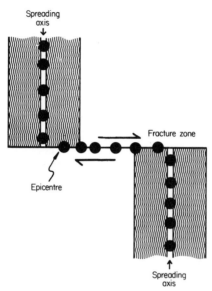

FIG. 35.25. Seismicity near a fracture zone which displaces the crest of a spreading ridge. The shaded region represents the new surface area created during some arbitrary time interval. The arrows indicate the direction of movement of adjacent blocks along the active part of the fault.

observed seismicity. In the ten events Sykes examined on the Mid-Atlantic Ridge and East Pacific Rise the direction of motion was found to be consistent with transform faulting, and a subsequent analysis of other earthquakes produced similar results (Sykes, 1970a). He thus provided evidence of crustal generation at ridge crests which is quite independent of the magnetic anomaly patterns.

The first-motions of earthquakes on parts of the oceanic ridges not displaced by fracture zones are generally characterized by a large component of normal faulting, with the axis of tension roughly perpendicular to the trend of the ridge crest. Thus these focal mechanisms also indicate a tectonic environment which is clearly compatible with the injection of new crustal material at the ridge axis. Frequently, the earthquakes beneath the ridge axis occur in swarms (Sykes, 1970b). One such swarm, consisting of several hundred events of magnitude 4·0–5·5, was detected in the Gulf of California in March, 1969 (Thatcher and Brune, 1971). Others, on the Mid-Atlantic Ridge and East Pacific Rise, have been described by Sykes (1970b).

Although earthquake swarms have recently been reported on fracture zones close to their intersection with a ridge crest (Tatham and Savino, 1974), nearly all those detected teleseismically occur beneath the ridge axis. Earthquake swarms may thus be directly associated with magmatic and hydrothermal processes which are presently active beneath the spreading centres. A recent series of closely spaced events under the axis of the Galapagos Ridge (Macdonald and Mudie, 1974) was located in an area well known for its current hydrothermal activity (Section 35.6.1). However, the underlying causes of stress build-up, leading to a high level of seismic activity, may be complex. Normal faulting in individual events of earthquake swarms on the Mid-Atlantic Ridge has been detected by Sykes (1970b). In another study, based on long-period surface waves from swarms in the NE. Pacific, many events were found to be characterized by strike-slip motion (Tatham and Savino, 1974). Sudden bursts of seismic activity on spreading ridges may, like earthquake swarms in continental areas, be due to several tectonic processes.

An investigation of the relationship between the seismicity of ridge crests and local physiography and structure has been made possible with the deployment of bottom seismometers and free-floating sonobuoys. An event of magnitude $m = 0$, or less, can now be located with a precision of a few hundred metres. Teleseismic recording, on the other hand, has a location precision of only about 20 km at best, and a much higher threshold of detection. Recent observations with the bottom instruments and the sonobuoys clearly indicate that the earthquake activity on ridge crests is very localized. An array of seismometers placed on the crest of the Mid-Atlantic Ridge at 45°N. detected about 30 events per day in the median valley, and few or none in the rift mountains (Francis and Porter, 1973). Reid and Macdonald (1973) recorded about 10 earthquakes per day on the Ridge at 37°N. (the FAMOUS* area; see Section 35.6.4) using sonobuoys and, again, found the seismic activity to be confined in its distribution. Spindel et al. (1974) have presented results to show that earthquake foci in the FAMOUS area tend to cluster beneath the walls of the median valley, suggesting an association with normal faulting on the valley walls rather than with intrusive processes under the ridge axis. (Section 35.6.4). In these studies, the depth of the events was found to be shallow, usually less than 6 km.

In marked contrast to the shallow seismicity of the oceanic ridges, the

* French–American Mid-Ocean Undersea Study.

seismic activity beneath the trenches and island arcs spans a depth range of a few hundred kilometres, a feature which was appreciated over fifty years ago by Turner (1922). The earthquakes occur within an inclined zone, which dips at an angle of about 45° beneath the island arc, reaching a maximum depth of 700 km. Recent observations indicate that the seismic region, often called the Benioff zone, is usually only a few tens of kilometres wide (Fig. 35.12).

The study of the distribution and focal mechanisms of earthquakes along the Benioff zone has played an important role in the formulation of the plate theory of tectonics. Perhaps the most significant investigation is that of the earthquakes in the vicinity of the Tonga-Kermadec Trench by Oliver and Isacks (1967). The frequency and amplitude of shear waves recorded at several local stations clearly show that seismic attenuation is appreciably lower for energy travelling through the seismic zone than it is for that outside it, whereas the propagation paths from shallow earthquakes to stations on the oceanic side of the trench are also ones of relatively small attenuation. These observations are compatible with the concept that the oceanic crust and part of the underlying upper mantle (the high-Q lithosphere) penetrate the highly attenuative lithosphere, a structure that forms an important element of the general tectonic synthesis described by Isacks et al. (1968) (Fig. 35.12). Structural sections similar to that at Tonga have emerged from studies of other trenches, although in some areas the complex configuration of the physiography and discontinuities in the seismic zone have demanded some modifications to the simple model (Isacks and Molnar, 1971).

The stress distribution in the lithosphere, as determined from first-motions, is consistent with this picture of underthrusting. The principal stresses are parallel to the slab and are indicative of tensional conditions at intermediate levels and of compressional ones at depths greater than about 300 km. In the Tonga region, Isacks and Molnar (1971) have found the descending Pacific plate to be under continuous compression from depths of 80–700 km, a result which has important implications when the driving forces of plate motion are considered (Section 35.8).

The maximum depths of earthquakes vary considerably from one region to another. Those trenches which have a relatively restricted depth range of seismic activity, such as the Aleutian (about 250 km), appear to be young features. The length of the seismic zone seems to be controlled by the rate at which the lithosphere descends and by the time which has elapsed since underthrusting began. The 700 km depth limit of earthquakes beneath major trenches, such as the Tonga–Kermadec and Japan, may represent either the upper boundary of a zone of mechanical resistance (the mesosphere, Fig. 35.12) or the level at which the temperature of the sinking plate becomes too high for elastic deformation.

35.6.3. GRAVITY

The pendulum method used by Vening Meinesz, Worzel and others for measuring gravity from submarines (see Worzel, 1965) was superseded in the 1950s when continuously recording gravity meters were developed for surface vessels. The surface observations have confirmed the conclusion, based on the pendulum data, that the ocean basins are in broad isostatic equilibrium. With the exception of the regions near trenches, the mean level of the free-air gravity anomaly is close to zero. On spreading ridges it is therefore clear that isostatic balance is maintained in the presence of a 2 km elevation of the sea floor.

Seismic refraction work has shown that the thickness and structure of the oceanic crust and the seismic velocity at the Moho do not vary greatly across the ridge, except near the ridge crest, although changes of crustal structure with age have been reported (Section 35.5.1). From the point of view of gravity variations these are minor and are far from sufficient to account for the compensation of the ridge, which must therefore take place within the mantle (Talwani et al., 1965). The compensation is best considered by examining the gravity anomaly over the Mid-Atlantic Ridge in Fig. 35.26. In this example, the Bouguer anomaly, which solely reflects changes in density below the sea floor, is plotted. The magnitude of the gravity low is directly related to the mass deficiency that compensates for the elevation of the ridge. The relatively steep gradients of the Bouguer curve require that a large part of the mass deficiency lies at relatively shallow depths, of the order of a few tens of kilometres. Because the gravity curve represents a potential field, however, the limits of the region of low density material cannot be uniquely determined. One solution suggested by Talwani et al. (1965), is shown in Fig. 35.26. They propose that the anomalous, low velocity mantle detected seismically beneath the ridge crest extends beneath the flanks as a low density tongue within the normal upper mantle material. As the depth of the sea floor increases, so the thickness of the low density material decreases. In this and other models, compensation occurs at depths less than 50 km, an assumption compatible with surface wave observations. At present, we know little about the nature of the mass deficiency.

Satellite data and compilations of surface gravity profiles reveal the presence of long-wavelength (> 1000 km) variations in the gravity field over spreading ridges. For example, between the Equator and 30°N. in the Atlantic the free-air values averaged over 5° squares are close to zero (Talwani and Le Pichon, 1969), whereas further north they are generally positive. Over the Reykjanes Ridge, free-air anomalies range from + 25 to + 60 mgl (Talwani et al., 1971).

An important clue to the origin of the long-wavelength changes in gravity

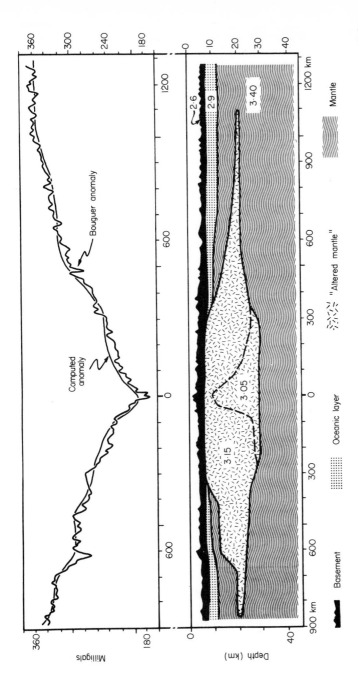

FIG. 35.26. Bouguer gravity anomaly across the Mid-Atlantic Ridge near 32°N., together with one structural model to account for the gravity low. Densities in g cm⁻³ are indicated. The Bouguer anomaly has been calculated assuming two-dimensionality and a density contrast of 1·57 g cm⁻³ between the upper crust and seawater. (After Talwani et al., 1965.)

is provided by their relation to topography. Sclater *et al.* (1975) have noted that the anomalies are generally positive in areas which are relatively shallow for their crustal age, as on the Reykjanes Ridge. The "elevation" anomalies (Section 35.6.1) appear to be linearly related to the gravity variations. As density differences within a 100 km-thick lithosphere can only account for the short-wavelength (< 500 km) components of the gravity field, it might be suggested that the topography in regions of elevation anomalies is totally uncompensated and kept at its present depth by distortion of the lithosphere under compressive stress. However, it can be shown (McKenzie, 1967) that the required shear stresses are large enough to cause failure in the plate. It appears that the only feasible mechanism to maintain the necessary dynamic stresses is convection in the upper parts of the mantle. The possible form of convective motions has been discussed by McKenzie *et al.* (1974) and by McKenzie and Weiss (1975) (Section 35.8).

It is only over the regions of crustal consumption (i.e. the trenches) that large departures from isostatic equilibrium occur in the oceans. The free-air gravity anomalies in these areas are strongly negative, typically -100 to -300 mgl, but reach -380 mgl over the Puerto Rico Trench. The nature and the distribution of the low density material are still a matter for debate, partly because deep seismic control is so poor. On many gravity profiles the gravity minimum occurs over the lower part of the landward wall of the trench. Such a displacement from the trench axis is consistent with the presence of low density sediments on the landward side and can be accounted for by the scraping of sediments from the oceanic plate as it descends into the mantle. However, the presence of sediments alone cannot explain the size of the observed anomalies and other causes of mass deficiency must be found. The large positive gravity anomalies associated with the island arc and the ocean floor just seaward of the trenches are important in this context With the exception of that associated with the Puerto Rico Trench, these positive anomalies balance out the main negative anomaly, and so give rise to a regional anomaly which is slightly positive. The total mass imbalance associated with crustal consumption is, therefore, not large. Studies by Hanks (1971) and by Watts and Talwani (1975) have shown that the positive anomalies associated with the basement rise seawards of many trenches can be explained by a simple upward flexure of the lithosphere. This implies that the total gravity effect of the downgoing slab is small and confined to the trench and the island arc. Thus, either the density contrast between the lithosphere and the adjacent mantle is small (less than 0.05 gm cm^{-3}) or the slab is regionally compensated at depth.

The small influence of descending oceanic lithosphere on the regional gravity field means that the configuration of the slab cannot be deduced from gravity data alone. Watts and Talwani (1975) have suggested that the regional gravity high over the trench–island arc system in the western Pacific may be associated with the process of crustal extension behind the island arcs (Section 35.3.2), rather than with the descending plate.

35.6.4. MAGNETIC ANOMALIES AND THE ZONE OF CRUSTAL ACCRETION

One aspect of sea-floor spreading which has received much attention is the distance over which new crust is generated at the ridge crest. Surveys, such as that by Loncarevic *et al.* (1966), indicate that a simple model involving dyke injection precisely along the ridge axis does not account for the variability of the magnetic anomalies. Matthews and Bath (1967) approached the problem by assuming that fresh material is not only injected axially, but is also distributed about the median line with a normal probability of a certain standard deviation. They carried out a series of calculations in which normally distributed, pseudo-random numbers, with different standard deviations, were used to determine the location of each intrusion of new crustal material. It is clear that the zone of accretion is only a few kilometres wide. With a standard deviation of approximately 5 km they were able to match the amplitudes of observed profiles closely; almost all of the central block consists of normally magnetized crust whereas the adjacent blocks are contaminated with material of opposite polarity to such an extent that the amplitudes of the simulated anomalies closely correspond to those seen aboard ship. Harrison (1968), employing a dyke-injection model to explain anomalies on the East Pacific Rise and Reykjanes Ridge, found the width of emplacement to be 3 km or less. If the percentage of extrusives associated with the dykes is large, the width could be considerably less than this, because the randomness of the lava flows tends to smear out the reversal boundaries in a similar manner to a scattered injection process.

In addition to the surface magnetic measurements, near-bottom total field observations have been made to study the fine structure of the anomalies at ridge crests (Spiess and Mudie, 1970). Being much closer to the anomaly source, these profiles reveal short-wavelength information which is undetectable at the surface (Fig. 35.27). Unfortunately, near-bottom systems are rare and for this reason only a few areas have been examined in detail. A study of part of the Gorda Ridge has been described by Atwater and Mudie (1973). In crossing the polarity reversals seen on the surface magnetometer profiles they found that the transition zone is 1·5–3·5 km wide. This is an important result because the extent of the transition region is a measure of the width of crustal emplacement. To simulate the observed anomalies it is necessary to assume that most of the new magnetic material was added within 2 km of the spreading centre. A similar result has been obtained on the Galapagos spreading centre by Klitgord *et al.* (1975).

The prediction that crustal accretion takes place in a zone 5 km, or less, in width receives strong support from observations made during a recent intensive study of the crest of the Mid-Atlantic Ridge near 37°N. (The FAMOUS area; Fig. 35.28). The rift valley in this region is about 30 km wide and 2 km deep and contains a central inner rift, 1·5–3 km wide, the floor of

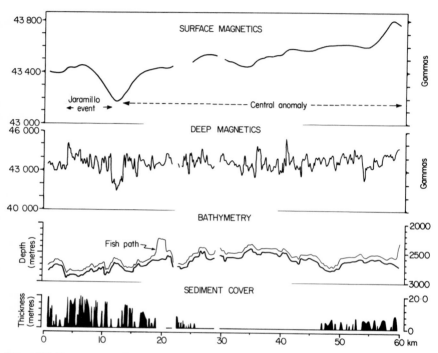

FIG. 35.27. East–west profile across the crest of the East Pacific Rise showing the magnetic anomalies and sediment distribution recorded with an instrument package towed about 100 m above the sea bottom. Note the high frequency character of the anomalies and their relationship to the surface magnetic profile. Small topographic irregularities are important in producing field fluctuations of less than 3 km wavelength (Klitgord *et al.*, 1975). Sediments are absent from the ridge crest. They gradually thicken on the ridge flanks, with increasing crustal age. (Fig. 24 of Spiess and Mudie (1970); reproduced by permission.)

which is depressed a further 100–400 m. Remarkably detailed evidence for crustal extension in the area has been obtained using submersibles, sidescan sonar and narrow-beam echo-sounders (Heirtzler and Bryan, 1975; Laughton and Rusby, 1975). Several major, high-angle normal faults, up to 50 km in length, have been mapped on the flanks of the rift valley as well as many minor ones. It is the narrow inner rift, however, which has the greatest

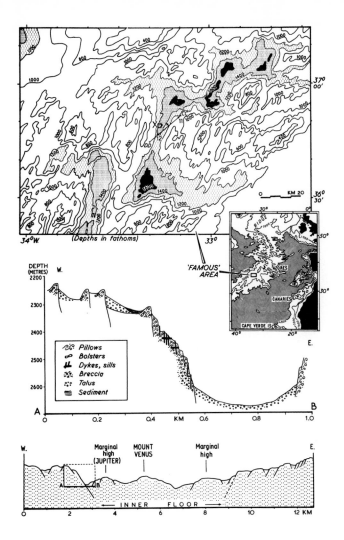

FIG. 35.28. UPPER: Bathymetric map of the FAMOUS area, at the crest of the Mid-Atlantic Ridge south of the Azores. (Adapted from Heirtzler and Bryan, 1975, with the permission of the authors and *Scientific American Inc.*) Note that the rift valley is displaced along two E.–W. fracture zones near 36°30′N. and 37°00′N. X shows the location of the photograph in Fig. 35.29. The sonograph in Fig. 35.30 was recorded along track CC–DD. The sea floor was insonified to the NW.

MIDDLE: Sketch showing the surface geology observed during submersible dive up the scarp bounding the western side of the inner floor of the rift valley (no vertical exaggeration). (After Arcyana, 1975. Copyright © 1975 by the American Association for the Advancement of Science.)

LOWER: Profile showing bathymetry in region of submersible dive. The section is located close to X in the upper part of the diagram.

E. J. W. JONES

density of active tectonic features, and it is this region where major crustal accretion appears to occur. The inner rift is strongly fissured and is made up of lavas often showing remarkable forms (Fig. 35.29; see Heirtzler and Bryan, 1975; Arcyana, 1975; Ballard, 1975). The most recent eruptive activity appears to be confined to a number of discontinuous median ridges, about

FIG. 35.29. Pillow lavas photographed from the submersible *Alvin* at position X in Fig. 35.28 (36°49.05′N., 33°16·15′W.). Water depth 2746 m. (Photograph kindly provided by J. R. Heirtzler and J. D. Phillips, Woods Hole Oceanographic Institution.)

100–250 m high and up to 10 km in length (Figs. 35.28 and 35.30). In form, they bear a close resemblance to the moberg ridges of Iceland, which are the products of the extrusion of basaltic lavas beneath ice. The median ridges, such as Mount Venus in Figs 35.28 and 35.30, appear to be built up by localized fissure eruptions.

The lack of continuity of the ridges may be a result of each interval of extrusive activity being followed by a period of disruption and subsidence. In a preliminary model, published by Moore *et al.* (1974), continued subsidence produces a tilting of the lava sequence towards the axis of the rift valley, former centres of median ridges subsiding to the level of intrusive

activity. (Fig. 35.31). The junction between the pillow basalts and the intrusive section is correlated with the *Layer 2A/2B* seismic boundary discussed
in Section 35.5.1.

There is some doubt as to whether the magma chamber is as shallow as

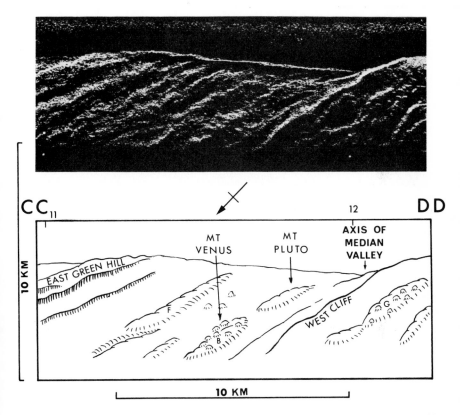

FIG. 35.30. Oblique acoustic view of median valley to NW. of track CC–DD in Fig. 35.28.
Reflections from acoustic targets are white areas on the sonograph. Scanned range, 6·7 km. An
interpretative drawing by Laughton and Rusby (1975) is included. Note the lineated appearance
of the topography. Within the median valley echoes are received from low volcanic ridges, such
as Mount Venus and Mount Pluto. Along the valley sides reflections from long, normal faults,
such as West Cliff, can be seen. (Figure 15 of Laughton and Rusby, 1975. Reproduced by kind
permission of the authors and Pergamon Press.)

Moore *et al.* (1974) have proposed. Teleseismic data certainly indicate the
existence of a local region of low Q beneath the crests of spreading ridges
(Molnar and Oliver, 1969; Francis, 1969), but the upper level of the anomalous region cannot be defined very precisely. Other evidence is conflicting.

For example, Fowler and Matthews (1974) failed to find a low velocity zone, indicative of a region of partial melting, on a long-range seismic refraction profile shot in the rift valley of the FAMOUS area.

However, Francis and Porter (1973) have postulated a shallow region of low velocity on the ridge crest at 45°N. to account for variations in seismic

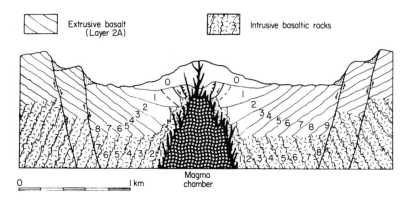

Fig. 35.31. Preliminary structural model of the inner rift valley in the FAMOUS area. Relative ages of basaltic units are shown by numbers 0–9 (0, youngest). (After Moore *et al.*, 1974. Reproduced by permission.)

propagation with direction. They found that a bottom seismometer in the rift valley recorded far more earthquakes than did one placed in the mountainous region to the east. They suggested that this difference can be accounted for by a shadow zone created by a low velocity layer beginning about 3 km under the median valley. A similar conclusion has been reached by Solomon and Julian (1974). Evidence for a low velocity channel under the crest of the East Pacific Rise near 105°W. has been presented by Orcutt *et al.* (1976). Inversion of travel times from a pattern of closely spaced shots received by bottom seismometers indicates a low velocity layer, some 1–2 km thick, starting at a depth of about 2 km. Observations on crust about three million years old failed to detect this shallow anomalous zone, indicating that it is a very localized feature of the ridge crest.

One interpretation of the present data is that shallow (depths less than 3–5 km) magma chambers are short-lived and only exist along parts of the ridge system at any given time, whereas the main region of anomalous mantle is a persistent feature. More micro-earthquake studies and additional seismic experiments carried out in the manner described by Orcutt *et al.* (1976) will help to resolve the question of crustal magma chambers, a problem which is of considerable petrogenetic significance.

35.6.5. ASYMMETRICAL SPREADING

Along most parts of the oceanic ridge system the magnetic anomalies are disposed symmetrically about the ridge crest, reflecting the symmetry of crustal generation at constructional plate boundaries. Such symmetry is, perhaps, not entirely unexpected Because the strength of the lithosphere is strongly temperature-dependent, plate rupture and the injection of fresh crustal material would tend to occur where the plates are hottest and, therefore, weakest (Morgan, 1971). The site of the accretion process would thus be closely confined to the zone of the most recent dyke injection which, as has been seen in Section 35.6.4, is less than 5 km wide. Surveys carried out over the past ten years have revealed, however, that crustal accretion is not always a symmetrical process. Asymmetrical spreading has been reported south of Australia (Weissel and Hayes, 1971, 1974), near the Galapagos Islands (Hey *et al.*, 1973), on the Kolbeinsey Ridge between the Jan Mayen Ridge (M in Fig. 35.7) and Greenland (Johnson *et al.*, 1972), and in the NE. Pacific (Elvers *et al.*, 1973).

The best-documented example of asymmetrical spreading occurs along a 300-km section of the South-east Indian Ocean Ridge, between Australia and Antarctica (Weissel and Hayes, 1971, 1974). This region, known as the Australian–Antarctic Discordance, is separated from normal parts of the ridge by N.–S. fracture zones. Within the anomalous zone, the morphology of the sea floor, with its N.–S. grain, is distinctly different (Fig. 35.32). Magnetic anomalies are linear, but lack symmetry about the ridge axis (Fig. 35.32). Using the reversal time-scale described by Heirtzler *et al.* (1968), Weissel and Hayes (1974) have shown that the spreading rate on the northern limb $(4\cdot0 \text{ cm year}^{-1})$ is appreciably faster than that on the southern $(3\cdot0 \text{ cm year}^{-1})$. Because there are no discontinuities in the total rate of separation of the Australian and Antarctic plates, the geometry requires that the length of the transform faults bounding the discordance should change with time.

Hey *et al.* (1973) have attributed the asymmetrical pattern of magnetic anomalies in the Galapagos region to many small southward jumps of the ridge crest, which tend to keep the spreading centre close to the Galapagos hot-spot. Ridge jumps also appear to be important in the Atlantic (Vogt *et al.*, 1971). Discrete displacements of the ridge axis do not seem to have occurred within the Australian–Antarctic Discordance. The magnetic anomalies indicate that asymmetrical spreading has been a continuous process down to a scale of 10 km or less (Hayes, 1976). The lack of symmetry may arise from long-term thermal contrasts between the adjacent plates. Hayes (1971) has suggested that the thermal gradients in the Antarctic plate are higher than those in the Australian plate. Such a difference would tend

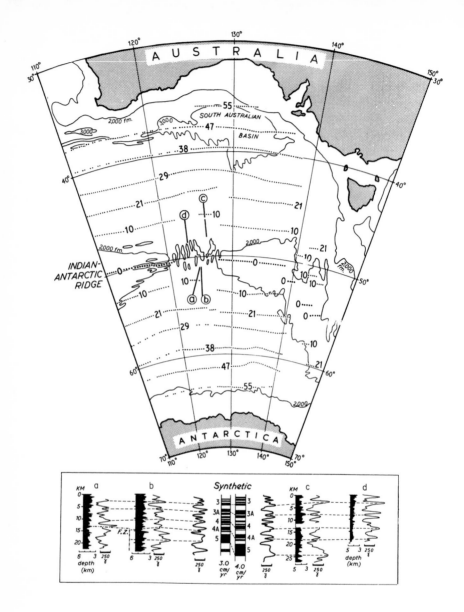

FIG. 35.32. The Australian–Antarctic Discordance, a region of asymmetrical sea-floor spreading. The numbers on the dotted lines refer to ages in millions of years of selected magnetic anomalies. Bathymetry is in fathoms (1 fathom = 1·83 m). Note the N.–S. grain of the topography near the ridge crest and the difference in the spacing of the magnetic anomalies of the same age to the north and south. (After Weissel and Hayes, 1974. Diagram reproduced with the authors' permission. Copyright © American Geophysical Union.)

to maintain the region of highest temperature gradient, where splitting and accretion takes place, a short distance south of the axis of the most recent injection of crustal material. Thus, new crust is added at a slower rate onto the hotter Antarctic plate. The thermal contrasts may be brought about by a local downward movement of mantle material beneath the anomalous zone (Weissel and Hayes, 1974), a flow which could explain why the depth of the sea floor is greater than would be expected from the crustal age. It would also account for the long-wavelength gravity low, apparent from satellite data (Gaposchkin and Lambeck, 1971), over the discordance (see also Sections 35.8, 35.9). These tentative suggestions have yet to be rigorously tested.

35.7. DEEP-SEA SEDIMENTS AND PLATE MOTIONS: PALAEOCEANOGRAPHY

In most regions, the oceanic basement is covered with a sedimentary layer varying in thickness from a few metres to several kilometres. The sediments in the basins close to the continents have mostly been derived from the adjacent land masses, a large proportion of the material reaching its site of deposition in dense, sediment-laden bottom currents; either fast-moving turbidity currents or slower, geostrophic currents (Heezen et al., 1959, 1966; Davies and Gorsline, 1976). The areas seaward of the abyssal plains and marginal trenches are usually covered with pelagic clays and organic oozes. The clays, generally brown in colour, cover large expanses of the deep-ocean floor, particularly in the Pacific. The oozes are composed of the remains of planktonic organisms and can be classified according to the dominant form present (see Chapter 24). The most widely distributed of these is the calcareous *Globigerina* ooze, which covers nearly half the Earth's surface. Radiolarian and diatom oozes are important siliceous deposits found in the high productivity areas of the equatorial Pacific and the convergence zones.

It has long been realized that deep-sea sediments are sensitive indicators of the oceanic environment and that they thus hold important information relating to the geological evolution of the ocean basins (Kuenen, 1950). If the question of the variation in thickness of the sedimentary layer is considered, the problem must arise as to whether or not this is consistent with the formation of the underlying basement by the process of sea-floor spreading. The answer to this has been provided by the many thousands of kilometres of seismic reflection profiles recorded over the past 15 years. These show a general absence of sediments over the crests of spreading ridges and a gradual thickening of the pelagic cover with increasing distance down the ridge flanks (Fig. 35.33; see also Ewing and Ewing, 1970). Such a sediment distribution is clearly compatible with the pattern of crustal ages predicted by spreading from a ridge axis. The rate at which the sediments thicken, however, is quite variable because of differences in spreading velocity, surface productivity and terrigenous influence.

In some areas, systematic changes in sediment thickness have been important in determining the locations of relict spreading centres. Thin

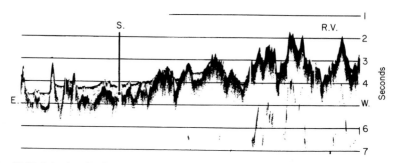

FIG. 35.33. Seismic reflection profile, recorded by Professor J. L. Worzel and the author aboard R. V. *Vema*, across the crest of the Mid-Atlantic Ridge along latitude 42°20′N. The profile runs in an E.–W. direction and is 330 km long. The rift valley (R.V.) is well defined. The sediment layer (S.) thickens away from the ridge axis. Two-way travel time is expressed in seconds. In water, 1 s ~ 730 m and, in sediments, 1 s ~ 1000 m. (Section is reproduced by kind permission of the Lamont–Doherty Geological Observatory.)

sediments covering a shallow basement surface in the central part of the Labrador Sea led Drake *et al.* (1963) to suggest that the basin opened by sea-floor spreading along a median rift. The presence of a region of thin sediments between the Norwegian continental margin and the Jan Mayen Ridge (Eldholm and Windisch, 1974) delineates a spreading axis which was active during the early stages of the opening of the Norwegian Sea. Avery *et al.* (1968) have demonstrated the symmetry of magnetic anomalies about this line.

An indication of large-scale horizontal motion of the sea floor has been obtained from the palaeomagnetism of some of the early Atlantic sediment cores recovered by the drilling vessel D.V. *Glomar Challenger* (Sclater and Cox, 1970). At Site 10 (Fig. 35.34) the magnetic inclination of Upper Miocene sediments is close to that expected from the present latitude (32°52′N). Deeper in the succession, however, the inclination decreases with age. In Fig. 35.34 the palaeolatitude of the drilling-site is plotted against time. With the exception of one sample, which appears to be magnetically unstable, the palaeolatitude shows a remarkably systematic variation, considering the many changes that can affect the weak magnetism of sediments after their

deposition. The shift in latitude is close to that expected from the North
American Cretaceous palaeomagnetic data of Grommé *et al.* (1967), and
thus is consistent with the view that the western Atlantic and the North
American continent form part of the same rigid plate. Although similar
analyses have been possible elsewhere in the oceans (Blow and Hamilton,
1975), the samples from the D.V. *Glomar Challenger* have often proved to be
too disturbed by drilling, or too unstable magnetically, for palaeomagnetic
studies.

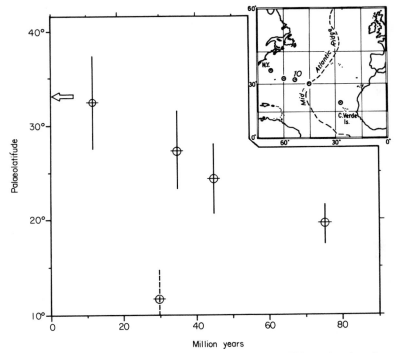

FIG. 35.34. The palaeolatitude of DSDP Site 10 (32°52′N., 52°13′W.) as a function of age. The
palaeolatitudes (with error bars indicated) have been deduced from the magnetic inclination of
sediment cores ranging in age from Maestrichtian to Miocene. The arrow marks the present
latitude of Site 10. (After Sclater and Cox, 1970).

Other evidence of plate movement can, however, be derived directly from
major lithological changes observed in the drill cores. This is possible
because the kind of sediment deposited on the ocean floor is closely related
to the water depth. Murray and Renard (1891) were the first to appreciate
that *Globigerina* ooze is generally restricted to the shallower parts of the

ocean basins, those regions less than about 4500 m deep. They correctly ascribed this feature to the rapid dissolution of the calcareous tests of the foraminifera below this critical depth known as the calcite compensation depth (CCD) (see Chapter 29). Below the CCD, carbonate-free abyssal clays or siliceous oozes generally accumulate.

Drilling results from areas presently below the CCD reveal that the abyssal clays are underlain by at least one thick layer of calcareous sediment. In the western Pacific two layers of carbonate sediment have been found, one resting directly on the basement and the other sandwiched between two clay units (Fig. 35.35; see also Heezen et al., 1973). At first sight, the transition from clay to a carbonate might be accounted for by past fluctuations in the depth of the CCD. However, the age of the abyssal clay/carbonate contacts is not the same everywhere and the contacts are, in fact, systematically time-transgressive. In the western Pacific they increase in age northwards; in the Atlantic they generally become older with increasing distance from the crest of the Mid-Atlantic Ridge.

These apparently simple stratigraphic relationships can readily be interpreted in terms of sea-floor spreading. At the present time, the crests of the active ridges lie above the CCD so that a carbonate ooze is the first sediment to be deposited on newly formed oceanic basement. It is not until the sea floor moves below the CCD that siliceous oozes or abyssal clays form the principal bottom deposit. Because of lateral motion of the crust, therefore, the carbonate/clay boundary becomes older with increasing crustal age.

The second layer of calcareous sediments encountered during drilling in the NW. Pacific is also a time-transgressive unit (Fig. 35.35). It has been interpreted as a deposit formed beneath the equatorial belt of high surface productivity where the CCD is depressed as far as the sea floor. (Heezen et al., 1973; Winterer, 1973). A model to explain the stratigraphy is shown in Fig. 35.36. The basal calcareous section is deposited on a ridge crest, running roughly E.–W., which is located south of the Equator. As the crust subsides below the CCD, abyssal clays begin to accumulate. Northward movement brings the plate into the equatorial belt, where the layer of clay is first covered by siliceous sediments and then by a thick carbonate ooze. Abyssal clay deposition resumes when the sea floor moves out of the equatorial region. As a result of the plate motions the upper section of carbonates thus becomes older northwards. The ages of the equatorial carbonate layer determined at 14 drilling sites are given in Fig. 35.36. The pattern of ages is consistent with the northward motion of the Pacific plate inferred from magnetic anomalies. Furthermore, the observed thicknesses of the main lithological units are close to those predicted from the rate of northward movement of the crust, the deposition rates and the width of the productivity

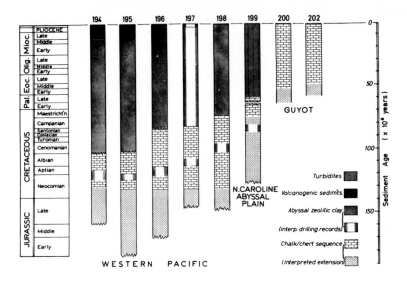

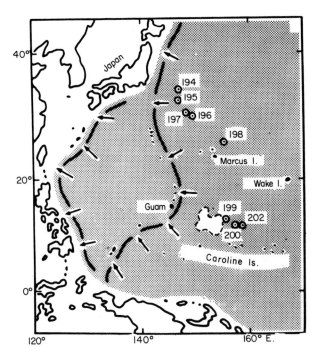

FIG. 35.35. Stratigraphy of eight holes drilled by the *Glomar Challenger* in the western Pacific. Note that the lithological units are time-transgressive. (After Heezen *et al.*, 1973. Diagram reproduced with the permission of the publishers of *Nature*.)

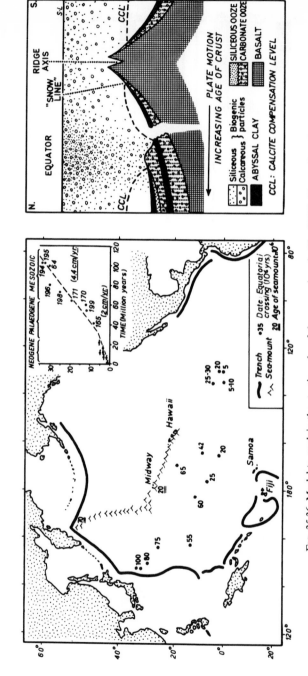

Fig. 35.36. Model to explain the stratigraphy of the successions in Fig. 35.35. (For explanation see text.)

zones. (Heezen *et al.*, 1973). The model has been discussed further by Berger (1976) who has also considered fluctuations in the depth of the CCD during the Tertiary.

Other changes within the oceanic sedimentary column also appear to reflect plate motions. From a study of depositional regimes over the past 80 million years, Jones *et al.* (1970) found evidence for a major alteration in the circulation pattern of the North Atlantic in late Cretaceous–early Tertiary time. They presented seismic and hydrographic data to show that in several areas of the North Atlantic sedimentation rates are largely governed by the movement of bottom waters of Arctic origin. Depositional rates often change rapidly over distances of a few kilometres, with the result that distinctive sedimentary features, such as the Blake Outer Ridge (Ewing and Ewing, 1964; Heezen *et al.*, 1966) and the Feni Ridge (Jones *et al.*, 1970), are built up. Although common in the Tertiary, current-controlled deposits appear to be absent from seismic recordings of sediments of Cretaceous age, implying that thermohaline circulation was not initiated until early Tertiary time.

The early Tertiary was thus a time of major oceanographic change, one to which Jones *et al.* (1970) have attributed the opening of the Norwegian and the Labrador Seas and the formation of the basin between the Rockall Plateau (L in Fig. 35.7) and Greenland, since these plate movements then

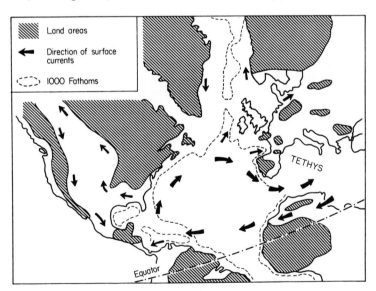

FIG. 35.37. Surface currents of the North Atlantic in late Cretaceous time. (After Jones *et al.*, 1970. Diagram reproduced by permission of the American Geophysical Union.) 1 fathom = 1·83 m.

allowed exchange of deep water between the Arctic and the Atlantic. With the creation of major basins in the north and the restriction or elimination of connections to the Tethys and the Pacific, the surface circulation altered during the Lower Tertiary from the form indicated for the late Cretaceous in Fig. 35.37 to one more closely resembling the present pattern. The change in shape of the main North Atlantic gyre appears to be expressed in the sedimentary column by unconformities on the North American continental margin. Similar arguments have been used to interpret the oceanic sedimentary sequences of the South Pacific and Indian Ocean (Kennett *et al.*, 1972; Davies *et al.*, 1975). With much more data from continuously cored drilling sites available, detailed analyses are now possible.

35.8. CAUSES OF SEA-FLOOR SPREADING

The process of sea-floor spreading is intimately associated with the forces that move the lithosphere and several driving mechanisms have been proposed for these.

(i) Expansion of the Earth: horizontal displacements of the lithosphere are caused by global expansion resulting from a decrease in the universal gravitational constant, G (Heezen, 1962; Hoyle, 1972).

(ii) Pull from the subduction zones: plates are in motion because they are being pulled towards the subduction zones by the sinking slabs of dense lithosphere (Elsasser, 1969).

(iii) Push from the crests of spreading ridges: plates are driven by forces arising from the hydrostatic pressure created by the elevation of the spreading ridges (Lliboutry, 1972) and their associated mantle inhomogeneities (Artyushkov, 1973).

(iv) Viscous drag by convection currents on the base of lithosphere: the movement of the lithosphere is primarily maintained by convection in the mantle, and motion occurs because viscous forces couple the plates to the moving mantle below (McKenzie, 1969).

Of these suggestions, only the last appears to be adequate to explain large-scale plate motions. The rate of increase of the Earth's radius required to account for the area of the present ocean floor is about $2\,\mathrm{cm\,year^{-1}}$, a value which is too large to be compatible with palaeomagnetic data (Ward, 1963) or with allowable variations of G. A recent determination of $\dot{G}$, from changes in the length of the day (Hoyle, 1972), gives a figure of $1 \cdot 5 \times 10^{-2}$ $\mathrm{km^2\ year^{-1}}$ for the rate of production of oceanic crust. This is about two orders of magnitude too low (McKenzie *et al.*, 1974). Le Pichon (1968) has pointed out that if the generation of oceanic crust was entirely compensated by a change in the Earth's radius then the pattern of spreading would require the equatorial bulge to increase at a rate which is unacceptably large. However, despite these fundamental objections, the expansion hypothesis has recently been resurrected (Owen, 1976).

Several observations indicate that Elsasser's driving mechanism is not the primary cause of plate motions. Thus, the opening of the South Atlantic cannot be explained by gravitational sinking alone since the African and American plates are not being subducted except in local areas. Furthermore, the rates at which plates descend into the mantle appear to increase with their size rather than with the length of the descending slab (McKenzie, 1969). The state of stress in the lithosphere is also inconsistent with Elsasser's suggestion. If the lithosphere acts as a stress guide in the way he proposes then the interiors of plates should be in tension. However, there is strong seismological evidence to show that, except on the ridge crests and near the seaward walls of trenches, plates are, in fact, under compression. A large proportion of earthquakes occurring well away from plate boundaries are characterized by a predominance of compressional thrust faulting, with the principal stress uniform in orientation over wide areas (Sykes and Sbar, 1973). Isacks and Molnar (1971) have been able to determine that in the region of the Tonga Trench the descending plate is under continuous compression from depths of 80–700 km, with the principal stress aligned parallel to the seismic zone. Such a stress field can only be produced if the lithosphere is being pushed into the mantle.

Although the ridge-push mechanism can account for the compression of the lithosphere, it can be eliminated as the major cause of plate movement. If it did provide the main forces, then those small plates which have an extensional boundary should move faster than those of large area since the driving force must be independent of spreading rate or of the size of the plate to which new lithosphere is added (McKenzie, 1972). However, the opposite effect is observed, the fastest relative motion occurring between the largest plates (Le Pichon, 1968). The ridge-push hypothesis can also be rejected because it fails to satisfy energy requirements. McKenzie (1972) has noted that the rate of work done by forces resulting from the elevation of the oceanic ridges (about 2×10^{18} joules year^{-1}) is considerably less than the rate of release of elastic energy from shallow earthquakes (about 6×10^{18} joules year^{-1}).

Mantle convection appears to be the only means of maintaining movements of major plates. However, there is little agreement on the form it takes since many of the physical parameters necessary to derive the flow are poorly known. Because of the complexity of present plate motions, the early view that ridge axes are situated over the rising limbs of large convection cells, with trenches marking the descending flow, has been discarded.

Two models for convection in the mantle have recently been discussed by Morgan (1971, 1972a, b) and by McKenzie and Weiss (1975). Morgan has suggested that flow is governed by narrow, pipe-like plumes of hot material which spread out radially in the upper mantle and produce the driving stresses at the base of the lithosphere. The return flow into the deeper mantle is diffuse. Mantle plumes, presently less than 20 in number, are recognized by localized volcanic outpourings such as those on Hawaii and Iceland. The plumes do not necessarily bear a direct relationship to the plate surfaces, though there is a tendency for them to occur near ridge crests.

To account for present plate motions in the light of numerical and laboratory experiments on convecting fluids, McKenzie and Weiss (1975) have built up a more elaborate model in which convection takes place on two horizontal scales, at depths less than 650 km. The large-scale flow, with an extent of several thousand kilometres, drives the major plates, whereas the smaller-scale motion, extending over a few hundred kilometres, is responsible for the greater part of the heat transfer.

35.9. Concluding Remarks

Observations made over the past two decades have firmly established sea-floor spreading as the dominant process involved in the formation of the ocean basins. Although the timing and geometry of the movement in areas where magnetic anomalies are ill-defined or absent are the subjects of current debate, the history of spreading in most parts of the major oceans is now well determined. The present uncertainties surrounding the evolution of the oceanic crust largely arise because its composition is only known in broad outline. Recent discussions have focused on the results of investigations of ophiolite complexes (Moores and Vine, 1971). There is considerable doubt, however, whether ophiolites represent typical oceanic crust. With the exception of the Bay of Islands complex, their "*Layer 3*" thickness is smaller, by as much as a factor of five, than in normal oceanic crust. Moreover, a large part of the extrusive sections of some complexes, such as Troodos and Vourinos, are composed of andesitic basalts and andesites, suggesting that they were formed close to an island arc rather than in an open-ocean environment, comparable to the central Atlantic or South Pacific (Miyashiro, 1973). Thus, ophiolite studies do not provide a substitute for the direct examination of the oceanic basement by drilling.

The source layer of the linear magnetic anomalies, on which the correctness of the concept of sea-floor spreading initially depended, is still poorly defined despite intensive investigation. Measurements on drill cores suggest that its thickness is much greater than the 400–500 m obtained by anomaly inversion (Section 35.5.3). At several sites drilled by the D.V. *Glomar Challenger* since early 1974, the magnetization of the upper few hundred metres of the basement has proved to be so complex that, if the results had been available 15 years ago, it is perhaps doubtful whether Vine and Matthews would have had the courage to publish their initial paper. Other important problems, which are still largely unsolved, include the origin of the seismic layering in the oceanic crust, the reasons for its variations with spreading rate and crustal age, and the relation between crustal structure and mantle inhomogeneities. The relative roles of compositional and thermal gradients, fracturing processes and metamorphic reactions in this context can obviously be clarified by drilling deep into the oceanic basement. The results of the present International Phase of Ocean Drilling (IPOD) are thus awaited with immense interest.

A major problem which cannot be solved by drilling is the nature of the convective motion that maintains sea-floor spreading. Few geophysical measurements can be related to mantle convection so that the models proposed by Morgan (1972a, b), McKenzie and Weiss (1975) and others cannot be rigorously tested at the present time. Focal mechanisms of intraplate earthquakes (Sykes and Sbar, 1973) cannot yet be interpreted in terms of the forces acting on the base of the lithosphere because the cause of stress in the lithosphere itself is not well understood. Sources of stress not directly associated with the driving mechanism may be involved in intraplate deformation (Turcotte and Oxburgh, 1973). The rate of heat loss from the lithosphere is, in theory, an important indicator of mantle motions, but it is not clear how much of the variability in heat flow arises from processes taking place within the lithosphere. Convective heat transfer in the oceanic crust is important in this respect (Section 35.6.1).

Regional gravity anomalies offer valuable indications of convective motions (Section 35.6.3). The ascending parts of a convection cell give rise to a gravity anomaly partly because of the higher temperatures and partly because of the upward deformation of the surface over the rising current. There has been some uncertainty, however, over the sign of the resulting gravity anomaly because the two effects act in opposite directions, while being of the same order of magnitude. A recent series of numerical experiments, involving simple two-dimensional models of a convecting fluid, have conclusively shown that the gravity effect is dominated by surface deformation at all Rayleigh numbers so that the anomaly is positive over regions where mantle material is ascending (McKenzie *et al.*, 1974). Following earlier work by Talwani and Le Pichon (1969), Sclater *et al.* (1975) have established that positive elevation anomalies in the North Atlantic do indeed occur in areas where the regional gravity anomaly is positive (Section 35.6.3). Although this study has not been fully extended to other regions it is evident that valuable indications of sub-lithospheric flow can be obtained from existing gravity and bathymetric information.

ACKNOWLEDGEMENTS

I thank Dr Drummond H. Matthews, F.R.S. and my research students, Philip Bishop, Donald Goddard and Clement Mgbatogu, for critically reading the manuscript of this chapter. I am also grateful to Colin Stuart for drawing the diagrams.

REFERENCES

Ade-Hall, J. M. (1964). *Geophys. J.R. Astron. Soc.* **9**, 85.
Anderson, R. N. and Hobart, M. A. (1976). *J. Geophys. Res.* **81**, 2968.
Arcyana. (1975). *Science, N.Y.* **190**, 108.
Artyushkov, E. V. (1973). *J. Geophys. Res.* **78**, 7675.
Atwater, T. (1970). *Bull. Geol. Soc. Am.* **81**, 3513.
Atwater, T. and Mudie, J. D. (1973). *J. Geophys. Res.* **78**, 8665.
Avery, O. E., Burton, G. D. and Heirtzler, J. R. (1968). *J. Geophys. Res.* **73**, 4583.
Ballard, R. D. (1975). *Nat. Geogr.* **147**, 604.
Barazangi, M. and Dorman, J. (1969). *Bull. Seism. Soc. Am.* **59**, 369.
Barker, P. F. (1972). *Earth Planet. Sci. Lett.* **15**, 123.
Barrett, D. L. and Aumento, F. (1970). *Can. J. Earth Sci.* **70**, 1117.
Ben-Avraham, Z., Bowin, C. and Segawa, J. (1972). *Nature, Lond.* **240**, 453.
Berger, W. H. (1976). *In* "Chemical Oceanography" (J. P. Riley and R. Chester, eds), Vol. 5, pp. 265–388. Academic Press, London and New York.
Blow, R. A. and Hamilton, N. (1975). *Nature, Lond.* **257**, 570.
Bonatti, E., Honnorez, J. and Ferrara, G. (1970). *Earth Planet. Sci. Lett.* **9**, 247.
Bonatti, E., Honnorez, J. and Ferrara, G. (1971). *Phil. Trans. R. Soc. A*, **268**, 285.
Bott, M. H. P., Browitt, C. W. A. and Stacey, A. P. (1971). *Mar. Geophys. Res.* **1**, 328.
Brune, J. N., De la Cruz, S., Bradner, H., Lomnitz, C., Villegas, C., Reid, I., Reichle, M., Nava, A. and Silva, P. (1972). *Geofisica Internacional*, **12**, 201.
Bullard, E. C. (1954). *Proc. R. Soc. A*, **222**, 408.
Bullard, E. C., Maxwell, A. E. and Revelle, R. (1956). *Adv. Geophys.* **3**, 153.
Bullard, E. C., Everett, J. E. and Smith, A. G. (1965). *Phil. Trans. R. Soc. A*, **258**, 41.
Cann, J. R. (1970). *Nature, Lond.* **226**, 928.
Cann, J. R. (1974) *Geophys. J. R. Astron. Soc.* **39**, 169.
Cann, J. R. and Funnell, B. M. (1967) *Nature, Lond.* **213**, 661.
Chase, C. G. (1972). *Geophys. J.R. Astron. Soc.* **29**, 117.
Christensen, N. I. (1972). *J. Geol.* **80**, 709.
Christensen, N. I. and Salisbury, M. H. (1975). *Rev. Geophys. Space Phys.* **13**, 57.
Corliss, J. B. (1971). *J. Geophys. Res.* **76**, 8128.
Cox, A. (1969). *Science, N.Y.* **163**, 237.
Cox, A., Doell, R. R. and Dalrymple, G. B. (1964). *Science, N.Y.* **144**, 1537.
Davies, T. A. and Gorsline, D. (1976). *In* "Chemical Oceanography" (J. P. Riley and R. Chester, eds), Vol. 5, pp. 1–80. Academic Press, London and New York.

Davies, T. A., Weser, O. E., Luyendyk, B. P. and Kidd, R. B. (1975). *Nature, Lond.* **253**, 15.

Davis, E. E. and Lister, C. R. B. (1974). *Earth Planet. Sci. Lett.* **21**, 405.

Dick, H., Heirtzler, J. R., Dmietriev, L. and others (1976). *Nature, Lond.* **262**, 768.

Dietz, R. S. (1962). *In* "The Crust of the Pacific Basin" (G. A. Macdonald and H. Kuno, eds), *Geophys. Mon.* **6**. American Geophysical Union.

Dietz, R. S. and Holden, J. C. (1970). *Scient. Am.* **223**(4), 30.

Drake, C. L., Campbell, N. J., Sander, G. and Nafe, J. E. (1963). *Nature, Lond.* **200**, 1085.

Duffield, W. G. (1924). *Mon. Not. R. Astron. Soc. Geophys. Suppl.* **1**, 161.

Eldholm, O. and Windisch, C. C. (1974). *Bull. Geol. Soc. Am.* **85**, 1661.

Elsasser, W. M. (1969). *In* "The Application of Modern Physics to the Earth and Planetary Interiors" (S. K. Runcorn, ed.), pp. 223–246. Wiley-Interscience, London.

Elvers, D., Srivastova, S. P., Potter, K., Morley, J. and Seidel, D. (1973). *Earth Planet. Sci. Lett.* **20**, 211.

Ewing, J. and Ewing, M. (1959). *Bull. Geol. Soc. Am.* **70**, 291.

Ewing, J. and Ewing, M. (1970). *In* "The Sea" (A. E. Maxwell, ed.), Vol. 4(1), pp. 1–51. Wiley-Interscience, New York.

Ewing, M. and Ewing, J. (1964). *In* "Studies on Oceanography", pp. 525–537. Tokyo University Press, Japan.

Ewing, M. and Press, F. (1955). *Geol. Soc. Am. Spec. Pap.* **62**, 1.

Foster, J. H. and Opdyke, N. D. (1970). *J. Geophys. Res.* **75**, 4465.

Fowler, C. M. R. and Matthews, D. H. (1974). *Nature, Lond.* **249**, 752.

Fox, P. J. and Opdyke, N. D. (1973). *J. Geophys. Res.* **78**, 5139.

Fox, P. J., Schreiber, E. and Peterson, J. J. (1973). *J. Geophys. Res.* **78**, 5155.

Francis, T. J. G. (1969). *Geophys. J. R. Astron. Soc.* **17**, 507.

Francis, T. J. G. and Porter, I. T. (1972). *Nature, Lond.* **240**, 547.

Francis, T. J. G. and Porter, I. T. (1973). *Geophys. J. R. Astron. Soc.* **34**, 279.

Gaposhkin, E. M. and Lambeck, K. (1971). *J. Geophys. Res.* **76**, 4855.

Gass, I. G. and Masson-Smith, D. (1963). *Phil. Trans. R. Soc. A*, **255**, 417.

Gass, I. G. and Smewing, J. D. (1973). *Nature, Lond.* **242**, 26.

Grindley, G. W. (1974). *In* "Mesozoic-Cenozoic Orogenic Belts" (A. M. Spencer, ed.), p. 387. Geological Society of London.

Grommé, C. S., Merrill, R. T. and Verhoogen, J. (1967). *J. Geophys. Res.* **72**, 5661.

Grow, J. A. and Atwater, T. (1970). *Bull. Geol. Soc. Am.* **81**, 3715.

Hall, J. M. (1976). *J. Geophys. Res.* **81**, 4223.

Hall, J. M. and Ryall, J. (1976). *In* "Initial Reports of the Deep Sea Drilling Project", Vol. 37, Ch. 16. U.S. Government Printing Office, Washington, D.C.

Hanks, T. C. (1971). *Geophys. J.R. Astron. Soc.* **23**, 173.

Harrison, C. G. A. (1968). *J. Geophys. Res.* **73**, 2137.

Hayes, D. E. (1971). *Abstr. Am. Geophys. Union Fall Mtg.*

Hayes, D. E. (1976). *Bull. Geol. Soc. Am.* **87**, 994.

Hayes, D. E. and Rabinowitz, P. D. (1975). *Earth Planet. Sci. Lett.* **28**, 105.

Hecker, O. (1903). *Veroff. preus.; Geodat. Inst.* No. 11.

Hecker, O. (1908). *Zentral-bureau der Internationalen Erdmessung, N.F.* No. 16.

Heezen, B. C. (1962). *In* "Continental Drift" (S. K. Runcorn, ed.), pp. 235–288. Academic Press, New York and London.

Heezen, B. C. and Ewing, M. (1961). *In* "Geology of the Arctic" (G. O. Raasch, ed.), pp. 622–642. Toronto University Press, Canada.

Heezen, B. C. and Tharp, M. (1967). Map of Indian Ocean Floor. *Nat. Geogr.* **(October)**.

Heezen, B. C. and Tharp, M. (1968). Map of Atlantic Ocean Floor. *Nat. Geogr.* **(June)**.
Heezen, B. C. and Tharp, M. (1969). Map of Pacific Ocean Floor. *Nat. Geogr.* **(October)**.
Heezen, B. C. and Tharp, M. (1971). Map of Arctic Floor. *Nat. Geogr.* **(October)**.
Heezen, B. C., Tharp, M. and Ewing, M. (1959). *Geol. Soc. Am. Spec. Pap.* **65**.
Heezen, B. C., Hollister, C. D. and Ruddiman, W. F. (1966). *Science, N.Y.* **152**, 502.
Heezen, B. C., MacGregor, I. D., Foreman, H. P., Forristal, G., Hekel, H., Hesse, R., Hoskins, R. H., Jones, E. J. W., Kaneps, A., Krasheninnikov, V. A., Okada, H. and Ruef, M. H. (1973). *Nature, Lond.* **241**, 25.
Heirtzler, J. R. and Bryan, W. B. (1975). *Scient. Am.* **233**(2), 78.
Heirtzler, J. R. and Hayes, D. E. (1967). *Science, N.Y.* **157**, 185.
Heirtzler, J. R. and Le Pichon, X. (1965). *J. Geophys. Res.* **70**, 4013.
Heirtzler, J. R., Le Pichor, X. and Baron, J. G. (1966). *Deep-Sea Res.* **13**, 427.
Heirtzler, J. R., Dickson, G. O., Herron, E. M., Pitman, W. C., III and Le Pichon, X. (1968). *J. Geophys. Res.* **73**, 2119.
Heirtzler, J. R., Veevers, J. V. and others (1974). *In* "Initial Reports of the Deep Sea Drilling Project", Vol. 27, pp. 89–127. U.S. Government Printing Office, Washington D.C.
Herron, E. M. (1972). *Bull. Geol. Soc. Am.* **83**, 1671.
Herron, E. M., Dewey, J. F. and Pitman, W. C., III (1974). *Geology,* **2**, 377.
Hess, H. H. (1962). *In* "Petrologic Studies: a volume in honor of A. F. Buddington", pp. 599–620. Geological Society of America.
Hess, H. H. and Maxwell, J. C. (1953). *In* "Proceedings of the Seventh Pacific Science Congress, New Zealand", **2**, 14.
Hey, T. N., Johnson, G. L. and Lowrie, A. (1973). *EOS* (*Am. Geophys. Union Trans.*) **54**, 244.
Hill, M. N. (1957). *Physics Chem. Earth,* **2**, 129.
Hollister, C. D., Ewing, J. I. and others (1972). *In* "Initial Reports of the Deep Sea Drilling Project", Vol. 11, pp. 75–104. U.S. Government Printing Office, Washington, D.C.
Houtz, R. and Ewing, J. (1976). *J. Geophys. Res.* **81**, 2490.
Hoyle, F. (1972). *Quart. J. R. Astron. Soc.* **13**, 328.
Irving, E. (1966). *J. Geophys. Res.* **71**, 6025.
Isacks, B. and Molnar, P. (1971). *Rev. Geophys. Space Phys.* **9**, 103.
Isacks, B., Oliver, J. and Sykes, L. R. (1968). *J. Geophys. Res.* **73**, 5855.
Jeffreys, H. (1970). "The Earth", 5th Edn. Cambridge University Press, England.
Johnson, G. L., Southall, J. R., Young, P. W. and Vogt, P. R. (1972). *J. Geophys. Res.* **77**, 5688.
Jones, E. J. W., Ewing, M., Ewing, J. I. and Eittreim, S. L. (1970). *J. Geophys. Res.* **75**, 1655.
Karasik, A. M. (1968). *In* "Geophysical Methods of Prospecting in the Arctic". Vol. 5, pp. 8–19. Leningrad.
Karig, D. E. (1971). *J. Geophys. Res.* **76**, 2542.
Keen, C. and Tramontini, C. (1970). *Geophys. J. R. Astron. Soc.* **20**, 473.
Kennett, J. P., Burns, R. E., Andrews, J. E., Churkin, M., Davies, T. A., Dumitrica, P., Edwards, A. R., Galehouse, J. S., Packham, G. H. and van der Lingen, G. J. (1972). *Nature, Phys. Sci.* **239**, 51.
Klitgord, K. D. and Mudie, J. D. (1974). *Geophys. J. R. Astron. Soc.* **38**, 563.
Klitgord, K. D., Mudie, J. D., Larson, P. A. and Grow, J. A. (1973). *Earth Planet. Sci. Lett.* **20**, 93.

Klitgord, K. D., Huestis, S. P., Mudie, J. D. and Parker, R. L. (1975). *Geophys. J. R. Astron. Soc.* **43**, 387.

Kuenen, P. H. (1950). "Marine Geology". John Wiley and Sons, London.

Langseth, M. G. and Von Herzen, R. P. (1970). *In* "The Sea" (A. E. Maxwell, ed.), Vol. 4(1), pp. 299–352. Wiley-Interscience, New York.

Larson, R. L. (1975). *Geology*, **3**, 69.

Larson, R. L. and Chase, C. G. (1972). *Bull. Geol. Soc. Am.* **83**, 3627.

Larson, R. L. and Hilde, T. W. (1975). *J. Geophys. Res.* **80**, 2586.

Larson, R. L. and Ladd, J. W. (1973). *Nature, Lond.* **246**, 209.

Larson, R. L. and Pitman, W. C., III (1972). *Bull. Geol. Soc. Am.* **83**, 3645.

Laughton, A. S. (1971). *Nature, Lond.* **232**, 612.

Laughton, A. S. and Rusby, J. S. M. (1975). *Deep-Sea Res.* **22**, 279.

Laughton, A. S., Whitmarsh, R. B. and Jones, M. T. (1970). *Phil. Trans. Roy. Soc. A*, **267**, 227.

Le Pichon, X. (1968). *J. Geophys. Res.* **73**, 3661.

Le Pichon, X. (1969). *Tectonophys.* **7**, 385.

Le Pichon, X., Bonnin, J. and Francheteau, J. (1973). "Plate Tectonics". Elsevier, Amsterdam.

Lister, C. R. B. (1970). *J. Geophys. Res.* **75**, 2648.

Lister, C. R. B. (1972). *Geophys. J.R. Astron. Soc.* **26**, 575.

Lliboutry, L. (1972). *J. Geophys. Res.* **77**, 3759.

Loncarevic, B. D., Mason, C. S. and Matthews, D. H. (1966). *Can. J. Earth Sci.* **3**, 327.

Lowrie, W. (1974). *J. Geophys./Z. für Geophys.* **40**, 513.

Macdonald, K. C. and Mudie, J. D. (1974). *Geophys. J.R. Astron. Soc.* **36**, 245.

Mason, R. G. (1958). *Geophys. J.R. Astron. Soc.* **1**, 320.

Mason, R. G. and Raff, A. D. (1961). *Bull. Geol. Soc. Am.* **72**, 1259.

Matsumoto, T. and Kimura, T. (1974). *In* "Mesozoic and Cenozoic Orogenic Belts" (A. M. Spencer, ed.), pp. 513–541. Geological Society of London.

Matthews, D. H. (1971). *Phil. Trans. R. Soc. A* **268**, 551.

Matthews, D. H. and Bath, J. (1967). *Geophys. J. R. Astron. Soc.* **13**, 349.

Maxwell, A., Von Herzen, R. P., Hsu, K., Andrews, J. E., Saito, T., Percival, S. F., Milow, E. D. and Boyce, R. E. (1970). *Science, N.Y.* **168**, 1047.

Maynard, G. L. (1970). *Science, N.Y.* **168**, 120.

McKenzie, D. P. (1967). *J. Geophys. Res.* **72**, 6261.

McKenzie, D. P. (1969). *Geophys. J.R. Astron. Soc.* **18**, 1.

McKenzie, D. P. (1972). *In* "The Nature of the Solid Earth" (E.C. Robertson, ed.), pp. 323–360. McGraw-Hill, New York.

McKenzie, D. P. and Morgan, W. J. (1969). *Nature, Lond.* **224**, 125.

McKenzie, D. P. and Parker, R. L. (1967). *Nature, Lond.* **216**, 1276.

McKenzie, D. P. and Sclater, J. G. (1969). *Bull. Volc.* **33**, 101.

McKenzie, D. P. and Sclater, J. G. (1971). *Geophys. J. R. Astron. Soc.* **24**, 437.

McKenzie, D. P. and Weiss, N. (1975). *Geophys. J. R. Astron. Soc.* **42**, 131.

McKenzie, D. P., Roberts, J. M. and Weiss, N. O. (1974). *J. Fluid Mech.* **62**, 465.

Melson, W. G. and Thompson, G. (1971). *Phil. Trans. R. Soc. A*, **268**, 423.

Menard, H. W. (1967). *Science, N.Y.* **157**, 923.

Menard, H. W. (1969). *Earth Planet. Sci. Lett.* **6**, 275.

Menard, H. W. (1973). *J. Geophys. Res.* **78**, 5128.

Miyashiro, A. (1972). *Am. J. Sci.* **272**, 629.

Miyashiro, A. (1973). *Earth Planet. Sci. Lett.* **19**, 218.

Miyashiro, A., Shido, F. and Ewing, M. (1970). *Earth Planet. Sci. Lett.* **7**, 361.

Miyashiro, A., Shido, F. and Ewing, M. (1971). *Phil. Trans. R. Soc. A*, **268**, 589.
Moberly, R. (1972). *In* "Studies in Earth and Space Sciences: A memoir in honor of Harry Hammond Hess" (R. Shagam, R. B. Hargraves, W. J. Morgan, F. B. Van Houten, C. A. Burk, H. D. Holland and L. C. Hollister, eds), pp. 35–55. Geological Society of America.
Molnar, P. and Oliver, J. (1969). *J. Geophys. Res.* **74**, 2648.
Moore, J. G., Fleming, H. S. and Phillips, J. D. (1974). *Geology*, **2**, 437.
Moores, E. M. and Vine, F. J. (1971). *Phil. Trans. R. Soc. A*, **268**, 443.
Morgan, W. J. (1968). *J. Geophys. Res.* **73**, 1959.
Morgan, W. J. (1971). *Nature, Lond.* **230**, 42.
Morgan, W. J. (1972a). *Bull. Am. Assoc. Pet. Geol.* **56**, 203.
Morgan, W. J. (1972b). *In* "Studies in Earth and Space Sciences: A memoir in honor of Harry Hammond Hess" (R. Shagam, R. B. Hargraves, W. J. Morgan, F. B. Van Houten, C. A. Burk, H. D. Holland and L. C. Hollister, eds), pp. 7–22. Geological Society of America.
Murray, J. and Renard, A. F. (1891). "Report on Deep-sea Deposits Based on Specimens Collected During the Voyage of H.M.S. *Challenger* in the Years 1872–1876". H.M.S.O., London.
Nowroozi, A. A. (1973). *Bull. Seism. Soc. Am.* **63**, 441.
Oliver, J. and Isacks, B. (1967). *J. Geophys. Res.* **72**, 4259.
Opdyke, N. D. (1962). *In* "Continental Drift" (S. K. Runcorn, ed.), pp. 41–65. Academic Press, New York and London.
Orcutt, J. A., Kennett, B. L. N. and Dorman, L. M. (1976). *Geophys. J. R. Astron. Soc.* **45**, 305.
Owen, H. G. (1976). *Phil. Trans. R. Soc. A*, **281**, 223.
Palmason, G. (1967). *In* "Iceland and Mid-Ocean Ridge, Publication 38" (S. Bjornsson, ed.), pp. 111–127. Geoscience Society of Iceland, Reykjavik, Iceland.
Parker, R. L. and Oldenburg, D. W. (1973). *Nature, Phys. Sci.* **242**, 137.
Phillips, J. D. and Forsyth, D. (1972). *Bull. Geol. Soc. Am.* **83**, 1579.
Pitman, W. C., III and Hayes, D. E. (1968). *J. Geophys. Res.* **73**, 6571.
Pitman, W. C., III and Heirtzler, J. R. (1966). *Science, N.Y.* **154**, 1164.
Pitman, W. C., III and Talwani, M. (1972). *Bull. Geol. Soc. Am.* **83**, 619.
Pitman, W. C., III, Larson, R. L. and Herron, E. M. (1974a). Map of Magnetic Lineations of the Oceans. Geological Society of America.
Pitman, W. C., III, Larson, R. L. and Herron, E. M. (1974b). Map of the Age of the Ocean Basin. Geological Society of America.
Poehls, K. A. (1974). *J. Geophys. Res.* **79**, 3370.
Poehls, K. A., Luyendyk, B. P. and Heirtzler, J. R. (1973). *J. Geophys. Res.* **78**, 6985.
Prothero, W. A. (1974). *Bull. Seism. Soc. Am.* **64**, 1251.
Purdy, G. M. (1975). *Geophys. J. R. Astron. Soc.* **43**, 973.
Raff, A. D. and Mason, R. G. (1961). *Bull. Geol. Soc. Am.* **72**, 1267.
Reid, I. and Macdonald, K. C. (1973). *Nature, Lond.* **246**, 88.
Reyment, R. A. and Tait, E. A. (1972). *Phil. Trans. R. Soc. B*, **264**, 55.
Ribando, R. J., Torrance, K. E. and Turcotte, D. L. (1976). *J. Geophys. Res.* **81**, 3007.
Roberts, D. G. (1975). *Phil. Trans. R. Soc. A*, **278**, 447.
Runcorn, S. K., ed. (1962). *In* "Continental Drift", pp. 1–40. Academic Press, New York and London.
Rutland, R. W. O. and Walter, M. R. (1974). *In* "Mesozoic and Cenozoic Orogenic Belts" (A. M. Spencer, ed.), pp. 491–500. Geological Society of London.
Sclater, J. G. and Cox, A. (1970). *Nature, Lond.* **226**, 934.

Sclater, J. G. and Francheteau, J. (1970). *Geophys. J. R. Astron. Soc.* **20**, 509.
Sclater, J. G., Anderson, R. N. and Bell, M. L. (1971). *J. Geophys. Res.* **76**, 7888.
Sclater, J. G., Ritter, U. G. and Dixon, F. S. (1972). *J. Geophys. Res.* **77**, 5697.
Sclater, J. G., Lawver, L. A. and Parsons, B. (1975). *J. Geophys. Res.* **80**, 1031.
Sclater, J. G., Crowe, J. and Anderson, R. N. (1976). *J. Geophys. Res.* **81**, 2997.
Scrutton, R. A. (1972). *Geophys. J.R. Astron. Soc.* **27**, 259.
Shor, G. G., Menard, H. W. and Raitt, R. W. (1970). *In* "The Sea" (A. E. Maxwell, ed.), Vol. 4(2), pp. 3–27. Wiley-Interscience, New York.
Smith, A. G. and Hallam, A. (1970). *Nature, Lond.* **225**, 139.
Solomon, S. C. and Julian, B. R. (1974). *Geophys. J.R. Astron. Soc.* **38**, 265.
Spiess, F. N. and Mudie, J. D. (1970). *In* "The Sea" (A. E. Maxwell, ed.), Vol. 4(1), pp. 205–250. Wiley-Interscience, New York.
Spindel, R. C., Davis, S. B., Macdonald, K. C., Porter, R. P. and Phillips, J. D. (1974). *Nature, Lond.* **248**, 577.
Spooner, E. T. C. and Fyfe, W. S. (1973). *Contr. Mineral. Petrol.* **42**, 282–304.
Sutton, G. H., Maynard, G. L. and Hussong, D. M. (1971). *In* "The Structure and Physical Properties of the Earth's Crust" (J. G. Heacock, ed.), *Geophys. Mon.* **14**, pp. 193–209. American Geophysical Union.
Sykes, L. R. (1967). *J. Geophys. Res.* **71**, 2981.
Sykes, L. R. (1970a). *Bull. Seism. Soc. Am.* **60**, 1749.
Sykes, L. R. (1970b). *J. Geophys. Res.* **75**, 6598.
Sykes, L. R. and Sbar, M. L. (1973). *Nature, Lond.* **245**, 298.
Talwani, M. and Le Pichon, X. (1969). *In* "The Earth's Crust and Upper Mantle" (P. J. Hart, ed.), pp. 341–351. American Geophysical Union.
Talwani, M., Le Pichon, X. and Ewing, M. (1965). *J. Geophys. Res.* **70**, 341.
Talwani, M., Windisch, C. C. and Langseth, M. G. (1971). *J. Geophys. Res.* **76**, 473.
Tatham, R. H. and Savino, J. M. (1974). *J. Geophys. Res.* **79**, 2643.
Thatcher, W. and Brune, J. N. (1971). *Geophys. J.R. Astron. Soc.* **22**, 473.
Thompson, G. (1973). *Chem. Geol.* **12**, 99.
Turcotte, D. L. and Oxburgh, E. R. (1973). *Nature, Lond.* **244**, 337.
Turner, H. H. (1922). *Mon. Not. R. Astron. Soc. Geophys.* Suppl. 1, 1.
Vacquier, V., Raff, A. D. and Warren, R. E. (1961). *Bull. Geol. Soc. Am.* **72**, 1251.
Van Andel, Tj. H. and Bowen, C. O. (1968). *J. Geophys. Res.* **73**, 1279.
Vening Meinesz, F. A. (1948). "Gravity Expeditions at Sea, 1923–1938". J. Waltmann, Delft, Holland.
Vening Meinesz, F. A. (1952). *Geologie Mijnb.* **14**, 373.
Vine, F. J. (1966). *Science, N.Y.* **154**, 1405.
Vine, F. J. (1968). *In* "History of the Earth's Crust" (R. A. Phinney, ed.), pp. 73–89. Princeton University Press, U.S.A.
Vine, F. J. (1970). *Nature, Lond.* **227**, 1013.
Vine, F. J. and Hess, H. H. (1970). *In* "The Sea" (A. E. Maxwell, ed.), Vol. 4(2), pp. 587–622. Wiley-Interscience, New York.
Vine, F. J. and Matthews, D. H. (1963). *Nature, Lond.* **199**, 947.
Vine, F. J. and Moores, E. M. (1972). *In* "Studies in Earth and Space Sciences: A memoir in honor of Harry Hammond Hess" (R. Shagam, R. B. Hargraves, W. J. Morgan, F. B. Van Houten, C. A. Burk, H. D. Holland and L. C. Hollister, eds), pp. 195–205. Geological Society of America.
Vine, F. J. and Wilson, J. T. (1965). *Science, N.Y.* **150**, 485.
Vogt, P. R. and Ostenso, N. A. (1966). *J. Geophys. Res.* **71**, 4389.
Vogt, P. R. and Ostenso, N. A. (1967). *Nature, Lond.* **215**, 810.

74 E. J. W. JONES

Vogt, P. R. and Ostenso, N. A. (1970). *J. Geophys. Res.* **75**, 4925.
Vogt, P. R., Avery, O. E., Schneider, E. D., Anderson, C. N. and Bracey, D. R. (1969). *Tectonophys.* **8**, 285.
Vogt, P. R., Johnson, G. L., Holcombe, T. L., Gilg, J. G. and Avery, O. E. (1971). *Tectonophys.* **12**, 211.
Von Herzen, R. P. and Anderson, R. N. (1972). *Geophys. J.R. Astron. Soc.* **26**, 427.
Ward, M. A. (1963). *Geophys. J. R. Astron. Soc.* **8**, 217.
Watts, A. B. and Talwani, M. (1975). *Bull. Geol. Soc. Am.* **86**, 1.
Wegener, A. (1912). *Petermanns Mitt.* **58**, pp. 185–195, 253–256, 305–309.
Weissel, J. K. and Hayes, D. E. (1971). *Nature, Lond.* **231**, 518.
Weissel, J. K. and Hayes, D. E. (1974). *J. Geophys. Res.* **79**, 2579.
Wheeler, J. O., Charlesworth, H. A. K., Monger, J. W. H., Muller, J. E., Price, R. A., Reesor, J. E., Roddick, J. A. and Simony, P. S. (1974). *In* "Mesozoic and Cenozoic Orogenic Belts" (A. M. Spencer, ed.), pp. 591–623. Geological Society of London.
Whitmarsh, R. B. (1973). *Nature, Lond.* **246**, 297.
Williams, C. A. (1975). *Earth Planet. Sci. Lett.* **24**, 440.
Williams, D. L., Von Herzen, R. P., Sclater, J. G. and Anderson, R. N. (1974). *Geophys. J. R. Astron. Soc.* **38**, 587.
Wilson, J. T. (1963a). *Nature, Lond.* **197**, 536.
Wilson, J. T. (1963b). *Nature, Lond.* **198**, 925.
Wilson, J. T. (1965). *Nature, Lond.* **207**, 343.
Winterer, E. L. (1973). *Bull. Am. Assoc. Pet. Geol.* **57**, 265.
Worzel, J. L. (1965). "Pendulum Gravity Measurements at Sea: 1936–1959". John Wiley and Sons, New York.

Chapter 36

Sea-floor Sampling Techniques

TED C. MOORE, JR. and G. ROSS HEATH

Graduate School of Oceanography, University of Rhode Island, Kingston, R.I., U.S.A.

36.1. INTRODUCTION

The early sailors knew much of the surface ocean, its currents, tides and waves; of the atmosphere, its winds and clouds; and of the firmament, with its passing stars and planets. However, they knew or cared little of the ocean bottom except when it came close to the hulls of their ships. The nature of the bottom was of concern only in shallow waters when it was necessary to determine whether or not the sea floor would hold an anchor without hopelessly entangling it.

Later these early navigators extended their measurements into deeper waters. Soundings of several hundred metres were made as early as the sixteenth century, and by the late-nineteenth century piano wire had been substituted for fibre rope to aid in these deeper measurements. Probably the first systematically collected samples of the sea floor were taken with sounding leads. A small recess in the bottom of the lead was filled with tallow, to which a sample of the bottom sediment adhered. Knowledge of the type of sediment recovered was used as an aid to navigation and as an indication of the nature of fishing grounds. In addition to navigation and fisheries, one of the earliest needs for exact knowledge of the sea floor was brought about by the laying of trans-oceanic cables. Cable-laying operations, which began in the mid-nineteenth century, were tremendous feats of engineering. They required not only innovations in the design of the cable itself and in the types of ships that deployed the cables, but also a fairly good knowledge of the submarine topography and of the types of deposits that lay on the sea floor. Each deep sounding made for these trans-oceanic surveys was an arduous and time consuming task. But a compilation of the results first revealed such gross topographical features of the deep-sea floor as the Mid-Atlantic Ridge. Samples recovered along with these soundings were as eagerly received by the scientists of that day as the moon-rocks have been in modern times. By serendipity, these samples proved much more interesting to the nineteenth century scientists than a collection of moon-rocks would have done, for in addition to volcanic and other mineral debris they revealed an abundance of fossil organisms. These materials greatly increased the contemporary knowledge and appreciation of the importance of planktonic life tc deep-sea sedimentary deposits, and of the relationships of these deposits to the geological record on land.

Thanks to the extensive sampling programmes of early oceanographic expeditions, such as that of the H.M.S. *Challenger*, the nature of what had been an almost totally obscure part of our globe rapidly became known, and known with surprising accuracy and completeness. It is true that several misconceptions and misinterpretations were fostered in the early days of exploration. For example, Huxley (Murray and Renard, 1891) originally thought that coccoliths were the skeletal elements of a deep-ocean creature named *Bathybius*. When *Bathybius* was found to be a calcium sulphate precipitate caused by the addition of alcohol to the contents of the specimen jars, however, the coccoliths were restored to their rightful status in the phytoplankton. These occasional mistakes were not of major importance; the breakthrough had been made. Samples of the deep-sea floor had been recovered and they had literally opened another world to scientific investigation.

Purely intellectual interest in the sea floor has been greatly reinforced by such pragmatic needs as the laying of submarine cables, navigation and fisheries, as well as later demands for knowledge of the sea floor resulting from the advent of submarine warfare and the development of offshore oil fields. At first the only scientific question posed was: "What is the nature of the deep-sea floor?". Once the diverse nature of deep-sea sediments was appreciated, however, the questions began to multiply: "What are the distributions of these various sediment types; how fast do they accumulate; how are they related to the physical, biological and chemical characteristics of the oceans; and finally, how has the character of the sediments changed with time and, from these changes, what can be deduced about oceanographical changes?". To answer these questions, marine geologists have usually sampled the sea floor with specific purposes in mind. Accordingly, techniques and equipment have been developed that are aimed at solving particular problems.

Many of the tools used in early explorations of the sea floor were of crude and simple design. They were scoops, dredges and pipes lowered on long lines from surface ships. In the last 100 years the designs of many of these samplers have become quite sophisticated, but they still recover samples from the sea floor in essentially the same simple way. The biggest changes are to be seen in the retrieval system. Where once miles and miles of manila line were required to probe the deep-sea bed, samples are now recovered on thin steel lines or even by free-vehicle samplers. Manila has been replaced by steel, and the steam- and sail-driven nineteenth century research ships have given way to specially designed research vessels, submersibles and drilling ships.

This chapter is not intended to provide an exhaustive list of all the sampling devices and systems ever used by marine geologists. Rather, it considers the reasons for different types of sampling programme, then summarizes the problems associated with carrying them out and finally describes the sampling devices and sampling strategies which have been designed to overcome these problems.

36.2. SAMPLING PATTERNS IN SPACE: SURFACE SEDIMENTS

At the major core repositories, requests for surface sediment samples from deep-sea cores greatly outnumber those for sub-surface samples. These requests reflect the degree of interest in one of the oldest and still one of the most active areas of research in marine geology; the definition of present-day patterns of sediment distribution and the determination of their relationships to the modern oceanic environment. Among the various scientific problems included in this broad field of investigation are the determination of:

(i) the influence of physical, chemical and biological features of near-surface waters on marine sediments;

(ii) the influence of vertical variations of physical, chemical and biological variables on sediment composition;

(iii) the effects of the flow regime and the physico-chemical characteristics of bottom waters on deep-sea sediments;

(iv) the influence of benthic fauna on the sediment;

(v) the relationship of sediment composition, and surrounding physical and chemical conditions, to the geotechnical properties of deep-sea sediments.

These general categories have been subdivided many times according to particular geographical locations, minerals, chemical components, size fractions or microfossil groups to be studied.

Each particular study, however, encounters many problems that are common to all. One of the first questions that usually arises in such investigations concerns the scale of the pattern under investigation and the detail with which it must be defined. How many samples will be required and how should they be spaced? If the sedimentary properties of interest are related to the oceanography of near-surface waters, they tend to have very broad geographical patterns. Thus, samples spaced thousands of kilometres apart will often reveal the general patterns of sediment distributions. The H.M.S. *Challenger*, which took less than 300 samples at depths in excess of 1000 fathoms, not only sampled all the major sediment types of the pelagic realm, but also did much to define their distribution patterns.

Imbrie and Kipp (1971) made one of the first attempts to establish quantitatively the relationship between sea-surface conditions and microfossil assemblages preserved on the sea floor. They were able to relate the broad distribution patterns of sedimentary foraminiferal assemblages (Fig. 36.1) to sea surface temperatures through a regression equation which gave a correlation coefficient, r, of greater than 0·95 and a standard error of the estimate of about $\pm 2°C$. This was achieved using only 61 surface sediment samples from the Atlantic and western Indian Oceans—about one sample per million square kilometres of ocean area.

If the influence of more restricted oceanographical features, such as currents or zones of divergence, are to be defined, sample spacings of no more than a few hundreds of kilometres are usually needed. Thus, in a study similar to that described above, but using the foraminiferal assemblages of 161 sediment surface samples from the North Atlantic (about one sample per 300 000 km^2), Kipp (1976) was able to clearly map the influence of the Gulf Stream and the North Atlantic Drift (Fig. 36.2). Similarly, Kennett and Watkins (1975) were able to map the influence of the deep circum-Antarctic flow and its associated bottom currents on the deep-sea deposits in the south-eastern Indian Ocean by using 187 sediment cores supplemented with data from 143 camera stations. These data allowed them to delineate both the lateral extent of erosion and the direction of bottom flow. The average sample density in their study was approximately one sample per 50000 km^2.

In a study of sediment dispersal patterns in the Panama Basin, a sample density of about one per 17 000 km^2 was sufficient to clearly define the effects

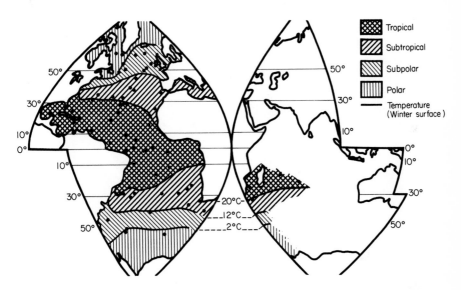

FIG. 36.1. Dominant assemblage map showing distribution of four foraminiferal assemblages in the Atlantic and Indian Oceans; derived from 61 samples (from Imbrie and Kipp, 1971).

of large topographical features and bottom flow on the distribution of sand-sized material (Kowsmann, 1973; Yamashiro, 1975), silt-sized sediments (van Andel, 1973), biogenic debris (Moore *et al.*, 1973) and clay minerals (Heath *et al.*, 1974).

In general, the sampling density required for any study of surface sediments is determined by the variability or gradients in the processes which control the distribution of the sediment type in question. The examples cited above indicate that those properties which are primarily water-mass related can be defined by very widely spaced samples. If, however, the high gradient regions at water-mass boundaries or beneath major current systems are to be sharply defined, samples should be at least an order of magnitude more closely spaced.

Another example of the resolution attainable by different sampling densities comes from successive studies of the wind-blown quartz distribution in the North Pacific. In an early study of this continentally-derived component of deep-sea sediments, the general pattern of a broad, mid-latitude maximum in quartz concentrations could be derived for the North Pacific from a map containing about 100 points (Rex and Goldberg, 1958; Arrhenius,

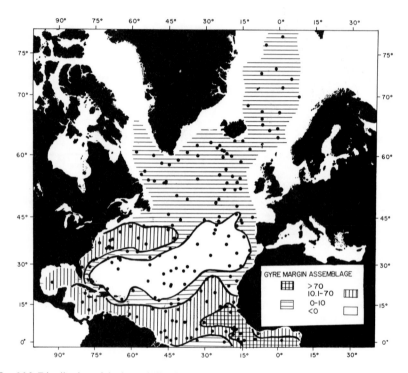

FIG. 36.2. Distribution of the foraminiferal gyre margin assemblage that delineates the equatorial and tropical circulation of the North Atlantic. (Based on samples shown; after Kipp, 1976.)

1963; Fig. 36.3). A later study with roughly five times the number of samples (Fig. 36.4; Heath *et al.*, 1973b) better defines the distribution of quartz in surface sediments and differentiates zones of ice transport, jet stream transport and trade wind transport. The significance of the pattern seen in Fig. 36.3 and its relation to zonal winds and arid areas was much more clearly shown by averaging values over five degree latitudinal strips and displaying the data in a meridional plot, perpendicular to the major climatic and wind gradients (Fig. 36.5).

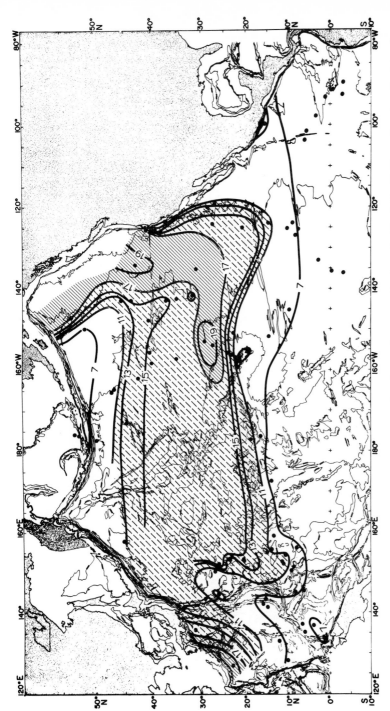

FIG. 36.3. Distribution of quartz concentrations in approximately 90 surface sediment samples from the North Pacific Ocean (based on the data of Rex and Goldberg, 1958). Heavy contours are quartz concentrations in weight per cent on a carbonate-free basis. Dots indicate sample locations. Light contours are bathymetry at 1000 m depth intervals (after Chase, 1975). Note that ill-defined band of high values at mid-latitudes and compare with the detail in Fig. 36.4. Diagonal solid lines encompass samples with more than 17% of quartz; diagonal dashed lines encompass samples with more than 11% of quartz.

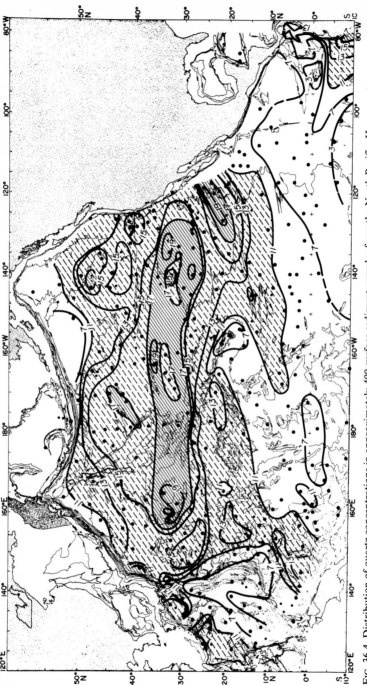

FIG. 36.4. Distribution of quartz concentrations in approximately 400 surface sediment samples from the North Pacific. Heavy contours are quartz concentrations in weight per cent on a carbonate-free and opal-free basis. Shading as in Fig. 36.3. Contours in the eastern tropical Pacific are based on data by Moore et al. (1973). Large dots indicate sample locations (data produced by the CLIMAP Project). Small dots indicate sample locations from Rex and Goldberg (1958) which are unlikely to have large concentrations of opal and therefore can be used in this data set. Note the narrowly defined mid-latitude band of high values which is distinctly separated from the lobe of high values associated with the NE. trade winds.

In a similar way, transects of closely spaced samples across oceanic gradients are often the most efficient way to define the changes in sediment properties. However, in such instances, as with areal sampling, increased sample density does not necessarily improve the correlation between a sediment property and its inferred oceanographical control. Plots of carbonate concentrations in surface sediments versus water depth provide good

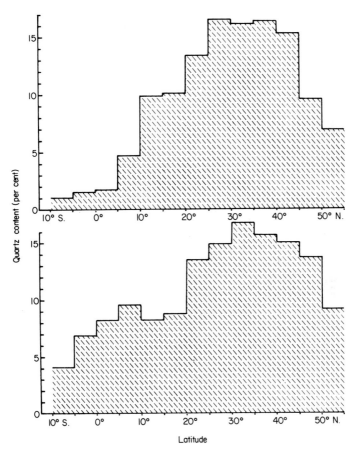

Fig. 36.5. UPPER: Plot of average quartz concentrations versus latitude showing mid-latitude quartz maximum (after Rex and Goldberg, 1958).

LOWER: Plot of average quartz concentration from Fig. 36.4. The main difference between the upper and lower plots lies in the sharpness of the mid-latitude maximum and the apparent separation of a secondary maximum associated with the NE. trade winds. The latitudinal averaging of the scant data shown in Fig. 36.3. brings out the dominant signal–the mid-latitude maximum. When data are sufficient to resolve detailed geographical patterns, however (Fig. 36.4), latitudinal averaging will obscure longitudinal variation in these patterns.

examples of this phenomenon. In one of the early studies of this relationship, Bramlette (1961) showed that for limited regions, and over water depths of about 3–5 km, the carbonate concentrations decrease monotonically with increasing depth. He also indicated that this carbonate–depth line appeared to be different for different regions of the oceans. As more data were collected, the plots of carbonate versus depth became much more complicated (see e.g. Lisitzin, 1970; Moore *et al.*, 1973). No longer was the relationship simply monotonic, but rather each depth was characterized by sediments having a fairly wide range of carbonate concentrations. In this case, increased sampling has begun to "see" effects other than the simple effect of depth (pressure) on carbonate preservation. These effects, which include slumping, dilution of carbonate with terrigenous debris and geographical variations in the rates of supply and dissolution of carbonate, may be constant with depth over restricted areas such as that studied by Bramlette (1961). Over wide regions, however, the relationships change (Berger, 1971; Berger and Winterer, 1974) so that only a carefully designed sampling programme, which takes the possible complicating factors into account, can extract the more basic depth-carbonate relationship.

In general, the selection of sample spacing is determined by the answers to two questions:

(i) "how large are the patterns which it is desired to define?" and
(ii) "of how large an area is each sample representative?"

If the distribution of the property in question is rather even and the patterns are quite broad, as are those of the foraminiferal assemblages of the Atlantic (Imbrie and Kipp, 1971), the samples can be widely spaced. If the distribution of the property is heterogeneous, so that one sample is not likely to adequately represent the samples from a large surrounding area, then, even if the average pattern lacks steep gradients, there may be considerable unexplained detail or "noise" in the data. Some sort of averaging, such as the latitudinal averaging of quartz values by Rex and Goldberg (1958), may be required to bring out the general pattern.

In high-energy environments near to shore, a high density of samples is required if all the complexities of the depositional patterns are to be resolved. Broad general patterns can still be recognized, but scatter in the data is often quite large (see e.g. Hollister, 1973). For such detailed studies, when the characteristics of the bottom sediments reflect only the environment immediately surrounding the site of deposition, the accurate location of a sampling station becomes a very critical part of the sampling programme. Recent technical advances in electronic and satellite navigation systems (Talwani, 1970; Shepard, 1973) allow the marine geologist to locate a sampling ship, absolutely to within a few hundred metres on the surface of the globe, and relatively to within a few metres from one sampling site to another. Often the important consideration, however, is to locate the sample with respect to surrounding bottom topography. A narrow-beam echo-sounder is the basic tool used to achieve this goal. In relatively shallow water such a tool can adequately show the details of bottom topography. Depending on the frequency of the sound pulses used, some sub-bottom penetration can be achieved and the sedimentary layering thus revealed may be used as an added tool in interpreting the sedimentary processes which are important in the study area. Detailed bathymetric maps (see e.g. Uchupi, 1969) and, in shallow and relatively clear water, aerial photographs (see e.g. Ball, 1967) aid in planning sampling schemes and in determining the sediment distribution patterns of very complex near-shore areas.

Even in the deep ocean, sediment distribution patterns may show considerable detail that does not appear to bear any relation to the broad oceanographical patterns which are usually thought of as controlling pelagic sedimentation. A study of a small area in the central Pacific highlights the problem of areal representativeness. Over an area of some $250 \, km^2$, the variations in particle size and mineralogy were scarcely greater than were the analytical errors (Heath et al., 1970). Thus, a single core would constitute a representative sample of the entire area for such measurements. In contrast, the thickness of Quaternary sediments, which varied by more than two orders of magnitude (Moore and Heath, 1966), and the Mn/Fe ratio in ferro-manganese nodules which varied from 2·6 to 5·4 (Calvert et al., 1977), are as variable on scales of hundreds of metres as they are on scales of hundreds of kilometres. Such local variability explains why, for example, global maps of sedimentation rates or ferro-manganese nodule composition show such "noisy" patterns.

The scattered recovery of outcropping Tertiary sediments in the central Pacific could not be explained until the topography of the deep sea could be mapped in detail and samples located with respect to this topography (Moore, 1970; Johnson and Johnson, 1970; Johnson and Parker, 1972). In early studies, the representation of the deep-sea floor by surface echo-sounders was very generalized. In deep waters the cone of sound emitted by conventional echo-sounders is reflected from such a large area that topographical details of the sea floor are lost in an envelope of overlapping hyperbolae (Krause, 1962). Even if a narrow-beam echo-sounding system is used, drift of the ship as the sampler is lowered will introduce a few hundred metres uncertainty in the position of the sample relative to the mapped topography. Acoustic transponders placed on the sea floor have done much to alleviate this problem of relating detailed surveys and sample locations in the deep sea. The time which elapses between the enquiry by the survey ship and the receipt of a response from three or more bottom transponders allows the ship to be accurately located relative to the transponder array. Sampling devices and deeply towed surveying instruments also can be navigated within the same array (Spiess *et al.*, 1966). Thus, not only can the topography be accurately depicted by towing an echo-sounder close to the bottom, but the sediment samples can also be precisely located relative to the topography.

One of the other major problems in studies of surface sediments is the geological time interval represented by each surface sample. Surface sediment samples are usually no less than 1 cm thick and, therefore, must represent a time average. In some respects this averaging is a boon to the marine geologist. It smooths the daily, seasonal and annual patchiness in the vertical and horizontal distributions of planktonic species and other oceanographic variables into much more even patterns. Thus, the distributions of micro-fossil species in samples collected from the sea floor show much more co-herent patterns than do comparable data from plankton tows. In comparing sediment and averaged oceanographic data, however, it must be remembered that the time-scales of averaging are quite different. Oceanographical measurements rarely extend back more than about 40 years. A sediment sample 1 cm thick from the deep sea, on the other hand, can represent a few hundred to a few thousand years of deposition. Furthermore, bioturbation of the bottom sediments effectively homogenizes the near-surface sediments over the upper 3–50 cm (Section 29.5.5; see also Arrhenius, 1963; Berger and Heath, 1968; Guinasso and Schink, 1975; Berger *et al.*, 1977; Peng *et al.*, 1977). If the properties of the surface sediments are to be correlated with the oceano-graphical conditions in near-surface waters, the assumption must be made that the average of the oceanographical conditions over the past few decades is a good estimate of those over the past several hundred years.

36.3. Sampling Patterns in Time: Stratigraphical Studies

In the seventeenth century Steno clearly spelled out the law of superposition. Subsequently, geologists have relied on the fact that the upper layers of a sequence of sedimentary rocks are younger than the lower layers in order to piece together the history of the Earth and the evolution of life. Because of tectonic movements, the erosion and destruction of sedimentary sections, as well as the paucity of exposures and preserved fossils, the construction of such a history has not been a simple task.

Early geologists recognized the marine origins of many deposits now exposed on land, and surmised that oceanic sediments should contain a more complete record of Earth history than the more ephemeral deposits they saw on land. A natural extension of this idea was that the deep oceans contained a complete record of earth history; i.e. that debris washed from the continents and the fossils of pre-existing marine life had settled aeon after aeon onto the sea floor and that there they remained, layer upon layer, as pages in a great history book. If, by some means, a complete sequence of these layers could be recovered from various parts of the World Ocean, this history could be read. Unhappily, now that this dream is technically feasible, we have learned that a complete record of Earth history cannot be found in the oceans. There are two reasons for this. First, the continual creation and destruction of the sea floor due to plate tectonics (see Chapter 35; Hess, 1962; Morgan, 1971) has destroyed deep-sea deposits older than about 200 million years. Thus, only the latest one-tenth of the history of life on this planet is preserved on the sea floor. Second, erosion, redeposition and dissolution of deep-sea deposits have exposed sediments ranging in age from 1 to 100 million years at the sea floor (Riedel and Funnell, 1964; Riedel, 1967; Burckle et al., 1967). The gaps in the record conceal many historical details but, inasmuch as these same gaps form coherent patterns in time and space (see e.g. Rona, 1973; Davies et al., 1975; Moore et al., 1978), they contain information on bottom currents and bottom-water chemistry that are just as much a part of the history of the oceans as are the properties of the sediments which remain.

The palaeoceanographer, biostratigrapher, geochemist, mineralogist, ocean engineer and geophysicist share common difficulties in interpreting the record of sediments laid down in ancient oceans. These problems include the determination of:

(i) the duration of the sedimentary record,
(ii) its resolution and
(iii) its completeness or freedom from disturbance. Most investigators seek sedimentary sections that are (a) undisturbed, (b) complete, (c) as long as possible, (d) well dated and (e) as highly resolved as possible. Unfortunately, such sections probably do not exist over most of the ocean basins.

One of the obvious compromises which must be made is the length of the record versus its resolution. Because corers have limited penetration capabilities, the longer the span of time represented in the recovered section, the lower the resolution. The length and resolution of the record (in terms of time) depend on the accumulation rate of the sediment, which varies widely in the deep ocean. In the truly pelagic regions, rates range from less than 1 m(m.y.)^{-1} (metre per million years) in areas of "red" clay deposition to almost 100 m(m.y.)^{-1} in highly productive regions where the siliceous and calcareous tests of micro-organisms rapidly accumulate. The average accumulation rate for such fossiliferous pelagic sediments is about 20 m(m.y.)^{-1}. This is a little lower than the median accumulation rate for the Atlantic derived from the data of Ericson et al. (1961) and a little higher than most of the values for carbonate sediments of the equatorial Pacific. A sample 1 cm thick from a core with a 2 cm $(10^3 \text{ years})^{-1}$ (i.e. 20 m$(\text{m.y.})^{-1}$) accumulation rate obviously represents at least 500 years of deposition. For a 1 cm sample to represent only 100 years of deposition, the sediment must have accumulated at a rate of at least 10 cm $(10^3 \text{ years})^{-1}$ (i.e. 100 m$(\text{m.y.})^{-1}$). Accumulation rates in areas of hemipelagic sedimentation range from about 40 m$(\text{m.y.})^{-1}$ up to a few hundred metres per million years. Although such rates allow shorter depositional events to be resolved, the improvement is not directly proportional to the increased accumulation rates. Organisms which live on and in the sea floor mix the near-surface sediments by burrowing through them and feeding on them. Such bioturbation degrades or blurs the sedimentary record. Observations on the mixing of short-lived and fall-out radionuclides into the sediments (Arrhenius, 1963; Noshkin and Bowen, 1973), and on the thickness of the upper sediment layer giving a constant radiocarbon age (Peng et al., 1977), suggest that the thickness of this mixed layer can range from about 3 to almost 50 cm (see Chapter 40). The mixing depth tends to be greater in rapidly accumulating hemipelagic sediments than in pelagic clays, thus reducing the anticipated increase in resolution expected from the more rapidly deposited sections. In general, mixing, rather than sample thickness, limits the temporal resolution of the sedimentary record.

The degree of homogenization produced by the burrowing of organisms, and its effect on depositional events, is best indicated in cores by the final distributions of sedimentary particles that originally were deposited by an instantaneous event such as a volcanic eruption (Ruddiman and Glover, 1972) or a microtektite fall (Glass, 1969; see also Section 29.5.5). Mathematical models of the mixing process (see e.g. Berger and Heath, 1968; Guinasso and Schink, 1975) simulate the observed distributions of microtektites and ash beds to some extent, and allow the amount of distortion and blurring that is imposed on the record of deposition to be appreciated. The next step is the development of a deconvolution procedure which will "unmix" the bioturbated signal and restore the record, at least partially, to its original character (Ruddiman et al., 1976).

In regions where the oxygen content of the bottom waters is very low, mixing by burrowing organisms is much less of a problem. Anoxic sediments are found in basins in which the bottom waters are rarely renewed (e.g. in the Santa Barbara Basin, the Cariaco Trench and many fjords), or where a strongly developed oxygen minimum layer impinges on the continental rise (e.g. along the western coasts of Africa and South America). The benthic population that can survive under such conditions does not greatly disturb the original depositional layering of the sediment. In fact, if the delivery of sediments to such regions varies seasonally, annual varves are commonly preserved (Hulsemann and Emery, 1961; Soutar, 1971). Although such depositional environments are not representative of typical oceanic conditions, their sediments offer a unique view of the amount of information that is initially present in the stratigraphical record. Because accumulation rates of anoxic sediments are usually high (300–500 m(m.y.)$^{-1}$), annual, and even seasonal, events can be resolved (Calvert, 1966; Soutar, 1971). Unfortunately, the length of record obtainable in such sediments by normal coring techniques (10–20 m) is not great enough to recover more than a few thousand years of oceanographical history.

In general, the rapidly accumulating sections that are sought for high temporal resolution are found in high-energy environments in which the slow rain of pelagic debris is greatly supplemented by sediments which are transported laterally by near-bottom currents. Because the influx of laterally transported material is often episodic, sedimentary sequences in such environments tend to have highly variable accumulation rates as well as gaps due to erosion or non-deposition. Thus, it is difficult to establish a time-scale for such records. This leads to compromise between high resolution in terms of the time represented by the thickness of a single sample, and low accuracy in terms of the correspondence between a series of samples and an accurate time-scale.

Sample spacing, as well as sample size, governs the stratigraphical resolution obtainable from a series of measurements. The selection of a sampling interval down a core is entirely comparable to a decision on how far apart to space samples in a study of surface sediments. In a down-core study, the investigator is usually looking for patterns in time; in the surface sediment study, for patterns in space. The preceding discussion of bioturbation and mixing of the sedimentary record implies that sampling more closely than some lower limit yields no additional information. Data on the variability of mixing in deep-sea sediments are still rather sparse, so that the relatively large amount of detail which can be resolved from cores that have accumulated very slowly is sometimes surprising. In a core from the North Pacific (BNFC-43PG2, Scripps Institution of Oceanography) with an accumulation rate of only about 1 cm (10^3 years)$^{-1}$, a sample spacing of 2 cm gave an oxygen isotope record with significantly more detail than a 4 cm spacing (Shackleton, 1977). Thus, the limiting sample spacing should be determined directly for each core, rather than estimated from models of bioturbation.

The upper limit of sample spacing depends on the duration or wave length of the phenomenon under investigation. All too often, samples are taken at some regular interval down a core where the magnitude of the interval reflects the length of the core and the number of samples that an investigator feels like running, rather than a logical evaluation of the probable rate at which sediment has accumulated and the time resolution actually desired. If a particular cyclicity in some variable is being investigated, the sample spacing (in time) should be at least one-half the wavelength of the cycle (corresponding to the Nyquist frequency; Pisias *et al.*, 1973). If a unique event is being sought within the stratigraphical record, then the sample spacing should be such that the feature will be defined by at least three points (Ledbetter and Ellwood, 1976). With less points, it may be difficult to demonstrate that an event, or feature, is missing from the record. Thus, preliminary stratigraphical work is usually needed to establish an approximate time-scale for a core. A sampling plan can then be based on the assumption of a constant accumulation rate. If this assumption proves to be incorrect, a revised sampling plan may be required.

Tying these stratigraphical sections together in regional and global net-works of correlated sequences is a difficult task; it is a necessary first step, however, in reconstructing past oceanographical conditions in space and time. Detailed stratigraphical correlations should be based on one or more continuously fluctuating parameters which have unbounded upper and lower values (for example, concentrations which do not approach 0 or 100%). The fluctuations in such parameters should be distinctive; they should vary more with time than with space and they should be measurable in all cores on small samples. If such a perfect stratigraphical parameter could be found and its fluctuations tied feature by feature and point by point to an accurate time-scale, then increases or decreases in the accumulation rate of a few centi-metres per thousand years, as well as gaps in the record of a few thousand years, could easily be detected. Curves of carbonate concentrations are often used for stratigraphical correlations and to establish time-scales for Quater-nary sections. Like any concentration measurement, however, the curves lose all character when concentrations approach 0 or 100%. Furthermore, the carbonate content of sediments is highly variable over the ocean basins at any given time, as are the concentrations of quartz, opal and the clay minerals. The stratigraphical use of the abundance of microfossil assemblages, or of individual species, encounters this same problem. The correlative tool that has come closest to meeting all stratigraphical criteria is the oxygen isotopic composition of the calcareous tests of benthic foraminifera (see Chapter 29 and Shackleton and Opdyke, 1973). Calcareous forms are not ubiquitous over the ocean basins, but they are sufficiently widespread to make possible very accurate inter-regional and inter-ocean correlations.

Over time-scales of several millions years, and for somewhat less detailed correlations, two additional basic stratigraphical tools are available for the assignment of a time-scale to the sedimentary record; magnetic stratigraphy and biostratigraphy. Magnetic stratigraphy offers the great advantage that it is globally applicable; most biostratigraphical zonations are useful over restricted, albeit wide, regions. However, because magnetic stratigraphy relies solely on the normal, or reversed, character of the remanent magnetization of the sediments, biostratigraphical control is necessary to ensure that the core section is free of hiatuses. The youngest five million years of the magnetic reversal record is well controlled by radiometric dating (Watkins, 1972). Older reversals have only been dated by correlation to dated biostratigraphical zonations (Burckle, 1972; Theyer and Hammond, 1974).

Biostratigraphical correlations and absolute age assignments in many pelagic sections are aided by the availability of more than one microfossil zonation. Thus, radiolarian and coccolith zonations add resolution to, and provide at least a partial check on, the foraminiferal zonation in many Deep-Sea Drilling Project cores. Existing zonations allow a time resolution of the order of ± one million years, at least for the last ten million years (van Andel *et al.*, 1975). This does not mean that the geological time-scale is known to this degree of accuracy; only that examination of well-preserved, well-sampled microfossils permits a gap of this magnitude to be detected. Thus, compared to the resolution obtainable in the Pleistocene with oxygen isotope stratigraphy, zonal stratigraphy is at least two orders of magnitude less precise.

Now that most marine sedimentary sections can be dated rather precisely (Berggren and van Couvering, 1974), systematic gaps in the record can be much more clearly defined. For the most recent few tens of thousands of years, missing sections can be identified by detailed local studies of groups of cores using the quantitative stratigraphical techniques discussed previously. For intervals of several millions of years, missing biostratigraphical zones or parts of zones (sub-zones) clearly indicate the presence of gaps in the records. In sections of severe, but incomplete, dissolution of microfossils, or in sections of almost unfossiliferous clays, the existence of a hiatus in the record is often very difficult to prove. Such sections provide good examples of the previously mentioned sampling problem in which it is more difficult to establish the absence of an event than its presence. In practice, investigators tend to define gaps, or hiatuses, in the sedimentary record rather arbitrarily in order to be able to map the temporal and spatial distributions of such features (see Chapter 35). Thus, Moore *et al* (1978) defined a hiatus as any portion of a sedimentary section where the accumulation rate was less than 1 m(m.y.)$^{-1}$. Studies such as this have shown that gaps are concentrated in certain parts of the marine record (Fig. 36.6; Kennett *et al.*, 1972; Edwards, 1973; Rona, 1973; Davies *et al.*, 1975; Moore *et al.*, 1978). The older the sedimentary section, the more likely it is to contain hiatuses. Thus, the Cretaceous–Tertiary gap (65 million years B.P.) is found in about 90% of the sections sampled by the Deep-Sea Drilling Project, whereas in the uppermost Pliocene interval (2–3 million years B.P.) hiatuses are found in only about 20% of the sections (Fig. 36.6; Moore *et al.*, 1978; Moore and Heath, 1977).

Regardless of how sophisticated sampling techniques become in the future these inherent deficiencies in the marine sedimentary record will prevent the recovery of complete sequences of sediments from all portions of the deep sea.

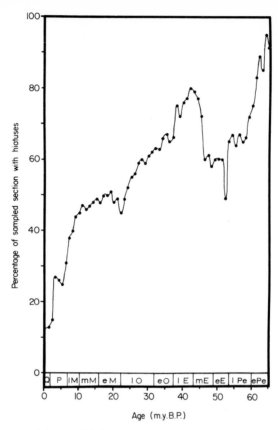

Fɪɢ. 36.6. Percentage of the sampled deep-sea section represented by hiatuses (after Moore and Heath, 1977). Data compiled from Deep-Sea Drilling Project Legs 1–33. Time-scale based on Berggren (1972).

36.4. Tʜᴇ Sᴀᴍᴘʟɪɴɢ Dᴇᴠɪᴄᴇꜱ

Many different sampling devices have succeeded the early tallow and weight recovery method (Table 36.1). In selecting the particular device to be used, however, there are many considerations in addition to those discussed above. Some of these are more fiscal than physical or scientific in nature. For example, budgetary considerations and the cost of each sample in terms of equipment manufacture, ship time and manpower may have a large influence on the sampling programme. To lower a sampling device and retrieve a sample from the sea can take from two to five hours of station time depending on the depth of water, the winch characteristics and the complexity of rigging and unrigging the sampling device. For a ship costing about $5000 (U.S.) per day, not an abnormally high rate for oceanographic vessels, the cost of each deep-sea sample is at least $400 in ship time alone. Because of this, the devices used in the deep sea usually represent a compromise between several ideals so that the samples which are collected can be used for a number of scientific purposes. Because of this, and because they better preserve the near-surface layering, corers, rather than grabs or dredges, are usually used to collect sediment samples from the floor of the deep ocean.

36.4.1. DREDGES

Dredges are amongst the oldest devices used to collect samples from the deep sea. Today, however, their use is usually limited either to studies of areas of basalt or indurated sediment outcrops, or to the recovery of larger masses of coarse components of marine deposits such as manganese nodules or reef debris, for which mechanical damage is not a serious problem. Although rock corers (very heavy corers with hardened steel barrels) and *in situ* rock drills (Brooke and Gilbert, 1968; Ade-Hall *et al.*, 1973) can recover oriented samples of rock from areas where a thin veneer of sediment defeats dredging attempts, the dredge remains the only really effective way to recover large volumes of rock from the sea floor. Dredges, which usually have rect-angular steel mouths, are attached to the ship's wire with a steel bail and to a chain bag which retains fragments larger than about 10 cm in diameter. The dredge has to be pulled along the bottom behind the ship with a considerable scope of wire (usually twice the depth of water). Thus, if samples are to be located relative to specific bathymetric features, care has to be taken to determine the interval and depth range for which the dredge was on the bottom. For this, a recording tensiometer, good record-keeping and a knowledge of the bottom topography are required (Aumento, 1970). A dredging operation is time-consuming and often results in the temporary anchoring of the ship when the dredge snags on rocks.

36.4.2. GRAB SAMPLERS

Large volumes of loose debris (such as manganese nodules) can also be collected from the sea floor using free-fall grab samplers (Table 36.1). These devices are easy to use, have minimal requirements for supporting shipboard equipment, can be located much more accurately than dredges and sample an accurately known area. For near-shore work, various types of wire-line grab samplers (Fig. 36.7) have proved to be very useful. They can recover large sample volumes from both coarse and fine sediments. Usually, the only serious problem encountered is when debris prevents the complete closure of the jaws. In most models the jaws are spring-loaded or held closed

TABLE 36.1

Types of sea-floor sampling devices and their general characteristics

Core type	Required lifting capacity (kg)	Cross-section (cm²)	Corer I.D. (cm)	Max. length (cm)	Disturbance	Use and remarks	Key references
Grab samplers	110–250	≤400	—	—	Negligible in centre	Surface samples primarily for lithological and petrographical analyses. Layering usually preserved.	van Veen (1936), Jonasson and Olausson (1966), Emery and Champion (1948), La Fond and Dietz (1948)
Free-fall grab	—	900	—	—	—	Samplers fitted with position-finding transmitter.	Kollwentz (1973)
Gravity Open-barrel	200–2500	17·7–51·5	4·75–8·1	300	Shortening of sediment	Upper few metres of sediment and surface samples.	Hvorslev and Stetson (1946), Emery and Dietz (1941)
Phleger	70–100	4·9–11·3	2·5–3·8	36	Slight deformation	Surface sample and pilot core.	Phleger (1951)
Hydroplastic	68–136	51·5–80·1	8·1–10·1	400	Negligible	Undisturbed surface samples for engineering and mass physical properties.	Richards and Keller (1961)
Multiple (3–5 barrels)	250–600	4·9–11·3	2·5–3·8	48	Negligible	Surface samples to test reliability of very small samples to represent comparatively large areas.	Fowler and Kulm (1966)
Free-fall	82	12·6–33·2	4·0–6·5	122	Negligible	Short cores, working from small boats. Hydrodynamic release with glass sphere floats for recovery.	Sachs and Raymond (1965), Moore (1961)

					Dependence upon piston set-up		
Piston	200–2500	17·7–51·5	4·75–8·1	3400		Down-core studies.	Kullenberg (1947, 1955), Swedish Committee (1961), Zenkovitch (1955), Richards (1961)
Large volume Sphincter	320–1000	113	12	1200	Negligible	Wide diameter piston corer.	Kermabon et al. (1966)
Kasten	—	225	15 × 15	1500	Negligible	Undisturbed down-core studies.	Kogler (1963)
Box	385	600 (2500)[a]	20 × 30 (50 × 50)[a]	60	Negligible	Large rectangular sample at water–sediment interface.	Bouma and Marshall (1964), [a]Hessler (personal communication, 1976)
Large diameter	180	176·6	15	600	Negligible	Enlargement of Richards and Keller (1961) PUC corer.	McManus (1965)
Giant piston	3200–4000	126·6	12·7	3000	Negligible	Long cores in abyssal depths.	Hollister et al (1973)
Vibro-corer	100	19·6–62·2	5·0–8·9	1000	—	Disturbs unconsolidated formations. Suitable for semi-consolidated formations.	Kudinov (1957)
Hydrostatic/Gas	50–100	11·3	3·8	600	—	Dependence upon sediment type, layering distorted. Operating depth <90 m.	Mackereth (1958), Rosfelder (1966)
Deep-sea drilling	Capacity 373 tonnes	9·6–19·6	3·5–5·0	—	—	Drilling tower on board ship. Discontinuous coring. Hard formation rotary bits with drilling fluid.	JOIDES (1965)

by weights and are not designed to be watertight. Thus, some loss of water and fine sediment is inevitable. When a piece of coarse debris is caught between the jaws, however, the washing away of fine sediment and the loss of sample is severe. The cost of grabs is moderate. In shallow water, the smaller versions can be raised and lowered by hand and used from the smallest rowing boat. In such areas, if the jaws fail to close properly, another sample can easily

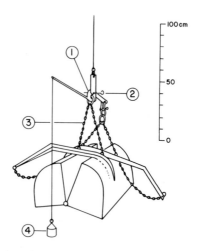

FIG. 36.7. Grab sampler. Release hook (1) is held by locking pin (2) until strain is taken on the hoisting wire. The device is lifted by the support chains (3), which are released when the tripping weight (4) touches bottom. After the unit has tripped, hoisting wire will close the jaws of the sampler. In some models, closure is assisted by springs or weights on the jaws.

(and economically) be taken near the same location, thereby overcoming the grab's chief disadvantage.

The ease of use, the effectiveness in all sediment types and the large sample volumes collected are the main advantages of grab samplers. The fact that some near-surface structure is often preserved in the recovered sediment is a slight additional advantage.

36.4.3. LARGE-VOLUME CORERS

If the preservation of the original structure is a major consideration in a study, the box corer is an appropriate sampling device (Fig. 36.8). Box corers effectively sample both coarse and fine sediments. In most models the coring weights can be adjusted to prevent over-penetration and to ensure that near-surface layering is preserved. Other large-volume samplers such as the Sphincter and Kasten corers (Table 36.1) also recover near-surface layers in a fairly undisturbed condition if the weight of the corer is adjusted to suit the penetration characteristics of the sediments being sampled. Of these, the Kasten corer (Fig. 36.9) yields the longest core. Even so, few Kasten cores longer than 10 m have been taken in the deep sea. Cores longer than about 3 m usually have to be hoisted on board ship in a horizontal position, leading to disturbance of the core top unless the supernatant water is first drained off and the upper sedimentary layers are in some way stabilized (for example with a layer of expanding plastic foam).

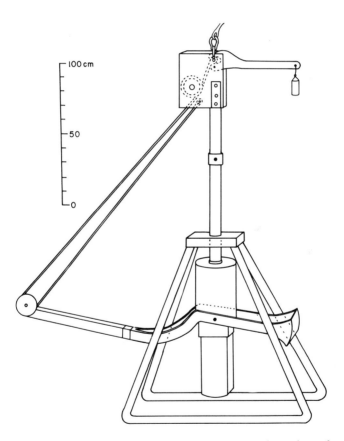

FIG. 36.8. Box corer. The sampling tripod is lowered to the bottom where release of tension on the hoisting wire allows the device to trip and the weighted rectangular box to sink into the sediment. The depth of penetration may be adjusted by adding or removing weights above the sampling box. In some models, hydraulic damping reduces the speed of penetration and aids in the recovery of the uppermost sedimentary layers. Retrieving the sampling device causes the forked arm to pivot underneath the bottom of the sample box to close it.

The smaller models of the Sphincter and Kasten corers can be used from a hydrographic winch, but even then good preservation of the surface sediment layers requires that the core be kept vertical. Thus, a fairly large stable working vessel is desirable. The box corer requires a heavier winch and is more un-wieldy than comparable Sphincter or Kasten corers, but it does have its own tripod platform which reduces the problem of keeping the core upright once it is on deck.

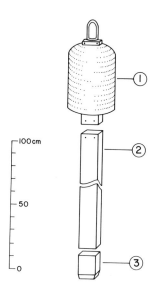

FIG. 36.9. Kasten corer. A heavy weight-stand (2700–5400 kg) drives the square barrel (2) into the sediment. In some models, the mass of the weight (1) is adjustable. In others, a piston core weight-stand is employed through the use of barrel adapters. The barrel is formed of two right-angle pieces which are joined longitudinally by screws or pop-rivets. Barrels are commonly 3 m long. Additional length may be obtained by coupling two or more barrels together. The core catcher (3) contains two overlapping rectangular flaps which are spring loaded. During descent, the flaps are held in the open position by latches with external tabs. These tabs are tripped upon pull-out.

The great advantage of these devices for stratigraphical studies is that they allow very thin down-core samples to be taken which, because of the large diameter of the corer, still give a large volume of sample material. This advantage of large volume can also be a disadvantage when it comes to recovering the core. The weight of the recovered core is large—several tens to more than a hundred kilograms, depending on its length. Thus, special provisions are necessary to handle the heavy, cumbersome samples on the rolling deck of a standard oceanographical vessel.

36.4.4. *In situ* SAMPLERS

Studies which focus on just the uppermost skin of bottom sediments, and on the interactions that take place across the sediment–water interface, impose very specific requirements on sample recovery. First, this skin must be recovered relatively intact; second, the cross-sectional area of the sample should be large compared to its thickness so that a sufficient volume of sample is available for study. Saidova (1968) took advantage of the epibenthic fauna, which feed from the uppermost surface layer of pelagic sediments, to sample these elusive deposits. Her examination of their gut contents showed, for example, that delicate calcite tests of foraminifera reach the sea floor even in the deepest trenches. Thus, the absence of such tests beneath the calcite compensation depth must result from dissolution at the sea floor rather than in the water column.

The recovery of the intact undisturbed surface of the sea floor is no mean feat. Even if the sampling device itself accomplishes the task, the hoisting of the sample on board, the draining of the surplus water from it, storage, transportation to the laboratory and all the handling required at each step in the operation can easily destroy what some well-designed sampling device has recovered. In addition, serious geochemical and geotechnical problems may be created by punching a hole in the sediment, and retrieving a core at 100 m min^{-1} through a temperature change of some $20°C$ and a pressure drop of 400 atmospheres. Because of these problems, investigators of sediment–water interactions and near-surface gradients in the sediment are turning to *in situ* systems which perform the required measurements on the sea floor. These devices make chemical, biological and geotechnical measurements which are either possible only *in situ* or more accurate if made in the natural environment (see e.g. McDonald *et al.*, 1972; Richards *et al.*, 1972; Barnes, 1973; Smith *et al.*, 1976; Sayles *et al.*, 1976).

In situ measurements tend to be expensive, as the equipment is complex and time-consuming to set up and recover. As yet such instruments have not been deployed in sufficient numbers to resolve spatial patterns. Rather, they provide a few direct measurements of processes taking place, on or within, the sea floor; these can be used to test conclusions based on shipboard or laboratory measurements.

Manned and unmanned submersibles have added much to our knowledge of the nature of the deep sea and of deep-sea deposits. One of the most sophisticated of the unmanned instruments is the deep-towed "fish" developed at the Scripps Institution of Oceanography. This vehicle can be towed within a few metres of the sea floor at 4–6 km h^{-1}. The instrument package contains echo-sounders, sidescan sonar, water samplers, television, magnetometers, camera and a 4 kHz sub-bottom reflection profiling system. It yields a very detailed picture of both the sea floor and shallow acoustic reflectors. The current version cannot retrieve samples off the sea floor, however, only manned submersibles have accomplished the feat of visually selecting a sampling site and retrieving a rock or sediment sample from the deep-ocean floor. The great advantage of knowing precisely the nature of the physical setting of a sample is obvious and for certain studies, such as those concerned with the processes taking place within a spreading ridge or fracture zone (Ballard et al., 1975), it is all important. Because of the rather myopic view from submersibles, however, a good base map of the area under investigation is a prerequisite for any efficient sampling programme from such a vessel.

The need for sophisticated electronics, complex transponder navigation systems and support ships when operating either manned or unmanned submersibles makes their operation very expensive. When the cost of pre-dive site surveys is added to this, it becomes clear that such operations can only be justified if the goals are very clearly defined and are attainable through studies of specific, quite small and well-surveyed areas. Sight-seeing trips on these vehicles are too expensive for most scientific budgets.

In contrast, randomly located observations by means of bottom photographs are relatively inexpensive and have added much to our knowledge of the nature of bottom sediments and the processes of the deep sea which affect them. For example, the interaction between the marine benthos and the sea-floor sediments is clearly revealed by photographs of the ocean floor (see e.g. Paul, 1976). Similarly, information on bottom currents can be fairly easily derived from pictures of the depositional structures and biological features of those areas of the sea floor which are affected by current flow. Finally, bottom photographs, seismic reflection records of the sub-bottom sedimentary layers and detailed maps of bottom topography, when taken together with evidence from bottom samples, have clearly defined the major pathways of bottom flow in the deep ocean by revealing areas of erosion, sediment dispersal paths (Burckle and Stanton, 1975; Kolla et al., 1976a), bed forms and other depositional features (Heezen and Hollister, 1971).

36.4.5. GRAVITY CORERS

The various modifications of the simple gravity corer are perhaps the easiest corers of all to use and represent one of the oldest basic designs of marine sampling gear. The cost of the equipment is minimal; there are practically no moving parts to jam or fail and, depending on the overall weight of the device, a small hydrographic winch is usually adequate for lowering and raising the corers. The hydroplastic-type corer is particularly attractive in this respect (Fig. 36.10). The diameter can be large, while at the same time the wall thickness of the barrel can be relatively small. Such cores are, therefore,

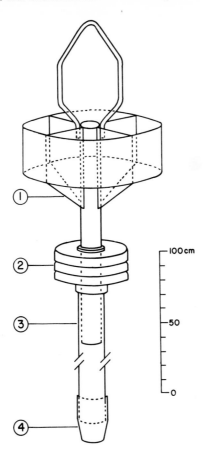

FIG. 36.10. Hydroplastic corer. This is one of many basic designs of the gravity corer. These samplers are driven into the sediment by the weight of the device. The amount of weight (2) is usually adjustable. In this model, the barrel (3) is made of plastic (plastic liner from a piston corer is sometimes used). The barrel can be attached to the weight-stand with screws or hose clamps. The barrel length is variable, but usually less than 3 m. The nose piece (4) contains an annulus of slightly curved, spring-metal fingers which prevent the loss of the core sample upon pull-out. The prevention of sample loss is also aided by a check valve at the top of the core which allows water to flow through the barrel during descent, but closes upon retrieval.

suitable for geotechnical studies, in which reduced disturbance of the sediment by wall friction is desired, and for other studies (pore waters, rare minerals, etc.) in which large volumes of sediment are required. In addition, because the plastic tubing used for the barrel is comparatively inexpensive, the recovered sediment can be stored in this and a new barrel used for each lowering operation.

The disadvantage of the gravity corer is its rather limited penetration. Thus, while it is entirely adequate for most studies of surface sediments, it is unsuitable for many stratigraphical purposes. Increasing the weight of the corer increases its penetration up to a point, but the same design characteristics which cause relatively high wall friction and shallow penetration lead to disturbance of the recovered sediments through compression of the section (Ross and Riedel, 1967). Another common form of disturbance in gravity cores results when they are not retrieved quickly after being pulled out of the sediment. Rolling of the ship can cause the corer to bounce one or more times along the sea floor. Thus, the uppermost layers will be repeated lower down the core (see Fig. 36.13c.)

36.4.6. PISTON AND TRIGGER WEIGHT CORERS

For most stratigraphical studies, the piston corer and trigger weight (or pilot) corer combination has become the most widely used and successful means of sampling the deep-sea floor (Fig. 36.11). Early designs of this combination used a simple weight to trip the free-fall mechanism, but it was soon realized that two cores could be taken at no extra cost in ship time and a short gravity-type corer was substituted as the trigger weight. At first, the trigger weight corers were of small (2–3 cm) diameter, similar to the Phleger corer (Table 36.1). However, when it became evident that these gravity corers provided a much better sample of the surface sediment than did the piston corer, larger versions were introduced. Because of the better coring characteristics of the wider diameter barrels, these devices not only provide a larger volume of surface sample, but they also give a less compressed, longer core of sediment. More recently, multiple-barrelled corers (Fig. 36.12; Fowler and Kulm, 1966) have been used as trigger weights. The multiple corer not only collects more surface sediment, but it also allows the small-scale spatial variability (of the order of a few centimetres) between individual barrels to be evaluated and compared with the variability between the piston and gravity cores (separated by about one metre).

The piston in a piston corer creates a negative pressure above the sediment as the barrel moves past it. This pressure counteracts friction between the sediment and barrel wall, leading to deeper penetration than is possible with gravity corers (Kullenberg, 1947). Most oceanographical ships are equipped to take 12–18 m piston cores with little difficulty. Cores of up to 30 m in length have been taken successfully from conventional oceanographical vessels using standard piston coring gear.

In addition to a pressure wave which precedes the nose of a piston corer and blasts away a few centimetres of surface sediment (Ewing et al., 1967), other factors such as differing core diameters, adjustment of the piston seals

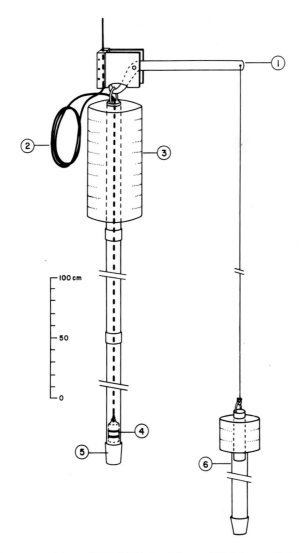

FIG. 36.11. Piston corer. A heavy (2700–5400 kg) weight-stand (3) drives a cylindrical steel barrel up to 25 m into the sediment. The main coring device is released when the trigger weight (usually a gravity-type corer) (6) hits the bottom. The trigger arm (1) then releases the piston corer which free-falls into the sediment. The length of line on the trigger weight corer is equal to the length of the piston corer barrel (usually 6–18 m) plus the length of free-fall wire (2) (usually 3–5 m). The nose of the piston corer should, therefore, reach the sea floor at precisely the end of the free-fall. The winch stops lowering when the corer trips. After free-fall, the wire halts the piston (4) at the sediment surface and the corer falls past the piston. The resulting reduction of pressure in the barrel allows greater penetration than can be obtained with a gravity corer of comparable dimensions. The core catcher (5) can be of either a spring-finger or flap type.

and the length of the free-fall line can cause differences in recovery between the piston corer and its accompanying trigger weight corer (Ross and Riedel, 1967). Disturbance within the sedimentary section recovered by piston corers can occur in several ways. Wall drag often bows the sedimentary layers. This is clearly visible in layered sediments (Fig. 36.13b), but may be invisible in sediments with homogeneous colour and lithology. Similarly, if the corer does not penetrate to the full length of the barrel, sediment is sucked in from the region around the core nose as the piston is pulled to the top of the barrel

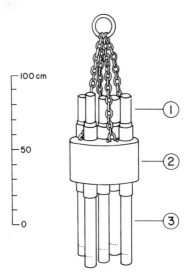

FIG. 36.12. Multiple corer. This device mounts up to five short core barrels (3) on a single weight-stand (2) in another variation of the gravity corer design. This design includes spring-finger core catchers at the bottom of the barrels and check valves (1) at the top. It can be used by itself or as a trigger weight with piston corers.

during recovery. Vertical striping in the sediments at the bottom of piston cores usually serves to identify such "flow-in" material (Fig. 36.13a). "Flow-in" can be eliminated by the use of a detachable piston which frees the hoisting line from the piston after penetration. Some investigators welcome "flow-in" sections, however, as they provide large volumes of sediment from well below the sea floor.

For piston corers which do not contain plastic liners, the sediment must be extruded from the core barrel by means of a pump. This may cause further disturbance. Even in a corer with a liner, however, malfunctions commonly disturb the recovered section. The suction pressure due to the piston readily

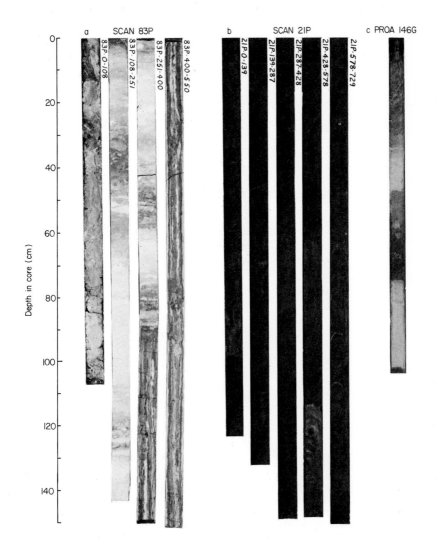

Fig. 36.13. (a) Piston core SCAN 83P. Note the vertical striations beginning about 90 cm down from the top of the third section. This sort of striping identifies the portion of the core sucked up by the piston action. Note also the disturbed nature of the uppermost section.

(b) Piston core SCAN 21P. Note the bowing of layers in the piston core in the lower two sections.

(c) Gravity core PROA 146G. Note the apparently repeated section below about 50 cm. Such repeated sections can occur when the corer strikes the bottom more than once.

(Core photographs courtesy of T. Walsh and W. R. Riedel, Scripps Institution of Oceanography.)

collapses the wall of a defective liner. Furthermore, leakage at joints in the string of liners leads to water-filled gaps in the recovered sedimentary sections.

The advantages of the piston corer have been married to those of the large-diameter corer in the giant piston corer (Fig. 36.14; Hollister *et al.*, 1973) to achieve maximum penetration and recover large-volume, minimally

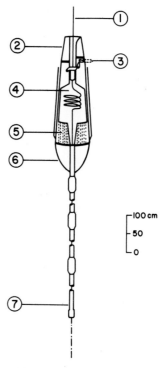

FIG. 36.14. Giant piston corer. This device works like a normal piston corer (Fig. 36.11) but has been modified to obtain very long, undisturbed sections. The main wire (1) has to have at least a 12 000 kg working strength. The weight-stand (weighing approximately 5 600 kg) includes instrument compartments (2), a compartment for the free-fall loops of wire (4), and removable (5) as well as permanent (6) weights. The corer may be tripped either acoustically or mechanically (3). The barrel (7) is usually 30–40 m long (composed of 10–14 coupled segments) and has an inside diameter of 11·4 cm.

disturbed cores. This corer is capable of recovering sections which are more than 30 m thick. Recent models have core barrels in which the wall thickness tapers from 0·6 cm at the core nose to 1·9 cm just below the weight-stand. They also include a core-nose velocimeter and accelerometer, and core-head

3·5 and 12 kHz acoustic transducers to record the acoustic structure of the sediment at the coring site. Because the weight of this device exceeds the capabilities of most regular oceanographical vessels, and is close to the limit of the working load of most deep-sea wires, equipment losses have been high.

36.4.7. FREE-FALL CORERS

Free-fall (or boomerang) corers (Fig. 36.15) have proved to be most effective where a sampling plan calls for a closely spaced transect of cores across a previously defined topographical feature. The only shipboard requirements

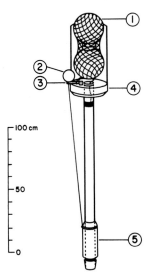

FIG. 36.15. Free-fall corer. This corer is deployed without an attached wire and depends on two hollow glass spheres (1) to float the 1 m core to the surface. A hollow rubber ball (2) is wedged into an opening in the steel cylinder which encloses the floats. This ball prevents accidental tripping of the float release (3) during launching. After launching, the ball collapses under pressure and slips out of the opening. The ballast weight (5) keeps the corer vertical while falling through the water column. The weight (5) is attached to the float release mechanism (3) by a flexible wire. Upon penetration of the sediment, the ballast weight is forced to slide up the barrel, thus allowing the float release mechanism to pivot and release the floats. The rising floats close a valve at the top of the core liner and pull the liner from the barrel. The coring device (4) remains on the sea floor.

for such a study are accurate navigation and a good echo-sounder. Two people can easily toss the corers over the side while the ship is underway, and they can retrieve the floatation devices and cores upon their return. The corers fall at about 0·5 km min^{-1}; thus, lateral displacement of the corer by current drift is practically non-existent. The location of the core is taken to be that of the ship at the time that the corer went over the side. Round-trip travel time is usually about 15 minutes per 1000 m of depth; this is not much faster than a normal gravity corer. However, there is no way in which a string of accurately located, closely spaced gravity cores can be taken up the side of a hill with the ease and low cost with which several free-fall corers can be dropped over the side. Thus, the advantages of this device lie in its minimal requirements for shipboard equipment and in its ability to take closely spaced samples which are accurately located relative to each other and to the bottom topography. However, it has disadvantages which include:

(i)   a lack of penetration (the cores are approximately 1 m long);
(ii) the necessity to recover it at night when the strobe beacon in the float is visible;
(iii) a loss rate of 10–20% which is considerably higher than for more conventional, wire-line coring devices;
(iv) an initial unit cost which is considerably greater than that of a simple gravity corer; and
(v)   the time required to rig and check each corer, as well as to service the flotation and beacon components, which is considerably greater than that required to rig a gravity corer capable of retrieving a sample of similar size.

36.4.8. DEEP-SEA DRILLING

Even using the giant piston corer, the maximum length of a continuous sequence of fossiliferous pelagic sediments that can be recovered spans only a few million years. If older sections are to be sampled from conventional oceanographical ships, the areas in which they outcrop must be found. The location of such outcrops is difficult in the absence of surveys using both low and high frequency acoustic profilers. The low frequency (20–100 Hz) profilers can trace layers tens or even hundreds of metres below the sea floor to areas of outcrop, whereas the high frequency (3·5 kHz) profilers can distinguish true outcrops from cases where the older deposits are mantled by a few metres of recent sediments. A series of cores taken across an outcrop can easily recover a sequence of sediments spanning several million years. As discussed previously, the difficulty lies not so much in the recovery of old sediments but in the correlation of the recovered cores both with one another, and with the reflection profiler records. In unlithified sediments the free-fall corer is the device with which it is easiest to locate samples relative both to each other and to an area of outcrop. The length of record that is recovered in each core, however, is very short. In the central equatorial Pacific, the area of outcrop is so large that randomly located, relatively short piston corers and gravity corers have produced a rather comprehensive collection of

Cenozoic sediments (Riedel, 1971; Funnell, 1971; Saito *et al.*, 1974), even though individual outcrops are so small that a systematic sampling of the same sequence would be extremely difficult.

The use of outcropping sediments and a pieced-together record to reconstruct the geological history of an area encounters two difficulties, however. Firstly, outcrops by their very nature are associated with areas of erosion, lateral transport, mixing and reworking of sediments. Thus, the sediments recovered from these areas are often reworked and badly mixed, which in turn tends to result in an inaccurate stratigraphy and its associated problems. Secondly, the tropical Pacific is not a biogeographically, chemically or lithologically homogeneous area. Thus, the piecing together of a few composite sections is not adequate to resolve the palaeoceanographical history of an ocean basin. What are needed are many sections, spaced according to the criteria discussed in Section 36.2, and containing as complete a stratigraphical record as possible.

The Deep-Sea Drilling Project (DSDP) (van Andel, 1968) has provided a means for recovering such sections. The drilling ship used for this project, the D.V. *Glomar Challenger*, can maintain its position relative to a transponder on the sea floor and extend its drill down through almost 7000 m of sea water and sediment to recover the sedimentary column and even some of the underlying basaltic layer over most of the ocean basins (Fig. 36.16). The use of this tool has enabled marine geologists to study the longest most-nearly complete sections that have ever been recovered from the deep sea. The individual sites drilled by this ship, which are located in all oceans except the Arctic Ocean, now number well over 400. At most of these sites, a significant portion of the sedimentary section has been recovered for study. Of course, these DSDP sections are still not the perfect records dreamed of by nineteenth and early twentieth century geologists because of the gaps mentioned previously (Fig. 36.6).

The age of the sediments immediately above the basaltic basement (see e.g. van Andel and Bukry, 1973) and the distribution of biogenic sediments (Tracey *et al.*, 1971; Berger, 1973; van Andel *et al.*, 1975) are consistent with the theory of plate tectonics (Chapter 35). As feared, the record of earth history that can be sampled in the oceans is very short. The average age of the ocean crust is about 60 million years (Berger and Winterer, 1974), and the oldest sediments recovered thus far are late Jurassic (~ 140 million years B.P.) in age.

In addition to these difficulties, which set natural limits on the comprehensiveness of the marine geological record, technical problems further limit the usefulness of some sections collected by the DSDP. Prior to DSDP Leg 8, for example, massive and light set diamond bits were most commonly used in drilling. These bits could not penetrate the chert layers encountered

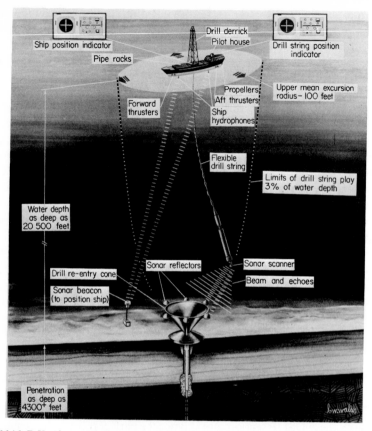

FIG. 36.16. D.V. *Glomar Challenger* uses "dynamic positioning" to hold station above a sonar sound source placed on the ocean bottom while drilling. Two tunnel thrusters forward and two thrusters aft, together with the vessel's two main propellers, are computer controlled to hold position without anchors in water depths up to 6100 m so that drilling and coring can be accomplished. When a drill bit is worn out it is possible to retract the drill string, change the bit and return to the same bore hole through a re-entry funnel placed on the ocean floor. High-resolution scanning sonar is used to locate the funnel and to guide the drill string over it. The artist's concept shows a sonar beacon used for "dynamic positioning" and a sonar scanner at the end of the drill string searching for the sonar reflectors on the re-entry cone. The relative position of bit and funnel are displayed at the surface on a Drill String Position Indicator Scope. Re-entry was developed by the DSDP when the Project was stopped short of its scientific goals at many sites in the Atlantic and Pacific Oceans at which blunting of the bits by hard chert layers forced early abandonment of bore holes.

in Mid- to Upper Eocene intervals; therefore, older intervals were rarely sampled. Roller-cone bits through which cores could be taken were tried first on Legs 7 and 9 and were subsequently routinely used to drill through chert layers. Such bits also recover a higher percentage of basaltic sections

(43 %) than do the diamond bits (31 %). The introduction of re-entry cones on DSDP Leg 15 further enhanced the ability of the Project to core thick sedimentary sequences by allowing the drill string to be pulled out of the hole in order to replace a worn bit, and then to re-enter the same hole to continue coring.

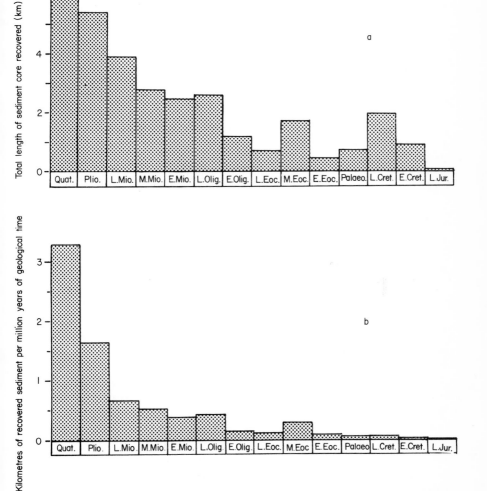

FIG. 36.17. (a) Thickness of sediments representing various geological time intervals recovered by DSDP Legs 1–33.
   (b) Thickness of sediments representing one million year intervals (grouped by geological time intervals) recovered by DSDP Legs 1–33. Durations of geological units from Berggren (1972).

T.

Percentage recovery ((metres recovered/metres cored) × 100) tabul

Ocean basin or sea	Late Jurassic	Early Cretaceous	Late Cretaceous	Palaeocene	Early Eocene	Midd Eocer
Global	45·9	28·7	37·4	38·7	40·6	50·4
Pacific	50·0	16·5	34·4	36·3	43·5	49·0
Atlantic	45·5	31·9	41·0	44·4	38·1	54·8
Indian	58·3	49·8	46·1	39·2	43·4	48·8
Caribbean	0·0	9·2	33·9	34·5	21·7	56·9
Mediterranean	0·0	27·3	0·0	0·0	0·0	0·0

Many problems associated with the recovery of sediments that are penetrated, however, have not been solved. The most obvious and pervasive deficiencies of DSDP cores are their incomplete recovery and the disturbed nature of soft sediments. Drilling, pumping and coring occur intermittently as a hole is drilled. During all three operations, the drill pipe moves up and down in the hole in response to the rolling and pitching of the drilling ship. These motions are partly responsible for incomplete filling of the core barrel and disturbance of the recovered sediments.

When the drill bit encounters lithified sediments or rock, drilling fluid (sea-water) is circulated down the pipe in order to prevent it from being jammed in the hole by rock fragments. Unfortunately, this circulating water also washes away any soft sediments which are interlayered with the lithified deposits. Thus, in older, more indurated sections, the percentage of core recovered tends to decrease (Table 36.2). Recovery of Recent through Lower Oligocene deposits is approximately constant at about 60%; in Eocene and Palaeocene deposits, in which chert layers are commonly encountered, recovery drops to 40–50%; and in Cretaceous deposits, in which limestones and cherts are found interlayered with clays, recovery decreases to about 30–40%. This difficulty in recovering old sections, combined with problems such as the age distribution of the oceanic crust, results in a marked decrease in total recovery as a function of age (Fig. 36.17).

Mechanical mixing may also mar sediment recovery. Because the hole is not cased, shallow sediments can slump down and mix with older material recovered from the deeper part of the section. When such slumping and mixing is severe, it becomes extremely difficult to establish the exact age of the recovered samples. Slumping of sediments is most common in sections which contain thick sandy layers and, if it is particularly severe, it can completely jam the drill pipe in the hole. Even for rather soft biogenic oozes of Upper Palaeogene and Neogene age, recovery is far from complete. That some *ocean and geological age. (Data from DSDP Legs 1–33.)*

Late ocene	Early Oligocene	Late Oligocene	Early Miocene	Middle Miocene	Late Miocene	Pliocene	Quaternary
49·7	59·6	68·7	64·0	59·8	60·9	63·0	63·4
47·4	57·6	67·8	66·3	61·2	64·5	62·3	61·6
60·8	71·0	83·0	63·0	41·1	72·3	69·7	62·2
54·6	60·7	68·5	55·8	62·3	60·4	64·1	70·8
56·4	62·8	46·6	52·8	61·0	45·4	65·2	62·3
0·0	0·0	0·0	0·0	29·4	16·3	50·1	59·8

intervals are completely missed, even in continuously cored sections with 100% recovery, is indicated by the distribution of biostratigraphical boundaries in the recovered cores. Through DSDP Leg 34, about 4000 m of sediment have been recovered in contiguous cores with 100% recovery. If this sediment actually represented perfect recovery of the section cored, the biostratigraphical zonal boundaries encountered should be uniformly divided amongst the six 1·5 m sections which make up each 9 m core. A few boundaries would fall between cores, but nowhere near the number actually found (Fig. 36.18). This compilation shows that there is a slight deficiency of boundaries in the top sections (1) of the cores, and too many boundaries in Section 6. The distribution of boundaries in Sections 2–5 is approximately uniform. Roughly 30% of the boundaries occur between contiguous cores. These results support an earlier study (based on DSDP Legs 1–8 only; Moore, 1972), which showed that of the 9 m cored, 3·8 m was lost. Thus, only about 60% of a continuously cored section is actually recovered even though the core barrel is full. The most obvious explanation for this somewhat surprising result lies in the size and shape of the drill-bit face relative to the opening to the core barrel. The ratio of the bit face to core area is approximately 20/1. Although much of the sediment is pushed aside, it appears that some of it is pushed inwards and enters the core barrel together with the correctly cored material as the bit moves ahead. Obviously, the bit does not act like a perfect funnel, however, or the core barrel would be filled before it had penetrated even one metre and the layering which is preserved in many DSDP cores would be totally destroyed.

A core barrel which extends in front of the bit has been tested in soft sediments at a few sites (e.g. on Legs 15 and 16) in the hope of reducing disturbance of the sediments and improving core recovery. The overall recovery was not improved by this technique—in the opinion of some of the chief scientists this was because recovery and disturbance is more a function of the up and down motion of the ship than of the coring technique or equipment used (Heath *et al.*, 1973a).

In "spot" cores (short intervals sampled as a hole is drilled ahead), the distribution of biostratigraphical boundaries is much like that in continuously cored sections (Fig. 36.18). Slightly fewer boundaries occur in Core-section 1 and slightly more in the lowermost section (plus core catcher) than in the intermediate sections of the cores. This pattern indicates that the recovered sequence is only roughly representative of the 9 m interval that was cored.

It is clear from these results that a continuous, undistorted record of the

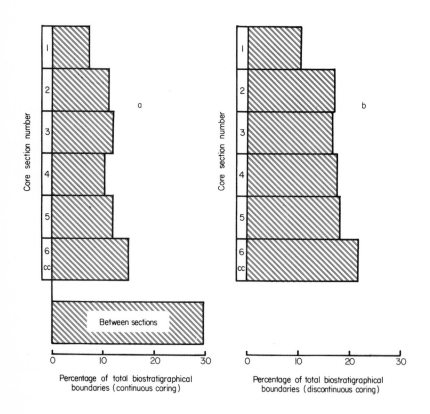

Percentage of total biostratigraphical
boundaries (continuous coring)

Percentage of total biostratigraphical
boundaries (discontinuous coring)

FIG. 36.18. (a) Distribution of biostratigraphical boundaries within DSDP core sections and between cores at intervals which were continuously sampled with 100% recovery (cc, core catcher; total boundaries observed = 325).

(b) Distribution of biostratigraphical boundaries in DSDP core sections that were taken in discontinuously sampled intervals. For cores in which recovery was less than 100% sections were renumbered so that the lowest one was always Section 6 (cc, core catcher; total boundaries observed = 1488).

Cenozoic and late Mesozoic has yet to be recovered from the deep sea. Furthermore, using only the DSDP material it will be very difficult to determine the exact periodicity of fluctuations in any biological or lithological variable if their peak to peak durations are less than about one million years, for example those in the Clipperton Oceanic Formation (Tracey et al., 1971). This does not mean, however, that the DSDP material is so badly disturbed and distorted that it is useless—far from it. Investigators need only be aware that the lower limit of resolution in a series of DSDP cores is nowhere near as great as that attainable in a piston core. The great usefulness of the DSDP material lies in the length of record it represents and in the broad geographical coverage which it provides. Together, these attributes fit the requirements for many temporal and spatial studies, thereby allowing us to follow the evolution of the oceans from Mesozoic to modern times.

36.5. RESULTS

36.5.1. SURFACE SEDIMENT STUDIES

Despite the efforts of marine geologists over the 100 years since the H.M.S. *Challenger* expedition, the distributions of the various properties and components of pelagic sediments are far from being completely mapped. Even the maps in which we have some confidence are of very recent origin and, in some cases, are based on surprisingly few data points (Table 36.3). For some of the parameters, such as carbonate and topography, data have been accumulated gradually over many years; for others, such as opal and quartz, virtually all of the data have been produced in the last decade. Although the list in Table 36.3 is not exhaustive, it does include many of the major data sets which have been published. There are obvious gaps in these data. Some of the gaps, such as those in the planktonic faunal and floral distributions, are rapidly being filled (see e.g. Cline and Hays, 1976). Others, however, languish until research into them is motivated.

Perhaps what is even more surprising to a marine geologist is the almost total lack of up-to-date, global compilations of physical, chemical and biological oceanographical data. When marine geologists try to correlate the properties of surface sediments with average modern oceanographical conditions, they find it is nearly impossible on a global basis. Recently, a group of global climatic modellers sought a compilation of mean monthly sea-surface temperatures for the world oceans to use as input to mathematical simulations of modern climate. Even for such basic data they found that they had either to assemble the information themselves (e.g. Alexander and Mobley, 1974), or to piece together a host of regional maps. Monthly or

TABLE 36.3

Quantitative measurements of components of marine surface sediments from large regional studies

Component	Oceanic area	References
Topography	World	Chase (1975)
Carbonate	Pacific	Berger *et al.* (1976)
	Atlantic	Biscaye *et al.* (1976)
	Indian	Kolla *et al.* (1976b)
Opal	World	Lisitzin (1972)
Clay minerals	World	Griffin *et al.* (1968), Rateev *et al.* (1969)
	Arctic	Carroll (1970)
	Indian	Venkatarathnam and Biscaye (1973), Goldberg and Griffin (1970)
Quartz	Pacific	Rex and Goldberg (1958), Heath *et al.* (1973b)
	Indian	Kolla and Biscaye (1977)
Ice-rafted debris	North Pacific	Conolly and Ewing (1970)
	Antarctic	Lisitzin (1960), Keany *et al.* (1976)
Manganese nodules	World	Horn *et al.* (1972a, b), Monget *et al.* (1976)
Foraminifera	Antarctic	Bé (1969)
	Atlantic	Kipp (1976), Imbrie and Kipp (1971)
	Indian	Bé and Tolderlund (1971)
	Pacific	Parker and Berger (1971)
Radiolaria	Antarctic	Lozano and Hays (1976)
	Atlantic	Goll and Bjorklund (1971, 1974)
	Indian	Nigrini (1967)
Coccoliths	Atlantic	McIntyre and Bé (1967)
	Pacific	McIntyre *et al.* (1970), Geitzenauer *et al.* (1976)

seasonal maps of other basic oceanographical data, such as sea-surface salinity, have yet to be compiled on a global basis.

36.5.2. STRATIGRAPHICAL STUDIES

Literally thousands of piston cores have been recovered from the ocean basins for study by marine geologists. Commonly, more than half of the cores from a given area are found to be unsuitable for a particular study. They may be too short, contain gaps in the record, be badly disturbed by the coring process, show widely varying sedimentation rates, show extreme dissolution of the biogenic components or simply contain a mixed, or homogenized, record. Of the cores that are acceptable, most do not penetrate Pleistocene deposits. In consequence, the geological studies of such cores have concentrated on very early diagenetic processes or on glacial–interglacial fluctuations in palaeoceanographical and palaeoclimatic conditions (see

Chapter 29). Early studies of the marine Pleistocene record stressed three main problems:

(i)   the general nature (intensity and duration) of climatic fluctuation,
(ii) the correlation of the marine record to the classical glacial sequence on land, and
(iii) the chronology of these fluctuations.

Stratigraphies were developed to permit many cores to be correlated (e.g. in the Caribbean and Atlantic; Ericson and Wollin, 1968; Kennett and Huddleston, 1972). However, when these rather simple stratigraphies were interpreted in terms of climatic fluctuations or extended beyond the range of their applicability, they led to the surprising and apparently erroneous conclusion (CLIMAP, 1976) that the Pleistocene warm and cool intervals were out of phase in the Atlantic and Pacific Oceans (Ericson and Wollin, 1970; Wollin et al., 1971; CLIMAP, 1976). One of the earliest, and currently still the best, method used to define the intensity and duration of Pleistocene climatic fluctuations is the determination of the oxygen isotopic composition of the calcareous tests of foraminifera (Emiliani, 1955, 1966; Shackleton and Opdyke, 1973, 1976). These data are thought to record changes in the global ice volume. During glacial times the lighter oxygen isotope (^{16}O) was preferentially deposited in continental ice masses, leaving an excess of ^{18}O in the oceans; this was incorporated in the calcite tests of the foraminifera. Such isotopic variations not only provide an important record of climatic variations, but also constitute a stratigraphical tool which is globally applicable and (within the mixing time of the oceans) synchronous throughout the ocean basins (Shackleton and Opdyke, 1973).

The detail of this record, which has been recognized in cores from all the major oceans of the world, far exceeds any comparable climatic record that has been sampled on land. Thus, one of the hopes of nineteenth century geologists has been fulfilled—the marine Pleistocene section has become the standard with which the less complete land records are being compared.

Variations in other biological and lithological characteristics of sediment cores have also been used to characterize the oceanographical fluctuations of the ice-ages. The new palaeontological technique of Imbrie and Kipp (1971) allows palaeontologists to estimate changes in the temperature of near-surface waters during the Pleistocene, an important complement to the isotopic studies.

In the same way as changes in the fossils of the microplankton reflect variations in the conditions in the near-surface waters overlying the sample site, so do changes in the dissolution of carbonate material record changes in the chemical nature of the bottom waters. When such changes are determined in the same core, the ice volume, surface water and bottom water records can be accurately compared, allowing slight phase shifts in the individual records to be interpreted in terms of the response of each part of the climatic and oceanic system to climatic change (Luz and Shackleton, 1975; Ninkovich and Shackleton, 1975; Hays *et al.*, 1976; Moore *et al.*, 1977).

The determination of a reliable chronology for these climatic fluctuations has given rise to much debate, particularly for that part of the section which is too old to be dated by radiocarbon (more than about 50000 years), but which is younger than the first magnetic reversal at 700000 years B.P. Other radiometric dating techniques have been applied to deep-sea cores (Broecker *et al.*, 1968; Mathews, 1973; Chappell, 1974; Ku *et al.*, 1974) in attempts to resolve this problem. The weight of the evidence appears to support an age of 125000 years B.P. for Emiliani's isotope stage 5e (the beginning of the last full interglacial). The ages of older glacial events have been estimated by interpolation between this age and the base of the Brunhes–Matuyama boundary, assuming a uniform sedimentation rate in a long core from the western Pacific (Shackleton and Opdyke, 1973).

Spectral analysis of the oxygen record (as well as other records) of two cores from the Antarctic Ocean (Hays *et al.*, 1976) showed peaks at frequencies corresponding to periodic variations in the Earth's orbit which affect the global distribution of solar radiation (i.e. variation in the tilt of the Earth's axis and in the eccentricity and precession of the Earth's orbit). When the calculated record of variations in the Earth's orbit was compared to measured fluctuations in the core records, a nearly consistent phase relationship was found which indicates that changes in the Earth's orbit slightly preceded changes in the ocean. Only slight adjustments to the time-scale used for the core (less than 5% of the down-core ages) were required to reconcile these phase relationships throughout the past 350000 years. This work illustrates a step-wise technique which first establishes as good a time-scale as possible with radiometric measurements and then uses calculated orbital parameters to fine-tune this time-scale. Theoretically, such an approach could be used for any part of the marine record in which known periodic changes in the amount or distribution of solar energy have acted as pacemakers to changes in the oceans.

The development of such a comparatively detailed chronology for the pre-Pleistocene record has not yet been achieved. A considerable improvement in the Cenozoic time-scale has been made during the past decade, however, as more and more terrestrial fossiliferous sections have been dated radiometrically, and as these sections have been painstakingly correlated to the marine record by both biostratigraphical and magnetostratigraphical techniques (Berggren, 1969a, b; 1971, 1972; Berggren and van Couvering, 1974; Ryan *et al.*, 1974). The resolution of biostratigraphy itself has been greatly increased by the definition of more detailed zonations (Martini and Worsley,

1970; Bukry, 1973) and by the establishment of reliable stratigraphies for microfossil groups such as the radiolarians (Riedel and Sanifilippo, 1970; Moore, 1971; Foreman, 1975), nannofossils (Bukry, 1973) and diatoms (Burckle, 1972; Schrader, 1973).

Our present understanding of the Cenozoic and Mesozoic marine record is comparable to our understanding of the Pleistocene record 25 years ago. Even where major oceanographical changes have been pinpointed in a stratigraphical sense (e.g. the marked cooling of the oceans and depression of the calcite compensation depth at the Eocene–Oligocene boundary; Kennett and Shackleton, 1976), the exact causes of these changes are inadequately understood.

The oxygen isotopic record preserved in the tests of foraminifera has again led the way in defining the magnitude and general character of oceanographical changes in the Mesozoic and Cenozoic (Douglas and Savin, 1975; Savin et al., 1975; Shackleton and Kennett, 1975). Prior to the mid-Miocene, when a major ice-cap began to develop on Antarctica, fluctuations in the isotopic records recorded by benthic and planktonic foraminifera primarily resulted from temperature changes in the deep and near-surface waters. After the mid-Miocene, however, the isotopic variations largely resulted from changes in the volume of glacial ice on land, as discussed previously. Little is known of the quantitative changes in the microfossil assemblages prior to the Pleistocene, other than information derived from biostratigraphical studies (such as the information on extinctions, first appearances and rates of evolution of species given by Bramlette and Martini (1964) and Berggren (1969a)). Lithological changes in Cenozoic deposits are a little better known, however. Thus, the rates of accumulation of carbonate (van Andel et al., 1975) and biogenic opal (Leinen, 1977) in the tropical Pacific have changed by factors of two to three in the last 40 million years, with maxima in carbonate accumulation during the Oligocene and Middle Miocene, and maxima in opal accumulation during the late Eocene and Middle Miocene.

These and other studies of DSDP material have taken advantage of the new microfossil stratigraphies, and the good areal coverage provided by drilling sites in some ocean basins, to reconstruct sedimentary patterns of the past. Using the theories of plate tectonics, the geographical positions (van Andel et al., 1975) and palaeodepths (Sclater et al., 1971; Berger, 1973a) of DSDP sites can be estimated throughout their depositional histories. These reconstructions have provided the necessary data to determine changes in the calcite compensation depth (Berger, 1973b; van Andel and Moore, 1974; van Andel, 1975) and in the rate of calcite dissolution (Heath et al., 1977) with time. In addition, our understanding of temporal changes in accumulation rates (van Andel et al., 1975; Leinen, 1977) and abundance of hiatuses (Moore and Heath, 1977; Moore *et al.*, 1978) has been greatly enhanced by our ability to view palaeogeographical reconstructions of the ocean basins at critical times in the past. It is unfortunate that the times of greatest changes are often poorly represented in the stratigraphical record. Thus, the abundance of hiatuses in the recovered sections (Fig. 36.6) is almost a mirror image of the thickness of sediment recovered for each age (Fig. 36.17). Poor representation of some intervals, such as the Cretaceous–Tertiary boundary, has been partially overcome by special sampling efforts on the part of DSDP scientists. In general, however, it is probably more realistic to study the details of critical transitions from one oceanographical stage to the next in the few good sections that are preserved (see e.g. Kennett and Shackleton, 1976), and to reconstruct the spatial patterns of the oceans just prior to and following the changes from the larger suite of available cores.

36.6. Conclusions

In the past decade, great technical advances have been made in our ability to sample the sea floor. Not only can it be dredged and cored, but man can physically descend to its depths and sample it. It is now possible to see into it with the aid of acoustical devices and, with the proper ship, it is feasible to drill all the way through it and sample at the same time. Still, scientists are not satisfied. Neither the preserved geological record nor the new tools which have been developed to sample it have fulfilled all that is required. It could be argued that additional effort is needed; large areas of the World Ocean have yet to be adequately sampled by the DSDP and many of these areas are known to be of critical importance (e.g. the Antarctic region). Perhaps new technology will alleviate some of the problems of sample disturbance and distortion. In contrast, however, it might be more productive to take a conservative approach and return to the more "traditional" sampling methods and tools.

The ocean floor will never yield a complete record of the Mesozoic. For this reason, a return must be made to the areas where the record exists on land and these exposures must be viewed with the new insights that oceanographical and palaeoceanographical studies have provided. The Cenozoic record in the deep sea is not complete and cannot be recovered from the sea floor undisturbed. Again, renewed study of marine Cenozoic deposits on land might aid greatly in piecing together the evolution of the oceans. Alternatively, it may be possible to recover less disturbed Cenozoic sections from the sea floor using piston corers and giant piston corers (carefully located in well-surveyed areas of outcrop) than are now gathered by the D.V. *Glomar Challenger.*

Finally, it may be argued that our ability to recover samples from the deep-sea floor has outstripped our ability to study them. Our libraries of archived sample material have grown faster than the resources and manpower available for their study. This may be only a temporary imbalance; but further efforts in sample collection might be more effective and more successful if they are preceded by additional studies of the cores we already have in hand.

REFERENCES

Ade-Hall, J., Aumento, F., Ryall, P. J. C., Gerstein, R. E., Brooke, J. and McKeowen, D. L. (1973). *Can. J. Earth Sci.* **10**, 679.

Alexander, R. C. and Mobley, R. L. (1974). "Monthly Average Sea-surface Temperatures and Ice-pack Limits on a 1° Global Grid". Advanced Research Projects Agency Report R-1310-ARPA.

Arrhenius, G. (1963). *In* "The Sea" (M. N. Hill, ed.), Vol. 3, pp. 655–727. Wiley–Interscience, New York.

Aumento, F. (1970). *Can. J. Earth Sci.* **7**, 534.

Ball, M. M. (1967). *J. Sed. Petrol.* **37**, 556.

Ballard, R. D., Bryan, W. B., Heirtzler, J. R., Keller, G., Moore, J. G. and van Andel, Tj. H. (1975). *Science, N.Y.* **190**, 103.

Barnes, R. O. (1973). *Deep-Sea Res.* **20**, 1125.

Bé, A. W. H. (1969). *Am. Georg. Soc. Antarct. Map Folio Ser.* **11**, 9.

Bé, A. W. H. and Tolderlund, D. S. (1971). *In* "The Micropaleontology of Oceans" (B. M. Funnell and W. R. Riedel, eds), pp. 105–149. Cambridge University Press, England.

Berger, W. H. (1971). *Marine Geology*, **11**, 325.

Berger, W. H. (1973a). *Nature, Lond.* **236**, 392.

Berger, W. H. (1973b). *Bull. Geol. Soc. Am.* **84**, 1941.

Berger, W. H. and Heath, G. R. (1968). *J. Mar. Res.* **26**, 134.

Berger, W. H. and Killingley, J. S. (1977). *Science, N.Y.* **197**, 563.

Berger, W. H. and Winterer, E. L. (1974). *In* "Pelagic Sediments on Land and under the Sea" (K. J. Hsu and H. C. Jenkyns, eds), *Spec. Pub. Int. Assoc. Sediment.* **1**, 11–48.

Berger, W. H., Adelseck, C. G., Jr. and Mayer, L. A. (1976). *J. Geophys. Res.* **81**, 2617.

Berger, W. H., Johnson, R. F. and Killingley, J. S. (1977). *Nature, Lond.* **269**, 661.

Berggren, W. A. (1969a). *Micropaleontology*, **15**, 351.

Berggren, W. A. (1969b). *Nature, Lond.* **224**, 1072.

Berggren, W. A. (1971). *In* "The Micropaleontology of Oceans" (B. M. Funnell and W. R. Riedel, eds), pp. 693–810. Cambridge University Press, England.

Berggren, W. A. (1972). *Lethaia*, **5**, 195.

Berggren, W. A. and van Couvering, J. A. (1974). *Paleogeog. Paleoclim. Paleoecol.* **16**, 216 pp.

Biscaye, P. E. (1965). *Bull. Geol. Soc. Am.* **76**, 803.

Biscaye, P. E., Kolla, V. and Turekian, K. K. (1976). *J. Geophys. Res.* **81**, 2595.

Bouma, A. H. and Marshall, N. F. (1964). *Marine Geology*, **2**, 81.

Bramlette, M. N. (1961). *In* "Oceanography", *Am. Assoc. Adv. Sci. Washington Pub. No.* **67**, pp. 345–366.

Bramlette, M. N. and Martini, E. (1964). *Micropaleontology*, **10**, 291.
Broecker, W. S. and Ku, T. L. (1969). *Science, N.Y.* **166**, 404.
Broecker, W. S., Thurber, D. L., Goddard, J., Ku, T., Mathews, R. K. and Mesolella, K. J. (1968). *Science, N.Y.* **159**, 297.
Brooke, J. and Gilbert, R. L. G. (1968). *Deep-Sea Res.* **15**, 483.
Bukry, D. (1973). *In* "Initial Reports of the Deep Sea Drilling Project", Vol. 15, pp. 685–703. U.S. Government Printing Office, Washington D.C.
Burckle, L. H. (1972). *In* "First Symposium on Recent and Fossil Marine Diatoms" (R. Simonsen, ed.), pp. 217–249. Verlag von J. Cramer, Bremerhaven, Germany.
Burckle, L. H. and Stanton D. (1975). *In* "Third Symposium on Recent and Fossil Marine Diatoms" (R. Simonsen, ed.), Proceedings VIII, pp. 283–292. Verlag von J. Cramer, Bremerhaven, Germany.
Burckle, L. H., Ewing, J., Saito, T. and Leyden, R. (1967). *Science, N.Y.* **157**, 537.
Calvert, S. E. (1966). *J. Geol.* **74**, 546.
Calvert, S. E., Price, N. B., Heath, G. R. and Moore, T. C., Jr. (1978). *J. Mar. Res.* **36**, 161.
Carroll, D. (1970). *J. Sed. Petrol.* **40**, 788.
Chappell, J. (1974). *Bull. Geol. Soc. Am.* **85**, 553.
Chase, T. E. (1975). "Topography of the Oceans". Scripps Institution of Oceanography, La Jolla, U.S.A.
CLIMAP (1976). *Science, N.Y.* **191**, 1131.
Cline, R. M. and Hays, J. D., eds (1976). "Investigation of Late Quaternary Paleoceanography and Paleoclimatology", *Geol. Soc. Am. Mem.* **145**, 464 pp.
Conolly, J. R. and Ewing, M. (1970). *In* "Geological Investigations of the North Pacific" (J. D. Hays, ed.), *Geol. Soc. Am. Mem.* **126**, pp. 219–231.
Davies, T. A., Luyendyk, B. P. and Weser, O. E. (1975). *Nature, Lond.* **253**, 15.
Douglas, R. and Savin, S. (1975). *In* "Initial Reports of the Deep Sea Drilling Project", Vol. 32, pp. 509–520. U.S. Government Printing Office, Washington D.C.
Edwards, A. R. (1973). *In* "Initial Reports of the Deep Sea Drilling Project", Vol. 21, pp. 641–692. U.S. Government Printing Office, Washington D.C.
Emery, K. O. and Champion, A. R. (1948). *J. Sed. Petrol.* **18**, 30.
Emery, K. O. and Dietz, R. S. (1941). *Bull. Geol. Soc. Am.* **52**, 1685.
Emiliani, C. (1955). *J. Geol.* **63**, 538.
Emiliani, C. (1966). *J. Geol.* **74**, 109.
Ericson, D. B. and Wollin, G. (1968). *Science, N.Y.* **162**, 1227.
Ericson, D. B. and Wollin, G. (1970). *Science, N.Y.* **167**, 1483.
Ericson, D. B., Ewing, M., Wollin, G. and Heezen, B. C. (1961). *Bull. Geol. Soc. Am.* **72**, 193.
Ewing, M., Hayes, D. E. and Thorndike, E. M. (1967). *Deep-Sea Res.* **14**, 253.
Foreman, H. P. (1975). *In* "Initial Reports of the Deep Sea Drilling Project", Vol. 32, pp. 579–676. U.S. Government Printing Office, Washington D.C.
Fowler, G. A. and Kulm, L. D. (1966). *Limnol. Oceanogr.* **11**, 630.
Funnell, B. M. (1971). *In* "The Micropaleontology of Oceans" (B. M. Funnell and W. R. Riedel, eds), pp. 507–534. Cambridge University Press, England.
Geitzenauer, K. R., Roche, M. B., McIntyre, A. (1976). *In* "Investigation of Late Quaternary Paleoceanography and Paleoclimatology" (R. M. Cline and J. D. Hays, eds), *Geol. Soc. Am. Mem.* **145**, pp. 423–428.
Glass, B. P. (1969). *Earth Planet. Sci. Lett.* **6**, 409.
Goldberg, E. D. and Griffin, J. J. (1970). *Deep-Sea Res.* **17**, 513.
Goll, R. M. and Bjorklund, K. R. (1971). *Micropaleontology*, **17**, 434.

Goll, R. M. and Bjorklund, K. R. (1974). *Micropaleontology*, **20**, 38.

Griffin, J. J., Windom, H. and Goldberg, E. D. (1968). *Deep-Sea Res.* **15**, 433.

Guinasso, N. L. and Schink, D. R. (1975). *J. Geophys. Res.* **80**, 3032.

Hays, J. D., Imbrie, J. and Shackleton, N. J. (1976). *Science, N.Y.* **194**, 1121.

Heath, G. R., Moore, T. C., Jr., Somayajulu, B. L. K. and Cronan, D. S. (1970). *J. Mar. Res.* **28**, 225.

Heath, G. R., Bennett, R. H. and Rodolfo, K. S. (1973a). *In* "Initial Reports of the Deep Sea Drilling Project", Vol. 16, pp. 3–17. U.S. Government Printing Office, Washington D.C.

Heath, G. R., Opdyke, N. D., Dauphin, J. P. and Moore, T. C., Jr. (1973b). *Geol. Soc. Am. Abst.* **5**, 662.

Heath, G. R., Moore, T. C., Jr. and Roberts, G. L. (1974). *J. Geol.* **82**, 145.

Heath, G. R., Moore, T. C., Jr. and Dauphin, J. P. (1976). *In* "Investigation of Late Quaternary Paleoceanography and Paleoclimatology" (R. M. Cline and J. D. Hays, eds), *Geol. Soc. Am. Mem.* **145**, pp. 393–409.

Heath, G. R., Moore, T. C., Jr. and van Andel, Tj. H. (1977). *In* "The Fate of Fossil Fuel Carbon Dioxide" (N. Anderson and A. Malahoff, eds), pp. 605–625. Plenum Press, New York and London.

Heezen, B. C. and Hollister, C. D. (1971). "The Face of the Deep". Oxford University Press, Oxford and New York.

Hess, H. H. (1962). *In* "Petrologic Studies: a volume in honor of A. F. Buddington", pp. 599–620. Geological Society of America.

Hollister, C. D. (1973). "Atlantic Continental Shelf and Slope of the United States— Texture of Surface Sediments from New Jersey to Southern Florida", Geol. Survey Prof. Paper 529–M. U.S. Government Printing Office, Washington D.C.

Hollister, C. D., Silva, A. J. and Driscoll, A. (1973). *Ocean Engng* **2**, 159.

Horn, D. R., Horn, B. M. and Delach, M. N. (1972a). "Ferromanganese Deposits of the North Pacific", *Tech. Rept.* **1**, NSF GX33616. Office for the International Decade of Ocean Exploration, National Science Foundation, Washington D.C.

Horn, D. R., Horn, B. M. and Delach, M. N. (1972b). *In* "Ferromanganese Deposits on the Ocean Floor" (D. R. Horn, ed.), pp. 9–18. Office of the International Decade of Ocean Exploration, National Science Foundation, Washington D.C.

Hülsemann, J. and Emery, K. O. (1961). *J. Geol.* **69**, 279.

Hvorslev, M. J. and Stetson, H. C. (1946). *Bull. Geol. Soc. Am.* **57**, 935.

Imbrie, J. and Kipp, N. (1971). *In* "Late Cenozoic Glacial Ages" (K. K. Turekian, ed.), pp. 71–181. Yale University Press, New Haven and London.

Johnson, D. A. and Johnson, T. C. (1970). *Deep-Sea Res.* **17**, 157.

Johnson, D. A. and Parker, F. L. (1972). *Micropaleontology*, **18**, 129.

JOIDES (1965). *Science, N.Y.* **150**, 709.

Jonasson, A. and Olausson, E. (1966). *Marine Geology*, **4**, 365.

Keany, J., Ledbetter, M., Watkins, N. and Huang, T. C. (1976). *Bull. Geol. Soc. Am.* **87**, 873.

Kennett, J. P. and Huddleston, P. (1972). *J. Mar. Res.* **2**, 38.

Kennett, J. P. and Shackleton, N. J. (1976). *Nature, Lond.* **260**, 513.

Kennett, J. P. and Watkins, N. D. (1975). *Science, N.Y.* **188**, 1011.

Kennett, J. P., Burns, R. E., Andrews, J. E., Churkin, M., Jr., Davies, T. A., Dumitrica, P., Edwards, A. R., Galehouse, J. S., Packham, G. H. and van der Lingen, G. J. (1972). *Nature, Lond.* **239**, 51.

Kermabon, A., Blavier, P., Curtis, V. and Delanze, H. (1966). *Marine Geology*, **4**, 149.

Kipp, N. G. (1976). *In* "Investigation of Late Quaternary Paleoceanography and Paleoclimatology" (R. M. Cline and J. D. Hays, eds), *Geol. Soc. Am. Mem.* **145**, pp. 3–41.

Kogler, F. C. (1963). *Meyniana*, **13**, 1.

Kolla, V. and Biscaye, P. E. (1977). *J. Sed. Petrol.* **47**, 642.

Kolla, V., Moore, D. G. and Curray, J. R. (1976a). *Marine Geology*, **21**, 255.

Kolla, V., Bé, A. W. H. and Biscaye, P. (1976b). *J. Geophys. Res.* **81**, 2605.

Kollwentz, W. (1973). *In* "The Origin and Distribution of Manganese Nodules in the Pacific and Prospects for Exploration" (M. Morgenstein, ed.), pp. 85–92. Hawaii Institute of Geophysics, Hawaii.

Kowsmann, R. O. (1973). *J. Geol.* **81**, 473.

Krause, D. C. (1962). *Hydrogr. Rev.* **39**, 65.

Ku, T. L., Kimmel, M. A., Easton, W. H. and O'Neil, J. J. (1974). *Science, N.Y.* **183**, 959.

Kudinov, E. I. (1957). *U.S.S.R. Acad. Sci.* **25**, 143.

Kullenberg, B. (1947). *Svenska Hydrogr.-Biol. Kommn. Skr.* **1**, 46pp.

Kullenberg, B. (1955). *K. Vet. O. Vitterh. Samh. Handl.* **6**, 17pp.

La Fond, E. C. and Dietz, R. S. (1948). *J. Sed. Petrol.* **18**, 34.

Ledbetter, M. T. and Ellwood, B. B. (1976). *Geology*, **4**, 303.

Leinen, M. (1977). *Geochim. Cosmochim. Acta*, **41**, 671.

Lisitzin, A. P. (1960). *Deep-Sea Res.* **7**, 89.

Lisitzin, A. P. (1970). *In* "Scientific Exploration of the South Pacific" (W. S. Wooster, ed.), pp. 89–132. National Academy of Sciences, Washington D.C.

Lisitzin, A. P. (1972). "Sedimentation in the World Ocean", *Soc. Econ. Paleo. Min. Spec. Pub.* **17**, 218 pp.

Lozano, J. and Hays, J. D. (1976). *In* "Investigation of Late Quaternary Paleoceanography and Paleoclimatology" (R. M. Cline and J. D. Hays, eds), *Geol. Soc. Am. Mem.* **145**, pp. 303–336.

Luz, B. and Shackleton, N. J. (1975). *In* "Dissolution of Deep Sea Carbonates" (W. V. Sliter, A. W. H. Bé and W. H. Berger, eds), *Cush. Found. Foram. Res. Spec. Pub.* **13**, 142.

Mackereth, F. J. H. (1958). *Limnol. Oceanogr.* **3**, 181.

Martini, E. and Worsley, T. (1970). *Nature, Lond.* **225**, 289.

Mathews, R. K. (1973). *Quatern. Res.* **3**, 147.

McDonald, V. J., Olson, R. E., Richards, A. F. and Keller, G. H. (1972). *In* "Ocean '72, International Conference on Engineering in the Ocean Environment". Institute of Electrical and Electronic Engineers, New York.

McIntyre, A. and Bé, A. W. H. (1967). *Deep-Sea Res.* **14**, 561.

McIntyre, A., Bé, A. W. H. and Roche, M. B. (1970). *Trans. N.Y. Acad. Sci. Ser. II* **32**, 720.

McManus, D. A. (1965). *Deep-Sea Res.* **12**, 227.

Monget, J. M., Murray, J. W. and Mascle, J. (1976), "A World-wide Compilation of Published Multicomponent Analyses of Ferromanganese Concretions", *Tech. Rept.* **12**, Manganese Nodule Project. Office of the International Decade of Ocean Exploration, National Science Foundation, Washington D.C.

Moore, D. G. (1961). *J. Sed. Petrol.* **31**, 627.

Moore, T. C., Jr. (1970). *Deep-Sea Res.* **17**, 573.

Moore, T. C., Jr. (1971). *In* "Initial Reports of the Deep Sea Drilling Project", Vol. 8, pp. 727–775. U.S. Government Printing Office, Washington D.C.

Moore, T. C., Jr. (1972). *Geo Times* **17**, 27.

Moore, T. C., Jr. and Heath, G. R. (1966). *Nature, Lond.* **212**, 983.

Moore, T. C., Jr. and Heath, G. R. (1977). *Earth Planet. Sci. Lett.* **37**, 71.
Moore, T. C., Jr., Heath, G. R. and Kowsmann, R. O. (1973). *J. Geol.* **81**, 458.
Moore, T. C., Jr., Pisias, N. G. and Heath, G. R. (1977). *In* "The Fate of Fossil Fuel Carbon Dioxide" (N. Anderson and A. Malahoff, eds), pp. 145–165. Plenum Press, New York and London.
Moore, T. C., Jr., van Andel, Tj. H., Sancetta, C. and Pisias, N. (1978). *Micropaleontology*, **24** (in press).
Morgan, W. J. (1968). *J. Geophys. Res.* **76**, 1959.
Morgan, W. J. (1971). *Nature, Lond.* **230**, 42.
Murray, J. and Renard, A. F. (1891). "Report on Deep-sea Deposits Based on Specimens Collected During the Voyage of H.M.S. *Challenger* in the Years 1872–1876". H.M.S.O., London.
Nigrini, C. (1967). *Bull. Scripps Instn Oceanogr.* **11**, 125 pp.
Ninkovich, D. and Shackleton, N. J. (1975). *Earth Planet. Sci. Lett.* **27**, 20.
Noshkin, V. E. and Bowen, V. T. (1973). *In* "Radioactive Contamination of the Marine Environment", Int. Atom. Energy Agency Symp. 158/45, pp. 671–686.
Parker, F. L. and Berger, W. H. (1971). *Deep-Sea Res.* **18**, 73.
Paul, A. Z. (1976). *Nature, Lond.* **263**, 50.
Peng, T.-H., Broecker, W. S., Kipphut, G. and Shackleton, N. (1977). *In* "The Fate of Fossil Fuel Carbon Dioxide" (N. Anderson and A. Malahoff, eds). Plenum Press, New York and London.
Phleger, F. B. (1951). *Geol. Soc. Am. Mem.* **46**, 88 pp.
Pisias, N. G., Dauphin, J. P. and Sancetta, C. (1973). *Quatern. Res.* **3**, 3.
Rateev, M. A., Gorbunova, Z. N., Lisitzin, A. P. and Nosov, G. L. (1969). *Sedimentology*, **13**, 21.
Rex, R. W. and Goldberg, E. D. (1958). *Tellus*, **10**, 153.
Richards, A. F. (1961). *U.S. Navy Hydro. Off. Tech. Rept.* **63**, 70 pp.
Richards, A. F. (1972). *In* "The Second International Ocean Development Conference", **2**, 1329–1346. Preprints, Tokyo, Japan.
Richards, A. F. and Keller, G. H. (1961). *Deep-Sea Res.* **8**, 306.
Richards, A. F., McDonald, V. J., Olson, R. E. and Keller, G. H. (1972). *Am. Soc. Test. Mat. Spec. Pub.* **501**, 55.
Riedel, W. R. (1967). *Science, N.Y.* **157**, 540.
Riedel, W. R. (1971). *In* "The Micropaleontology of Oceans" (B. M. Funnell and W. R. Riedel, eds), pp. 567–594. Cambridge University Press, England.
Riedel, W. R. and Funnell, B. M. (1964). *Quat. J. Geol. Soc. Lond.* **120**, 305.
Riedel, W. R. and Sanfilippo, A. (1970). *In* "Initial Reports of the Deep Sea Drilling Project", Vol. 4, pp. 503–575. U.S. Government Printing Office, Washington D.C.
Rona, P. A. (1973). *Nature, Lond.* **244**, 25.
Rosfelder, A. M. (1966). *J. Ocean Tech.* **1**, 53.
Ross, D. A. and Riedel, W. R. (1967). *Deep-Sea Res.* **14**, 285.
Ruddiman, W. F. and Glover, L. K. (1972). *Bull. Geol. Soc. Am.* **83**, 2817.
Ruddiman, W. F., Dicus, R. L. and Glover, L. K. (1976). *Geol. Soc. Am. Prog. Abst.* **8**, 1079.
Ryan, W. B. F., Cita, M. B., Rawson, M. D., Burckle, L. H. and Saito, T. (1974). *Riv. Ital. Paleont. Stratigr.* **80**, 631.
Sachs, P. L. and Raymond, S. O. (1965). *J. Mar. Res.* **2–3**, 44.
Saidova, Kh. M. (1968). *Dokl. Akad. Nauk SSSR*, **182**, 453.
Saito, T., Burckle, L. H. and Hays, J. D. (1974). *In* "Studies in Paleo-oceanography" (W. W. Hay, ed.), *Soc. Econ. Paleo. Min. Spec. Pub.* **20**, pp. 6–36.

Savin, S. M., Douglas, R. G. and Stehli, F. G. (1975). *Bull. Geol. Soc. Am.* **86**, 1499.
Sayles, F. L., Mangelsdorf, P. C., Jr., Wilson, T. R. S. and Hume, D. N. (1976) *Deep-Sea Res.* **23**, 259.
Schrader, H. J. (1973). *In* "Initial Reports of the Deep Sea Drilling Project", Vol. 18, pp. 673–797. U.S. Government Printing Office, Washington D.C.
Sclater, J. G., Anderson, R. N. and Bell, M. L. (1971). *J. Geophys. Res.* **76**, 7888.
Shackleton, N. J. (1977). *Phil. Trans. R. Soc. B*, **280**, 169.
Shackleton, N. J. and Kennett, J. P. (1975). *In* "Initial Reports of the Deep Sea Drilling Project", Vol. 29, pp. 801–806. U.S. Government Printing Office, Washington D.C.
Shackleton, N. J. and Opdyke, N. D. (1973). *Quatern. Res.* **3**, 39.
Shackleton, N. J. and Opdyke, N. D. (1976). *In* "Investigation of Late Quaternary Paleoceanography and Paleoclimatology" (R. M. Cline and J. D. Hays, eds), *Geol. Soc. Am. Mem.* **145**, pp. 449–464.
Shepard, F. P. (1973). "Submarine Geology", 3rd Edn. Harper and Row, New York and London, 517 pp.
Smith, K. L., Clifford, C. H., Eliason, A. H., Walden, B., Rowe, G. T. and Teal, J. M. (1976). *Limnol. Oceanogr.* **21**, 164.
Soutar, A. (1971). *In* "The Micropaleontology of Oceans" (B. M. Funnell and W. R. Riedel, eds), pp. 223–230. Cambridge University Press, England.
Spiess, F. N., Loughridge, M. S., McGehee, M. S. and Boegeman, D. E. (1966). *Navigation, Los Ang.* **13**, 154.
Swedish Committee on Piston Sampling (1961). *Swedish Geotech. Inst. Proc.* **19**, 45 pp
Talwani, M. (1970). *Geoexplor.* **8**, 151.
Theyer, F. and Hammond, S. R. (1974). *Earth Planet. Sci. Lett.* **22**, 307.
Tracey, J. I., Jr., Sutton, G. H. and others (1971). *In* "Initial Reports of the Deep Sea Drilling Project", Vol. 8, pp. 17–41. U.S. Government Printing Office, Washington D.C.
Uchupi, E. (1969). *Trans. N.Y. Acad. Sci. Ser. II*, **31**, 56.
van Andel, Tj. H. (1968). *Science, N.Y.* **160**, 1419.
van Andel, Tj. H. (1973). *J. Geol.* **81**, 434.
van Andel, Tj. H. (1975). *Earth Planet. Sci. Lett.* **26**, 187.
van Andel, Tj. H. and Bukry, D. (1973). *Bull. Geol. Soc. Am.* **84**, 2361.
van Andel, Tj. H. and Moore, T. C., Jr. (1974). *Geology*, **2**, 87.
van Andel, Tj. H., Heath, G. R. and Moore, T. C., Jr. (1975). "Cenozoic History and Paleoceanography of the Central Equatorial Pacific Ocean", *Geol. Soc. Am. Mem.* **143**, 134 pp.
van Veen, J. (1936). "Onderzoekingen in de Hoofden." Landsdrukkerij, The Hague, 252 pp.
Venkatarathnam, K. and Biscaye, P. E. (1973). *Deep-Sea Res.* **20**, 727.
Watkins, N. D. (1972). *Bull. Geol. Soc. Am.* **83**, 551.
Wollin, G., Ericson, D. B. and Ewing, M. (1971). *In* "The Late Cenozoic Glacial Ages" (K. K. Turekian, ed.), pp. 199–214. Yale University Press, New Haven and London.
Yamashiro, C. (1975). *In* "Dissolution of Deep-Sea Carbonates" (W. V. Sliter, A. W. H. Bé and W. H. Berger, eds), *Cush. Found. Foram. Res. Spec. Pub.* **13**, 151.
Zenkovitch, L. A. (1955). *Trudy Inst. Okeanol.* **12**, 5.

Chapter 37

Suspended Matter in Sea-water

WILLIAM M. SACKETT

Department of Oceanography, Texas A and M University,
College Station, Texas, U.S.A.

37.1. Introduction

In this chapter the terms suspended solids, suspended matter and suspensoids will be used to describe sea-water particulates and are defined as "those solid particles which spend a reasonable length of time suspended in sea water". The time constants of the most important oceanic processes are at least a few days. Therefore, according to the various settling laws, the largest particles having densities appreciably greater than unity, which will remain in the water column for such a period, are about 1000 µm in diameter.

127

Particles with diameters larger than this will not remain in suspension long enough to undergo significant reactions with the surrounding medium. In contrast, smaller particles ranging down to colloidal sizes (diameters 1·0–0·01 μm) will, according to their theoretical settling times, require tens of years or even orders of magnitude longer to reach the bottom. Such particles remain suspended sufficiently long for them to undergo reactions which may influence both the particles and their aqueous environment.

A knowledge of the spatial and temporal distribution, composition and lateral and vertical rates of movement of suspensoids is important in many areas of the marine sciences for several reasons.

(i) The suspended matter is the manifestation in the water column of that fraction of the sediments which is being deposited particle by particle. Thus, suspensoids are a true representation of the material which is being deposited in a given ocean basin at a particular time.

(ii) Suspensoids are the carriers by which many chemical species are transferred from the surface to the deeps. The transfer mechanisms involved include sorption phenomena, which can bring about the downward transport of many trace components (including pollutants such as lead, mercury, DDT, PCBs etc.), chemical reactions, such as the oxidation of the organically bound nutrient elements, and solution processes acting on calcareous and siliceous particles falling through the water column.

(iii) Suspensoids affect the transparency of water and so exert a control on the depth of the euphotic zone.

(iv) Suspensoids are the substrate, and often the food, for a significant community of micro-organisms which act as mediators for many of the chemical processes taking place in the sea.

(v) Earthquakes, and/or slumps, are thought to sometimes cause the instantaneous suspension of enormous amounts of sediment. This produces a great increase in the density of the water mass into which it is injected, with the result that it flows rapidly down-slope as a "turbidity" current transporting significant amounts of sediments very quickly to the ocean basins and frequently changing the temperature and salinity of the bottom waters.

Suspensoids may be classified according to their sources, mechanisms of formation or their chemical compositions. However, for the purposes of this review they will initially be classified as having either a detrital or an authigenic origin. Detrital suspensoids include terrigenous particles transported to the sea by river run-off or winds, re-suspended materials (which give rise to nepheloid (cloudy) layers and turbidity currents) and particles resulting from underwater eruptions. Although most detrital suspensoids are, in fact, inorganic, organic particles of terrestial origin are commonly found in coastal environments. Authigenic suspensoids are those organic and inorganic particles produced in the sea by biological or inorganic chemical processes. They include bacteria and living nannoplankton, faecal pellets and other marine-derived organic agglomerates together with inorganic particles, e.g. calcium carbonate polymorphs and barium sulphate produced by organisms or by purely inorganic precipitation. Colloidal hydrous metal oxides formed in estuaries, or environments such as the hot brine regions at the bottom of the Red Sea, would also be classified as authigenic suspensoids.

37.2. HISTORICAL DEVELOPMENTS

Early work on suspensoids was motivated by biological considerations and stemmed largely from an interest in the effect of particles on the transparency of sea-water, since the latter was known to determine the depth of light penetration which, in its turn, governs the depth distribution of primary producers. The earliest measurements of transparency and light scattering were made by Petterson (1934a, b). Surprisingly, these were *in situ* measurements but, because of the primitive state of electronic instrumentation at that time, they were not very accurate and the technique was only of value for coastal waters. The Tyndall effect was first used for the direct determination of the concentration of suspended matter by Kalle (1939)—see Table 37.1 which presents a summary of the significant developments in the study of suspended matter in sea-water.

The concentration of total suspended matter (TSM) in sea-water is ideally determined by separation by filtration or centrifugation followed by weighing. However, before significant measurements could be made, the small amounts (usually $< 100\ \mu g\,l^{-1}$) and fine particles sizes ($\sim 1\ \mu m$) found in the open ocean required the development of special filtering media. Thus, the development of membrane filters in the 1940s led to the first published study of TSM concentrations by Armstrong and Atkins (1950) who used them to measure the distribution of TSM in the English Channel. Subsequently, Atkins *et al.* (1954) and Armstrong (1958) reported on various aspects of TSM in the seas around Great Britain.

The first major study of the distribution of particles in the ocean was reported by Jerlov (1953). This study, which covered some regions of the Atlantic, Pacific and Indian Oceans and of the Red, Caribbean and Mediterranean Seas, utilized light-scattering measurements on discrete samples. Because of uncertainties in instrument calibration and the relevant light-scattering theory, only relative values of particle distributions were obtained, but the observed trends did confirm certain intuitive beliefs commonly

TABLE 37.1

Chronological summary of some selected sea-water suspensoid studies

Investigator	Brief description of study	Method(s)[a]
Kalle (1939)	First determination of particulate concentration by the Tyndall effect.	LSDS
Armstrong and Atkins (1950)	First reported use of membrane filters to determine concentrations of particles.	MFG
Jerlov (1953)	Spatial distribution of TSM (mostly organic) in Mediterranean and Red Seas, Atlantic, Pacific and Indian Oceans; first report of near-bottom nepheloid layers.	LSDS
Lisitzin (1959)	Summary of ten years of Russian work on spatial distributions and compositions of suspensoids.	MFG, CG
Sackett and Arrhenius (1962)	Specific determination of suspended aluminosilicates by aluminium normalization.	MFC
Ewing and Thorndike (1965)	Beginning of routine use of *in situ* nephelometry; confirmation of near-bottom nepheloid layer.	LSI
Jacobs and Ewing (1969 a, b)	Suspended matter concentration in major oceans and qualitative description of composition.	CG
Gibbs, ed. (1974)	State of the art collection of papers published in "Suspended Solids in Sea Water".	All
Brewer *et al.* (1976); Spencer *et al.* (1976)	GEOSECS Atlantic longitudinal sections published.	MFC, LSI, MFC

[a] LSDS, light scattering on discrete samples;
MFG, membrane filtration and gravimetry;
CG, centrifugation and gravimetry;
MFC, membrane filtration and chemical analysis;
LSI, light scattering *in situ*.

held by marine scientists. One such belief was that there was a decrease in suspensoid concentrations with depth in the water column and with distance from estuarine and coastal source regions. An unexpected result of this study was the recognition of particle maxima in the deep waters of various ocean basins. These maxima, which were often found in close proximity to the bottom, were attributed by Jerlov to the re-suspension of sediments by bottom currents. One comment made by him is particularly significant in view of subsequent studies of the thicknesses and the distributions of deep-sea sediments:

Fine material in suspension can be carried over long distances by currents. In depressions and basins where transporting agents are absent the material will accumulate.

Another of his observations, which is equally significant in the light of later work on metalliferous deposits in the Red Sea, is that:

The existence of red or brown particles could be ascertained . . . close to the bottom (181 m) at Station 247 in the Red Sea a red suspension was found.

Thus, Jerlov's paper pre-empted some of the most important discoveries in marine geochemistry in the following 20 years.

During the 1950s, the U.S.S.R. began a massive programme for the collection and analysis of suspended matter, using membrane filtration and industrial centrifugal separators, during which some 18 000 samples were collected. This work, which has been summarized by Lisitzin (1959, 1960), indicated that the amount of suspended matter in the oceanic water column ranges between 600 and 5000 g m^{-2} and that, on an average, the suspensoids consist of 10–30% of organic matter and 70–90% of mineral matter.

The aluminosilicates found in marine sediments contain $\sim 10\%$ of aluminium, and therefore the concentration of particulate aluminosilicates in the water column may be estimated by multiplying the concentration of membrane filter-retained aluminium by a factor of ten. Sackett and Arrhenius (1962) introduced an aluminium-specific spectrofluorimetric technique for this purpose, and this has been subsequently used by other investigators (e.g. Feely et al., 1971). Feely et al. (1974) were able to divide suspended matter into its major compositional groups by determining TSM, carbon (both organic and inorganic) and aluminosilicates in membrane filter-retained particulates. This differentiation between the various types of suspensoids is important for understanding the origin of the near-bottom maxima in suspensoid concentrations.

An important series of publications on suspended matter in the oceans by a group from the Lamont-Doherty Geological Observatory (L-DGO) under the direction of the late Maurice Ewing began to appear in 1963. The first of these (Groot and Ewing, 1963) was a very brief paper which reported the presence of 2500 μg l^{-1} of suspended matter in a sample taken at a depth of 4030 m in the Atlantic. This was followed by a more detailed paper by Ewing and Thorndike (1965), reporting that background haze in bottom photographs of the deep-sea floor strongly suggested the presence of cloudy water due to sediment re-suspension, and this was supported by the existence of ripple and scour marks. These observations led them to develop a photographic in situ light-scattering meter, a "nephelometer" (Section 37.3.2.2), with which they were able to show that light scattering increased markedly near the bottom in a transect of stations seawards from the Hudson Canyon off the eastern coast of the USA; an increase which they attributed to the presence of a nepheloid layer. Subsequently, Jacobs and Ewing (1965) used centrifugation, weighing and X-ray diffraction techniques and found that two samples of suspended matter from the Caribbean had mineral compositions similar to those of the underlying sediments. Over the next few years a considerable effort was made by this group to obtain additional data on the concentration of suspended matter in the Atlantic, Pacific and Indian Oceans and the Caribbean Sea and the Gulf of Mexico (Jacobs and Ewing, 1969a; see Table 37.2). With regard to this, it should be noted that

TABLE 37.2

Summary of weight–concentration data for suspended particulate matter for whole water column in the major oceans (after Jacobs and Ewing, 1969a)

Water body	Number of samples	Range ($\mu g\,l^{-1}$)	Mean ($\mu g\,l^{-1}$)
N. Atlantic	88	0·5–247	49
S. Atlantic	52	1·5–197	51
N. Pacific	145	0·5–152	37
S. Pacific	78	4·5–86	30
Indian	37	9–177	72
Caribbean	25	0·5–139	40
Gulf of Mexico	11	12·5–193	66

their Indian Ocean measurements are an order of magnitude lower than those reported by Lisitzin (1960). Other workers, e.g. Bassin *et al.* (1972), have suggested that the techniques which were used by Jacobs and Ewing (1969a) would not collect TSM quantitatively; however, the mean values reported in Table 37.2 for the Atlantic are also significantly higher than those found by subsequent workers using improved methods. This problem will be discussed in more detail in Section 37.3. In a further paper, Jacobs and Ewing (1969b) showed that a mica-rich mineral assemblage, possibly originating from the Amazon River, is carried by currents through the Caribbean Sea into the Gulf of Mexico where it is mixed with montmorillonite-rich material from the Mississippi before it is discharged back into the Atlantic.

Manheim *et al.* (1970) have made an interesting and comprehensive study of the suspended matter in the surface waters of the Atlantic continental margin from Cape Cod to the Florida Keys. In general, the TSM decreased systematically from $>4000\ \mu g\,l^{-1}$ near the coast to $<100\ \mu g\,l^{-1}$ at $\sim 50\,km$ off the coast. In the various estuaries studied, the TSM showed a systematic decrease with increasing salinity. In most samples organic matter usually comprised 40–90% of the TSM, the percentage increasing with distance from the coast. Grain sizes decreased markedly away from the coast as would be predicted. The authors concluded "that the continent is presently contributing little detritus to the continental slope and deeper regions of the Atlantic Ocean".

Since 1970, numerous studies have been published describing the spatial and temporal distributions, compositions, settling rates, etc., of suspended matter. The following discussion will concentrate on the improvements in instrumentation and technique which have made much of the pre-1970 work obsolete. Many of the most important recent papers are to be found in Gibbs (1974).

37.3. METHODS

37.3.1. INTRODUCTION

The concentration of suspensoids in sea-water has been determined either by weighing the material after separation by filtration or centrifugation, or by using optical phenomena *in situ*. Information about the amounts of certain components of suspended matter on the filter may be obtained by determining a particular chemical element; thus the weights of alumino-silicates or organic matter can be obtained from determinations of aluminium and carbon respectively. *In situ* optical methods provide the best information on the real distribution and nature of suspensoids because particulate sampling techniques may alter their *in situ* characteristics by processes such as agglomeration and disaggregation. In addition, optical methods may provide closely spaced observations which can give detailed information on the distribution of suspensoids—information that cannot be obtained by discrete sampling. However, optical phenomena depend on the sizes and shapes, as well as the number, of particles with the result that the interpretation of data from *in situ* optical measuring devices is complex.

37.3.2. *In situ* METHODS

The only practicable *in situ* methods for the investigation of sea-water suspensoids are based on optical phenomena, such as light absorption and light scattering (nephelometry) which have been used for analytical purposes in the laboratory since the early 1900s (see e.g. Yoe, 1928). These two techniques have been used by oceanographers to study sea-water suspensoids; the first-named technique being terms transmissometry rather than turbidimetry, and the second retaining the traditional name of nephelometry.

37.3.2.1. *Transmissometry*

A transmissometer consists basically of a light source and a light sensor mounted coaxially on a frame which can be lowered through the water. Suspended particles in the pathway between the two components absorb and scatter light, the total attenuation being proportional to the amount of suspended matter. In near-shore waters, the presence of significant concentrations of certain dissolved species may also lead to the absorption of the source radiation and thus result in erroneous suspensoid data. Calibration curves for the instruments are constructed from light attenuation measurements made on samples having known (gravimetrically determined) TSM concentrations.

Tyler *et al.* (1974) have presented an excellent summary of the theory and use of beam transmissometers for oceanographical measurements. A schematic representation of one such instrument designed by Petzold and Austin (1968) is presented in Fig. 37.1. The instrument features:

(i) a folded light path which makes the instrument more compact and easier to handle, and allows the source, the sensor and the electrical connections all to be placed in the same watertight compartment;

(ii) a direct light path from lamp to detector to allow the stability of the source to be monitored; and

(iii) a cylindrically restricted beam of light which allows all the projected rays to fall on the photodetector, thus reducing errors due to forward scattering.

The ideal operating range of this instrument was found by Drake (1974)

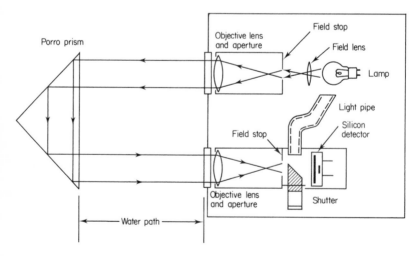

FIG. 37.1. Schematic diagram of folded-path transmissometer (after Tyler *et al.*, 1974).

to be $\sim 0.5-\sim 5$ mg TSM l^{-1} for suspended matter in the submarine canyons off southern California. Because open-ocean levels of TSM are usually <0.5 mg l^{-1}, the usefulness of this type of instrument is limited to the measuremeant of high loads of TSM, such as those encountered in submarine canyons and various other continental shelf and slope environments.

37.3.2.2. Nephelometry

The measurement of light scattering is preferred to that of light absorption for the *in situ* study of suspended matter in the deep sea because of its greater sensitivity. Descriptions of the theory, instrumental calibration and other aspects of this technique are to be found in papers by Smith *et al.* (1974), Gordon (1974), Zaneveld (1974) and Fry (1974).

In essence, the scattering of light by suspended particles is controlled by their concentration, sizes and shapes, and by the difference between their refractive index and that of water. Most scattering takes place at low angles in a forward direction relative to the source beam. The Mie theory (Mie, 1908) can be used to calculate scattering angles for spherical particles of different sizes; however, in nature, the wide spectrum of sizes and shapes only allows semi-quantitative information to be obtained from light-scattering data. Instrumental calibration is usually accomplished empirically by examination of water samples which have been analysed for TSM, gravimetrically.

To date, most of the *in situ* measurements of suspended matter have been made with the L-DGO-Thorndike nephelometer (LTN) (Thorndike, 1975; Thorndike and Ewing, 1967; Eittreim *et al.*, 1969). This instrument is illustrated diagrammatically in Fig. 37.2. Light from a continuous source passes through a slit perpendicular to the movement of a constantly moving strip of 35-mm film. The centre of the film receives light transmitted through an

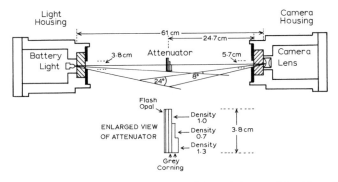

FIG. 37.2. Schematic diagram of the L-DGO-Thorndike nephelometer (after Thorndike, 1975).

attenuator which serves to monitor the source intensity, whereas the film on either side of the centre receives scattered light. Any given point on the film receives the summation of light from a range of angles, with the region nearest the centre corresponding to the smallest scattering angles and *vice versa*. Before the instrument is lowered, the film is calibrated in a sensitometer to provide an area with twelve steps of log-exposure differing from one another by a factor of 0·2. These steps are used to produce a graph defining the film contrast which is needed when the exposed and developed film is processed through a recording densitometer. The final output is log S, where S is the ratio of the film exposure E produced by the scattered light to that produced by the directly attenuated light E_D. This ratio, which is sometimes normalized so that the clearest water is assigned an optical density of zero, compensates for a number of experimental variables. It depends only on the light scattered by suspended particles over the range of angles from about 8 to 24 degrees.

The depths of the various exposures are found both by the deflection of a light spot produced by a bourdon-tube pressure sensing system and by correlating the time at which a particular wire length was paid out with time marks on the film. A typical arrangement for obtaining continuous light-scattering data down to the sea bottom and collecting a seabed sample is depicted in Fig. 37.3. The assemblage shown includes:

(i) a pinger, which enables the unit to be positioned very close to the bottom;

(ii) a Nansen bottle, which is used for collection of water samples for measurement of salinity and dissolved oxygen (its associated reversing thermometers permit the temperature to be determined accurately); and

(iii) a 30 litre Niskin bottle for the collection of large-volume water samples for the determination of TSM. Additional sampling bottles may be added for the collection of samples at intermediate depths.

Another type of nephelometer which is used routinely is depicted in Fig. 37.4. It is entirely battery powered, consisting of a flashing light source and a photomultiplier tube, the output of which is a measure of the intensity of the scattered light and is recorded on an internal strip-chart. The original design (Sternberg *et al.*, 1974) was adequate for studies of the relatively turbid water along the continental margins, but for the clearer waters of the Caribbean it was necessary to increase the sensitivity; this was accomplished by decreasing the angle between the source and the detector from 90 degrees to 45 degrees and adding a more sensitive step to the electronic detection system (Bassin, 1975). Both the original design, used by oceanographers at the University of Washington, and the modified version employed by both N. J. Bassin and R. A. Feely on Texas A and M University cruises have proved successful. Generally, the instrument was placed above an STD probe so that

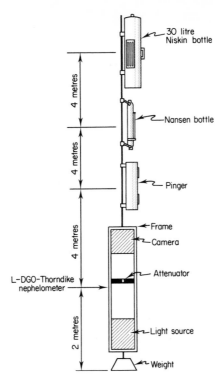

FIG. 37.3. The sampling system suspended on a hydrowire (after Feely *et al.*, 1974).

simultaneous temperature and salinity data could be obtained. Stopwatch annotations were made on the STD printout to enable these data to be correlated with the light-scattering data obtained with the nephelometer. Water samples for calibration purposes were collected by means of separate

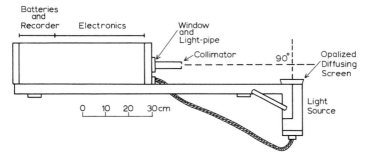

FIG. 37.4. Simplified schematic diagram of the University of Washington nephelometer.

hydrocasts. A comparison of the nephelometric data with that obtained by filtration techniques yielded correlation coefficients of 0·73 for the University of Washington instrument (Baker et al.. 1974), and 0·62 (Bassin, 1975) and 0·85 (Feely, 1975) for the modified version used by the Texas A and M University group. For comparative purposes it may be noted that Biscaye and Eittreim (1974) have reported a value of 0·91 for the L-DGO-Thorndike nephelometer.

A new type of nephelometer is being used with limited success in the GEOSECS programme. It features a laser source and a real-time output coupled to an STD system (Meade et al., 1975).

37.3.3. GRAVIMETRIC METHODS

Separation and subsequent weighing of suspended matter from seawater is the only satisfactory method for the determination of the absolute concentration of TSM. This separation can be accomplished either by centrifugation or filtration techniques. For example, Jacobs and Ewing (1969a, b) used a continuous centrifuge, producing 9000 g at the periphery of the bowl, in their studies of the concentration and composition of suspensoids in the major oceans. Bassin et al. (1972) attributed the low TSM concentrations reported by Jacobs and Ewing (1969a) for the Gulf of Mexico and Caribbean Sea to their failure to use a sufficiently high gravitational force when centrifuging, and estimated that at least 100000 g is necessary to quantitatively separate a clay mineral such as montmorillonite. They also pointed out that the density of particulate organic matter is too near to that of sea-water for it to be satisfactorily removed by centrifugation. Furthermore, losses of particulates occur because of the difficulty of quantitatively transferring the small amount of material (~ 10 mg) collected from 200 litre samples from the bowls of the centrifuge to a container for drying and weighing purposes. All of these factors would contribute to an underestimation of the concentration of suspended matter and to a discrimination against the elemental and mineralogical contributions from the smallest and least dense components. Much of the work reported by Lisitzin (1959) was also carried out using centrifugation and is probably open to the same objections.

The most commonly used technique at the present time for the determination of TSM is filtration through pre-weighed membrane filters. Three filtration techniques are commonly used:

(i) the water is transferred from a deep-sea sampler, such as a 30 litre Niskin bottle, to a plastic carboy from which the water is fed by suction through a filter into a vacuum reservoir;

(ii) the water can be fed directly from the Niskin bottle to a filter which may

be located on a lower deck of the ship to provide a considerable hydrostatic head;

(iii) as (ii), but the Niskin bottle is pressurized to about 40 p.s.i.

The rapidity of processing is in the order (iii) > (ii) > (i). The chances of
contamination are increased if an intermediate container is used, as in (i).
Handling and contamination problems are minimized in procedures (ii) and
(iii), although the P.V.C. tubing which is generally used may cause problems
in organic studies.

Although Millipore® membrane filters were used in most of the earlier
work in the U.S.A., the majority of workers concerned with TSM now prefer
Nuclepore® filters as these have several important advantages (see below;
see also Section 19.4). With both types of filter the amount of water which can
be filtered in a reasonable length of time (1–2 hours) varies from 1 to 30
litres depending on the amount and type of TSM present.

Millipore membrane filters are composed of a mixture of cellulose acetate
and cellulose nitrate and have a porosity of 70–80%. They retain on their
surfaces all rigid particles larger than the listed nominal pore sizes which
range from 0·025 µm to 14 µm, Type HA ones with a nominal pore size
of 0·45 µm being used in most oceanographical studies. These filters suffer
from the disadvantage that they contain water-extractable material and this
necessitates a considerable amount of pre-washing (Banse *et al.*, 1963).

Nuclepore filters are made by bombarding thin polycarbonate film
with high-speed charged particles in a nuclear reactor. The areas which have
been damaged in this way are then subjected to a chemical etching process
that can be controlled to give precise pore sizes ranging from 0·1 to 8·0 µm.

The advantages of Nuclepore over Millipore filters have been excellently
summarized by Biscaye and Eittreim (1974):

(1) Nuclepore filters weigh less than an equivalent Millipore filter (20 mg versus
90 mg) so that, in obtaining the weight of suspended particulate matter from a
20 liter sample of clean water (containing perhaps tens of micrograms), a small-
difference measurement is made on a smaller total weight. (2) Millipore filters
are coated with a wetting agent which washes out upon use, creating a weight
difference of the order of the suspended particulate matter burden. We have not
detected such a weight loss in Nuclepore filters. (3) Due to their lower surface
area and possibly the lack of a wetting agent, Nuclepore filters achieve relative
humidity equilibrium much more rapidly and suffer much less change in weight
due to change in weighing-room relative humidity, reducing (but not eliminat-
ing) the degree of relative-humidity control required in the weighing-room or
weighing chamber. (4) Because of the difference in the nature of the filter, the
surface of Millipore filters have enormously more visual 'texture' than do
Nuclepore filters. For scanning electron microscopy work, therefore, at a given
magnification one can distinguish the presence of small particles much more
readily on a Nuclepore filter. (5) If any analytical work on trace elements is to

be done on the samples, Nuclepore filters have a significantly lower blank for a number of elements than do the Millipore Type HA filters (Derek Spencer, personal communication).

To determine TSM with an estimated precision of $\pm 5\%$ at concentrations greater than $20\,\mu g\,l^{-1}$ it is necessary, according to Biscaye and Eittreim (1974), to control the humidity during the pre- and post-weighing operations in order to prevent atmospheric contamination and to wash the filter free of sea-salt using filtered doubly distilled water. As the TSM concentrations decrease the estimated error increases, and is about $\pm 25\%$ at the $4\,\mu g\,l^{-1}$ level.

37.3.4. OTHER METHODS

It was shown above that some of the major components of TSM can be estimated from its elemental composition. Thus, Sackett and Arrhenius (1962), Feely et al. (1974) and Feely (1975) have used particulate aluminium concentrations to obtain the proportion of the aluminosilicate fraction in samples of TSM by making the assumption that the aluminium is exclusively present in the aluminosilicates, of which it comprises 10%. Although individual clay minerals have widely varying aluminium contents, Sackett and Arrhenius (1962) based their assumption on the fact that the aluminosilicates of marine sediments contain an average of 10% aluminium, i.e. aluminosilicate equals aluminium × 10. Particulate iron has also been used as a measure of aluminosilicates in TSM (see e.g. Betzer and Pilson, 1970a, 1971). However, the proportions of iron in the common clay minerals are much more variable than those of aluminium, leading to considerable uncertainties in particulate iron-derived TSM aluminosilicate values. Another drawback of this method, which was pointed out by Betzer and Pilson (1970b) among others, is that contamination with iron may arise from hydrowire and Nansen bottles (Betzer and Pilson, 1970b). For example, the mean particulate iron concentrations in samples collected from the Atlantic Ocean and Caribbean Sea were found to be $0.18\,\mu g\,l^{-1}$ for surface water and $0.26\,\mu g\,l^{-1}$ for deep water when metal-free samplers were used; however, even with Teflon®-coated Nansen bottles the observed concentrations at the same locations and depths were 15–86 times greater. Nonetheless, the general usefulness of particulate iron measurements as an indicator of aluminosilicate concentrations is evidenced by the finding of up to six times more particulate iron in near-bottom nepheloid layers than in the overlying deep water, presumably resulting from the re-suspension of aluminosilicate-rich sediment.

The Coulter Counter® has been used with considerable success for determining the concentrations and sizes of particles in sea-water (see e.g. Sheldon and Parsons, 1967; Sheldon *et al.* 1972; Brun-Cottan, 1971; Zeitzschel, 1970; Carder and Schlemmer, 1973). In this instrument, an electrical field is applied across an aperture through which a given volume of the sea-water sample passes. Particles in it modify the electrical field and thus give rise to pulses which are counted electronically. The pulses are proportional to the sizes of the particles, and thus the number of particles in a certain size range may be determined either by varying the amount of amplification of the electrical disturbance created by the particles, or by changing the strength of the current across the aperture. Because the Coulter Counter cannot reliably detect particles smaller than 1 μm, and also because of uncertainty about the densities of suspended particles, the mass of TSM cannot be obtained reliably by Coulter counting techniques.

37.4. SPATIAL AND TEMPORAL VARIATIONS OF TOTAL SUSPENDED MATTER

A large amount of reliable data on the spatial and temporal variations of TSM concentrations in various ocean basins was accumulated during the early part of the present decade by the use of the improved nephelometric and gravimetric techniques described above. For convenience this data will be considered ocean by ocean.

37.4.1. ATLANTIC OCEAN

The most comprehensive TSM studies in the Atlantic Ocean were those carried out in connection with the GEOSECS (Atlantic) Expedition in 1972 and 1973 in which as many as 40 depths were sampled at most of the 121 stations worked. Using the best techniques available, a large amount of excellent data was obtained for a longitudinal section from $\sim 75°$N. to $\sim 55°$S. in the western Atlantic Ocean. The TSM concentrations in this section have been described by Brewer *et al.* (1976) (see Fig. 37.5) and some of their general observations are summarized below.

(i) Total suspended matter concentrations $> 100\ \mu g\,kg^{-1}$, and sometimes $> 200\ \mu g\,kg^{-1}$, are found in surface waters in biologically productive regions at high latitudes. High concentrations are also found in bottom waters of the Denmark Straits, the Iceland–Scotland overflows and the Antarctic Bottom Water. The rapid movement of these bottom-water masses apparently maintains a high concentration of re-suspended sediments.

(ii) Low concentrations ($< 20\ \mu g\,kg^{-1}$) occur in the middle of the water

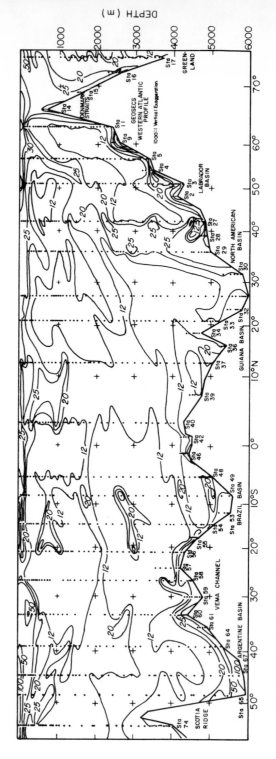

FIG. 37.5. Longitudinal distribution of total suspended matter concentrations (µg kg^{-1}) in the western Atlantic Ocean (after Brewer et al., 1976).

column in the Sargasso Sea and the subtropical areas of the South Atlantic.

(iii) Total suspended matter maxima of $\sim 25\,\mu g\,kg^{-1}$ are observed at about 10°N. and 10°S. at the same depth as the equatorial oxygen minima which occur between 500 and 1000 m. These TSM maxima are presumably the result of the presence of organic particles.

(iv) A plume of relatively high TSM concentrations ($\sim 25\,\mu g\,kg^{-1}$) was found to extend ~ 1 km from the bottom at latitudes between 35°N. and 40°N. The authors related this to the boundary of a cyclonic deep-water gyre in the North Atlantic.

(v) The only feature of the TSM distributions which cannot be correlated with hydrography is a maximum of $\sim 25\,\mu g\,kg^{-1}$ at a depth of 3000 m between 10°S. and 20°S. However, it may be associated with the weak oxygen minimum and the silica maximum which exist at that depth and location.

Even though many stations were worked in the GEOSECS study, the sampling programme was nevertheless not sufficiently extensive to provide a thorough understanding of the factors controlling the distribution of TSM in the Atlantic. However, it should provide a base-line with respect to which past and future studies may be compared.

The presence of a strong light-scattering layer of suspended sediments in water at depths greater than 3000 m near the continental margins off North America has been established using the LTN instrument (Thorndike, 1975). This layer, which is thought to be a permanent feature, increases in thickness from ~ 500 m off Cape Hatteras to ~ 2000 m in the western basins to the south. The appearance of this nepheloid layer suggested that it is produced by sediment re-suspension and possibly also be sporadic and steady-state density-flow-type turbidity currents.

In a similar manner, the re-suspension of sediments is thought to be responsible for the thick nepheloid layer in the Argentine Basin (Ewing et al. 1971). An LTN study of the deep waters of the Western Atlantic Trough from 40°N. to 55°S. indicated that the background turbidity varied by a factor of three, and was apparently related to biological productivity in the overlying water column (Eittreim and Ewing, 1974). The intensity of the near-bottom nepheloid layer decreases from both north and south towards the Equator, and is controlled by the velocity of bottom currents and possibly by the proximity of sources of terrigenous material, such as the Amazon River.

Biscaye and Eittreim (1974) used both photographic nephelometric and gravimetric techniques to make an intensive temporal study over a 19-day period at three locations in the North American Basin; two on the lower slope of the Blake–Bahama Outer Ridge and one on the Hatteras Abyssal Plain. All the TSM profiles showed the existence of high concentrations

($\leqslant 200\ \mu g\ l^{-1}$) in the benthic boundary layer, i.e. the zone of frictional interaction between the moving bottom water and the seabed. At these locations this layer was generally ~ 200 m thick and had a negative temperature gradient consequential on adiabatic conditions similar to those observed by Amos et al. (1971). Repeated measurements at each of these locations over the 19-day period showed significant variations in the TSM concentrations. At two stations, only 110 km apart, there were systematic variations with a period of about one week, although they were out of phase by about one day. The authors suggested that these temporal variations could be explained by the injection of clear Antarctic Bottom Water into the rapidly southward-flowing Western Boundary Undercurrent which is considerably more turbid.

Another recent study of the TSM distribution in a transect across the Atlantic Ocean was that by Betzer et al. (1974) at 11°N. A near-bottom nepheloid layer ~ 100 m thick was found in the Western, but not in the Eastern, Basin of the North Atlantic. This difference was attributed to the presence in the Western Basin of the rapidly northward-moving Western Boundary Undercurrent which was absent from the Eastern Basin. The increases in TSM which these authors found in the bottom water over the Mid-Atlantic Ridge apparently originated from the Ridge itself. High TSM concentrations at four locations on the African Continental Rise were attributed to terrigenous matter originating from the continental shelf off this part of Africa.

Biscaye and Eittreim (1977) have made an important contribution to our knowledge of the concentration and transportation of suspensoids in the nepheloid layer of the abyssal Atlantic Ocean. Data from over a thousand L-DGO-nephelometer stations were used to present the following parameters pictorially.

(i) The quantitative distribution of TSM (ranging from $<0.5\ \mu g\ l^{-1}$ to $>3.5\ \mu g\ l^{-1}$) in the clear water minimum which occurs at varying mid-depths in the water column.

(ii) The gross particulate standing crop which ranges from <50 to >3000 $\mu g\ cm^{-2}$.

(iii) The net particulate standing crop which has had the clear water component removed, and therefore shows the imprint of sediments re-suspended by boundary currents along the western margins of the Atlantic.

(iv) The total net re-suspended loads in various sections of the Atlantic Ocean which range from $\sim 2 \times 10^{6}$ tons in the equatorial Guyana Basin to 52×10^{6} tons in the North American Basin. The total suspended matter load in the Western Basins is about an order of magnitude higher than that in basins east of the Mid-Atlantic Ridge (111×10^{6} tons and 13×10^{6} tons respectively), a difference which can be attributed to the strong boundary current along the western margins of the Atlantic Ocean.

Other important studies of TSM concentrations in the Atlantic Ocean include one by Kullenberg (1974) for an upwelling area off the NW. African coast, and another by Gibbs (1974) which showed that, during times of high discharge rates, suspended matter from the Amazon River run-off produces surface TSM concentrations of $\geqslant 2 \cdot 0 \, \text{mg} \, \text{l}^{-1}$. These high concentrations extend seawards for $\sim 100 \, \text{km}$ and for $\sim 2000 \, \text{km}$ to the NW. along the coast of South America.

37.4.2. GULF OF MEXICO AND CARIBBEAN SEA

In considering the distribution of TSM in the Gulf of Mexico (GoM)–Caribbean Sea (CS) system it is important to take into account the water masses and circulation pattern. Wüst (1963) and Sturges (1965) have pointed out that the inflow of deep water into the Western Basins of the Caribbean occurs sporadically through the Windward Passage between Cuba and Haiti over a sill at a depth of 1600 m. This deep water spreads into the Eastern Caribbean Basin as well as into the Gulf of Mexico through the Yucatan Channel. Sturges (1965) has suggested that inflow into the Eastern Basins through the Jungfern Passage, which has an estimated sill depth of 1960 m, may have occurred in the past. At intermediate depths an important feature in this system is the core of Sub-Antarctic Intermediate Water (SIW) which is characterized by a salinity minimum at depths of 500–1000 m. This water mass enters through the passages on either side of St. Lucia Island in the eastern Caribbean and spreads westward to the Cayman Basin where it loses much of its identity as a result of mixing with water of higher salinity which enters through the Windward Passage. Wüst (1963) believed that North Atlantic Deep Water (NADW), characterized by high dissolved oxygen values, overflows through the Windward Passage into the Cayman Basin, and as a consequence the deep water passing through the Yucatan Channel into the Gulf of Mexico is a mixture of NADW and SIW. A third water mass, the Subtropical Underwater (formed at 20°N.–25°N. in the Atlantic), is characterized by its high salinity and is found at depths of ~ 100–200 m.

The surface current patterns and deep-water spreading in the Gulf of Mexico are largely determined by topography. Water enters from the Carribbean over the 1800 m sill of the Yucatan Straits and leaves through the Florida Straits, the controlling sill of which has a depth of 600 m. Leipper (1970) has suggested that there are seasonal changes in the surface current patterns. In general, a spring intrusion of water intensifies through the summer months and becomes a pronounced loop extending north-westward into the Gulf. In the late summer this loop sometimes becomes a separate eddy which detaches from the main current and drifts westwards. It thus seems probable that neither the water deeper than 2000 m, nor the near-bottom nepheloid layers in the Atlantic, contribute water or suspended matter to the GoM–CS system.

The Gulf–Caribbean system has probably been studied more intensively than any other oceanic region in the world, both in terms of the numbers of stations worked and the TSM samples collected per unit area, largely as a result of the activities of groups based at the Lamont-Doherty Geological Observatory and the Texas A and M University.

The first TSM values for the Gulf and Caribbean were reported by Jacobs and Ewing (1969a) who used a Gerard–Ewing water sampler which was attached to the wire above a piston coring device. This apparatus enabled TSM samples and cores to be taken simultaneously at the same location; however, difficulties could have arisen if the equipment was not properly rigged as serious contamination of the water samples could have resulted from sediment being re-suspended during the coring operation itself (Ewing et al. 1971). Nonetheless, the fact that the mean TSM concentrations listed in Table 37.3 are lower than those reported by most other investigators suggests that the recovery of TSM was not quantitative, probably because centrifugation techniques were used. For this reason it was not possible to establish the extent to which contamination had occurred.

Harris (1971, 1972) carried out the first intensive study of the spatial distribution of TSM in the GoM–CS system, and his findings (Table 37.3) suggested the following:

(i) The concentration of TSM in deep water in the Gulf of Mexico is about

TABLE 37.3

Total suspended matter concentrations in the Gulf of Mexico and Caribbean Sea

No. of Locations	No. of Samples	Depths (m)	Range (μg l^{-1})	Mean (μg l^{-1})	Investigator(s)
			Gulf of Mexico		
11	11	180–bottom	13–193	66	Jacobs and Ewing (1969a)
44	91	0–100	64–238	254	Harris (1971)
20	72	101–bottom	22–406	161	
1	8	900–bottom	39–296	110	Feely et al. (1974)
7	54	550–bottom	10–89	22	Feely (1975)
			Caribbean		
25	25	180–bottom	1–139	40	Jacobs and Ewing (1969a)
8	20	0–100	52–200	99	Harris (1971)
4	46	101–bottom	6–159	46	
20	47	0–105	32–679	147	Bassin et al. (1972)
19	170	110–bottom	1–207	38	Ichiye et al. (1972), Bassin (1975)

four times that found in the Caribbean Sea. This difference is to be expected because there is a major sediment source, the Mississippi River, in close proximity to the Gulf.

(ii) The concentration of TSM shows a bimodal distribution in the deep water of the western Gulf, and concentrations in 1968 ($\sim 300\,\mu g\,l^{-1}$) were about six times higher than they were in 1969 and 1970. The most obvious explanation for this is that the early results were erroneous because of experimental difficulties which were overcome in the later part of the programme. However, circulation in the Gulf is complex and the renewal of deep water in the western part may take place in a sporadic manner, and there is a real possibility that there was a large change in the TSM concentration between 1968 and 1970. This suggests that measurements of TSM may provide valuable information on both the complex circulation and the renewal of deep water in the western Gulf of Mexico.

(iii) The size distribution of particles, as measured by electron microscopy, was continuous between 5 μm and 0·05 μm and conformed to the relationship $N = 1·5 \times 10^7 D^{-2·52}$ where N is the number of particles and D is the mean particle diameter.

Carder and Schlemmer (1973) measured particle volume distributions on discrete samples of near-surface water using both Coulter Counter and light-scattering techniques and showed that suspended particle concentrations were dependent on the loop current. Entrainment of the particle-rich up-welled water along the Campeche Bank and Florida Shelf produced concentration patterns delineating the boundaries of the loop current circulation. High particle concentrations were also found in the Subtropical Underwater; these were apparently due to organic particles produced at the surface in the source region at 20°N.–30°N. in the central Atlantic. This water mass and its particle load are transported to the Gulf of Mexico at a depth of 150 m for hundreds of kilometres, with a possible transit time of five to ten years. In contrast, the concentrations of particles was relatively low in the centre of the loop current.

Feely and his co-workers have reported the TSM concentrations in 62 samples mainly collected within 1000 m of the bottom at eight stations in the Gulf of Mexico (see Table 37.3; Feely et al. 1974; Feely, 1974, 1975). Three of these stations were at the same location; of these three, two of them were worked one week apart in 1973 and gave very nearly the same low concentrations of TSM ($20–30\,\mu g\,l^{-1}$). In contrast, the third, sampled about a year and a half earlier, contained consistently higher concentrations ($\sim 60\,\mu g\,l^{-1}$) with an extremely high maximum of $296\,\mu g\,l^{-1}$ about 100 m above the bottom. The results of this interesting experiment illustrate the ephemeral nature of the nepheloid layer at this station, and suggest that similar temporal variations may exist at other locations in the World Ocean.

Concentrations of TSM in the Caribbean Sea have been determined by Bassin et al. (1972), Ichiye et al. (1972) and Bassin (1975), whose data are summarized in Table 37.3. The mean deep-water concentrations found by these investigators lie between $38 \, \mu g \, l^{-1}$ and $46 \, \mu g \, l^{-1}$. Since water from deeper than 2000 m does not enter the Caribbean from the Atlantic (see above), it seems probable that the reason that the mean TSM concentration is higher than that in the deep water of the Atlantic is the action of in situ processes in the Caribbean. One such process may be the scouring action of water moving over the sills of the various passages into the Caribbean, and indeed Ichiye et al. (1972) have attributed two mid-depth TSM maxima to this process. These authors also pointed out that Lisitzin (1972) and Klenova et al. (1962) have suggested that the mid-depth TSM maxima which they observed were a result of the scouring of ridge crests.

There have been several other studies of the distribution of certain chemical types of suspended matter in the Gulf of Mexico–Caribbean Sea system. These are discussed in Section 37.5.

37.4.3. PACIFIC OCEAN

Most recent TSM distribution studies in the Pacific have been carried out by investigators from Oregon State University (see e.g. Bearsdley et al. 1970; Pak et al. 1970; Carder et al. 1971; Pak, 1974). In these investigations, discrete samples were collected and determinations of the volume-scattering functions at 45 degrees were made using the Brice–Phoenix Model 2000 light-scattering photometer, as well as measurements of particle size distributions using a Coulter Counter (see Sheldon and Parsons, 1967). The latter measurements were usually made immediately after collection; this may, in fact, have been unnecessary as Carder and Schlemmer (1973) have found that samples can be preserved using Lugol's iodine solution for subsequent onshore measurement. There are two principal problems associated with the interpretation of this work: (i) much of the light scattering is caused by particles $< 1 \, \mu m$ in size which cannot be reliably measured by Coulter counting; (ii) it is necessary to assume a density for the particulate matter in order to obtain the weight of the TSM.

Descriptive studies carried out by this group include one on the Columbia River plume which they were able to trace seawards for about 200 km. Particle concentration appeared to be a conservative property of the plume water for about 100 km and this suggested that the residence time of the particles in the plume water is at least 10, and possibly as many as 30, days. In another study, Pak (1974) found that there are relatively high concentrations of TSM in the surface and the bottom water ($\sim 90 \, \mu g \, l^{-1}$) and a broad minimum ($\sim 40 \, \mu g \, l^{-1}$) at mid-depth in the Panama Basin in the equatorial Pacific. Relatively high concentrations are also present in the Equatorial Undercurrent, and probably arise from high biological production near the Galapagos Islands with subsequent transport to the area of the Panama Basin.

Important studies of the longitudinal distribution of TSM were made in the Pacific during the GEOSECS programme; however, detailed data from these were not available at the time of writing.

It is apparent from the TSM investigations described above that currents exert a great influence on the distribution of TSM at all depths. However, physical oceanographers have, as yet, made little use of TSM distributions as an aid to the study of the direction, velocity and continuity of currents in the deep sea.

37.5. CHEMICAL COMPOSITION OF TOTAL SUSPENDED MATTER

In recent years TSM studies have been extended to include chemical and isotopic aspects in addition to spatial distribution. The most abundant chemical component of TSM is organic matter, and because of its importance to biological phenomena and the relative ease with which it can be measured much more data have been accumulated on particulate organic matter (POM) than on TSM.

37.5.1. PARTICULATE ORGANIC MATTER (see also Chapter 13)

The term "organic matter" will be used here to denote only the reduced forms of carbon such as those making up the soft parts of plants and animals. Thus, that fraction of the TSM which is composed of calcareous and siliceous tests of organisms will be included with the inorganic fraction and will be discussed in Section 37.5.2. Ordinarily, POM is assessed from measurements of particulate organic carbon (POC) made using procedures such as that introduced by Menzel and Vaccaro (1964). Briefly, the latter involves filtration of the sea-water sample through an organic-free glass fibre filter (nominal pore size of $\sim 0.45 \, \mu m$), wet combustion of the organic matter using potassium peroxydisulphate and measurement of the CO_2 produced with an infra-red gas analyser. Particulate organic matter is evaluated by multiplying the POC by a factor of two, since carbon comprises $\sim 50\%$ of the ash-free dry weight of organic matter (Riley, 1970).

Studies of the spatial distributions of POC have been reported for the Atlantic (Menzel and Ryther, 1968), Pacific (Menzel, 1967), Gulf of Mexico

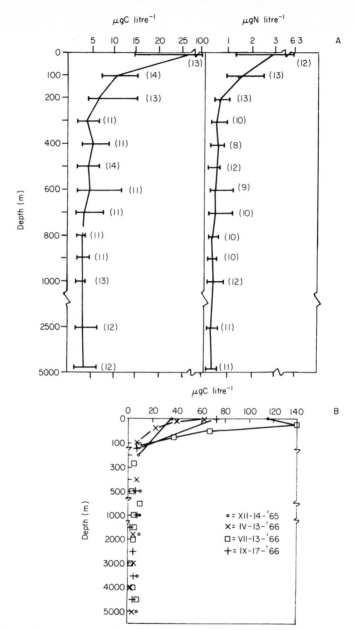

Fig. 37.6. (A). The vertical distribution of particulate organic carbon and nitrogen in the SW. Atlantic (Atlantis II Cruise 31, February–March, 1967). Bars indicate range; the numbers of observations are indicated in parentheses. (After Menzel and Ryther, 1968.) (B). The vertical distribution of particulate organic carbon at approximately 36°30′N., 67°50′W. in the Sargasso Sea. (After Menzel and Ryther, 1968.)

(Fredericks and Sackett, 1970) and the Antarctic (El-Sayed, 1968). In a typical open-ocean profile (Fig. 37.6A), the POC concentration in the surface layer is $\sim 100 \ \mu g \, l^{-1}$ and decreases sharply across the thermocline to values of only a few $\mu g \, l^{-1}$ at depths below 200 m. These low concentrations are not thought to be affected by horizontal and seasonal changes in the concentrations of POC in the surface water (Fig. 37.6B). Clearly, the POC in the

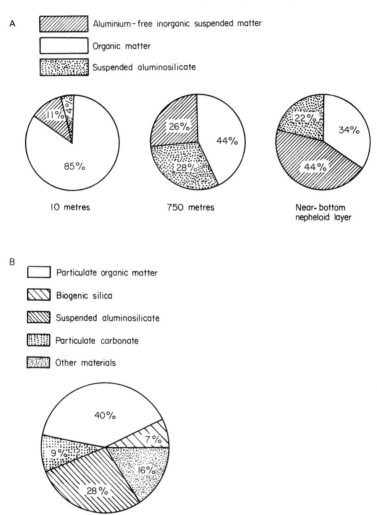

FIG. 37.7. (A). Percentage of total suspended matter attributable to particulate organic matter, suspended aluminosilicates and aluminium-free inorganic suspended matter at various depths in the water column (after Feely *et al.*, 1974). (B). Average composition of suspended matter in deep water in the Gulf of Mexico (after Feely, 1975).

euphotic zone is intrinsically related to primary production, and may reach levels of $1 \, \text{mg} \, l^{-1}$, or higher, in regions of upwelling and in other coastal environments with high productivity.

According to Feely et al. (1971, 1974) and Feely (1975), the percentage of POM in the TSM in the Gulf of Mexico varies from $\sim 90\%$ in the surface waters to $\sim 30\%$ in the near-bottom nepheloid layer (Fig. 37.7A). Since the percentage of organic matter in the sediments of the Gulf of Mexico is usually less than 1%, the relatively high concentrations of POM in the nepheloid layer observed by Feely et al. (1974) are not consistent with the concept that the particulate matter in this layer is comprised entirely of re-suspended sediments. In fact, the TSM is probably the most recently de-posited "soupy" material (which is not collected by most sediment sampling devices) that has undergone re-suspension. This material is probably more organic-rich than is the underlying more consolidated sediment which is ordinarily collected with gravity, piston and box corers (see Chapter 36).

If the average concentration of POC in deep water is assumed to be $5 \, \mu g \, l^{-1}$, as reported by Menzel (1967) and others, and the POC–POM conversion factor of two is applied, this gives an average POM concentration of $10 \, \mu g \, l^{-1}$ for such water. Using this average concentration it is apparent that POM constitutes $\sim 50\%$ of the TSM reported by Brewer et al. (1976) for deep water in the Atlantic.

The chemical composition of the POM in the euphotic zone probably approximates to that of plankton. However, beneath the euphotic zone there is likely to be a systematic change in POM composition with depth which reflects the relative stabilities of the various molecular species to bio-chemical degradation. The POM in the euphotic zone consists of a mixture of living micro-organisms and organic detritus (dead organisms). According to Menzel and Ryther (1964), the high light intensities and the depletion of nutrients in surface waters, particularly in the tropics, lead to the production of organisms relatively poor in chlorophyll, nitrogen and phosphorus. In contrast, near the compensation point, and presumably also in the surface water at higher latitudes, low light intensities and a plentiful supply of nutrients stimulate the production of organisms having high concentrations of chlorophyll, nitrogen and phosphorus. The composition of organic detritus depends on the extent to which the material has undergone decomposition and mineralization, both of these factors being functions of the age of the detrital material. According to Harvey (1960), Yentsch (1962) and Menzel and Ryther (1964), chlorophyll and phosphorus are quickly released in the euphotic zone on the death of micro-organisms, and this leads to the produc-tion of residues containing carbon and nitrogen, the ratio of which slowly changes with depth in favour of nitrogen (see Fig. 37.6A). Assays of living organic matter for adenosine triphosphate (ATP) indicated that at a station in the Pacific Ocean off Peru the percentage of living organic carbon in the POC ranged from ~90% in the euphotic zone, to ~4% at 200 m and down to <1% at 2400 m, (Holm–Hansen, 1970).

37.5.2. NON-ORGANIC PARTICULATE MATTER

The GEOSECS Expedition will eventually provide much valuable information on the chemical composition of the non-organic fraction of TSM, and an initial report has been published by Spencer et al. (1978). However, until the final results of this programme appear, the most complete chemical characterizations of the major sub-groups of the TSM in a deep-ocean basin will remain those reported by Feely et al. (1974) and Feely (1975) for deep-water samples from the Gulf of Mexico (Figs 37.7A and 37.7B). In these studies, the components of the TSM were classified as follows; organic matter, aluminosilicates, biogenic silica, calcium carbonate and "other material". The mean concentrations of these components in the deep water of the Gulf on Mexico were found to be 40%, 28%, 7%, 9% and 16% respectively (Feely, 1975).

Aluminosilicates and POM were determined as described in Section 37.3.4 and particulate calcium carbonate was estimated from particulate inorganic carbon. Biogenic silica was assumed to be the difference between total particulate silicon and the silicon present as aluminosilicate (following Grim and Johns (1954) an Si/Al ratio of 2·24/1 was assumed for the aluminosilicate); this is a useful but rather unreliable method of determining biogenic silicon. The value for "other material" was the difference between 100% and the sum of the other components. Each of these major non-organic groups will be considered individually below.

Transmission and scanning electron microscopy are being used to an increasing extent for the determination of the sizes, shapes, numbers and surface characteristics of particles in the non-organic fraction of TSM. According to Harris (1971) and Bassin (1975), electron micrographs of TSM collected in the deep water of the Gulf of Mexico and the Caribbean Sea indicated that, on a numerical basis, clay minerals are the most common components of suspensoids. Such clay minerals usually exist as thin (thickness about 1–2% of the diameter) irregular flakes and coacervate clusters, the latter being montmorillonites containing >90% trapped water. The only clay mineral which can be readily identified is kaolinite, the flakes of which are characterized by their hexagonal shapes. Coccoliths, the calcareous plates originating from coccolithophorides, are reported to be the most common identifiable biogenic component of TSM and siliceous diatom tests are also very abundant.

There have been numerous qualitative X-ray diffraction studies of the mineralogy of suspended aluminosilicates (see e.g. Lisitzin, 1959; Jacobs and Ewing, 1965, 1969b; Feely, 1975). Most workers have reported that the suite of clay minerals in suspension is the same as that found in the bottom sediments in the same region. Jacobs and Ewing (1969b) and Betzer *et al.* (1974) have suggested that particular suspended clay minerals which are peculiar to certain regions or basins may be useful diagnostic indicators of erosional processes that have occurred in an ocean basin upcurrent from the sampling location. Behairy *et al.* (1975) have examined the clay mineralogy of suspensoids collected by high-speed centrifugation of surface water from the eastern Atlantic Ocean, and found the average mineralogical composition of the clay to be illite $\sim 42\%$, kaolinite $\sim 25\%$, chlorite $\sim 19\%$ and montmorillonite $\sim 14\%$. Their data indicate that illite decreases and kaolinite increases towards low latitudes, reflecting the trends found for aeolian dusts and also for the soils of adjacent land masses. The authors concluded that these two clay minerals (and chlorite) in the deep-sea sediments of the Atlantic have a land-derived origin.

It was pointed out above that biogenous silica and carbonate particles comprise about 14% of the TSM in samples of deep water collected with Niskin bottles and similar devices. However, large particles (diameters $\sim 100 \, \mu m$), such as foram tests and faecal pellets, containing biogenic carbonate and siliceous material, fall more rapidly through the water column and the probability of sampling one of these particles in a 30 litre Niskin bottle is extremely low. Thus, the actual percentage of suspended biogenous silica and carbonate in TSM is probably significantly higher than that quoted above. This discrimination in sampling is responsible, in part, for the difference between the chemical compositions of a sediment and that observed for the TSM in the overlying water.

A considerable number of elemental analyses of TSM were carried out during the GEOSECS programme. In these investigations (see e.g. Spencer *et al.* 1978) the samples were filtered through 0·6 μm pore size Nuclepore filters and, after washing, the filters were pelletized and irradiated in a thermal neutron flux of $\sim 10^{13}$ neutrons $cm^{-2} s^{-1}$. The gamma ray spectra which resulted from the radionuclides induced by the irradiation were then analysed to give the amounts of La, Au, Mn, Hg, Cr, Sb, Ag, Sc, Fe, Zn, Co, Ba, Ti, Sr, Mg, Ca, V, Al, Cu and I. A very large amount of data was generated in this study and were mainly reported in the form exemplified by Fig. 37.8. Certain elemental concentrations can be correlated with the presence of particular minerals in the TSM on the filters. Thus, the amounts of Al, and also those of La, Ti, Fe and Sc, reflect the amounts of aluminosilicates; Ca, Sr and Ba may be regarded as measures of biogenous calcium carbonate, strontium and barium sulphates, respectively. Spencer *et al.* (1978) suggested that the near-surface maxima of Ca, Sr and Mg in the TSM are the result of biological assimilation in the euphotic zone, whereas the near-surface Al maximum in the Atlantic, particularly south of 40°N., is apparently due to the fall-out of dust transported by the NE. trade winds.

Chester and Stoner (1975) have described the trace element composition of 27 large samples of TSM collected from surface water (0–5 m) by high-speed centrifugation techniques on a track from Great Britain to Japan and back. A summary of the trace metal compositions of the dried (40°C) and homogenized samples, determined by emission spectrographic techniques, is

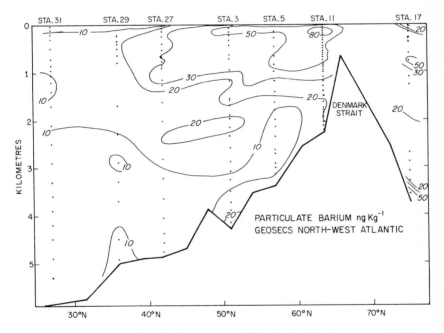

FIG. 37.8. Particulate barium distribution in the NW. Atlantic (after Spencer *et al.*, 1978).

presented in Table 37.4. From a comparison of these values with those reported for marine sediments and plankton, the authors deduced that: "Mn, Co and V are probably mainly present in continentally derived material and authigenic precipitates; Pb and Zn are probably located largely in plankton; and Cu and Ba are partitioned between the three components."

Chesselet *et al.* (1976) have used scanning electron microscope and electron microprobe techniques to examine TSM and have reported the consistent occurrence of barite crystals (mean size ∼1 μm) in the TSM from the Atlantic. At depths less than 3 km, the barite occurs as discrete crystals,

TABLE 37.4

Trace element concentrations in the particulate material from the surface layer (concentrations in ppm) (after Chester and Stoner, 1975)

Oceanic region[a]	Mn	Cu	Co	Ga	V	Ba	Pb	Zn
N. Atlantic (10)	145	74	11	4	38	191	52	159
S. Atlantic (4)	85	52	16	3	69	72	72	260
Indian Ocean (6)	385	202	14	3	60	77	44	231
China Sea (7)	1501	107	12	8	86	166	63	232
Overall average (27)	529	109	13	5	63	126	58	220

[a] Numbers in parentheses indicate number of samples.

whereas below this depth barium was found to be bound in complex aggregates. High concentrations of particulates were associated both with areas of high primary production and a region near the Mid-Atlantic Ridge, which suggests that particulate barium may have a biogenic and/or a volcanogenic origin. Similar techniques have also shown that particulate strontium is enriched in the TSM in near-surface water (Spencer *et al.* 1978) and it is probably present as celestite (strontium sulphate) in the tests of *Acantharia*. Such tests do not occur in deep water, probably because they dissolve rapidly in the upper parts of the water column. Evidence for the presence of other types of particles in TSM has been obtained with the scanning electron microscope: however, it has not been possible to identify some of these, and others have proved to be contaminants introduced during sampling and processing.

37.6. SETTLING RATES OF SUSPENDED MATTER

Many chemical species are transported through the water column either adsorbed onto settling particles or incorporated into their structures. Such particles have a wide range of sizes, shapes and densities which cause the individual particles to have a large spectrum of settling velocities. A further complication is that some particles may, as they sink, change in size and form; this will affect their rates of settling. Thus, in the near-surface water the particles may behave in accordance with Stokes' law. However, they may soon be eaten and so become incorporated in a much larger faecal pellet which has a faster settling velocity. This pellet may in its turn, either be eaten and eventually form a larger pellet, or be decomposed bacterially and thus return the particles to approximately their original size. Deeper in the water column, the particles may undergo partial dissolution which will cause the particle settling velocity to decrease. In the deepest water, the particles may be dissolved completely and therefore not be deposited at all. The actual pathway from the surface to the bottom for an average particle with a diameter of a few micrometres may be more complex than the above sequence suggests, and the time taken for a given particle to reach the bottom may be considerably different from that predicted theoretically. The only valid way of determining settling velocities is to measure them by methods based primarily on the distribution and partitioning of radionuclides. Several studies involving this approach are summarized below.

Initial estimates of the settling velocities for particles smaller than 100 μm may be obtained from Stokes' law which may be represented as follows:

$$V = d^2(P_s - P_f)\mathbf{g}/18\eta$$

where V is the settling velocity in cm s^{-1}, d is the particle diameter in cm (assuming the particles are spheres), P_s is the particle density (about 1·5 g cm^{-3}), P_f is the fluid density (1·03 g cm^{-3}) for sea-water, η is the fluid viscosity ($\sim$0·015 poises at 10°C) and $\mathbf{g}$ is the acceleration due to gravity (981 cm s^{-2}). Particles larger than 100 μm in diameter settle at a velocity lower than that predicted by this law because of resistance generated by the turbulence associated with them as they sink. The settling velocities of particles of various shapes are plotted against the radii of the equivalent spheres in Fig. 37.9 (Lerman et al., 1974). For particles having the same volume, the settling velocities increase in the order; cylinders > spheroids > discs. Thus, clay particles, which most nearly approximate to discs in shape, should have the lowest settling velocities. Thus, the diagram suggests that a particle with a volume equivalent to a sphere with a 1 μm radius will have a settling velocity ranging over two orders of magnitude from 3×10^{-5} cm s^{-1} (0·026 m day^{-1}) for a disc, to $\sim 2 \times 10^{-3}$ cm s^{-1} (1·7 m day^{-1}) for an elliptical cylinder. It would therefore take these particles 400 and 6 years, respectively, to fall through a 3800 m water column (the average depth of the ocean).

Brun-Cottan (1976) has pointed out that particle density is as important as shape in determining settling rates, and that there is a paucity of data on in situ densities which take account of both the porosities of particles and their bound water content. Furthermore, he has noted that Stokes' law is not applicable to particles with diameters of <4–5 μm. Smaller particles may behave conservatively and move horizontally over large distances in a particular water mass. According to the author, this implies that sedimentation from the upper layers of the water column takes place via large particles, or aggregates, in the 50–100 μm size range, and that the small particles which are found in sediments "reach the bottom by mechanisms and routes other than those determined for fast-settling Stokes particles". As suggested earlier, the mechanism for the formation of a large proportion of pelagic deposits may involve initial settling via large faecal pellets which then disaggregate through bacterial activity either in deep water, or on the

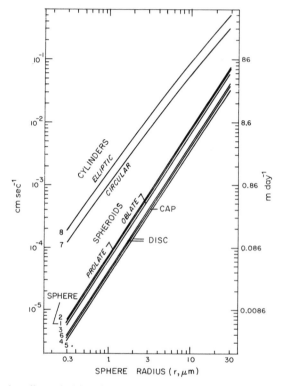

FIG. 37.9. Stokes' settling velocities of particles of different shapes (after Lerman *et al.*, 1974). The horizontal axis is the radius of the sphere equal in volume to the particle of different shape. (1) Sphere; (2) prolate spheroid, $c/a = 1.5$, settling parallel to the polar axis c (a is the equatorial radius); (3) oblate spheroid, $c/a = 0.5$, same settling as in (2); (4) hemispherical cap, wall thickness is 10 % of the outer radius; (5) disc, having a thickness/radius ratio of 0.1, settling broadside; (6) disc, as (5), settling edgewise; (7) circular cylinder, height/radius ratio of 4; (8) elliptical cylinder, with semi-axis ratio of 0.5 and α-axis/height ratio of 4, settling parallel to α-axis.

bottom, to yield the small-sized particles ordinarily found in deep-sea deposits.

The sinking rates of faecal pellets have been the subject of several investigations (Table 37.5). Measured values range from $15 \, \text{m day}^{-1}$ to $860 \, \text{m day}^{-1}$. Sinking rates calculated from data on nuclear bomb-derived

TABLE 37.5

Summary of the means (italic) and ranges (in parentheses) of zooplankton faecal pellet volumes and sinking rates found by various investigators (modified from Turner, 1977)

Source of faecal pellets	Food	Faecal pellet volume ($\mu m^3 \times 10^6$)	Equivalent sphere radius (μm)	Sinking rate (m day^{-1})	Number of measurements	Reference
Pontella meadii	*S. costatum*	*8·2* (1·7–26·0)	*125* (74–184)	*54·0* (22–133)	22	Turner (1977)
Pontella meadii	*S. costatum*	*11·7* (1·9–34·0)	*141* (77–201)	*88* (18–153)	27	Turner (1977)
Pontella meadii	*Nitzschia* sp.	*3·7* (1·1–6·9)	*96* (64–118)	*51* (23–125)	28	Turner (1977)
Pontella meadii	*D. salina*	*8·3* (1·3–15·9)	*126* (68–156)	*77* (15–117)	8	Turner (1977)
Pontella meadii	*G. monilata*	*38·6* (18·9–45·8)	*209* (165–222)	*666* (555–860)	6	Turner (1977)
Unknown	—	—	—	*159·8*[a] (36–376)	41	Smayda (1969)
Unknown	—	*0·21*[b] (0·11–0·33)[b]	*37* (30–43)	*116* (72–210)	7	Smayda (1971)
Acartia clausii	Microflagellates	—	(36–84)	*43*	—	Osterberg *et al.* (1963)
Euphausia pacifica	*S. costatum*	— (1·6–15·5)[d]	— (73–155)	— (53–411)	75[d]	Fowler and Small (1972)
Meganyctiphanes norvegica, Euphausia krohnii, Nematoscelis megalops	Juvenile *Artemia* or *Phaeodactylum tricornutum*	— (3·0–110·0)[d]	— (89–297)	*240* —	113[d]	Fowler and Small (1972)
Meganyctiphanes norvegica, Euphausia krohnii, Nematoscelis megalops	Natural diet					

[a] Mean calculated from data in Table 1 of Smayda (1969).
[b] Volumes calculated as cylinders from data in Table 3 of Smayda (1971).
[c] Mean calculated from data in Table 3 of Smayda (1971).
[d] Estimated from Fig. 1 of Fowler and Small (1972).
[e] For *E. krohnii* only.

fission products are much lower than this (<2 m day^{-1}) suggesting that transport by faecal pellets is relatively unimportant. However, such rates are probably the weighted average of an initial fast rate of about a 100 m day^{-1} and a slow rate of a few tenths of a metre per day for the particles produced subsequently by disaggregation in the water column.

Feely et al. (1971) have used an empirical approach to derive an estimate of the settling velocity of particulate aluminosilicates in the Gulf of Mexico. On the assumption of steady-state conditions, and using measured water column loads of particulate aluminium and an estimated mean sediment deposition rate based on radiometric data, they calculated that the residence time of particulate aluminosilicates is three years. If the particles were injected at the surface, this corresponds with a mean settling velocity of ~ 3 m day^{-1}. This is considerably higher than that predicted for clay-sized disc particles (see above). Use of the same procedure, assuming an average TSM value for the Atlantic of $20\,\mu g\,l^{-1}$, a water column of 5000 m and a sediment deposition rate of $1\ g\,cm^{-2}\,(10^3\ \text{years})^{-1}$, gives a settling velocity of ~ 1.4 m day^{-1}, similar to the previous estimate.

Radionuclides, with their precise internal clocks, are extremely useful in the Earth Sciences for determining the chronology and/or the rates of many different processes. However, for measurements to be of any value it is essential that the half-life of the nuclide is within about an order of magnitude of the time which has elapsed since the particular process was initiated. Carbon-14 ($t_{\frac{1}{2}} = 5730$ years), for example, cannot be used to date either rocks older than a million years or a twentieth century tree. Thus, for suspended particles with estimated settling velocities of metres per day and residence times in the water column of tens of years it would be necessary to use radionuclides with half-lives generally in the range from 10 days to 100 years. An exception to this generalization is that radionuclides, such as carbon-14 and plutonium-239 ($t_{\frac{1}{2}} - 24\,400$ years), which were introduced to the ocean in great quantities in the 1950s and 1960s as a result of the testing of atomic weapons have proved very useful in determining settling rates. The discussion below will consider firstly, the bomb-produced nuclides (e.g. ^{144}Ce, ^{147}Pm, ^{14}C, ^{55}Fe and ^{239}Pu), and secondly the naturally occurring radionuclides (e.g. ^{234}Th, ^{210}Pb and ^{210}Po) which are constantly being produced in the ocean from their soluble progenitors ^{238}U and ^{226}Ra (see also Chapter 18).

As early as 1965, Bowen and Sugihara (1965) reported that the relatively insoluble particulate fission products, such as ^{144}Ce and ^{144}Pm, move vertically in the water column at a greater rate than do the soluble species such as ^{90}Sr and ^{137}Cs. The authors attributed this phenomenon to an association of the insoluble species with other sinking particles. Mid-depth maxima of ^{144}Ce and ^{144}Pm were interpreted as being caused by either locally high concentrations of sinking particles, or the retention of these nuclides in plankton tissues. In one instance they found that:

> the Ce:Pm discrimination seen in the 1960, 700-m sample from 15°32'N., if adjusted for radioactive decay since 1958, is identical to that seen in the 1961, 100 m sample from the equator.

This suggests that the particles have settled about 600 m in one year, a settling velocity of about 1–2 m day^{-1}, although the authors themselves did not make such an extrapolation. Various workers, e.g. Slowey *et al.* (1965), have reported similar findings for these and other fission product activities but none of them have made an explicit attempt to use these activities to estimate particle settling velocities.

Plutonium-239 was introduced into the environment mainly by the atmospheric nuclear test series which took place during 1961 and 1962, although small amounts were introduced from the beginning of testing in 1954 as well as by the preliminary detonations in 1945. Pollution involving this nuclide results both from the dispersal of unreacted material from the nuclear devices and from plutonium produced by neutron capture by uranium-238. A smaller amount of plutonium-240 ($t_{\frac{1}{2}} = 6540$ years) was also produced at the same time, but since it is not possible to distinguish between the two by α-spectrometry, it is actually the sum of the plutonium-239 and, -240 α-activities that is usually reported. In addition to these nuclides, environmental contamination with plutonium-238 ($t_{\frac{1}{2}} = 87\cdot8$ years) occurred in 1964 when a satellite powered with this nuclide failed to orbit and burned up on re-entry. The spatial and temporal distributions of these isotopes of plutonium in the ocean are being measured in several laboratories and the data are yielding valuable information on particle settling velocities in the ocean.

In a brief report, Miyake *et al.* (1970) noted that by 1968 plutonium isotopes had penetrated to a depth of about 3000 m in the western North Pacific and Japan Sea. This indicates an average penetration rate of ~210 m year^{-1} (0·6 m day^{-1}). In a later, more detailed account, Bowen *et al.* (1971) pointed out that plutonium moves downwards in the water column at a much faster rate than do the "soluble" fall-out nuclides such as ^{90}Sr and ^{137}Cs, and suggested that plutonium becomes associated with sinking particles. Significant concentrations of plutonium were found in samples taken as deep as 2200 m in the Atlantic Ocean, an observation which would suggest a particle settling rate of 0·4 m day^{-1}.

Noshkin and Bowen (1973) have described the distribution of ^{239}Pu in several cores from the Atlantic Ocean. For example, one taken in 1969 in 4810 m of water contained significant amounts of plutonium. This may be translated to an average particle settling rate of 320 m year^{-1}, i.e. ~0·9 m day^{-1}. These authors have constructed models to predict the settling rate of plutonium associated with a mixed particle population in the water column and the plutonium distributions which they found suggested that 30% of it sinks at ~ 390 m year^{-1}, 40% at ~ 140 m year^{-1} and 30% at ~ 70 m year^{-1}, with a mean sinking rate of 195 m year^{-1}, i.e. ~ 0.5 m day^{-1}.

Labeyrie et al. (1976) have used ^{55}Fe ($t_{\frac{1}{2}} = 2.7$ years) to determine particle settling velocities in the water column. Most of the ^{55}Fe that is now measurable stems from nuclear bomb tests which were carried out in the atmosphere in 1961 and 1962. Measured distributions of this nuclide indicate a suite of iron-carrying particles, 20% of which sank at a rate of ~ 500 m year^{-1}, 60% at ~ 320 m year^{-1} and 20% at ~ 200 m year^{-1}. The mean settling velocities of the iron-carrying particles (350 m year^{-1}) is about double that of the plutonium-carrying particles mentioned above. The authors believed that their data are best explained in terms of "two populations of particles, of different sinking characteristics, one richer in iron, the other rich in plutonium"; however, it is probable that there is some overlap between the two populations.

The injection of "bomb" carbon-14 increased the surface water Δ^{14}C values for the total dissolved inorganic carbon at low and mid-latitudes from about $-50\permil$ (using the commonly accepted notation described by Broecker and Olson (1959); see Section 18.4.2) in pre-bomb times to about $+200\permil$ in the late 1960s and early 1970s. Carbon having such ^{14}C levels is being incorporated into the organic and calcium carbonate matrices of organisms living in the euphotic zone, and serves as a label for tracing the transport of this recently fixed carbon to the ocean depths. The use of "bomb" carbon-14 in this manner was described by Somayajulu et al. (1969) who determined the Δ^{14}C composition of two samples of particulate calcium carbonate collected at depths of about 2500 m and 3500 m in the Pacific. From this data they were able to estimate mean particle settling rates of 2·1 m day^{-1} and 1·6 m day^{-1}, respectively. Most of the particles concerned were reported to be less than 20 μm in size. These experimentally derived rates are similar to those of spheres with diameters of about 20 μm settling according to Stokes' law, and to those determined empirically by the method described by Feely et al. (1971).

Uranium-238, uranium-235 and thorium-232 are the parents of the three families of naturally occurring radionuclides shown in Fig. 37.10. In a closed system, after the elapse of sufficient time, the rate of decay of a radionuclide in one of these series becomes equal to its rate of formation and the number of atoms of that nuclide reaches a steady state, a condition known as secular equilibrium. Over a long period of time the only change in the series will be a decrease in the number of parent atoms and an increase in the number of atoms of the stable lead nuclides which terminate each series. In the ocean, uranium isotopes are present in solution as their stable carbonato complexes (e.g. $UO_2(CO_3)_2^{2-}$). When they decay, the thorium and other daughter

products tend to be adsorbed onto the surfaces of particulate material because, in contrast to uranium, they are rapidly hydrolysed. The settling of
these particles, with their associated radionuclides, often causes non-
equilibrium conditions to develop; thus, for example, surface waters may be
depleted in thorium daughters, whereas the intermediate water may contain

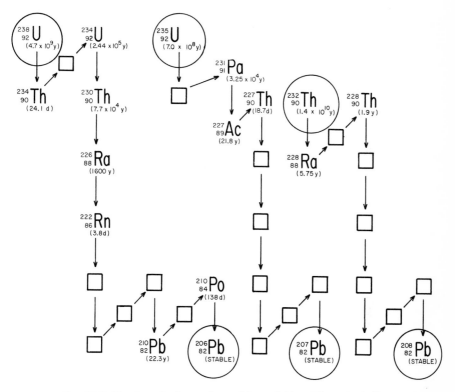

FIG. 37.10. Uranium–thorium series nuclides useful in settling rate studies.

an excess of these species. Thorium-232, which is the parent of the thorium-
232 series, does not occur in significant amounts in dissolved form in sea-
water. However, its immediate daughter (radium-228) is either released
from sediments and diffuses into the overlying water column, or enters the
ocean via river run-off; this gives rise to a similar disequilibrium between
thorium-232 and radium-228. Secular disequilibrium also exists between
radium-228 and its granddaughter thorium-228 because of the very short
oceanic residence time of the latter. The use of the uranium–thorium series
nuclides for determining settling velocities is discussed below.

Thorium-230 is produced by the decay of the uranium-234 present in solution in sea-water and it is quickly adsorbed onto suspended matter and deposited on the sea floor. According to Moore and Sackett (1964), thorium-230 has an oceanic residence time of 50 years. This corresponds with a particle settling velocity of $0.2 \, \text{m day}^{-1}$. Because of the difficulty of determining thorium-230 in the water column, this settling rate is, at best, only an order of magnitude estimate. As mentioned above, radionuclides with much shorter half-lives are more useful for measuring settling rates of about a metre per day.

The amount of a relatively short-lived member of the uranium–thorium series, such as ^{234}Th, ^{228}Th or ^{210}Pb, in a parcel of water may be described by the equation:

$$\underset{\text{(production rate)}}{\lambda_P N_P} = \underset{\text{(decay rate)}}{\lambda_D N_D} + \underset{\text{(scavenging rate)}}{k N_D} \tag{1}$$

where λ_P and λ_D are the radioactive decay constants of the parent and daughter respectively, N_P and N_D are the number of atoms of parent and daughter nuclides respectively and k is the scavenging coefficient, or removal rate, expressed in reciprocal time units. The mean life, or residence time, for this particular process is $1/k$. Since the activity A is equal to $N\lambda$ for the selected nuclide, the above equation may be rewritten

$$A_P = A_D(1 + k/\lambda_D). \tag{2}$$

Thus, if λ_D is known the residence time $(1/k)$ may be calculated from the measured activity ratio A_P/A_D. From these data approximate particle settling velocities may be estimated for particular depths of the water column.

The use of the short-lived nuclide ^{234}Th $(t_{\frac{1}{2}} - 24.1 \text{ days})$ for the measurement of particle settling velocities was first described by Bhat et al. (1969) who found $A_{234\text{Th}}/A_{238\text{U}}$ ratios of as low as 0.15 in near-shore surface waters. The ratio increased systematically further from shore and approached the equilibrium value of 1.00 several hundred kilometres offshore. Several vertical profiles showed activity ratios which were almost unity at the surface, but which decreased to 0.75 in the 100 m mixed layer, and increased again nearly to the equilibrium value of 1.00 in deep water. Use of Equation (2) and a value of $1/0.75$ for the ratio A_P/A_D gives a residence time of 100 days for the suspended material in the mixed layer. This corresponds to a particle velocity of about $1 \, \text{m day}^{-1}$. The authors have suggested that the relatively high activity ratios in the upper layers are the result of turbulence, and/or biological activity, which inhibit the rapid removal of ^{234}Th from surface water. In a later study, Matsumoto (1975) has reported that the residence time of particulate ^{234}Th in the upper 200 m of the water column is 139 days, corresponding to a particle settling velocity of $1.4 \, \text{m day}^{-1}$.

Aller and Cochran (1976) used a similar approach in a study of a near-shore environment in Long Island Sound and showed that $k = 1.4$ day^{-1}, which corresponds to a settling rate of ~ 10 m day^{-1}. This high settling rate is not surprising because the biological activity and the grain sizes are much greater than are those of the open-ocean environment studied by Bhat et al. (1969).

Thorium-228 derived from radium-228 is being removed from sea-water by adsorption onto particulate matter in an analogous fashion to thorium-234 derived from uranium-238. As has been noted above, radium-228 is produced in the sediment from its insoluble parent thorium-232 and diffuses into the overlying water column (see also Chapter 18). Thus, Moore (1969) has reported relatively high concentrations of ^{228}Ra in near-shore environments and has shown that it decreases systematically away from land. Broecker et al. (1973) have shown that the $A_{228\text{Th}}/A_{228\text{Ra}}$ ratio is also related to proximity to land, values generally increasing from ~ 0.1 in near-shore waters to ~ 0.4 in open-ocean environments. These ratios reflect the greater concentrations of TSM in near-shore environments and its greater role in the removal of ^{228}Th. By application of Equation (2), using a A_D/A_P ratio of 0.21, Broecker et al. (1973) have calculated a thorium residence time of 0.7 years. For a 100 m mixed layer, this corresponds to a particle settling velocity of ~ 0.4 m day^{-1}.

Thorium-234 and thorium-228 are produced in the water column from their soluble parents. The only other relatively short-lived radionuclide pair in the uranium–thorium series that may behave in a similar fashion is the actinium-227 ($t_{\frac{1}{2}} = 21.8$ years)–thorium-227($t_{\frac{1}{2}} = 18.7$ days) pair. However, to the author's knowledge the concentrations of actinium-227 in the water column have not been measured, although the chemistry of actinium suggests that it will not be readily adsorbed from sea-water by particulate matter.

Radon-222, which has a half-life and mean life of 3.8 and 5.5 days respectively, is produced by the radioactive decay of radium-226. The relatively long life of radon allows it to leak from continental rocks into the atmosphere where it is mixed upwards into the troposphere and transported laterally over the ocean where its decay products, of which lead-210 ($t_{\frac{1}{2}} = 22.3$ years) is the most important, are washed out by precipitation. Lead-210 and its daughter, polonium-210 ($t_{\frac{1}{2}} = 138$ days), serve as excellent tags for the determination of near-surface particle settling velocities. These two radionuclides are also present in deep water, in which case they are derived from radium-226 which is released from the underlying sediments and subsequently diffuses up the water column. Disequilibria between these two nuclides may be used to derive settling rates as discussed previously.

Rama et al. (1961) reported the first determinations of lead-210 in the ocean and found it to be in secular equilibrium with radium-226. Nevertheless, they were able to calculate an oceanic residence time of 2 years from estimates of the atmospheric fall-out of lead-210. For a 100 m mixed layer this would be a particle settling velocity of ~ 0.14 m day^{-1}; this value should be considered to be only very approximate. Subsequently, Goldberg (1963) reported that activities of lead-210 in several deep-water samples were only about one-half of those predicted on the assumption that secular equilibrium with radium-226 had been attained. Craig *et al.* (1973) confirmed the existence of a deep-water deficiency of lead-210 from profile measurements in deep water in both the Pacific and the Atlantic Oceans. Using a more sophisticated equation than Equation (2), they calculated that lead-210 has a residence time of 54 years. This again indicates a settling velocity of ~ 0.2 m day^{-1} if the uptake by particles is rapid relative to the settling time. Krishnaswami *et al.* (1975) have found $A_{210_{Pb}}/A_{226_{Ra}}$ ratios of ~ 9.3 in water deeper than 250 m in the Santa Barbara Basin. This ratio yields a settling rate of ~ 0.1 m day^{-1} for the 600 m water column. Bruland *et al.* (1974) have estimated that the residence time of ^{210}Pb in the highly productive surface waters in areas of upwelling is less than a month, equivalent to a particle settling velocity of at least several metres per day. By inserting a value of 0.15 for the ratio $A_{210_{Pb}}/A_{226_{Ra}}$ into Equation (2) and assuming an 800 m water column, a residence time of 6 years and a scavenging rate of ~ 0.4 m day^{-1} may be calculated.

The 138-day half-life of polonium-210 (granddaughter of lead-210) makes this is a convenient nuclide for the assessment of particle settling velocities. On the basis of estimates of both the atmospheric input and the *in situ* production of this nuclide by lead-210 decay, together with actual measurements of its concentration in the mixed layer, Shannon *et al.* (1970) have estimated that its residence time is 0.6 years, which corresponds to a particle settling velocity of ~ 0.5 m day^{-1}. Calculations based on lead-210 data gave a residence time of ~ 5 years and a particle settling velocity of 0.05 m day^{-1}. The authors pointed out that biological removal is particularly important for polonium-210 which, relative to sea-water, is concentrated by marine organisms by factors of 10^3–10^4. The affinity of organisms for polonium was demonstrated by the fact that the $A_{210_{Pb}}/A_{210_{Po}}$ ratios for sea-water, phytoplankton and zooplankton were ~ 2.0, 0.3 and ~ 0.08 respectively. Subsequently, Cherry *et al.* (1975) reported that the removal of polonium-210 from surface water by a mechanism solely involving zooplankton faecal pellet formation and sinking, requires 0.9 years, suggesting a particle settling velocity of ~ 0.3 m day^{-1} for a 100-m thick mixed layer.

Particle settling velocities, determined in the studies described above are summarized in Table 37.6. Measured velocities range from ~ 0.1–10 m day^{-1}; these are equivalent to the sinking velocities of spheres with diameters of ~ 2–20 μm, sinking according to Stokes' law. The highest rate was found in

TABLE 37.6

Summary of particle settling velocity studies (using radiometric methods)

Settling velocity (m day)	Radionuclide(s) and/or method	Reference
1–3	Via TSM residence time (T) where $T =$ TSM water column load/sediment deposition rate	Feely et al. (1971)
1–2	^{144}Ce and ^{144}Pm penetration in the DS[a]	Bowen and Sugihara (1965)
0·6	Plutonium nuclides penetration (DS)	Miyake et al. (1970)
0·4	Plutonium nuclides penetration (DS)	Bowen et al. (1971)
0·5	Plutonium nuclides penetration (DS)	Noshkin and Bowen (1973)
1·0	^{55}Fe penetration (DS)	Labeyrie et al. (1976)
1·6 and 2·1 ($CaCO_3$ part.)	^{14}C penetration (DS)	Somayaluju et al. (1969)
0·2	From estimated residence time of ^{230}Th	Moore and Sackett (1964)
1·0	^{234}Th–^{238}U disequilibria in the ML[b]	Bhat et al. (1969)
10 (near-shore)	^{234}Th–^{238}U disequilibria (ML)	Aller and Cochran (1976)
0·4	^{228}Th–^{228}Ra disequilibria (ML)	Broecker et al. (1973)
0·14	^{210}Pb residence time (ML)	Rama et al. (1961)
0·4	^{210}Pb–^{226}Ra disequilibrium in 800 m water column	Bruland et al. (1974)
0·2	^{210}Pb–^{226}Ra disequilibria (DS)	Craig et al. (1973)
0·1	^{210}Po–^{226}Ra disequilibria in Santa Barbara Basin	Krishnaswami et al. (1975)
0·05–0·5	^{210}Po and ^{210}Pb residence times (ML)	Shannon et al. (1970)

[a] DS, deep sea.
[b] ML, mixed layer.

a near-shore location in which the high biological activity and the large grain size of the sediment would certainly enhance the settling velocities. Some of the open-ocean areas with intermediate rates of about 1–2 m day^{-1} may either be regions in which the surface waters have a relatively high biological productivity, or deeper waters in which particles produced at the surface have been concentrated. An instance of the latter phenomenon was observed by Somayajulu et al. (1969) who found biogenous carbonate particles with diameters ranging from < 20 to > 100 µm in diameter at a depth of several thousand metres. As would be expected, some of the lowest settling rates, i.e. ∼2·0 m day^{-1}, occur in deep-sea environments.

Recently, Krishnaswami et al. (1976a, b) have investigated the composition of large quantities (< 2·0 g) of TSM collected from 2000–40000 litre samples of surface and deep water in the Pacific. Collections were made by pumping water (an *in situ* pumping system was used for deep samples) through either cotton-fibre or polyester-fibre filters. It was observed that at three stations ^{239}Pu reached a maximum concentration at water depths of 700–1500 m. Assuming that these maximum concentrations reflect the maximum injection of ^{239}Pu into the environment (which occurred in 1963) a settling rate of ~ 0.3 m day^{-1} was calculated by the authors. From the particulate ^{230}Th distribution they deduced a particle settling velocity of ~ 2 m day^{-1}; however, they pointed out that the "^{230}Th-based settling velocities are mean values for the entire oceanic column" and are not necessarily inconsistent with lower rates deduced from fall-out nuclides in the upper 1000 m.

Bacon *et al.* (1976) have described the distribution of ^{210}Pb and ^{210}Po in the dissolved and the particulate phases at 12 stations in the Atlantic and Pacific Oceans. Both nuclides were found to be present mainly as dissolved species at concentrations less than those which would have been present if secular equilibrium with respect to ^{226}Ra had been attained. The absence of significant vertical gradients in the concentrations of particulate ^{210}Pb from all but one profile would ordinarily suggest rapid settling velocities of the order of several metres per day. Because this explanation seemed inconsistent with the results of other studies, they suggested the alternative explanation that much of the ^{210}Pb is being removed at the sediment–water interface at topographical boundaries such as continental margins or mid-ocean ridges. In support of this argument, they pointed out that there was a significant correlation between the ^{210}Pb/^{226}Ra activity ratio and the horizontal distance to the nearest bottom feature. In contrast, ^{210}Po concentration distributions showed a deficiency of ^{210}Po with respect to ^{210}Pb in solution and an excess of it in particulate form was adduced as evidence for rapid *in situ* scavenging.

These last three papers provide a good deal of information on the radio-nuclide components of TSM and lead to conclusions generally similar to those derived from previous studies, with the exception that they suggest that the earlier settling rates based on lead-210 may be too high because of failure to take into consideration the uptake of this nuclide at the sediment–water interface along mid-depth topographical highs in the ocean.

Acknowledgements

The preparation of this manuscript was made possible by financial assistance provided by ERDA Contract No. AT(40-1) 3852. Secretarial assistance was afforded by Ms Ginger Franks. The author also wishes to thank Drs P. Biscaye, R. Feely and D. Spencer for their suggestions and comments.

REFERENCES

Aller, R. C. and Cochran, J. K. (1976). *Earth Planet. Sci. Lett.* **29**, 37.

Amos, A. F., Gordon, A. L. and Schneider, E. D. (1971). *Deep-Sea Res.* **18**, 145.

Armstrong, F. A. J. (1958). *J. Mar. Res.* **17**, 23.

Armstrong, F. A. J. and Atkins, W. R. G. (1950). *J. Mar. Biol. Ass. U.K.* **29**, 139.

Atkins, W. R. G., Jenkins, P. G. and Warren, F. J. (1954). *J. Mar. Biol. Ass. U.K.* **33**, 497.

Bacon, M. P., Spencer, D. W. and Brewer, P. G. (1976). *Earth Planet. Sci. Lett.* **32**, 277.

Baker, E T., Sternberg, R. W. and McManus, D. A. (1974). *In* "Suspended Solids in Sea Water" (R. J. Gibbs, ed.), pp. 155–172. Plenum Press, New York and London.

Banse, K. L., Falls, C. P. and Hobson, L. A. (1963). *Deep-Sea Res.* **10**, 639.

Bassin, N. J. (1975). Ph.D. Dissertation, Texas A and M University.

Bassin, N. J., Harris, J. E. and Bouma, A. H. (1972). *Marine Geology*, **12**, 171.

Beardsley, G. F., Pak, H., Carder, K. and Lundgren, B. (1970). *J. Geophys. Res.* **75**, 2837.

Behairy, A. K., Chester, R., Griffiths, A. J., Johnson, L. R. and Stoner, J. H. (1975). *Marine Geology*, **18**, 45.

Betzer, P. R. and Pilson, M. E. Q. (1970a). *J. Mar. Res.* **28**, 251.

Betzer, P. R. and Pilson, M. E. Q. (1970b). *Deep-Sea Res.* **17**, 671.

Betzer, P. R. and Pilson, M. E. Q. (1971). *Deep-Sea Res.* **18**, 753.

Betzer, P. R., Carder, K. L. and Eggimann, D. W. (1974). *In* "Suspended Solids in Sea Water" (R. J. Gibbs, ed.), pp. 295–214. Plenum Press, New York and London.

Bhat, S. G., Krishnaswami, S., Lal, D., Rama and Moore, W. S. (1969). *Earth Planet. Sci. Lett.* **5**, 483.

Biscaye, P. E. and Eittreim, S. L. (1974). *In* "Suspended Solids in Sea Water" (R. J. Gibbs, ed.), pp. 227–260. Plenum Press, New York and London.

Biscaye, P. E. and Eittreim, S. L. (1977). *Marine Geology*, **23**, 155.

Bowen, V. T. and Sugihara, T. T. (1965). *J. Mar. Res.* **23**, 123.

Bowen, V. T., Wong, K. M. and Noshkin, V. E. (1971). *J. Mar. Res.* **29**, 1.

Brewer, P. G., Spencer, D. W., Biscaye, P. E., Hanley, A., Sachs, P. L., Smith, C. L., Kadar, S. and Fredericks, J. (1976). *Earth Planet. Sci. Lett.* **32**, 393.

Broecker, W. S. and Olson, E. A. (1959). *Am. J. Sci. Radiocarbon Suppl.* **1**, 111.

Broecker, W. S., Kaufman, A. and Trier, R. (1973). *Earth Planet. Sci. Lett.* **20**, 35.

Bruland, K. W., Koide, M. and Goldberg, E. D. (1974). *J. Geophys. Res.* **79**, 3083.

Brun-Cottan, J. C. (1971). *Cah. Océanogr.* **23**, 193.

Brun-Cottan, J. C. (1976). *J. Geophys. Res.* **81**, 1601.

Carder, K. L. and Schlemmer, F. C., II (1973). *J. Geophys. Res.* **78**, 6286.

Carder, K. L., Beardsley, G. F. and Pak, H. (1971). *J. Geophys. Res.* **76**, 5070.

Cherry, R. D., Fowler, S. W., Beasley, T. M. and Heyraud, M. (1975). *Mar. Chem.* **3**, 105.

Chesselet, R., Jedwab, J., Darcourt, C. and Dehairs, F. (1976). *Am. Geophys. Union Trans.* **57**, 255.

Chester, R. and Stoner, J. H. (1975). *Nature, Lond.* **255**, 50.

Craig, H., Krishnaswami, S. and Somayajulu, B. L. K. (1973). *Earth Planet. Sci. Lett.* **17**, 295.

Drake, D. E. (1974). *In* "Suspended Solids in Sea Water" (R. J. Gibbs, ed.), pp. 133–153. Plenum Press, New York and London

Eittreim, S. L. and Ewing, M. (1974). *In* "Suspended Solids in Sea Water" (R. J. Gibbs, ed.), pp. 213–226. Plenum Press, New York and London.

Eittreim, S. L., Ewing, M. and Thorndike, E. M. (1969). *Deep-Sea Res.* 16, 613.

El-Sayed, S. Z. (1968). *Am. Geog. Soc. Folio* 10, 1.

Ewing, M. and Thorndike, E. M. (1965). *Science, N.Y.* 147, 1291.

Ewing, M., Eittreim, S. L., Ewing, J. and LePichon, X. (1971). *Physical Chem. Earth*, 8, 49.

Feely, R. A. (1974). Ph.D. Dissertation, Texas A and M University.

Feely, R. A. (1975). *Mar. Chem.* 3, 121.

Feely, R. A., Sackett, W. M. and Harris, J. E. (1971). *J. Geophys. Res.* 76, 5893.

Feely, R. A., Sullivan, L. and Sackett, W. M. (1974). *In* "Suspended Solids in Sea Water" (R. J. Gibbs, ed.), pp. 281–294. Plenum Press, New York and London.

Fowler, S. W. and Small, L. F. (1972). *Limnol. Oceanogr.* 17, 293.

Fredericks, A. D. and Sackett, W. M. (1970). *J. Geophys. Res.* 75, 2199.

Fry, E. S. (1974). *In* "Suspended Solids in Sea Water" (R. J. Gibbs, ed.), pp. 101–109. Plenum Press, New York and London.

Gibbs, R. J., ed. (1974). *In* "Suspended Solids in Sea Water", pp. 203–210. Plenum Press, New York and London.

Goldberg, E. D. (1963). *In* "Radioactive Dating", p. 121. International Atomic Energy Agency, Vienna.

Gordon, H. R. (1974). *In* "Suspended Solids in Sea Water" (R. J. Gibbs, ed.), pp. 73–86. Plenum Press, New York and London.

Grim, R. E. and Johns, W. D. (1954). *Proc. 2nd Natl. Conf. Clay Mineralogy*, 2, 81–102.

Groot, J. J. and Ewing, M. (1963). *Science, N.Y.* 142, 579.

Harris, J. E. (1971). Ph.D. Dissertation, Texas A and M University.

Harris, J. E. (1972). *Deep-Sea Res.* 19, 719.

Harvey, H. W. (1960). "The Chemistry and Fertility of Sea Waters". Cambridge University Press, England, 240 pp.

Holm-Hansen, O. (1970). *In* "Organic Matter in Natural Waters" (D. W. Hood, ed.), pp. 287–300. *Occas. Pub. No.* 1, Inst. Mar. Sci., University of Alaska.

Ichiye, T., Bassin, N. J. and Harris, J. E. (1972). *J. Geophys. Res.* 77, 6576.

Jacobs, M. B. and Ewing, M. (1965). *Science, N.Y.* 149, 179.

Jacobs, M. B. and Ewing, M. (1969a). *Science, N.Y.* 163, 380.

Jacobs, M. B. and Ewing, M. (1969b). *Science, N.Y.* 163, 805.

Jerlov, N. G. (1953). *Rep. Swed. Deep Sea Exped.* 3, 73.

Kalle, K. (1939). *Annln Hydrogr. Berl.* 67, 23–30.

Klenova, M. V., Lavrov, V. M. and Nikolaeva, V. K. (1962). *Soviet Oceanogr.* 3, 18.

Krishnaswami, S., Somayajulu, B. L. K. and Chung, Y. (1975). *Earth Planet. Sci. Lett.* 27, 388.

Krishnaswami, S., Lal, D., Somayajulu, B. L. K., Weiss, R. F. and Craig, H. (1976a). *Earth Planet. Sci. Lett.* 32, 420.

Krishnaswami, S., Lal, D., Somayajulu, B. L. K. and Craig, H. (1976b). *Earth Planet. Sci. Lett.* 32, 403.

Kullenberg, G. (1974). *In* "Suspended Solids in Sea Water" (R. J. Gibbs, ed.), pp. 195–202. Plenum Press, New York and London.

Labeyrie, L. D., Livingston, H. D. and Bowen, V. T. (1976). "Proceedings of the Symposium on Transuranium Nuclides", International Atomic Agency Proceedings, Series ST1/Publ. 410.

Leipper, D. F. (1970). *J. Geophys. Res.* 75, 637.

Lerman, A., Lal, D. and Decey, M. F. (1974). *In* "Suspended Solids in Sea Water" (R. J. Gibbs, ed.), pp. 17–47. Plenum Press, New York and London.

Lisitzin, A. P. (1959). International Oceanographic Congress Preprints, 470–471. American Association for the Advancement of Science, Washington D.C.

Lisitzin, A. P. (1960). *Oceanogr. Res.* **2**, 71.

Lisitzin, A. P. (1972). "Sedimentation in the World Ocean", *Soc. Econ. Paleo. Min. Spec. Pub.* **17**, 218 pp.

Manheim, F. T., Meade, R. H. and Bond, G. C. (1970). *Science, N.Y.* **167**, 371.

Matsumoto, E. (1975). *Geochim. Cosmochim. Acta,* **39**, 205.

Meade, R., Sachs, P. L., Manheim, F. T., Hathaway, J. C. and Spencer, D. W. (1975). *J. Sed. Petrol.* **45**, 171.

Menzel, D. W. (1967). *Deep-Sea Res.* **14**, 229.

Menzel, D. W. and Ryther, J. H. (1964). *Limnol. Oceanogr.* **9**, 179.

Menzel, D. and Ryther, J. (1968). *Deep-Sea Res.* **15**, 327.

Menzel, D. W. and Vaccaro, R. F. (1964). *Limnol. Oceanogr.* **9**, 138.

Mie, G. (1908). *Ann. Phys.* **25**, 377.

Miyake, Y., Katsuragi, Y. and Sugimura Y. (1970). *J. Geophys. Res.* **75**, 2329.

Moore, W. S. (1969). *Earth Planet. Sci. Lett.* **6**, 437.

Moore, W. S. and Sackett, W. M. (1964). *J. Geophys. Res.* **69**, 5401.

Noshkin, V. E. and Bowen, V. T. (1973). *In* "Radioactive Contamination of the Marine Environment", pp. 671–686. International Atomic Energy Agency, Vienna.

Osterberg, C., Carey, A. G. and Curl, H. (1963). *Nature, Lond.* **200**, 1276.

Pak, H. (1974). *In* "Suspended Solids in Sea Water" (R. J. Gibbs, ed.), pp. 261–270. Plenum Press, New York and London.

Pak, H., Beardsley, G. F. and Plank, W. (1970). Abstracts of Contributed Papers, the Ocean World Symposium, Tokyo, Japan.

Pettersson, H. (1934a). *Meddn Göteborgs Högsk. Oceanogr. Inst.* No. 7.

Pettersson, H. (1934b). *Meddn Göteborgs Högsk. Oceanogr. Inst.* No. 9.

Petzold, T. and Austin, R. (1968). *Scripps Inst. Oceanogr. Tech. Rep.* **68–9**, 5 pp.

Rama, Koide, M. and Goldberg, E. D. (1961). *Science, N.Y.* **134**, 98.

Riley, G. A. (1970). *In* "Advances in Marine Biology" (R. S. Russell and M. Young, eds), pp. 1–118. Academic Press, London and New York.

Sackett, W. M. and Arrhenius, G. (1962). *Geochim. Cosmochim. Acta,* **26**, 955.

Shannon, L. V., Cherry, R. D. and Orren, M. J. (1970). *Geochim. Cosmochim. Acta,* **34**, 701.

Sheldon, R. W. and Parsons, T. R. (1967). "A Practical Manual on the Use of the Coulter Counter in Marine Science". Coulter Electronic Co., Toronto, Canada, 66 pp.

Sheldon, R. W., Prakash, A. and Sutcliffe, W. H. (1972). *Limnol. Oceanogr.* **17**, 327.

Slowey, J. F., Hayes, D., Dixon, B. and Hood, D. W. (1965). *In* "Symposium on Marine Geochemistry". *Occas. Pub. No.* 3, University of Rhode Island.

Smayda, T. J. (1969). *Limnol. Oceanogr.* **14**, 621.

Smayda, T. J. (1971). *Marine Geology,* **11**, 105.

Smith, R. C., Austin, R. W. and Petzold, T. J. (1974). *In* "Suspended Solids in Sea Water" (R. J. Gibbs, ed.), pp. 61–72. Plenum Press, New York and London.

Somayajulu, B. L. K., Lal, D. and Kusumgar, S. (1969). *Science, N.Y.* **166**, 1397.

Spencer, D. W., Brewer, P. G., Biscaye, P. E., Sachs, P. L., Smith, C. L., Kadar, S. and Fredericks, J. (1978). *Earth Planet. Sci. Lett.* (in press).

Sternberg, R. W., Baker, E. T., McManus, D. A., Smith, S. and Morrison, R. A. (1974). *Deep-Sea Res.* **21**, 887.

Sturges, W. (1965). *J. Mar. Res.* **23**, 147.

Thorndike, E. M. (1975). *Ocean Engng,* **3**, 1.

Thorndike, E. M. and Ewing, M. (1967). *In* "Deep-Sea Photography", pp. 113–116. Johns Hopkins, Baltimore, U.S.A.

Turner, J. T. (1977). *Mar. Biol.* **40**, 249.

Tyler, J. E., Austin, R. W. and Petzold, T. J. (1974). *In* "Suspended Solids in Sea Water" (R. J. Gibbs, ed.), pp 51–59. Plenum Press, New York and London.

Wüst, G. (1963). *Deep-Sea Res.* **10**, 165.

Yentsch, C. S. (1962) *In* "Physiology and Biochemistry of Algae" (R. A. Lewis ed.), pp. 771–797. Academic Press, New York and London.

Yoe, J. H. (1928). *In* "Photometric Chemical Analysis". John Wiley, New York.

Zaneveld, J. R. V. (1974). *In* "Suspended Solids in Water" (R. J. Gibbs, ed.), pp. 87–100. Plenum Press, New York and London.

Zeitzschel, B. (1970). *Mar. Biol.* **7**, 305.

Chapter 38

Aerosol Chemistry of the Marine Atmosphere

WALTER W. BERG, JR.
National Center for Atmospheric Research, Boulder, Colorado, U.S.A.

and

JOHN W. WINCHESTER
Department of Oceanography, Florida State University, Tallahassee, Florida, U.S.A.

38.1. INTRODUCTION

This review covers information available up to the end of 1976 and as it was being prepared, there was a growing interest about the chemical composition of the atmosphere over the ocean and how it may be changing as a result of anthropogenic emissions into the air. In recognition of this interest, large-scale atmospheric chemical measurement programmes are being planned on both national and international levels. In several countries the programmes focus on the measurement of atmospheric trace substances in remote locations, especially oceanic sites far from land, in order to assess the extent of long-range transport of pollutants. The international Scientific Committee on Oceanic Research (SCOR), in co-operation with the U.S.

National Research Council, conducted a workshop in December 1975 in Miami, Florida, on "Tropospheric Transport of Pollutants to the Ocean" (TTPO, 1975). Subsequently, the World Meteorological Organization, in co-operation with the United Nations Environment Programme, conducted a workshop in October 1976 in Gothenburg, Sweden, on "Atmospheric Pollution Measurement Techniques" (APOMET, 1976). This chapter will review the status of our knowledge concerning the chemical composition of the marine atmosphere on which the measurement programmes may be based.

38.1.1. THE DETECTION OF POLLUTION OVER THE OCEAN

There are several reasons why improved compositional data are needed for atmospheric aerosols over the ocean. Some of these will be discussed below.

(i) There is need for a clearer understanding of basic atmospheric chemical processes in remote areas. At present, only limited data on concentrations and particle size distributions, relative to temporal and spatial variations, are available. Without a greatly increased programme for such measurement on a world-wide basis it is impossible to predict, even approximately, the consequences of man's modifications of the atmosphere by inputs of particles and gases. Nor, without further information, is it possible to assess the effects of the resulting impact on climate, on the composition of the ocean and ultimately on human health and welfare.

(ii) There is a distinct possibility that the concentration of atmospheric particulate matter is increasing globally. Changes in land-use practices, fossil fuel combustion, metallurgical operations and other technological activities may lead to a modification in the transfer of radiant energy by the scattering of solar radiation by particulate matter. In addition, the alteration of natural nucleation processes may affect the micro-structure of clouds, thereby changing the Earth's albedo and precipitation patterns. Of greatest importance is the systematic measurement of the major aerosol constituents which may regulate these processes. Particles in the aerodynamic radius range 0·5–1 μm are most important in scattering visible light; water-soluble ions in aerosols are most effective in upsetting natural nucleation processes. Therefore, measurements of components, especially sulphate, nitrate, chloride, ammonium, sodium, potassium, calcium and magnesium, and organic and inorganic forms of carbon (in the critical particle size range), may lead to an improved assessment of the climatic impact of large-scale air pollution.

(iii) There is a need for a better indication of the extent to which certain potentially biologically hazardous substances are rising above their natural atmospheric concentrations by transport from pollution sources. Attention has been focused on sulphuric acid, heavy metals and certain other trace elements because of their known effects at higher concentrations, either because acute exposure to them causes symptons of poisoning or because they are essential to nutrition. It is necessary, therefore, to be able to detect these substances at low levels, and also to be able to distinguish between the anthropogenic and natural components in the atmosphere, in order to evaluate the potential biological consequences of low-level chronic exposure to them on man and the biosphere.

Interest in the transport of anthropogenic trace substances from terrestrial sources to points of deposition on the sea surface has stimulated the investigation of aerosol composition over the ocean, especially for trace metals and organic materials. However, differentiation between natural and pollutant components is not straightforward and cannot, in general, be accomplished by measurements at one location and one time. Instead, a dozen or more alternative strategies may be adopted, many of which are listed below.

(a) Continuous measurements over extended periods of time in order to detect secular increases in concentration caused by pollution processes.

(b) Comparison of atmospheric concentrations in polluted regions with those in clean regions which are otherwise geographically similar.

(c) Comparison of concentrations in air masses having different flow trajectories, e.g. over clean and polluted areas or over land and sea.

(d) Comparison of particle size distributions of trace elements from natural and pollution sources.

(e) Comparison of aerosol composition with the summation of the expected natural and pollution source compositions, and the identification of reference elements for large natural sources.

(f) Use of global fluxes and well-characterized aerosol residence times to compare natural and pollution source strengths.

(g) Use of the historical record of aerosol deposition in the Greenland and Antarctic ice-caps and in marine sediments.

(h) Measurement of the stable isotope ratios of some elements, e.g. S, C, N, O and Pb , and the radioactivity of others, e.g. Rn and its daughter products.

(i) Comparison of trace element compositional relationships in aerosols with trace gas and organic compound concentrations in the same air.

(j) Comparison of concentrations of the same aerosol constituents in the Northern and Southern Hemispheres.

(k) Analysis of individual particles as well as of bulk particulate matter.

(l) Vertical profiling of atmospheric composition, e.g. measured from aircraft.

In regions in which the atmosphere is expected to be clean, such as over the ocean far from land, measurement techniques should be improved considerably over those required for studies of, for example, urban atmospheres. The techniques for sampling require the same degree of perfection as those for the subsequent chemical analysis, especially for atmospheric constitutents present at very low concentration levels. Contamination of samples from clean air locations before they reach the analytical laboratory is a possibility which must be scrupulously avoided.

For some substances, crustal reference elements for example, collection by filters will give useful information about total atmospheric concentrations. However, for other substances, such as sulphate or heavy metals derived from industrial pollution, it is necessary to determine their concentrations in certain particle size fractions. For this information suitable particle size classifying samplers are required. Sampling procedures, therefore should be selected according to the substance to be measured and the analytical technique to be used. Moreover, the procedures should be selected according to the requirements for remote background locations, which may be quite different from those for more polluted areas.

Sampler calibration should include not only the precise determination of the air flow rate, but also an assessment of the efficiency with which it collects particles from the air. The upper particle size limit for particles entering the sampler should be carefully defined. Some elements occur mainly in large particles and a large uncertainty is introduced by an imprecise definition of the collection efficiency for such particles. In some samplers, the upper size limit may depend on the wind speed, and this effect should be minimized.

Aerosol collection for the determination of certain elements is complicated by the fact that some of them occur to some extent in the gas phase. Moreover, some particles are affected by relative humidity, both with respect to their effective particle radius in the atmosphere and the degree to which they are retained by collection surfaces. These factors cause special sampling problems which must be considered in the design of an effective measurement programme.

Special consideration must be given to the siting of sampling equipment in the field and to the effect of local wind direction, wind speed and nuclei count in order to minimize local interferences with the objective of obtaining representative regional concentration measurements from a sampling station.

Because special requirements are necessary for the measurement of chemical constitutents of the atmosphere over the ocean, most laboratories have thus far only been able to mount investigations on a limited scale. It is anticipated that new co-operative studies of national or international scope will facilitate more comprehensive programmes.

38.1.2. THE OCEAN AS A SOURCE OF PARTICULATE MATTER

In evaluating the transport of particulate matter from source areas (on land or in the atmosphere) to the sea surface, predictions from transport models should be tested by measurements of particulate matter over the ocean. However, the aerosol above the water is a mixture of various components; some derived from the sea itself and some which have descended from aloft. The simple procedure of predicting the natural aerosol composition by a linear combination of the compositions of sea-water and material from terrestrial sources is not currently possible because source-to-atmosphere chemical fractionations, which may be substantial, are largely unknown. Moreover, the atmosphere itself may act as a source of some aerosol constituents by gas-to-particle conversion. It is also possible that the sea-spray aerosol can interact with other aerosol constituents so as to affect atmospheric chemical reactions and air-to-sea transfer rates. Consequently, a test of aerosol transport models by chemical measurements in the atmospheric boundary layer is by no means straightforward.

The importance of knowing the extent of fractionation effects at the sea–air interface, as well as the disconcerting lack of basic measurements which can lead to an unambiguous determination of the magnitude of these effects, have been emphasized in several recent reviews of sea surface chemistry. The following six references should be consulted concerning research published up to about November 1975: Working Symposium on Sea–Air Chemistry, Fort Lauderdale (WSSAC) (1972), International Symposium on the Chemistry of Sea–Air Particulate Exchange Processes, Nice (ISCSAPEP) (1974), MacIntyre (1974), Liss (1975), Winchester and Duce (1977) and Duce and Hoffman (1976).

A field measurement approach to establishing the composition of the natural marine aerosol is difficult because of interference from aerosol particles transported from land. However, the evaluation of any pollutant contribution to the marine aerosol requires a comparison of the observed concentration with the natural background concentration, which itself is the sum of sea surface and terrestrial components and may be enriched in one or more constituents by natural processes. Friedlander (1973) and later Gatz (1975) have developed a general mathematical formulation to account for aerosol composition in terms of the contributing sources, including fractionation factors when these are known. However, except for recent results obtained by Kowalczyk et al. (1978), discouragingly little success has been achieved for urban areas in fitting measured aerosol composition to the sum of inputs from several obvious sources and calculating their percentage contributions. Both the fact that fractionation factors have not been determined, and that aerosol measurement data are inadequate, lead to this frustration. The approach used by Kowalczyk et al. has apparently not yet been extended to the marine atmosphere far from land.

A laboratory measurement approach is difficult because all the relevant variables may not be controlled. The most serious of these is the possibility that the walls of the containing vessel for a model ocean in the laboratory may affect the extent of chemical fractionation when water is being transferred to the atmosphere as spray droplets. This difficulty may be partly overcome by generating spray droplets in the real ocean in a semi-controlled manner, using devices such as those which have been tested by Fasching et al. (1974), Morelli et al. (1974) and Berg and Winchester (1976).

An ideal case for field measurements, where a minimum of terrestrial source interference may be expected, is the determination of sea surface fractionation of chlorine, bromine and iodine. Duce et al. (1967) have documented the contents of these elements in marine aerosols and have demonstrated that iodine is 100–1000 times enriched in the atmosphere compared to sea-water, and that bromine may be slightly depleted to a degree which depends on particle size. For both elements the fractionation is relative to chlorine which is too often assumed to be relatively chemically unreactive over the sea surface. A detailed review of gas–particle relationships in the halogen elements is presented in Section 38.3.

Certain trace substances in sea-water have now been shown to be significantly enriched in aerosol droplets generated at the sea surface. Baylor et al. (1962) were apparently the first to observe that phosphate in aerosol droplets over the ocean may be strongly enriched compared to sea-water, and they established that this enrichment involved processes at the sea–air interface, rather than an influx of phosphate from land or some other source. MacIntyre and Winchester (1969) reproduced the qualitative features of the enrichment process in laboratory experiments which modelled some, though not all, of the relevant features of the real ocean. Recently, Van Grieken et al. (1974) have shown that radioactive Zn and Se tracers added to natural sea-water may be enriched in the aerosol, although not by the large factors found for phosphate. Similar experiments using trace metals in their natural form in sea-water (in contrast to the use of added tracers) appear not to have been made.

The sea surface may directly transfer materials concentrated at the air–water interface into the atmosphere as components of spray droplets; this is clearly demonstrated by measurements of the organic matter contents of marine aerosols. The work of Blanchard (1968) has shown that total organic matter in aerosols may comprise several per cent of the mass, as opposed to parts per million, or less, in bulk sea-water. Moreover, Blanchard and Syzdek (1972) have reported that living bacteria are found to be enriched in the droplets, apparently by a mechanism involving the scavenging of the water column by rising bubbles as well as the skimming of the surface during bubble bursting. Organic substances present in the marine aerosol are reviewed in Section 38.2.2. Specific inorganic substances may be carried with the organic material, but this problem has not been studied comprehensively.

For most constituents, measurement of the chemical composition of aerosols over the sea does not lead directly to the composition of the sea-spray component because of the possibility of admixture of material from the overlying air. However, by studying compositional variations within a suite of aerosol samples, collected as a function of particle size at an altitude of 40 m over Bermuda, Meinert and Winchester (1977) have shown that the contribution from the major sea-salt elements can be resolved, thus permitting a determination of any fractionation which may exist.

A finding of considerable significance in tropospheric pollution transport studies is that several trace metals and semi-metals occur at much higher concentrations in atmospheric particulate matter than can be accounted for by the unfractionated dispersion of soil or sea-water. This was apparently first recognized by Rahn (1971), whose subsequent results (Rahn, 1976a) indicate that these "anomalous atmospheric enrichments" are world-wide and may be due to a natural fractionation effect on land or at sea. Since many of these anomalous elements are of interest from the point of view of pollution, the extent and cause of the enrichments should be determined in order to evaluate the magnitude of the actual pollution contribution to the presumably natural background. Most studies have focused on either strictly continental or maritime sampling locations, and distinguishing between land or sea processes for the anomalies has therefore been difficult. However, Johansson et al. (1974) studied a coastal location in which sample sub-sets from continental and marine air flow conditions could be identified. By assuming Fe to be of terrestrial origin, their results showed for example, that the two anomalous elements Cu and Zn also appear to be largely of terrestrial, rather than marine, origin.

The atmosphere immediately over the ocean is a mixing zone as it is the zone of entry of pollutants into the sea, the zone of transfer of sea-water constituents to the atmosphere and the zone in which important gas–particle interactions may take place. Measurement of the contributions of oceanic and terrestrial sources to its overall chemical composition is not easily accomplished. Inference of its chemical reaction mechanisms through chemical sampling and analysis is also a complex task. The present status of each of these problems will be reviewed in the sections which follow.

38.2. Aerosol Composition Over the Ocean

38.2.1. ELEMENTAL CONSTITUENTS

38.2.1.1. Sea-salt

For many years considerable discussion has centred on the chemical composition of the marine aerosol relative to that of bulk sea-water. At locations over the World Ocean where terrestrial influences are minimal, the major aerosol elements exist in a particulate salt matrix which encompasses a spectrum of sea-salt particle sizes ranging from $<0.1\,\mu m$ to $>100\,\mu m$ in radius. Woodcock (1953, 1972) has shown that sea-salt particles exhibit a distinct maximum in number distribution below a $1\,\mu m$ radius. Figure 38.1 (Junge, 1963) indicates that both the number of particles and their upper size limit increase with increasing wind speed, probably as a result of increased wave action and bubble formation. In general, more than 90 % of the marine aerosol mass is found in particles of greater than $1\,\mu m$ radius (Junge, 1972).

The relative proportions of the major elements in the marine aerosol have often been found to deviate rather substantially from the virtually constant relative proportions of the major ions in sea-water, shown in Table 38.1 (Riley, 1975). In the literature, different investigators have reported element ratios, e.g. those of Na/Cl, Mg/Cl, S/Cl and K/Na, in marine aerosol and precipitation which agree in some cases and disagree in others with the sea-water ratios. A number of explanations have been advanced to account for these discrepancies (see e.g. Rossby and Egner, 1955; Junge and Werby, 1958; Gast and Thompson, 1959; Eriksson, 1959; Blanchard, 1968; Bloch and Luecke, 1968, 1970; MacIntyre and Winchester, 1969; Lazrus et al., 1970; Wilkniss and Bressan, 1971; Chesselet et al., 1972a, b; Junge, 1972; MacIntyre, 1974). These include: (i) intrusions of non-marine aerosols of continental or anthropogenic origins; (ii) ion fractionation during the formation of aerosol droplets at the sea–air interface; (iii) alteration of the ionic composition of the aerosol due to organic surface films and the selective volatilization of certain species from the sea surface; (iv) selective rain-out of particles of a certain size or composition; (v) particle-to-gas conversions in the atmosphere; (vi) adsorption of vapours, such as that of iodine, onto particles, and (vii) experimental error in the chemical analyses or the contamination of samples, especially if low concentrations are being determined. In spite of these explanations, many of the variations in chemical composition of the marine aerosol have not been adequately accounted for.

The bursting of bubbles, produced by the trapping of air in surface water by breaking waves, is believed to be the chief mechanism for the formation of particulate matter in the marine atmosphere. Direct generation of sea-spray may also be significant (Lai and Shemdin, 1974). In this section the production mechanisms and the particle size distribution of the marine aerosol will be reviewed; the subject of sea surface fractionation processes will not be considered as it has already been reviewed in the six basic references listed in Section 38.1.2.

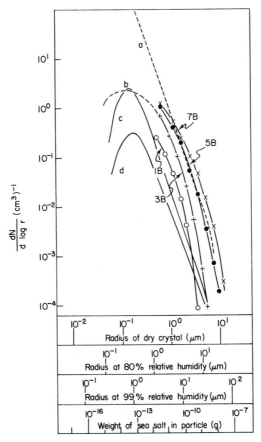

FIG. 38.1. Size distribution of sea-salt particles expressed as the number of particles per log radius interval, $dN/d\log r$. Curves 1B, 3B, 5B and 7B represent average data by Woodcock (1953) for wind forces 1, 3, 5 and 7 on the Beaufort scale. Curve a represents a typical continental distribution for comparison. Curve b is an extrapolation of Curve 3B on the basis of chloride analyses. Curves c and d are individual distributions obtained by Metnieks (1958) for Dublin and the west coast of Ireland. (After Junge, 1963).

Aitken (1881) was apparently the first to suggest that a large number of atmospheric particles might be produced by the oceans. Work by Köhler (1922, 1925) on the drop size and chlorinity of fog and cloud particles led him to conclude that most of the nuclei of the drops were formed from the ocean.

TABLE 38.1

Dissolved components of sea-water of salinity 35‰[a]

Component	Concentration $(g\,kg^{-1})$	Weight ratios X/Na	X/Cl
Major			
Na^+	10·773	1·0000	0·5569
Mg^{2+}	1·294	0·1201	0·0669
Ca^{2+}	0·412	0·0382	0·0213
K^+	0·399	0·0370	0·0206
Sr^{2+}	0·0081	0·00075	0·00042
B	0·004	0·0004	0·0002
Cl^-	19·344	1·7956	1·0000
SO_4^{2-}	2·712	0·2517	0·1402
Br^-	0·067	0·0062	0·0035
F^-	0·0013	0·00012	0·00007
HCO_3^-	0·142	0·0132	0·0073
Minor			
I	0·000064	6×10^{-6}	3×10^{-6}
Si	0·0029	0·00027	0·00015
Li	0·00017	$1·6 \times 10^{-5}$	$0·9 \times 10^{-5}$
Rb	0·00012	$1·1 \times 10^{-5}$	$0·6 \times 10^{-5}$
Ba	0·000021	2×10^{-6}	1×10^{-6}
C (org.)	0·00050	$4·6 \times 10^{-5}$	$2·6 \times 10^{-5}$
N	0·00067	$6·2 \times 10^{-5}$	$3·5 \times 10^{-5}$
P	0·000088	8×10^{-6}	5×10^{-6}

[a] Data from Riley (1975) and Turekian (1969).

Houghton (1932) suggested that the smaller drops found in coastal fogs had formed around sea-salt nuclei.

The first actual mechanistic approach to the problem was put forward by Jacobs (1937) when he suggested that bubbles breaking at the surface might be the source of sea-salt nuclei. He envisaged that the collapse of a high-velocity, upward-moving water jet would lead to the formation of many small droplets. Both Owens (1940) and Köhler (1941) reinforced Jacobs' work and provided evidence that droplets $< 10\,\mu m$ in radius could be produced by the ocean. Woodcock (1948), in studies on the red tide off the west coast of Florida, also backed up Jacobs' work by suggesting that bursting bubbles at the air–sea interface were the source of marine aerosols. Boyce (1951) produced interesting evidence which suggested that bubble and foam patches resulting from breaking waves were the most significant source of marine particles. The bubble-jet phenomenon, associated with a bubble bursting at the air–sea interface, was finally put on a solid, unerring data foundation when Woodcock *et al.* (1953) and Kientzler *et al.* (1954) produced high-speed camera photographs which showed that a wide size-range of sea-salt nuclei could be produced by sea-water droplets from the jets of bubbles ranging from 40 μm to 900 μm in diameter. As the bubbles reach the sea-water surface they collapse and rapidly produce high-speed jets of water which move upwards from the bottom of the bubble cavities. These jets rise rapidly, become unstable, and each fragment into two to five droplets, termed "jet drops". The jet drop diameters have been found to be approximately 10% of the diameters of the bubbles from which they were formed (Blanchard, 1963). In addition to sea-water, the jet drops formed from bubbles with diameters of 100–1000 μm also contain material skimmed off the top 0·05–0·5 μm of the water surface (MacIntyre, 1974). Smaller droplets (of the order of < 10 μm), termed "film drops", are also produced as a result of the bursting of the thin film of water on the top of the bubble which separates the air in the bubble from the atmosphere (Mason, 1957; Blanchard, 1963). Blanchard (1963) and Day (1964) have indicated that bubbles of < 300 μm bursting in sea-water do not appear to produce any film drops, whereas a 2000 μm bubble will produce a maximum of about 100. The bubble film cap thickness is of the order of 2 μm or more, hence the jet drop appears to be produced from a surface layer one to two orders of magnitude thinner than that which produces the film drops. The microtome effect (MacIntyre, 1968) is therefore different for the two classes of drops. A typical 500 μm diameter bubble bursting at the air–sea interface is shown diagrammatically in Fig. 38.2.

The major production of bubbles in the oceans is due to the formation of breaking waves or whitecaps. Blanchard and Woodcock (1957) have shown that in the vicinity of a breaking wave about 300000 bubbles m^{-2} s^{-1} break at the surface of the bubble-laden water. The resulting bubble spectrum is weighted in favour of the small bubbles, most of which are < 500 μm in diameter. Precipitation, as either rain or snowflakes, also produces a large number of bubbles, but quantitatively much less than do waves. The majority of precipitation-produced bubbles are < 100 μm in diameter (Blanchard and Woodcock, 1957).

Recent work by Lai and Shemdin (1974) suggests that in addition to bubble bursting there is another mechanism involved in aerosol production. They have postulated that sea-spray produced by the direct shearing of droplets from wave crests, by the bursting of bubbles produced by breaking waves and by the aerodynamic suction of droplets from the crests of capillary waves, may play a significant role in the mass transfer at the ocean surface. They are of the opinion that the production of droplets by bursting bubbles (Blanchard, 1963) does not account for the total production of marine aerosols. However, when considered together, bubble production and spray generation at the sea surface appear to account for the injection of particulate matter into the undisturbed marine atmosphere.

Those droplets with a diameter $<10\,\mu m$ form an aerosol when injected into the atmosphere and remain airborne for periods estimated to range from five to ten days (Martell and Moore, 1974; Tsunogai and Fukuda, 1974; Ketseridis *et al.*, 1976). Woodcock (1953, 1972) found that the number of

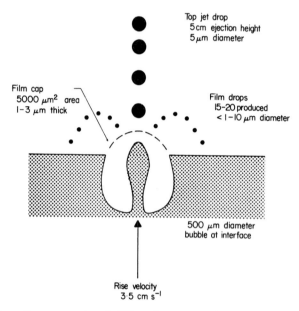

FIG. 38.2. Schematic representation of a 500 μm diameter bubble bursting at the air–sea interface, together with the atmospheric particles it produces. (After Winchester and Duce, 1977.)

particles and their mass median diameter increased with increasing wind speed and with decreasing altitude above the ocean (see Fig. 38.1). He observed sea-salt particles as small as 10^{-15} g, with radii of 0.1–$0.2\,\mu m$, at 90% relative humidity. At least 90% or more of the sea-salt aerosol mass is found in the giant particle size range (radius 1–$10\,\mu m$) (Woodcock, 1953; Duce *et al.*, 1967; Junge, 1972; Chesselet *et al.*, 1972a; Wilkniss and Bressan, 1972; Martens *et al.*, 1973). Less than 10% is carried by film particles in the particle size range, radius 0.1–$1\mu m$. Sea-state and atmospheric variables such as wind speed, relative humidity and ultra-violet light intensity appear to have some effect on the flux of several major elements, such as chlorine, across the air–sea interface, but do not appear to change the relative particle size distributions.

In the marine atmosphere, concentration levels of total particulate matter have been measured up to and exceeding $100 \, \mu g \, m^{-3}$, but generally concentrations are much lower, depending on such variables as wind speed, relative humidity and sea-state. Marine aerosol major element concentrations are generally of the order of $1 \, \mu g \, m^{-3}$. For example, chlorine, the principal element in the marine aerosol, is often in the range $1-10 \, \mu g \, m^{-3}$ near sea-level, for fair weather wind velocities ranging up to $10 \, m \, s^{-1}$. As the aerosol moves from the marine environment to continental areas, it has been well established that this concentration decreases rapidly (Rahn, 1971; Duce et al., 1973; Tsunogai, 1974; Okita et al., 1974; Hsu and Whelan, 1976). As the aerosol moves to increasing altitude above sea-level, the salt content has also been found to decrease (Durbin and White, 1961; Toba, 1965, 1966; Delany et al., 1973; Delany et al., 1974). Delany et al. (1973) showed, in fact, that the marine component of the tropospheric aerosol does not extend much beyond $2 \, km$ in altitude. Toba (1965) found that in the tropics the sea-salt concentration at $3 \, km$ over the ocean was only 1% of that near sea-level. In general, he found that the salt content decreased exponentially with altitude. Junge (1972) reasoned that this decrease is probably attributable to wash-out of the these hygroscopic particles in convective clouds and low-pressure storm situations. Data presented by Stensland and de Pena (1975) appear to support this conclusion. Their results show that under the best scavenging conditions only 20% of the initial aerosol mass remained in the sub-cloud layer after a $4 \, mm$ rainfall. The major chemical reactions in the marine atmosphere involving marine aerosols probably occur relatively close to the ocean surface.

The relative elemental composition of sea-salt may be changed during the transfer from bulk sea-water to the atmosphere. For example, measurements of the halogens in the marine atmosphere of Hawaii show a Br/Cl weight ratio which is approximately equal to the ratio in sea-water (0·00346), although there are measurable variations with particle size. For example, Fig. 38.3 (Moyers and Duce, 1972a) shows that there is a 50% deficit of bromine relative to chlorine in sub-micrometre particles (impactor stages D and E), but a relative increase toward the sea-water ratio at stage F. The weight ratio I/Cl, shown in Fig. 38.4 (Moyers and Duce, 1972b), is always far in excess of the sea-water ratio (3×10^{-6}), especially in the smallest particle sizes in which iodine, relative to chlorine, is enriched 1000-fold over the sea-water concentration. The cause of the iodine enrichment in marine aerosol droplets has been much discussed (see e.g. Komabayasi, 1962; Duce et al., 1963, 1965, 1967), and is believed to be the result of a combination of initial skimming of surface active organic matter, which may contain iodine, into the droplets during bubble bursting, and of absorption by the droplets of iodine vapours emanating from the sea surface. The bromine deficit relative to chlorine is small, but apparently real, and was used by Robbins (1970) as the basis of a mathematical model for gaseous evolution. Now that there is a greater appreciation of the reactivity of chlorine in marine aerosols (see Section 38.3), there is an urgent need for a re-evaluation of this model.

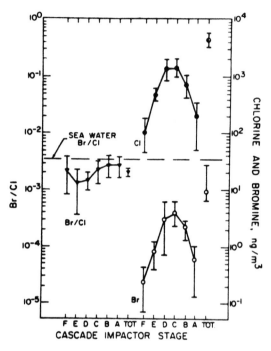

FIG. 38.3. Average size spectra for chlorine concentration, bromine concentration and Br/Cl ratio, using an Andersen impactor with minimum stage E cut-off diameter 0·5 μm (Moyers and Duce, 1972a).

38.2.1.2. Tropospheric aerosol

It has long been known that the atmosphere over the ocean, even far from land, contains dust particles of terrestrial origin. For example, Sahara dust transported across the Atlantic Ocean has been detected in samples from Florida (Junge, 1957) and Barbados (Delany et al., 1967) and has served as an elegant tracer of aerosol transport over long distances (Prospero and Carlson, 1972). Moreover, the phenomenon of aerosol transport from continental areas to the ocean raises the possibility that there is rapid transport of pollutants to the deep sea (Winchester and Desaedeleer, 1977), and prompts the need for new strategies for marine pollution monitoring and control (Goldberg, 1976). However, it also complicates the problem of rigorously resolving the contribution of trace substances from the sea surface to the overlying aerosol from measurements of their concentrations in aerosols themselves over the ocean. Unless this can be achieved, the mere presence of a pollutant over the ocean does not necessarily indicate its transport from land into the sea.

The most recent efforts to measure marine aerosol composition have been made at points far from land. In one such study (Meinert and Winchester, 1977) a series of aerosol samples, collected as a function of particle size at Bermuda during weather conditions ranging from stormy to calm, was found to contain major sea-salt constituents which varied roughly monotonically with the observed weather conditions. By means of a two-component model,

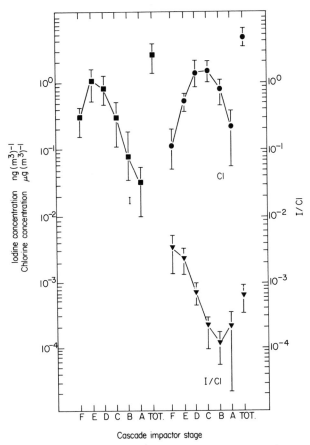

Fig. 38.4. Average particle size spectra of particulate chlorine concentration, particulate iodine concentration and particulate I/Cl ratio, using an Andersen impactor (Moyers and Duce, 1972b).

associations between S, Cl, K, Ca, Fe and Zn were established which permitted the tropospheric aerosol component to be resolved from a component derived from sea-water. The concentrations of five of these elements are shown for different particle size ranges in Fig. 38.5, together with the concentrations of the sea-water-derived components of S, K and Ca. Even in Bermuda, the sea surface cannot be assumed to be the sole source of these elements which are abundant in sea-water. In fact, sulphur in fine particles apparently

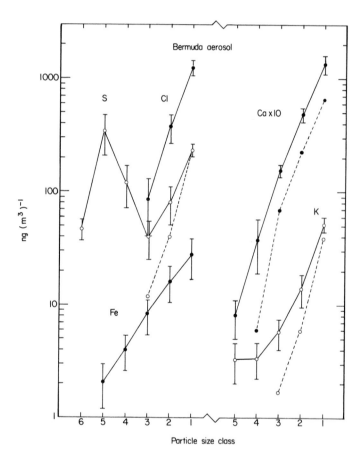

Fig. 38.5. Particle size distributions of S, Cl, K, Ca and Fe in Bermuda aerosol collected with a Battelle impactor and PIXE analysis (after Meinert and Winchester, 1977). Points connected by solid lines are geometric mean concentrations, with standard deviations of the means, for ten samples collected for 50–100 hours each under both stormy and calm conditions, October–November 1973. Dashed lines connecting points for S, K and Ca represent the sea surface-derived aerosol component which is assumed to contain negligible Fe. Particle size classes and aerodynamic radii (μm) are: (1) > 2; (2) 2–1; (3) 1–0·5; (4) 0·5–0·25; (5) 0·25–0·125; (6) <0·125.

is an essentially tropospheric constituent and is not directly related to a sea surface source.

The coastal zone in unpolluted areas offers the possibility of determining the directions of fluxes of atmospheric trace constituents from land to sea, or vice versa. Johansson *et al.* (1974, 1976), using sampling and elemental analysis techniques similar to those of Meinert and Winchester (1977), measured the elemental composition of the north Florida aerosol in a series of samples taken over several weeks, during which the direction of the air flow shifted from landward to seaward. Correlations of the average wind directions with the elemental compositions led to the conclusion that the flux direction was from sea to land for S and Cl in large particle size classes, but from land to sea for small particle size S and for Ti, Mn, Fe, Cu, Zn, Br and Pb in all particle size ranges. Seaward and landward fluxes of K and Ca partially offset each other in large particles, but for small particle K the land-to-sea direction predominated. A directional flux of V from sea to land was observed; this resulted from fuel-oil combustion emissions from a shore-based power plant. These findings suggest that fluxes to the ocean, as distinct from concentrations over the ocean, may be better determined from measurements in a boundary region, such as a coastal zone, which, on average, is influenced to a comparable degree by both terrestrial and sea surface sources, but for which the detailed influences are to a degree dependent on the direction of the air flow.

38.2.1.3. *Anomalous trace element enrichments*

A number of trace elements in the aerosol over the ocean occur in higher proportions, relative to either the major elements of sea-water or the most abundant elements in terrestrial rocks, than can be accounted for simply by dispersing soil dust or sea-spray. At present, the additional source processes which enrich the atmosphere with these elements have not been clearly identified. It appears that for some elements these anomalous processes may be mainly at the land–air, rather than the sea–air, interface, but it is not certain that this is so for all elements. The general nature of anomalous trace element enrichments in the atmosphere, first pointed out by Rahn (1971), has now been thoroughly reviewed by Rahn (1976a).

Aerosol composition in the marine atmosphere resembles that expected from the mixing of finely divided materials from large-scale sources, including soils from the weathering of rocks and sea-spray droplets, with the addition of trace gas-to-particle conversion products. Certain elements, including Na, Cl and S in particle sizes >1 μm aerodynamic diameter, occur in proportions which suggest a predominantly sea-spray origin. A large group of electropositive elements, including Fe, Al and the lanthanide elements, are found almost universally in proportions similar to those in average crustal rocks. However, other elements, such as S, Se, As, Sb, Cu, Zn, Cd, Hg and Pb, which occur in particles $< 1 \mu m$ in diameter, appear to be derived ultimately from the Earth's crust, but are found at anomalously high concentrations in the atmosphere relative to the previous group of directly crust-derived elements. The distinction between these two groups of apparently crust-derived elements can be appreciated by considering the relative elemental composition of the major rock types found at the Earth's surface.

A summary of selected element weight ratios in five of the most abundant rock types of the continental areas of the Earth, based on averages of geochemical data compiled by Turekian and Wedepohl (1961) is given in Table 38.2. In addition, the table gives the ratios in "average crustal rock" compiled by Mason (1966) and used by Rahn (1976a, b) in his comparisons with aerosol composition. Wedepohl (1969) has estimated that high-calcium and low-calcium granites together comprise 80–90 % of the igneous rocks exposed at the Earth's surface, and that the bulk of sedimentary rocks can be considered to be composed of shales, sandstones and carbonate rocks roughly in the proportions 10/2/1. Other igneous rock types, such as syenites, basalts and ultrabasics, are much less common than granites. Furthermore, clay and carbonate marine sediments, which may not be identical in composition to terrestrial sedimentary rocks, are not exposed to sub-aerial erosion and therefore are not available for the generation of aerosols. Metamorphic rocks may be considered to reflect the composition of their igneous or sedimentary precursors. It may be concluded, therefore, that the data in Table 38.2. give an indication of the most likely composition of wind-blown dust although this composition will obviously vary geographically according to the distribution of the various rock types. In most instances the average values given by Mason (1966) are close to the actual rock compositions.

Atmospheric aerosol particles are, in fact, generated by a complex of chemical and mechanical actions during erosion (Gillette, 1974). This may break down some rocks or minerals more easily than others and alter the atmospheric aerosol composition from that expected purely on the basis of the bulk composition of rocks. Moreover, the atmosphere may transport finer particles over longer distances than it does coarser particles, and the area of the Earth's surface used for averaging the contribution of different types of rock will therefore depend on the aerosol particle size which is being considered. However, the quantitative aspects of aerosol generation by erosion, and the transport and removal of particles over the Earth, are still incompletely understood. The data of Table 38.2 may therefore be used as a guide to the approximate composition of aerosol-generating rocks and as an indication of the variability in composition of different rock types.

For convenience, the data in Table 38.2 have been separated into two groups: (i) a "Group I" set of elements which includes the major rock-forming elements,

TABLE 38.2

Average element weight ratios in abundant crustal rocks[a]

Elements	Granitic rocks		Sedimentary rocks			Average Crustal rock
	High Ca	Low Ca	Shales	Sandstones	Carbonates	
Group I						
Al/Fe	2·8	5·1	1·7	2·6	1·1	1·6
Si/Fe	10·6	24·4	1·5	37·6	6·3	5·5
Si/Al	3·8	4·8	0·9	14·7	5·7	3·4
K/Fe	0·9	3·0	0·6	1·1	0·7	0·5
Ca/Fe	0·9	0·4	0·5	4·0	0·7	0·7
K/Ca	1·0	8·2	1·2	0·3	0·009	0·7
K/Rb	230	250	190	180	900	290
Ca/Sr	58	51	74	2000	500	97
Ti/Fe	0·11	0·08	0·10	0·15	0·11	0·09
Mn/Fe (× 100)	1·8	2·7	1·8	0·x	29	1·9
Ni/Fe (× 1000)	0·51	0·32	1·4	0·20	5·3	1·5
Zr/Fe (× 1000)	4·7	12	3·4	22	5·0	3·3
Group II						
Cu/Fe (× 1000)	1·04	0·70	0·95	—	1·05	1·10
Zn/Fe (× 1000)	2·0	2·7	2·0	1·6	5·3	1·4
Cu/Zn	0·50	0·26	0·47	—	0·20	0·8
As/Fe (× 1000)	0·064	0·11	0·28	0·10	0·26	0·036
As/Zn	0·032	0·038	0·14	0·06	0·05	0·026
Pb/Fe (× 1000)	0·51	1·3	0·42	0·71	2·4	0·26
Pb/Zn	0·25	0·49	0·21	0·44	0·45	0·19
Pb/As	7·9	12·7	1·5	7	9	7·2
S/Fe	0·010	0·021	0·051	0·024	0·316	0·005
S/Zn	5·0	7·7	25	15	60	3·7
S/As	160	200	185	240	3800	144

[a] Data from Turekian and Wedepohl (1961), Wedepohl (1969) and Mason (1966).

including Fe, and those minor elements which are generally thinly dispersed among the principal mineral components of rocks and (ii) a "Group II" set of elements which, relative to Fe, are low in overall abundance, but which may be segregated into accessory minerals, such as sulphides, and which have therefore been termed "chalcophilic" by geochemists. Both groups contain some elements which have relative abundances which differ only slightly between the different rock types (e.g. Ti/Fe and Pb/Zn) whereas others show considerable differences (e.g. Si/Fe and Ca/Fe). The atmospheric aerosol may therefore be expected to vary in composition in accordance with variability caused by the mixture of effective rock sources.

Rahn (1976a) has exhaustively reviewed the literature and has compiled one suite of 104 reports on the elemental composition of aerosols from marine and non-marine regions, and another listing 38 reports giving elemental distributions as a function of the mass median diameters of aerosol particles. He then computed an enrichment factor (EF) given by the expression

$$EF = \left(\frac{[\text{Element}]}{[\text{Al}]}\right)_{\text{aerosol}} \Bigg/ \left(\frac{[\text{Element}]}{[\text{Al}]}\right)_{\text{average crust}}$$

for each element using the data given by Mason (1966) for the average crustal composition. A grand average obtained from all the 104 reports, is shown as geometric mean enrichment factor in Fig. 38.6 from which it is apparent that the enrichment factors span a range from about unity to several thousand. There are 40 elements which cluster in an EF range from unity to five or six, and 30 which range from Cr $(EF = 8)$ to Se, C and N with factors of 3000–4000. Most of the data were taken from studies carried out in the Northern Hemisphere, and the marine samples were largely from the North Atlantic; however, similar values were found for samples from Antarctica. It appears, therefore, that the anomalies in trace element concentrations are of worldwide extent and occur over both the ocean and the land.

Fortunately, many investigators have measured the variation of elemental concentrations and enrichment factors with respect to aerosol particle size. The enriched elements tend to be more concentrated in particles in the smaller size ranges than do the elements with EF values around unity, a fact which appears to suggest that the processes for their atmospheric enrichments involve gas-to-particle conversions. In Fig. 38.7 the results from two Bermudan atmospheric aerosol samples are plotted as mass median diameter versus EF. In both samples, the "crustal" elements, which have EF values close to unity, are found in particles with aerodynamic diameters of several micrometres, whereas elements with high EF values are found mainly in particles with diameters of 1 μm or less.

In a practical sense it is important that the natural sources of these elements

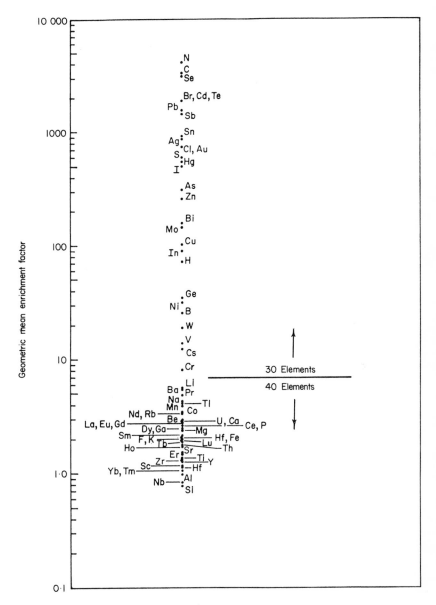

FIG. 38.6. Geometric mean enrichment factors of aerosols relative to the Earth's crustal average composition (Mason, 1966), using Al as the reference element (Rahn, 1976a).

be identified. The Florida coastal zone and Bermuda studies, by Johansson *et al.* (1974) and by Meinert and Winchester (1977) respectively, both indicated that for a few elements, including Zn, the concentrations correlated better with elements of terrestrial origin than with those found in sea-spray. However, research on the composition of the sea surface microlayer (reviewed

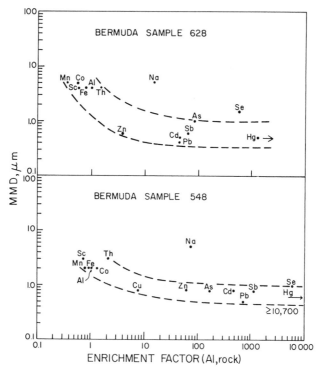

FIG. 38.7. Mass median diameter of Bermuda aerosol versus enrichment factor (Duce *et al.*, 1976; Rahn, 1976a).

in the six references cited in Section 38.1) has indicated that there are substantial enrichments of many of the anomalous elements at the air–sea interface. Because organic matter is also substantially enriched in spray droplets formed by bubble breaking, some of the elemental enrichment observed in the atmosphere may be due to sea-to-air fractionation effects. The anomalous elements are also amongst those which may be introduced into the atmosphere via anthropogenic mobilization such as smelting and high-temperature industrial processes. It is difficult to evaluate the extent of the atmospheric contamination from these emissions at locations far from their source unless the natural sources can be identified. Inasmuch as pollution transport to the ocean by atmospheric pathways may be rapid and long range, both the extent of sea-to-air transport of the same substances, and the concentrations at steady-state over the ocean, must be known in order to infer the downward fluxes into the sea.

38.2.2. ORGANIC MATERIAL

The identity and movement of organic compounds in the particulate or gas phase of the marine atmosphere are known in considerably less detail than are those of the elemental constituents discussed above. The chemical characterization of the marine aerosol in terms of organic constituents has only been made on a very general, and for the most part incomplete, basis. The quantification of the many organic compounds continues to be an enigma because of complex sample collection problems (e.g. gas–particle interactions), and the many analytical problems associated with the identification of extremely small quantities of organic species found in aerosol samples (see Chapter 39).

The organic fraction of the marine aerosol represents a complex matrix of ~ 150 compounds, including C_8–C_{22} fatty acids and esters, aliphatic hydrocarbons, a wide range of aromatic hydrocarbons and aromatic esters, bases such as pyridine and aniline, alcohols, ketones, carboxylic acids, amino acids, proteins, lipids and chlorinated hydrocarbons. At one time it was frequently assumed that the sea surface, through the action of bubble rupture at the air–sea interface, was the major source of these constituents. However, a large number of complicating factors, e.g. gas–particle reactions, volcanic emanations, influxes of wind-blown continental material and anthropogenic emissions, are now recognized to be potential sources of the particulate organic matter found in the marine atmosphere (Delany *et al.*, 1967; Folger, 1970; Parkin *et al.*, 1972).

Some recent research has concentrated on the characterization of the sea surface microlayer and on the transport processes necessary for the injection of organic compounds and marine micro-organisms into the atmosphere (see the six basic references cited in Section 38.1 and Chapter 10). Interest in the influx of micro-organisms into the marine atmosphere dates from long before research on the chemical composition of the marine aerosol developed, primarily as a result of interest in the spread of disease. The reviews by Blanchard and Parker (1977) and Duce and Hoffman (1976) give particularly good coverage in this area. Initial laboratory studies on the transfer of bacteria across the air–sea interface led to the investigation of the jet drops formed from organic monolayers at the ocean surface. During research on the bubble bursting phenomenon in the laboratory, Blanchard (1963) noted that some fraction of the organics in the microlayer was ejected into the atmosphere along with sea-salt. Bezdek and Carlucci (1974) examined the transport of insoluble monomolecular films to the atmosphere by bursting bubbles. They noted that ^{14}C-labelled stearic acid was incorporated into the jet drops and that the smaller droplets were more enriched in this material than were larger droplets. Hoffman and Duce (1976) made a careful laboratory study of the factors influencing the organic carbon content of marine aerosols. They noted that the organic-C/Na ratio of the laboratory-generated atmospheric sea-salt particles was dependent on four factors: (i) the quantity of organic carbon in the sea-water; (ii) the amount of surfactants in the sea-water; (iii) the distance the bubbles travelled in sea-water before bursting at the air–water interface; and (iv) the chemical and physical nature of the organic compounds in the sea-water. They concluded that the chemical form of the organic material plays a critical role in determining the amount of organic carbon on the sea-salt particles.

Substantially more work has been done to characterize the organic fraction of the aerosol in the natural marine atmosphere than in the laboratory. On the basis of studies of the non-volatile organic carbon content of atmospheric precipitation, Neuman et al. (1959) postulated that the sea surface supplies organic material to the atmosphere. Lodge et al. (1960) observed a mean concentration of $1.6 \mu g \, m^{-3}$ of benzene-soluble organic material in aerosol samples collected over the Pacific Ocean. The benzene-extractable organic material/salt mass ratio of 0.29 was equivalent to an enrichment factor of over 1000. Blanchard (1964) reported the presence of a compressed film of surface-active material on sea-salt particles collected on thin platinum wires along the east coast of Hawaii. Using a similar collection procedure, Blanchard (1968) later estimated that the surface-active film/salt mass ratio ranged from 0.3 to 0.7 on these Hawaiian particles. Interestingly, Blanchard (1968) noted that the organic material/salt ratio of the Hawaiian sea-salt aerosol showed changes from day to day, and could vary by a factor of four over a period of a week. The highest ratio was found in an air mass whose surface trajectory suggested that, prior to arriving at Hawaii, it had passed over a section of the Pacific rich in organic material (Woodcock et al., 1971; Blanchard and Syzdek, 1974). These results imply that the ocean is a major source of particulate organic constituents for the marine aerosol. Barger and Garrett (1970) collected aerosols from a 20 metre high tower on the north-eastern coast of Oahu, Hawaii, and found that the chloroform-extractable organic material (lipoids) in marine aerosols collected with a high-volume glass fibre filter system ranged from 0.7 to $6.3 \mu g \, m^{-3}$ (5–15% of the total particulate mass), with a lipoid/salt mass ratio of 0.07–0.14. They reported that the marine aerosol contained a mixture of surface-active compounds and non-polar hydrocarbons. They noted that the lipoid content appeared to increase with increasing salt content. Five fatty acids (C_{14}–C_{18}) were identified, and all were found to be in relative proportions which were similar to those in sea surface samples. Barger and Garrett (1970) then reasoned that because a transport mechanism existed for the transfer of organic matter from the ocean into the marine atmosphere, and since a strong chemical similarity appeared to exist between sea surface material and airborne organics, the sea must be the principal source of the organic constituents of marine air.

Barker and Zeitlin (1972) carried out a mass spectroscopic study of Hawaiian marine aerosols and observed that the organic coating materials in them are more concentrated in the smaller sized aerosols. Hoffman and Duce (1974) have made essentially the same observation for aerosol samples collected along the SW coast of Bermuda. They found that the content of organic carbon in the marine aerosol ranged from 1% to 19% of the mass of sea-salt in the particles, the percentage decreasing with increasing salt content. The concentration of total organic carbon, collected on glass fibre filters, varied from 0·15 to 0·47 μg m^{-3} (i.e. substantially less than that found in previous studies) with the organic-C/salt mass ratio varying from 0·01 to 0·19. It should be noted that in the previous work by Barger and Garrett (1970), described above, the lipoid content was found to increase with increasing salt content. This is the reverse of the results obtained by both Barker and Zeitlin (1972) and Hoffman and Duce (1974). Although several plausible explanations have been advanced to explain the discrepancy (Duce and Hoffman, 1976), none of them have been validated.

Barger and Garrett (1976) collected marine aerosol samples from shipboard in the Mediterranean Sea and the Pacific Ocean, and their results appear to verify the general concentration range of extractable organic material previously found in the marine aerosol. Total organic materials extracted by chloroform ranged from 0·08 to 4·02 μg m^{-3} of air, and the greatest concentration of any one methyl ester detected was 50 ng m^{-3}. In agreement with the observations of Garrett (1967), methyl palmitate was found to be the most prominent ester detected. A simple model of the marine aerosol was pictured by these authors as one of large lipid molecules mixed with as yet unidentified compounds having a wide range of molecular weights and shapes which make up the surface films associated with the particulate material in the marine atmosphere.

A rather different picture of the organic fraction of the marine aerosol, in terms of organic constituent complexity and source generation, may develop as the result of work by Ketseridis et al. (1976). From three Atlantic Ocean air masses and three continental European air masses (polluted to varying degrees) organic acids and phenols, organic bases, aliphatic hydrocarbons, aromatic hydrocarbons and neutral compounds were characterized. In all the areas sampled the organic fraction of the aerosol had a very complex composition; however, the major groups of organic constituents remained rather uniform. Some of the compounds identified in the gas chromatographic analysis of the ether-extractable particulate organic matter (EEOM) are shown in Table 38.3. The EEOM concentration in samples of clean air aerosol particles appeared to be rather uniform over both the North Atlantic and Europe. The average EEOM concentration was 5–10% of the total particulate matter (TPM), and ranged from 0·30 to 2·70 $\mu g\,m^{-3}$ STP of air with an average value of 1·20 $\mu g\,m^{-3}$. The total organic carbon (TOC) concentrations ranged from 0·33 to 1·60 $\mu g\,m^{-3}$, some samples having TOC concentrations higher than the corresponding EEOM concentrations, thus indicating that there was organic material in the samples which was not soluble in ether (e.g. plant debris and bituminous substances of high molecular weight). The TOC concentrations were remarkably constant, ranging from 2·0% to 3·9% of the TPM. On the basis of these data, Ketseridis et al. (1976) suggested (i) that most of the organic fraction of the clean air background aerosol found globally may simply be a diluted continental aerosol; or (ii) that in addition to the aerosols from polluted regions there may be a constant, but not necessarily major, production of organic particulate matter even outside continental areas (including contributions from sea surface films and gas-to-particle reactions of organic trace gases). The supposition made by many investigators that the ocean is the primary source of organic constituents for the marine aerosol may have to be critically re-evaluated. Further studies on the source and the compound and particle size composition of the organic particulate material found in the marine atmosphere are urgently needed.

38.3. GAS–PARTICLE RELATIONSHIPS IN THE MARINE ATMOSPHERE

38.3.1. CHLORINE, BROMINE AND IODINE

38.3.1.1. *Chlorine*

Chlorine occurs in the natural atmosphere in both the particulate and the gaseous forms. The ocean is considered to be the major source for particulate chlorine. However, the sources of gaseous chlorine are uncertain and the chemical speciations of its major components remain unknown. Because chlorine is the major elemental constituent in both sea-water and the marine aerosol, the formation and chemical nature of particulate and gaseous chlorine in the marine atmosphere will be discussed in detail below.

The chlorine content of marine particulate matter has generally been viewed relative to the Cl/Na ratio in sea-water. Analyses of particulate matter by different investigators show excesses, as well as deficits, of Cl in the marine aerosol. In order to view these data systematically the following nomen-

TABLE 38.3

Constituents of the ether-extractable organic fraction of particulate matter over the North Atlantic Ocean and the continent of Europe in 1973[a]

Aliphatic fraction	Aromatic fraction	Polar fractions	Organic acid fractions	Organic base fractions
(20–50 peaks)	(20–40 peaks)	(20–40 peaks)	(15–30 peaks)	(15–25 peaks)
18 peaks identified as C_{11}–C_{28} straight-chain paraffins	naphthalene methylnaphthalenes acenaphthene acenaphthylene fluorene anthracene phenanthrene methylanthracene fluoranthrene pyrene benzofluorene methylpyrenes benzo-(a)anthracene chrysene benzo(e)pyrene/benzo(a)pyrene perylene dibenzanthracene benzo(g,h,i)-perylene	indene methylindene carbazol coumarin anthrone benzanthrone flavone peri-naphthenone	16 peaks identified as C_9–C_{22} straight-chain fatty acids; even-numbered acids predominated	pyridine aniline quinoline iso-quinoline α-naphthylamine benzo(h)-quinoline acridine phenanthridine

[a] Gas chromatographic analysis of the fractions separated from the samples yielded 110–145 peaks per sample, but only a few compounds from each fraction were identified. The number of peaks did not differ appreciably between samples from different locations. Compiled from Ketseridis *et al.* (1976).

clature will be adopted (Duce *et al.*, 1972):

$$\text{Fractionation; } F_{Na}(X) = \frac{(X/Na)_{atm}}{(X/Na)_{seawater}}$$

$$\text{Enrichment; } E_{Na}(X) = F_{Na}(X) - 1$$

where $(X/Na)_{atm}$ is the weight ratio of any element to sodium in a marine aerosol (or rain sample, etc.), and $(X/Na)_{seawater}$ is the same elemental weight ratio in bulk sea-water. In the present discussion $X = $ chlorine.

After sea-water spray or droplets are ejected into the atmosphere there are essentially four ways in which the Cl/Na ratio in the particles can change: (i) particulate Cl may be lost by atmospheric conversion to gaseous Cl species; (ii) Cl may be gained by adsorption of gaseous Cl species; (iii) the particles may possibly gain Na from terrestrial dust; and (iv) the composition may change as a result of the selective wet or dry deposition of particles of a certain size or composition (Rossby and Egnér, 1955; MacIntyre, 1974). All four ways have been considered by investigators examining the marine atmospheric chemistry of chlorine, particular attention being focused on the gain, or loss, of Cl through some conversion mechanism.

Early work concentrated on the examination of the Cl/Na ratio in precipitation and laboratory-generated samples, very little work being done directly on actual marine aerosols. Cauer (1951), who carried out what was apparently the first laboratory examination of marine atmospheric chlorine chemistry, reported a rapid loss of Cl from an atomized aerosol. He concluded that 85 % of the Cl$^-$ escaped into the gaseous phase, probably as ClO_2, Cl_2 and HCl. Köhler and Bath (1952) atomized sea-water and collected the aerosols in bulk. Strangely enough, they reported the Cl/Na ratio as a little greater than the sea-water value of 1·80, suggesting that there was a Cl excess. However, they found that the Mg/Cl ratio was about equal to that of sea-water.

Initial results for fog and rain-water samples reported by Miyake (1948) showed that the Cl/Na ratio varied from 1·18 to 3·25 (Cl/Na = 1·80 in sea-water). Rossby and Egnér (1955) found Cl/Na ratios for precipitation samples along the Swedish coast to be always less than 1·80. They reported a rapid oscillation of the ratio as the distance from the coast increased. However, Gambell (1962) did not observe this effect for North American samples. Gorham (1955) measured the Na/Cl and Mg/Cl ratios in the rain-water over the Lake District in Great Britain and found that they agreed closely with those of sea-salt. The rather extensive precipitation data which now exist have been carefully reviewed by Junge (1963). In general, the results show that there is either no enrichment, or even a slight depletion, for chlorine with $E_{Na}(Cl)$ values down to $\sim -0·10$ (Junge and Werby, 1958; Eriksson, 1960; Junge, 1963). Two fairly recent studies support the conclusion that no Cl fractionation occurs. Seto *et al.* (1969) measured the Cl/Na ratio in Hawaiian rains between altitudes of 500 m and 1600 m at a distance of 11–27 km from the windward coast of the island of Hawaii. Within very narrow limits of accuracy the ratio was found to be the same as that of sea-water. Lazrus *et al.* (1970) examined the Cl/Na ratio in cloud-water at 1 km above sea-level in Puerto Rico and found virtually no difference in the ratio from that of sea-water.

Viewing all of the precipitation data together, it appears that they have generally been of high quality. Interpretation of the data, however, remains somewhat controversial when viewed in terms of Cl particulate chemistry over the ocean since it is complicated by factors including; (i) a very significant Cl gas phase that may, or may not, exchange with marine aerosols, (ii) wash-out and rain-out efficiencies which are a function of particle size and also particle composition, and (iii) intrusions of non-marine aerosols with continental and/or anthropogenic origins. These three complicating factors may significantly alter both the Cl/Na ratio and the interpretation of the Cl chemistry of the rain droplets in the marine environment.

When data for particulate chloride over the ocean are considered, it becomes apparent that there have been far fewer measurements of chlorine in particulate matter than in precipitation. Junge (1956) found a mean atmospheric Cl/Na ratio of 1·79 (confidence limits not given) in particles collected along the coast of Florida using a cascade impactor. For aerosols from the North Pacific sampled by filters, Tsunogai *et al.* (1972) have reported Cl/Na ratios of 3–6 (confidence limits not given), suggesting a two-fold enrichment of chlorine. These results represent the only case showing such a large enrichment. Chesselet *et al.* (1972b) found that over the North Atlantic, in aerosols apparently collected by cascade impactors, the Cl/Na ratio in particles was 1·72($\pm$0·03), suggesting a Cl loss of about 5%. Green (1972) found what he considered to be Cl losses for aerosols collected along the coast of California. He attributed these to the evolution of HCl or Cl_2 with an associated decrease in the particulate Cl/Na ratio. Martens *et al.* (1973) have reported a Cl/Na ratio of 1·57($\pm$0·16) for aerosol samples taken with an Andersen impactor in Puerto Rico. This approximately represents a 7–25% depletion in Cl. Pierson *et al.* (1974) found an average $E_{Na}(Cl)$ value of −0·09 between 1972 and 1973 in the Shetland Islands. Wada and Kokubu (1973) have reported that shipboard samples taken in mid-Pacific showed an $E_{Na}(Cl)$ of −0·03($\pm$0·03). Buat-Menard *et al.* (1974) produce strong evidence from aerosol data taken over the tropical and equatorial Atlantic for what they believe to be a definitive Cl loss balanced by an SO_4^{2-} enrichment. The mechanism which they believe to be involved in their sampling location is the evolution of gaseous HCl from the particulate phase. Lodge (1960) appears to have previously explored the $Cl^- - SO_4^{2-}$ relationship over the Pacific, but was unable to reach a definite conclusion about it. Gordon *et al.* (1977) have recently reported that the Cl/Na ratio between 75 m and 2000 m above the Bahamas is $1.83(\pm 0.04)$. The weights of Na and Cl in each sample were determined by neutron activation analysis and each element apparently showed no enrichment. The ratios also showed no variation with altitude. The problem of any Cl loss or enrichment still appears to be a controversial issue.

The variation of the Cl/Na ratio with particle size has been examined by several investigators. Rahn (1971) observed a general absence of Cl from the smallest particles ($< 1.0 \, \mu m$) collected by an Andersen cascade impactor at inland sites in northern Canada. Wilkniss and Bressan (1972) collected particulate samples from surface air over the Atlantic and Pacific Oceans by filtration. A mean Cl/Na ratio of $1.88(\pm 0.34)$ for all particles $\geqslant 2 \, \mu m$ in radius, and $1.40(\pm 0.26)$ for all particles $\geqslant 0.2 \, \mu m$ in radius has been calculated from their data. The latter value represents a 22% deficit, and comparison with the former value suggests that any conversion of Cl from particle to gas phase may occur preferentially on small particles. Martens *et al.* (1973) have reported large Cl losses, assuming an initial Cl/Na ratio equal to that of sea-water, from the smallest sea-salt particles collected by an Andersen impactor in ambient air from the San Francisco Bay area and from Puerto Rico. In all their samples, increasing loss of Cl with decreasing particle size was observed, over 90% of the Cl being lost from the smallest particles with radii of 0.2–0.4 μm, and less than 10% of it being lost from the large particles with radii $> 5 \, \mu m$. They found a linear correlation between the measured gaseous NO_2 concentration and the total Cl loss in the San Francisco samples, indicating a Cl loss of $0.06 \, \mu mol \, m^{-3}$ per $\mu mol \, m^{-3}$ of NO_2. The investigators felt that the size dependency of the Cl loss could be explained by a size-dependent gain of hydrogen ion resulting from the accretion of nitric acid vapour on the marine aerosol.

Recently, Meinert and Winchester (1977) showed an absence of particulate Cl in the 0.1–$0.5 \, \mu m$ radius range from samples of the North Atlantic marine aerosol taken by means of a cascade impactor in Bermuda. In addition, Johansson *et al.* (1974, 1976) have demonstrated a similar absence of particulate Cl in aerosols of the same size range from atmospheric samples taken in northern Florida with a cascade impactor.

Several important points concerning the atmospheric chemistry of chlorine may be drawn from the literature. Precipitation data are probably not instructive in the examination of the aerosol chemistry of Cl. In general, there appears to be a real variation of the Cl content with particle size in the marine aerosol. The deficit of Cl on sub-micrometre particles in the marine atmosphere is based on the decrease in the Cl/Na ratio with particle size compared to that of bulk sea-water. It should be noted that these conclusions are based mostly on data for samples collected with cascade impactor sampling devices.

The pathway for the production of gaseous chlorine in the marine atmosphere is unknown. In fact, the identification of the major gaseous chlorine compounds in the marine atmosphere has remained elusive. The gaseous Cl in uncontaminated air is believed to be derived mostly from sea-salt particles, although this has not been conclusively demonstrated. Other minor sources appear to include volcanic eruptions and emanations and, in arid areas, salt crystals from the soil (Eriksson, 1959). Valach (1967) has suggested that the major part of the atmospheric chlorine is in the form of gaseous HCl of volcanic origin. Duce (1973) showed convincingly, however, that probably only 10%, at most, of the gaseous Cl is volcanic. Cadle (1975) has estimated that the annual emission of HCl to the troposphere and stratosphere from volcanic eruptions is 7.5×10^5 tons; i.e. less than 1% of that believed to be derived from sea-salt particles. It has been traditionally assumed that most gaseous Cl exists as HCl, but recent work by Lovelock (1971, 1974), Lovelock *et al.* (1973), Su and Goldberg (1973), Grimsrud and Rasmussen (1975a) and Wilkniss *et al.* (1975) strongly suggests that gaseous Cl is organically bound in such compounds as CCl_3F, CCl_2F_2, CCl_4 and CH_3Cl.

Relatively few measurements exist for gaseous Cl in the marine environment. The first appear to have been made by Junge (1956, 1957) in the coastal regions of Florida and Hawaii in which he found 1.0–$3.9 \mu gCl\,m^{-3}$. Duce *et al.* (1965) also measured gaseous Cl on Hawaii and found levels ranging from 0.3 to $6.5 \mu gCl\,m^{-3}$, with a mean of $3.7 \mu gCl\,m^{-3}$. Buat-Menard (1970) found gaseous Cl levels over the North Atlantic to range from 1.6 to $11.6 \mu gCl\,m^{-3}$, with a mean of $3.4 \mu gCl\,m^{-3}$. Chesselet *et al.* (1972a) found a constant level of at least $2 \mu gCl\,m^{-3}$ of gaseous Cl above the North Atlantic ($45°N., 20°W.$) with maximum values of $7 \mu gCl\,m^{-3}$. Okita *et al.* (1974) reported gaseous Cl levels in 1973 along the coast of Japan (Tareyama) ranging from 5.3 to $9.9 \mu gCl\,m^{-3}$, with a mean of $6.4 \mu gCl\,m^{-3}$. Berg and Winchester (1977) found gaseous Cl levels in 21 samples from over the Gulf of Mexico ranging from 4.3 to $5.9 \mu gCl\,m^{-3}$. At higher elevations, the concentration levels appear to drop by about a factor of two (Junge, 1957; Duce *et al.*, 1965).

The published values of gaseous Cl concentrations cover a wide range and this may be a real reflection of their variations in the marine atmosphere. However, recent data on the organic nature of gaseous Cl species cast some doubt on the measurements since most of the collection schemes employed to sample these gas levels were based on the assumption that the gaseous Cl was an inorganic species (e.g. HCl or Cl_2). Bubblers and filter devices were designed to trap only inorganic gaseous chlorine. With the application of new analytical tools (e.g. gas chromatography with an electron capture detector, and mass spectrometry) and new sampling schemes, however, it has become apparent that there are substantial amounts of organochlorine compounds in the marine atmosphere.

Gaseous chlorine species were generally considered to be entirely inorganic in nature until Lovelock (1971), using a gas chromatograph with an electron capture detector, determined CCl_3F (Freon 11) and SF_6 levels in the atmosphere of Adrigole, County Cork, Ireland. He reasoned that both of these compounds were of anthropogenic origin and therefore could be used as effective tracers for recently polluted air masses. Subsequently, Lovelock et al. (1973) found CH_3I and CCl_4, together with CCl_3F, to be uniformly distributed both in the air and in the ocean surface along a cruise track from the United Kingdom to Antarctica and back. Interestingly, CH_3I and CCl_4 displayed no systematic variation with latitude, whereas CCl_3F was clearly more abundant in the Northern Hemisphere. Lovelock et al. (1973) reasoned that CCl_3F was anthropogenic in origin, but that CH_3I was probably the result of biological methylation in the ocean. A few sea-water measurements of CCl_4 down to 300 m showed that the CCl_4 concentration dropped exponentially with depth to 50 m. The authors reasoned that the sea is probably a sink for CCl_4 and that this compound seems very likely to be a product of natural chemistry, possibly from a complex series of reactions in the troposphere between methane and chlorine.

Su and Goldberg (1973) found CCl_3F (Freon 11) and CCl_2F_2 (Freon 12) at slightly higher concentrations than those observed by Lovelock at locations in or near San Diego, California. The dichlorodifluoromethane values were up to an order of magnitude less than those of trichlorofluoromethane. The authors reasoned that this difference might be due to the relative instability of the C-Cl bonds in CCl_3F.

Lovelock (1974) found seven atmospheric halocarbons over western Ireland and the North Atlantic (CCl_2F_2, CCl_3F, $CHCl_3$, CH_3CCl_3, CCl_4, $CHCl{=}CCl_2$ and $CCl_2{=}CCl_2$). He noted that virtually no difference has been observed in all the data collected up to that time for the abundance of CCl_4 between the Northern and Southern Hemisphere. In addition, he reported preliminary laboratory experiments which showed that the reaction in air between CH_4 and Cl_2 in glass vessels can produce significant quantities of CCl_4.

As a result of this pioneering work other investigators began to explore the troposphere and the stratosphere for halogenated aliphatic hydrocarbons. Molina and Rowland (1974) stimulated great interest in this research by expounding a theory that the influx of gaseous chlorine compounds (possibly originating near the ocean surface) into the stratosphere from the troposphere could lead to the destruction of ozone, catalysed by chlorine atoms. In order to investigate a wider range of halocarbons in the atmosphere, Grimsrud and Rasmussen (1975b) linked a mass spectrometer to a gas chromatograph. They searched for 19 simple halocarbons in the atmosphere near Pullman, Washington, and found seven chlorocarbon gases (Grimsrud and Rasmussen, 1975a). The most interesting result of their work was the fact that for the first time methyl chloride was detected in the atmosphere and was found to be present at more than twice the concentration, 530 ($\pm$ 50) parts per 10^{12} (or 841 ng m^{-3}) of any other halocarbon gas. Further work by Wilkniss et al. (1975) and Zafonte et al. (1975) established the ubiquitous nature of halo-carbon gases in the atmosphere.

Lovelock (1975) supported the observation of Grimsrud and Rasmussen (1975a) by finding appreciable levels of CH_3Cl in southern England in air which originated over the Atlantic Ocean. He postulated that the ocean was the major source for CH_3Cl and that the compound resulted from the reaction of methyl iodide with the chloride ion of sea-water. The chemistry of this possible reaction has been supported by calculations and experiments by Zafiriou (1975). Perhaps the most significant part of Lovelock's (1975) work, however, is the supposition that the transport of natural, ocean-produced CH_3Cl into the stratosphere can account for most of the present production of stratospheric Cl. The data base for this theory is extremely limited and contradictory at present, and the chemical pathways remain to be elucidated. An understanding of both the flux of Cl across the air–sea interface and the chemistry of the element near the sea surface is crucial for an understanding of the movement of gaseous Cl species in the marine atmo-sphere and their transport to higher altitudes. Clearly, more research is needed in this area.

The marine atmosphere is a natural source for gaseous chlorine, and many reactions and mechanisms have been suggested which would lead to its release from sea-salt particles. The principal pathways which have been proposed will be reviewed below.

The oxidation of chloride by ozone has been suggested by Cauer (1935, 1938) as being responsible for the pH values of precipitation. The first step in this oxidation is:

$$O_3(g) + Cl^-(aq) \rightarrow O_2(g) + ClO^-(aq) \tag{1}$$

followed by

$$ClO^-(aq) + 2H^+(aq) + Cl^-(aq) \rightarrow Cl_2(g) + H_2O(liq). \tag{2}$$

Yeatts and Taube (1949) have studied these reactions and found that the first step is the rate determining one. Erikson (1960) calculated the rate constant of this reaction for a sea-salt droplet using the Yeatts and Taube rate expression and an ozone partial pressure of 6×10^{-8} atm. He found that it would take about 1000 years for the Cl$^-$ to be converted to ClO$^-$ to any appreciable extent. This strongly suggests that the Cauer mechanism does not play a major role in the release of gaseous Cl_2. It also appears to be very likely that free gaseous Cl_2 would undergo rapid photolysis and, through a series of recombination steps, including possible hydration, lead to the formation of HCl (Zafiriou, 1974a). It is interesting to note that Reaction (1) probably occurs at an appreciable rate at altitudes of 20–50 km (Rowland and Molina, 1975) as part of the chain reaction initiated by the photolysis of chlorofluoromethanes.

Nitrosyl chloride (NOCl) has been mentioned as a possible gaseous form of chlorine in the marine atmosphere. Robbins *et al.* (1959) have suggested that nitrogen dioxide will react with moist NaCl at room temperature to produce nitrosyl chloride:

$$NaCl(s) + 2NO_2(g) \rightarrow NaNO_3(s) + NOCl(g) \tag{3}$$

which is a very easily hydrolysed or photolysed gas.

$$NOCl + H_2O \rightarrow HCl(g) + HNO_2(liq) \tag{4}$$

$$NOCl \overset{h\nu}{\rightarrow} NO + Cl. \tag{5}$$

Schroeder and Urone (1974) favour the reaction of NO_2 with NaCl and found Reaction (3) to be rapid for moist NaCl aerosols. However, at equilibrium an insignificant amount of both products was formed. Junge (1956) attempted, but failed, to find any nitrite in a study of sea-salt particles along the northeastern coast of the United States. Hence, it seems reasonable to suggest that nitrosyl chloride is not a significant gaseous carrier of chlorine in the marine environment.

Robbins *et al.* (1959) further proposed that atmospheric NO_2 may be hydrolysed in the vapour phase to form HNO_3:

$$3NO_2(g) + H_2O(g) \rightarrow 2HNO_3(g) + NO(g). \tag{6}$$

The equilibrium constant for this reaction is $0.004\ atm^{-1}$ at 300 K. With an extreme excess of water vapour, about 5% of the NO_2 is converted to nitric acid at room temperature. The resulting dilute acid could then either dissolve in a liquid sea-salt droplet or be adsorbed by a dry NaCl particle, thus lowering the droplet pH and leading to the release of HCl. Assuming that NaCl and HCl are 100% dissociated, the reactions may proceed as follows:

$$HNO_3(g) \rightarrow HNO_3(aq) \tag{7}$$

$$HNO_3(aq) + NaCl(aq) \rightarrow NaNO_3(aq) + HCl(aq) \tag{8}$$

$$HCl(aq) \rightarrow HCl(g) \tag{9}$$

where

$$K = \frac{P_{HCl}}{a_{H^+} \cdot a_{Cl^-}} = 10^{-6.36} \tag{10}$$

and the free energy change of the reaction (ΔG°) is $-8.68\ kcal\ mol^{-1}$ at 25°C (Eriksson, 1959 and 1960). Reaction (9) appears to have been first postulated by Eriksson (1955). Duce (1969) has considered this relationship at equilibrium and has shown that if all the gaseous Cl were present as HCl the partial pressure of HCl would be 10^{-9} atm and the equilibrium pH of the sea-salt aerosols would vary from 2·2 (for a sea-water composition of 0·55 M Cl$^-$) to 3·3 (if the salts in the droplet were concentrated up to 7·0 M Cl$^-$ by evaporations). For sea-salt aerosols these are rather low pH values. Junge (1963) has indicated that the ammonia concentration in the marine atmosphere (generally ranging from 2 to 5 μg m^{-3}) should be high enough to prevent a pH drop to these levels in salt droplets, and hence inhibit or stop Reaction (9).

Following the same reasoning used by Robbins et al. (1959), Eriksson (1960) has suggested that it is hygroscopic SO_3, formed in the atmosphere by the oxidation of SO_2, which dissolves in the sea-salt aerosols, forms H_2SO_4, lowers the pH and releases HCl to the marine atmosphere according to Reaction (9). This mechanism, however, suffers the same short-comings of the NO_2 hydrolysis pathway, i.e. it seems highly unlikely that the pH can become low enough for the particles to liberate HCl in accordance with Reaction (9) considered above.

Very few laboratory studies have been conducted to examine the particle-to-gas conversion of chlorine. Using an artificially produced marine aerosol and a limited number of determinations of filter samples without size differentiation, Chesselet et al. (1972a) found a Cl/Na ratio consistently close to the sea-water value. Wilkniss and Bressan (1972) employed a micro-ocean with a depth of 4 cm and a cascade impactor to examine the ion fractionation of F, Cl, Na and K, with decreasing sea-salt particle size. They found Cl/Na ratios significantly below the sea-water value of 1·80 for sea-salt particles < 1·0 μm in diameter.

Recent attention has been directed toward the study of photolytic reactions occurring at the air–sea interface. Petriconi and Papee (1972) have reasoned that highly reactive transient pernitrite intermediate oxidants can be formed photolytically in saline solutions containing NO_3^-. This has led them to suggest that these strong oxidizing agents may result in the release of halogen-containing gases from sea-water directly, or from atmospheric sea-salt particles. In another study, Zafiriou (1974b) has found that both kinetic calculations and flash photolysis experiments indicate that, in sea-water, OH radicals react rapidly and semi-quantitatively with bromide ion to yield dissolved bromine atoms, which complex with other halide ions. He suggested that similar processes may affect marine aerosols. Zafiriou (1975) has recently found that CH_3I reacts with sea-water chloride ion to form CH_3Cl. The latter compound has been reported by Grimsrud and Rasmussen (1975a) to be the major gaseous Cl species in a study of seven chlorocarbon gases in the atmosphere at Pullman, Washington. Zafiriou reasoned that CH_3Cl should be quite stable in sea-water and should exchange with the atmosphere. Once in the atmosphere it would probably be relatively photo-stable and, hence, have a comparatively long residence time. General support for this mechanism has been given by Dilling *et al.* (1975) who appear to have been the first to examine the evaporation rates and reactivities of various chlorinated compounds in dilute aqueous solutions. Methyl chloride (and other chlorinated hydrocarbons including CH_2Cl_2, $CHCl_3$ and CCl_4) evaporated rapidly from the surfaces of different test solutions. Less than 50% of the initial compound was left after only 29 minutes in all cases, and this was independent of sea-water salinity and contaminants such as clay, limestone, sand and peat moss. Recent work by Graedel and Allara (1976) also supports such a direct-emission hypothesis. They found that selected thermal and photochemical atmospheric reactions lead to the formation of negligible amounts of CCl_4, $CHCl_3$, CH_3I, CH_3CCl_3 and the chlorinated ethylenes in the troposphere. These workers noted that the atmospheric formation of CH_3Cl was extremely slow. The movement of CH_3Cl and other chlorinated hydrocarbons across the air–sea interface by direct emanation or droplet transfer thus appears to be quite possible.

A study conducted in the marine atmosphere over the Gulf of Mexico has been recently reported by Berg and Winchester (1976a, b). Measurements were made of: (i) the ambient levels of inorganic and organic gaseous chlorine; (ii) particulate chlorine as a function of particle size; and (iii) inorganic and organic gaseous chlorine, together with particulate chlorine, in a controlled atmosphere, in an *in situ* aerosol generation chamber. Their results showed that the total amount of chlorine in the marine atmosphere at the ocean surface may be relatively constant with time for at least several days. Both particulate chlorine and inorganic gaseous chlorine exhibit large changes in their concentration levels over short periods. The organic gaseous chlorine content was found to roughly equal the sum of the inorganic gaseous and particulate chlorine components, the content generally ranging from 50% to 70% of the total gaseous chlorine. Organic gaseous chlorine represented a constant 50% of the total atmospheric concentration of chlorine found over the sea surface, but the partition of inorganic chlorine between gaseous and particulate phases was variable. Data from the *in situ* aerosol generation chamber suggested that the ocean surface could be a source for organic gaseous chlorine. Particulate chlorine generated from the ocean surface as a result of bursting bubbles was observed to be the source of inorganic gaseous chlorine. Through the use of the aerosol generation chamber it was observed that the conversion process occurred within the chamber residence time of 30 minutes and was independent of any trace gases or atmospheric acid sources such as H_2SO_4 and HNO_3. The most widely accepted mechanism for the generation of gaseous chlorine, viz. the evolution of gaseous HCl by reaction with acidic droplets, may not, in fact, be the major gas production process in the marine atmosphere. In addition, data from samples collected with cascade impactors and total filter devices, run side by side, indicated that evolution of inorganic gaseous chlorine is to some degree dependent on the type of sampler used. This observation may call for a re-examination of the previously reported particulate and gaseous chlorine measurements.

38.3.1.2. Bromine

Our knowledge of the atmospheric chemistry of bromine compounds over the ocean is considerably more limited than that for chlorine. Sources of both gaseous and particulate species in the marine atmosphere include: (i) the ocean; (ii) long-range, wind-borne terrestrial sources; (iii) various atmospheric gas–particle conversion processes; and (iv) anthropogenic sources such as the automotive combustion of gasoline containing tetraethyl lead and ethylene dibromide. The bromine content of the marine atmosphere is the result of the mixing of local and long-range components and of the chemical reactions occurring during transport. The chemical form of gaseous bromine, like that of chlorine, remains unknown.

Rain and sea-spray in New Zealand were analysed for Br and Cl by Dean (1963) who reported that the average Br/Cl ratio was virtually the same as that found in sea-water (3.4×10^{-3}). However, his Br/Cl values ranged from 0.002 to 0.005, which suggested that in some instances the aerosol may be enriched with bromine and in others, depleted. Subsequently, Duce *et al.* (1963, 1965) made Br and Cl measurements on precipitation and atmospheric aerosol samples taken in Hawaii. They found that the Br/Cl ratios of aerosol samples were generally less than those of sea–water, whereas in the samples of rain-water the ratio was generally double that in sea water. They suggested that the apparent depletion of Br in the marine aerosol might be the result of the chemical oxidation of Br^- to Br_2, with the subsequent release of the latter to the gas phase.

Photochemical reactions, as pointed out by Zafiriou (1974a), probably play an important role in the chemistry of gaseous bromine species. In a first attempt to measure gaseous bromine, Duce *et al.* (1965) found that it was approximately equal to the concentration of the element in the particulate form. They suggested that the excess bromine in rain-water may be due to a chlorine loss from cloud- and rain-water, since both the Br/Cl and I/Cl ratios in rain appeared to increase with distance away from the sea-coast, whereas the I/Br ratio remained virtually constant. However, Seto *et al.* (1969) found that the Na/Cl ratio in rain-water, as a function of distance from the coastline, was essentially constant, whereas both the Br/Cl and I/Cl ratios appeared appreciably higher than the sea-water value and varied as a function of distance from the east coast of Hawaii (Fig. 38.8). It appears that: (i) no chlorine is lost from rain- and cloud-water; (ii) some bromine is lost from the marine aerosol; and (iii) some bromine is gained by rain- and cloud-water.

The Br/Cl ratio of Hawaiian marine aerosols as a function of particle size was examined by Duce *et al.* (1967) who observed a U-shaped curve for the Br/Cl versus particle size plot (see e.g. Fig. 38.3). They explained this as being due to the existence of two different sources for the particulate bromine. They also found that the Br/Cl ratio was lower than that of sea-water for all particle sizes.

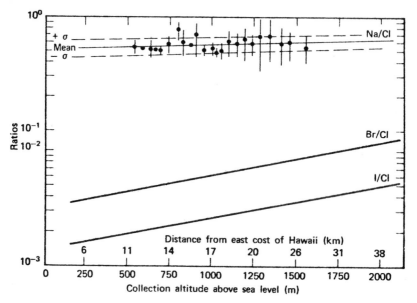

FIG. 38.8. Na/Cl, Br/Cl and I/Cl ratios in Hawaiian rain (Seto *et al.*, 1969). The Na/Cl line is not significantly different from the sea-water value.

Duce and Woodcock (1971) also observed that the Br/Cl ratio in the marine aerosol was substantially lower than that in sea-water. A plausible case for the loss of bromine from the aged marine aerosol can be made on the basis of their work (see Fig. 38.9). It is possible that bromine is lost from the marine aerosol and captured by rainfall, either on small condensation nuclei or in wash-out (MacIntyre, 1974).

The average particle size spectra for bromine and the Br/Cl ratio in the Hawaiian marine atmosphere are shown in Fig. 38.3 (Moyers and Duce, 1972a). The major mass of bromine was found on large particles with radii ranging between 1 and 5 μm; total particulate bromine concentrations averaged ~ 9 ng m^{-3}. The concentration of gaseous Br was found by these authors to be $\sim 50\,\mathrm{ng\,m}^{-3}$ in the Hawaiian marine atmosphere, or four to ten times that of the particulate bromine. Measurements of particulate Br in Antarctic aerosols (Duce *et al.*, 1973) showed the levels at a coastal site (McMurdo, $\sim 300\,\mathrm{km}$ from the coast) to be similar ($\sim 1\,\mathrm{ng\,m}^{-3}$) to those at the South Pole at an altitude of $2800\,\mathrm{m}$ ($0.5\,\mathrm{ng\,m}^{-3}$). At the South Pole, the gaseous bromine had a concentration approximately 20 times that of the

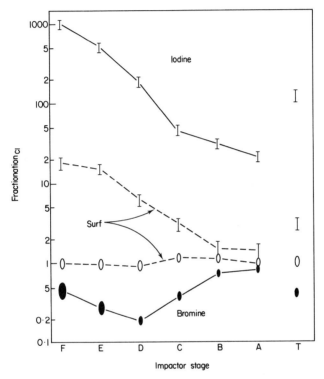

Fig. 38.9. Iodine and bromide fractionation with respect to Cl in fresh aerosol from surf (dashed lines) and in aged open-ocean particles (solid lines). (Data from Duce and Woodcock, 1971.)

particulate Br, with $F_{Cl}(\mathrm{Br})$ values generally greater than 12 for the total aerosol population.

Martens (1972) investigated bromine fractionation ($F_{Na}(\mathrm{Br})$), as a function of particle size at coastal and inland sampling sites in Puerto Rico and found his results to be in relatively good agreement with the Hawaiian data presented above. However, in four samples, particles ranging from 0.5 to 1.0 μm had $F_{Na}(\mathrm{Br})$ values ranging from 2 to 11. Martens suggested that the high fractionation values for Br on the smallest particles may be due to local anthropogenic influences, such as the presence of Br associated with Pb from the combustion of tetraethyl lead and ethylene dibromide in gasoline. The Br/Pb ratios measured on these small particles in Puerto Rico support this interpretation. The possibility also exists for the uptake of natural gaseous Br by the background, non-marine aerosol.

In general, it seems likely that no bromine fractionation occurs at the air–sea interface since knowledge of atmospheric processes can be used to explain apparent fractionation effects. The deviations of the Br/Cl and the Br/Na ratios found in marine aerosols from those of sea-water appear to be due primarily to gaseous bromine exchange with the marine and background aerosol, or to the addition of Br from the automotive combustion of ethyl fluid to small particles $< 1 \cdot 0 \, \mu m$ in radius. However, because the chemical speciation of gaseous bromine remains unknown, and since few reliable measurements of gaseous bromine in the marine atmosphere have been made, further research should be carried out. Because of possible gas–particle reactions, trace gas measurements of bromine compounds pose a particular sampling and analytical problem.

38.3.1.3. Iodine

The behaviour of iodine in the marine atmosphere appears to be distinctly different to that of either chlorine or bromine. A comparison of Fig. 38.3 with Fig. 38.4 shows that although the major mass of both chlorine and bromine is generally found on particles with radii $> 1.0 \, \mu m$, the major mass of iodine is generally on particles with radii of $0 \cdot 5 – 1 \cdot 0 \, \mu m$. As a result, the I/Cl ratio as a function of particle size is distinctly V-shaped and is suggestive of an iodine enrichment. Many investigators have shown that the I/Cl ratios in the marine aerosol and in precipitation samples taken in maritime air are generally 100–1000 times larger than they are in sea-water (see e.g. Komabayasi, 1962; Dean, 1963; Miyake and Tsunogai, 1963; Duce et al., 1963, 1965, 1967; Paslawska and Ostrowski, 1968a, b; Moyers and Duce, 1972b). Considerable evidence exists to support the ideas of an iodine enrichment in the marine aerosol, and of a chemical fractionation effect for iodine at the air–sea interface. A detailed review of the possible pathways has recently been given by Duce and Hoffman (1976) who have summarized the supporting data.

The ocean is generally believed to be the major source of the iodine found in the marine aerosol, but the exact fractionation mechanism at the sea surface remains unknown. It is widely believed (Duce and Hoffman, 1976) that the enrichment of iodine in the marine aerosol is attributable to either: (i) a gas-phase release of iodine, perhaps in the form of CH_3I or I_2, from the ocean surface with subsequent exchange with sea-salt particles; or (ii) an iodine enrichment at the sea surface resulting from bubble rupture and preferential injection of surface-active organic material enriched in iodine. It is probable that both mechanisms play a substantial role in iodine enrichment processes since evidence exists to support both pathways. However, the chemical speciation of the dominant gas-phase iodine compound, like those of chlorine and bromine, remains unknown, and the chemical pathways are not known exactly at present.

The total particulate iodine concentration in the marine atmosphere lies in the range 1–3 ng m^{-3} (Moyers and Duce, 1972b; Duce et $al.$, 1973). The particle size distribution is heavily weighted in favour of the small particle size fractions (Fig. 38.4).

Total gaseous iodine concentrations have been measured by Duce et $al.$ (1965), Moyers and Duce (1972b) and Duce et $al.$ (1973) in surface air near Hawaii and in Antarctica, and were generally found to range from 1 to 20 ng m^{-3}. The levels of gaseous iodine species are normally two-to-four times those of particulate iodine. It is possible that as the chemistry of the gas–particle interactions of iodine becomes better known, the absolute values and relative magnitudes of both gaseous and particulate iodine concentrations will be found to be significantly different. Particle-to-gas and gas-to-particle conversions within sampling devices may have appreciably affected previous atmospheric iodine measurements.

Early work by Miyake and Tsunogai (1963), Martens and Harriss (1970) and Seto and Duce (1972), using $^{131}I^{-}$ tracer in the laboratory, showed that I_2 could be released when a model ocean system was irradiated with light of wavelength 300–500 nm. The I_2 release has generally been attributed to the reaction:

$$2I^{-} + \tfrac{1}{2}O_2 + H_2O \xrightarrow{h\nu} I_2\uparrow + 2OH^{-}. \qquad (11)$$

However, Moyers and Duce (1972b) have pointed out that at the concentrations of I^{-} and IO_3^{-} present in sea-water, and at the concentration of iodine gas present in the atmosphere (assuming that it is gaseous I_2), a consideration of simple ionic thermodynamics leads to the conclusion that I_2 should be dissolving in the sea rather than being released. In addition, Moyers and Duce (1972b) have suggested from thermodynamic considerations of a system containing the inorganic species IO_3^{-}, I^{-}, $I_2(g)$ and $I_2(aq)$, that the marine aerosol may, in fact, be an efficient sink for atmospheric I_2. They have suggested that the rate of iodine uptake by the particles may be controlled by the rate at which IO_3^{-} is formed on the particles. Despite the fact that I_2 is known not to be the stable gas phase of iodine (because of the ease of photodissociation, Zafiriou (1974a)), other organic species such as CH_3I or HI could fulfil this role. Methyl iodide has recently been detected in appreciable quantities (~ 7 ng I m^{-3}) over the North Atlantic by Lovelock et $al.$ (1973), who have suggested that it is produced biologically in the ocean and subsequently released from the sea surface; it probably has a residence time of only a few days because of the ease with which it is photolytically destroyed. Recently, Zafiriou (1975) added to the iodine puzzle by showing, by means of reaction kinetic calculations and laboratory studies, that CH_3I in sea-water reacts with Cl^- to produce CH_3Cl at about the same rate that CH_3I escapes into the marine atmosphere. In addition, he showed that the reaction of CH_3I with Cl^- in the marine aerosol could not account for the observed high-iodine enrichments.

The effect of ultra-violet light on the production of gaseous iodine species has been studied by several investigators (e.g. Martens and Harriss, 1970; Seto, 1970; Seto and Duce, 1972). Seto (1970) observed a three-fold increase in I_2 formation and evaporation when he irradiated a model ocean with u.v. light. Seto and Duce (1972) found that, during bubbling of natural sea-water in the laboratory under u.v. irradiation, the gaseous iodine comprised 90% of the total iodine released, the remainder being particulate iodine. Without irradiation the percentage of gaseous iodine decreased to 75%. During similar experiments without bubbling, the release of gaseous iodine decreased sharply, but not to zero, and appeared to be at the same level both with, or without, u.v. light. From these experiments it seems likely that gaseous iodine is released from the ocean surface, in some form and that this release is enhanced by bubbling. Interestingly, the release of iodine from the sea surface may not be entirely dependent upon photochemical reactions.

From this brief consideration of atmospheric iodine over the ocean it should be evident that substantially more research is needed before the movement of this element through the marine environment can be fully elucidated. Major consideration should be given to the chemical identification of gaseous iodine species and to the delineation of gas–particle iodine exchange processes.

38.3.2. SULPHUR AND NITROGEN COMPOUNDS

38.3.2.1 *Sulphur*

The geochemistry of gaseous and particulate sulphur compounds over the oceans is complicated by multiple sulphur sources and by gas–particle interactions which have not yet been delineated. Probably the major sulphur species in the marine aerosol are $(NH_4)_2SO_4$, NH_4HSO_4, H_2SO_4 and various metallic sulphates and organic sulphates (Dinger *et al.*, 1970; Kellogg *et al.*, 1972; Environmental Protection Agency, 1975). Sulphur gases are likely to be mainly SO_2, with small amounts of H_2S and perhaps minor amounts of various organic sulphides such as carbon disulphide (CS_2), dimethyl sulphide (DMS), dimethyl disulphide (DMDS) and methyl mercaptan (CH_3SH). Very few of the gaseous sulphur species, however, have actually been measured over the ocean. Recent results by Prahm *et al.* (1976) suggest that the undisturbed marine atmosphere may contain only about $0.2\,\mu\text{g m}^{-3}$ SO_2 and $0.8\,\mu\text{g m}^{-3}$ sulphate (both values having tolerances of $\pm 50\%$). Cuong *et al.* (1974) reported SO_2 atmospheric concentrations of 0.04–$0.90\,\mu\text{g m}^{-3}$, and sulphate aerosol concentrations of 0.05–$1.65\,\mu\text{g m}^{-3}$ for the remote oceanic regions of the sub-Antarctic and Antarctic. Table 38.4 presents some representative SO_2 and sulphate values, together with computed SO_2/SO_4^{2-} ratios found over various continental and oceanic areas. To date, DMS and most other organic sulphides have not been detected in the marine atmosphere.

The major sources of sulphur species in the marine atmosphere are believed to include: (i) the release of gaseous sulphur compounds from the ocean surface, possibly H_2S under anaerobic conditions and organic sulphides under both aerobic and anaerobic conditions; (ii) the oxidation of sulphur gases such as SO_2, H_2S and, perhaps, organic sulphides to particulate SO_4^{2-}; (iii) the long-range transport of both gaseous and particulate sulphur compounds from anthropogenic sources; (iv) the ejection of particulate sulphates as a result of bubble rupture at the air–sea interface, leading to a possible sulphur enrichment in the marine aerosol; and (v) an influx of sulphate-containing particles from continental weathering sources. The contributions and spatial distribution of each of these sources are poorly known and are currently the subject of active research.

The possible release of gaseous sulphur compounds from the sea surface has been given recent attention because of its importance to the global sulphur cycle. Material balance estimates for sulphur on a global basis, which were started by Conway (1943) and now include more recent ones (see e.g. Eriksson, 1960, 1963; Junge, 1963; Robinson and Robbins, 1968; Kellogg *et al.*, 1972), have generally implied that sulphur entering the oceans via the rivers must be accounted for largely by re-emission to the atmosphere, ostensibly in the gaseous form as H_2S. Oceanic production estimates for H_2S have ranged up to 2×10^8 tons year^{-1} (Eriksson, 1960). However, there are no reports of successful attempts to measure this species in remote oceanic regions at the very low concentration levels (0.1 ppb or less) at which it would be expected to occur (Rasmussen, 1974). In addition, it seems likely that H_2S is readily oxidized by the dissolved oxygen in surface ocean waters before it can escape into the atmosphere. Even in intertidal flats with an abundance of anaerobic muds, the intervening aerobic surface waters probably prevent any substantial evolution of H_2S, at least in quantities sufficient to account for the amounts of excess sulphate estimated to be in circulation (Östlund and Alexander, 1963; Rasmussen, 1974). Lovelock *et al.* (1972) suggested that DMS, and possibly other organic sulphides, may play the role originally ascribed to H_2S. Dimethyl sulphide is known to be readily

TABLE 38.4

Sulphur dioxide and sulphate in aerosols

a. Aerosol and sulphur dioxide data, Torshavn, February 1975 ($ng\,m^{-3}$) [a]

Date	Atlantic				British Isles				Atlantic average 16–19	British average 25–28
	16	17	18	19	25	26	27	28		
SO_2–S	80	80	80	130	140	540	280	120	90	270
SO_4^{2-}–S_{cor}	30	90	200	230	710	1420	1350	810	140	1070
Sea–S	190	90	90	90	30	70	80	110	115	70
NO_3^-–N	1·1	1·2	0·8	0·8	2·5	3·2	3·5	3·5	1	3
NH_4^+–N	8	19	15	10	29	28	42	40	13	35
H^+	6	9·2	4·5	5·3	28·5	15·8	32·2	35·9	3	28
Pb	0·0	1·3	0·7	0·8	20·8	7·5	34·0	60·0	1	30
Br	5·4	2·1	2·4	2·2	3·6	2·1	3·5	7·5	3	4
Zn	0·7	2·1	0·0	0·4	126	55	209	391	1	19
Cu	0·4	0·5	0·0	0·0	0·7	0·3	1·1	1·5	0	1
Fe	3·7	7·9	0·0	0·0	47·6	26·5	57·7	96·1	4	57
Si	0·0	0·0	0·0	0·0	211	116	148	248	0	180
Mn	0·0	0·0	0·0	0·0	2·3	0·7	4·8	8·5	0	4
V	0·0	0·0	0·0	0·0	1·7	0·8	1·8	3·5	0	2

[a] Prahm *et al.* (1974).

b. Atmospheric concentration of sulphur dioxide and sulphate aerosols[a]

Location	SO$_2$	SO$_4^{2-}$	SO$_2$/SO$_4^{2-}$
	$\mu g SO_2\ m^{-3}$	$\mu g SO_4^{2-}\ m^{-3}$	SO$_2$/SO$_4^{2-}$
Antarctic (60° S.–70° S.) area	0·13	1·57	0·08
Sub-Antarctic (40° S.–60° S.) area	0·18	1·68	0·11
South Pacific Ocean (20° S.–40° S.)	0·12	1·15	0·10
North Atlantic Ocean (50° N.–8° N.)	—	2·33	—
Mediterranean Sea	2·27	8·43	0·27
Boulder, Colorado			
Ground	2·1	1·7	1·3
5·2 km above ground[b]	0·4	0·1	4·0
	$\mu g SO_2\ Nm^{-3}$	$\mu g SO_4^{2-}\ Nm^{-3}$	SO$_2$/SO$_4^{2-}$
Germany			
Ground to 1600 m	4–25	3–7	5–3
Unpolluted area at ≃2800 m[b]	~1	~4	0·25
Sweden	$\mu g S\ kg^{-1}$	$\mu g S\ kg^{-1}$	SO$_2$-S/SO$_4^{2-}$-S
400–2800 m unpolluted area (30 flights)[c]	1–7	0·1–1·6	10·4

[a] Cuong et al. (1974).
[b] Georgii (1970).
[c] Rodhe (1972).

produced by many living systems, especially marine algae. Rasmussen (1974) has noted that laboratory analyses for the sulphur gases in a wide variety of sea-water samples indicate that organic sulphur emissions dominate those of H_2S. On the basis of new, lower background sulphate and sulphur dioxide measurements, Prahm et al. (1976) reported that Robinson and Robbins (1970) may have seriously overestimated the global oceanic free-sulphur emission. They suggested that on a global scale the sea surface may be only a minor source for gaseous sulphur compounds. Further research is needed to determine if biogenic production of gaseous sulphur compounds in sea-water does actually occur on a large scale, and whether it constitutes a major input to the sulphur cycle.

The mechanisms involved in the production of particulate sulphate via the oxidation of gaseous sulphur compounds such as SO_2 and H_2S have been extensively studied, but are still poorly understood. Sulphur dioxide in the marine atmosphere may be oxidized to SO_3, which may be converted to a sulphuric acid aerosol, or it may react with ammonia in liquid droplets and so lead to the formation of $(NH_4)_2SO_4$. Alternatively the SO_2 may form sulphite ions which are then oxidized to sulphate. The resulting acid droplets may interact with other materials found in the atmosphere to form additional sulphate compounds. The major SO_2–sulphate conversion mechanisms that are believed to occur in the atmosphere are shown in Table 38.5 (Environmental Protection Agency, 1975). The SO_2–NH_3 liquid-phase reaction (Mechanism 3) has been studied both experimentally (van den Heuvel and Mason, 1963) and theoretically (Scott and Hobbs, 1967), and it may be a major mechanism for the formation of particulate ammonium sulphate over the oceans. Since this mechanism is an example of a gas-to-particle transformation process, it would be expected to lead to a large population of particles with radii appreciably less than 1 μm. Aerosol measurements made at many locations over the ocean using cascade impactors lend support to this idea, although other processes may also be involved (Meinert and Winchester, 1977; Berg and Winchester, 1976).

Comparison of the SO_2/SO_4^{2-} ratio over the ocean with the values reported for remote, relatively unpolluted land areas provides a rough indication of the extent to which SO_2 has been converted to SO_4^{2-}. Recent land values for the ratio have ranged from 1·3 to 10 (Georgii, 1970; Rodhe, 1972), whereas values in oceanic areas have ranged around 0·1 as shown in Table 38.4 (Cuong et al., 1974). This suggests a rapid conversion of SO_2 to SO_4^{2-} in the marine atmosphere, or a rapid removal of SO_2 from the atmosphere. On the basis of laboratory experiments, Beilke and Lamb (1974) and Spedding (1972) have suggested that the ocean could be a major sink for SO_2. Interestingly, Cuong et al. (1974) have, in fact, reported that the Mediterranean Sea is definitely a sink for SO_2.

TABLE 38.5

Mechanisms by which sulphur dioxide is converted to sulphates

Mechanism	Overall reaction	Factors on which sulphate formation primarily depends
1. Direct photo-oxidation	$SO_2 \xrightarrow[\text{water}]{\text{Light, oxygen}} H_2SO_4$	Sulphur dioxide concentration, sunlight intensity
2. Indirect photo-oxidation	$SO_2 \xrightarrow[\text{hydroxyl radical (OH·)}]{\substack{\text{Smog, water, NO}_x \\ \text{organic oxidants,}}} H_2SO_4$	Sulphur dioxide concentration, organic oxidant concentration, $OH^·$, NO_x
3. Air oxidation in liquid droplets	$SO_2 \xrightarrow{\text{Liquid water}} H_2SO_3$ $NH_3 + H_2SO_3 \xrightarrow{\text{Oxygen}} NH_4^+ + SO_4^{2-}$	Ammonia concentration
4. Catalysed oxidation in liquid droplets	$SO_2 \xrightarrow[\text{heavy metal ions}]{\text{Oxygen, liquid water,}} SO_4^{2-}$	Concentration of heavy metal (Fe, Mn) ions
5. Catalysed oxidation on dry surfaces	$SO_2 \xrightarrow[\text{carbon, water}]{\text{Oxygen, particulate}} H_2SO_4$	Carbon particle concentration (surface area)

Recently, PIXE cascade impactor data have been used to study the long-range transport of gaseous and particulate sulphur from anthropogenic and crustal weathering sources. Preliminary results (Darzi, 1977) show that sulphur in aerosols over the western North Atlantic is most abundant in particles having a radius $< 0.5\ \mu m$. These particles are most probably formed as a result of a gas-to-particle conversion process (see above) involving, among other compounds, gaseous SO_2 derived from pollution sources in urbanized land areas. Some of the sub-micrometre sulphur, however, may be formed as a result of bubble rupture at the air–sea interface (Berg and Winchester, 1976). In general, the sulphur in particles $> 1\mu m$ in diameter is also believed to be derived predominantly from the sea surface by bubble rupture and sea-spray. One exception showing the effect of continental weathering material has been reported by Buat-Menard et al. (1974) who found that the SO_4^{2-}/Na ratios in most of the samples which they collected off the coast of NW. Africa correlated remarkably well with the Ca/Na ratios. These authors suggested that a large fraction of the sulphur in the marine aerosol was in the form of gypsum ($CaSO_4 \cdot 2H_2O$) which was derived from the African desert areas.

The transport to oceanic locations of sulphur, probably as sulphate, from anthropogenic sources, continental dust, background aerosols, volcanic emissions and assorted gas-to-particle conversion processes has resulted in many investigators finding enrichment values ($E_{Na}(SO_4)$) substantially greater than zero for marine aerosols. Duce and Hoffman (1976) have critically reviewed the possible sulphur fractionation processes at the air–sea interface. In general, the sulphur fractionation which has been found is attributable to the sources mentioned above, and not to actual sea surface fractionation processes (Junge, 1963; Buat-Menard et al., 1974; Cuong et al., 1974).

Attention has recently been focused on isotopic studies of sulphur in marine atmospheric sulphate particles. By examining and comparing the $^{34}S/^{32}S$ ratios in carefully selected samples it should be possible to identify the sources of the sulphur (Dequasie and Grey, 1970; Manowitz et al., 1970; Grey and Jensen, 1972; Nielsen, 1974). Such source identification would be extremely important in understanding the marine air chemistry of sulphur compounds, although, as yet, little experimental work has been reported in this potentially fruitful area.

Normally, the deviation of the $^{34}S/^{32}S$ ratio of the samples from that of a standard is reported as follows:

$$\delta^{34}S = 1000 \left[\frac{(^{34}S/^{32}S)_x - (^{34}S/^{32}S)_{std}}{(^{34}S/^{32}S)_{std}} \right] \qquad (12)$$

where the subscripts x and std refer to sample and standard, respectively.

The estimated accuracy of the analysis is generally ∼ ±0·3 per mille (Ludwig, 1976). Figure 38.10 shows some typical sulphur isotope ratios, together with the results reported by Ludwig (1976). In general, sea-water sulphur usually has a very high sulphur isotope ratio; fossil fuel sulphur tends to have a smaller isotope ratio, and sulphur of bacteriogenic origin a still smaller one. Ludwig (1976) found that samples collected in unpolluted marine air south of San Francisco had sulphur isotope ratios that were significantly lower

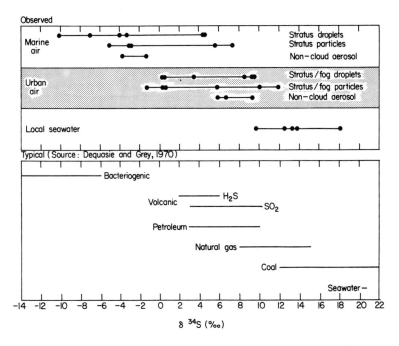

FIG. 38.10. Observed and typical sulphur isotope ratios (Ludwig, 1976).

than those of samples collected in nearby areas that were subject to urban pollution. The highest sulphur isotope ratios were found in the offshore sea-water. Ludwig has reasoned that these results suggest that the aerosol sulphur in the marine air has a bacteriogenic origin, since the low isotope ratios in the marine air cannot be explained as arising from a mixture of sea-water sulphur and pollutant sulphur because both tend to have higher isotope ratios. This result lends support to the idea of a release of gaseous compounds, such as H_2S and various organic sulphides, from the ocean surface (Lovelock, 1972), followed by various oxidation steps leading to a sulphate aerosol. Sulphur isotope research appears to be a promising avenue for investigation of the movement of sulphur through the natural marine environment.

38.3.2.2. Nitrogen

Nitrogen-containing constituents account for a major share of the gaseous and particulate compounds found in the marine atmosphere, yet remarkably little data can be found in the literature on their concentration levels, horizontal and vertical distributions, potential sources and gas–particle chemistry. Furthermore, despite the fact that organically bound nitrogen compounds are expected to be present at appreciable levels in the marine aerosol near the ocean surface, practically no data have been found to date on these compounds. Most of the information on the atmospheric geochemistry of nitrogen over the ocean has been developed from precipitation data, gas measurements, laboratory simulation experiments and isotopic studies of airborne nitrogen compounds.

In addition to molecular nitrogen, the major nitrogen-containing gaseous components in the marine atmosphere include NH_3, N_2O, NO_2 and possibly unknown organically bound nitrogen compounds. A much larger number of nitrogen compounds appears likely to occur in the gas phase, since nitrogen is able to form a very large number of covalently bonded compounds with oxygen. However, source and sink information on gaseous nitrogen components over the ocean is almost non-existent. The production of atmospheric ammonia is widely held to be biogenic (Georgii and Müller, 1974) and to be predominantly continental. More data are needed to firmly establish its global cycle. Background concentration levels at ground-level usually range from ~ 4–$20\,\mu gNH_3\,m^{-3}$ (Georgii and Müller, 1974). Probably the only presently known major gaseous, globally distributed, nitrogen compound believed to have an oceanic source is nitrous oxide. On the basis of daily analyses of air samples collected over a 2·5 year period, Schütz et al. (1970), have postulated that large parts of the Atlantic Ocean act as a source for this gas. This hypothesis was supported by data later presented by Hahn (1974), who found from sea-water measurements that the upper layers of the North Atlantic Ocean, and probably those of other oceans, act as a net source of atmospheric N_2O.

The global pathways of the important gaseous nitrogen compound, nitrogen dioxide, remain largely unknown. This compound is soluble in water with which it reacts, e.g. in rain-water, in accordance with the reaction:

$$2NO_2 + H_2O \rightarrow HNO_3 + HNO_2 \qquad (13)$$

This reaction has been held by many (e.g. Georgii, 1963) to lead to the observed nitrate concentrations found in rain-water, but more recent work by Georgii and Müller (1974) and by Cox (1974) suggests that perhaps more complicated reactions are involved. In terms of the World Ocean, however, the atmospheric chemistry of this component remains obscure.

Over the oceans, the marine aerosol is characterized predominantly by the presence of both the NH_4^+ ion (bound primarily by the SO_4^{2-} ion as $(NH_4)_2 SO_4$) and the NO_3^- ion. Precipitation data dating back to 1855 have been comprehensively surveyed by Eriksson (1952) and atmospheric nitrogen compounds in general have been reviewed by Georgii (1963) and Junge (1963). In the troposphere, ammonium sulphate exists either in crystalline form (at relative humidities $<81\%$), or as droplets (at relative humidities $>81\%$). Junge (1953) has shown that the crystals are mainly of relatively large size (diameter $0\cdot2$–$1\cdot0$ μm). Electron microscope studies by Heard and Wiffen (1969) indicated that the ammonium sulphate exists as separate and discrete particles which are not necessarily associated with other aerosol constituents. Very few atmospheric concentration values have been reported in the literature for nitrogen in the marine aerosol. Rahn (1976a), has made a comprehensive literature compilation of aerosol elemental data and found only eight concentration values, of which only one was for a marine aerosol. In this marine aerosol, from San Nicolas Island off the southern coast of California, Hidy et al. (1974) found 899 ngN m^{-3}. The other seven values were urban measurements which showed a mean of 3500 (±2300) ngN m^{-3}. Whether or not these values are representative of the marine aerosol in general remains to be demonstrated by further sampling and analysis.

Open-ocean source and sink information on the nitrogen components in the marine atmosphere remains poorly understood because of both a lack of measurements and an apparent lack of understanding concerning major gas–particle interactions. It has usually been assumed that ammonium sulphate must be formed by the reaction of ammonia with sulphur compounds in the atmosphere (Healy et al., 1970). A purely gas-phase set of reactions has been ruled out by Healy et al. (1970), who showed that for an appreciable amount of particulate $(NH_4)_2SO_4$ to be formed by gaseous interaction of SO_2 and NH_3 it was necessary to postulate a very high, and unlikely, concentration of SO_2 (>1000 μgSO_2 m^{-3}). Reactions involving SO_2 and NH_3 in the presence of solid particles or liquid water seem to be a more likely mechanism, and several conversion mechanisms have been postulated, although, to date, none have gained wide acceptance.

Van den Heuvel and Mason (1963) have directly demonstrated in the laboratory that ammonium sulphate is formed in water droplets suspended on a quartz grid in a current of air heavily loaded with NH_3 and SO_2. Ammonia promotes the oxidation by maintaining a high pH (Junge and Ryan, 1958), which is necessary since the species actually being oxidized is SO_3^{2-} rather than HSO_3^- or dissolved SO_2. If no alkali is present, the product is sulphuric acid, so the pH falls and the reaction stops. On the basis of the kinetic data of Van den Heuvel and Mason (1963) and the known equilibria, Scott and Hobbs (1967) have calculated the rate of production of $(NH_4)_2SO_4$ in a water droplet at 25°C. Further calculations, some by McKay (1969), based on the kinetic data of Fuller and Crist (1941) and Barron and O'Hern (1966), have indicated that: (i) a shorter reaction time was more probable than that calculated by Scott and Hobbs (1967); and (ii) the reaction must proceed considerably faster at lower temperatures because of the increased solubility of the gases. The results suggest that ambient $(NH_4)_2SO_4$ levels can build up in less than 1 hour in a thick mist (~ 100 $mgH_2O\,m^{-3}$). At a later time, if the relative humidity drops and the mist droplets dry out, the $(NH_4)_2SO_4$ crystals formed should have diameters ranging between 0·1 and 1·0 μm, corresponding to the sizes most prevalent in the atmosphere (Healy et al., 1970).

The above mechanism probably accounts for the formation of some of the ammonium sulphate present over the ocean, but at lower relative humidities this may not be the case. Georgii and Müller (1974) have, in fact, reported data which, in general, do not support the NH_3–SO_2–liquid-water hypothesis. At humidity levels well below 100%, any droplet phase would have to be stabilized by dissolved solids. Foster (1969) has considered the oxidation of SO_2, employing dissolved solids containing catalysts, in power-plant plumes. He has shown that at relative humidities greater than 70% the principal product is H_2SO_4, the catalytic effect being ascribed to iron oxide. The H_2SO_4 product will be rapidly converted to $(NH_4)_2SO_4$ as the plume mixes with ambient levels of ammonia (Cadle and Robbins, 1960). Over the ocean this mechanism may partially account for the high levels of small-particle sulphur (diameter $<1·0$ μm and probably bound as $(NH_4)_2SO_4$) that have been observed to travel great distances from anthropogenic sources (Darzi, 1977). In general, however, this mechanism probably does not account for a very significant part of the global load of $(NH_4)_2SO_4$ found over the ocean, since both the SO_2 and dust levels are orders of magnitude lower in the marine atmosphere than in a plume, except in very special cases.

Further research is vitally needed in this area of gas–particle conversion and interaction. Reactions occurring directly on the surface of the solid marine aerosol remain to be carefully examined and their relative importance evaluated. Urone et al. (1968) have observed SO_2 uptake at 50% relative humidity by such marine aerosol components as NaCl and $CaCO_3$. Reactions occurring directly on the sea-salt particle surface, however, remain to be delineated.

Isotopic measurements of atmospheric nitrogen compounds over the ocean may provide significant new information on the sources and sinks of the compounds. The isotopic composition of a particular chemical compound is related to its origin and is modified by various physical and chemical reactions involving the compound during its lifetime. Isotopic fractionation can occur under equilibrium conditions or for irreversible reactions. Normally, nitrogen isotope results are reported in units of per mille deviations from that of atmospheric molecular N_2 as shown in Equation (14):

$$\delta^{15}N = 1000 \left[\frac{(^{15}N/^{14}N)_x - (^{15}N/^{14}N)_{std}}{(^{15}N/^{14}N)_{std}} \right] \qquad (14)$$

where the terms are each similar to those shown in Equation (12). The precision for individual sample measurements has been estimated to be ± 0.2 per mille (Moore, 1974).

Relatively few isotopic studies of atmospheric nitrogen compounds have been undertaken, and almost no studies have been conducted exclusively for the marine atmosphere. However, recent work suggests that valuable information could be obtained on the chemistry of nitrogen compounds in and over the ocean from such studies. For example, Moore (1974) has reported variations in $^{15}N/^{14}N$ ratios for different samples; these are presented in Fig. 38.11, which also includes the published results of other

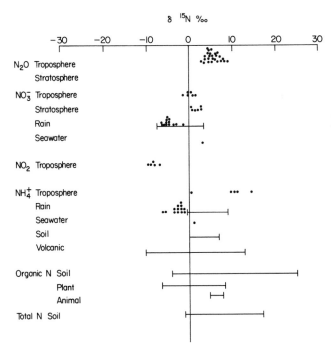

FIG. 38.11. Variation in $^{15}N/^{14}N$ ratios for various samples. The dots indicate values determined by the author. The bars represent ranges of values found by other workers. (From Moore, 1974.)

authors (Hoering, 1958; Cheng *et al.*, 1964; Volynets *et al.*, 1967; Delwiche and Steyn, 1970). Examination of these data leads to several significant observations concerning the origin of nitrogen compounds in the atmosphere. A comparison of the $^{15}N/^{14}N$ ratio in NH_4^+ in particulate matter collected on filters and in soil dust suggests that most of the particulate NH_4^+ in air results from the presence of entrained soil particles, and that the NH_4^+ in rain must come from another source, presumably NH_3 in the atmosphere. A comparison of the $^{15}N/^{14}N$ ratio for NO_3^- in rain, in aerosols and in soil again suggests that the NO_3^- in rain must originate from the NO or the NO_2 in the atmosphere. The role that the ocean plays with regard to the origin of these nitrogen compounds seems to be minimal. Mechanistically, it appears that the NH_4^+ and NO_3^- in rain result from the incorporation of NH_3 and NO_2 following the formation of water droplets (Moore, 1974). This conclusion is slightly at variance with some previous work and points to the fact that more research is needed before the marine air chemistry of nitrogen compounds is known with any significant degree of confidence.

Despite the fact that nitrogen-containing constituents account for a major share of the gaseous and particulate compounds found in the marine atmosphere, it is apparent from this review that remarkably little is known concerning their physical and chemical behaviour over the ocean. Traditionally, nitrogen compounds have been studied mainly in terms of urban pollution problems, or in relation to the stratospheric ozone budget. A new and detailed consideration of the nitrogen budget in the marine atmosphere would seem to be a highly productive area for future research.

ACKNOWLEDGEMENTS

This review was carried out in part with financial support from the U.S. Environmental Protection Agency, the National Institutes of Health and, for one of us (J.W.W.), the National Science Foundation through a visiting scientist appointment at the National Center for Atmospheric Research.

REFERENCES

Aitken, J. (1881). *Trans. R. Soc.* **30**, 337.
APOMET (collected papers) (1976). "Air Pollution Measurement Techniques", Special Environmental Report No. 10, W.M.O. No. 460. World Meteorological Organization, Geneva.
Barger, W. R. and Garrett, W. D. (1970). *J. Geophys. Res.* **75**, 4561.
Barger, W. R. and Garrett, W. D. (1976). *J. Geophys. Res.* **81**, 3151.
Barker, D. R. and Zeitlin, H. (1972). *J. Geophys. Res.* **77**, 5076.

Barron, C. H. and O'Hern, H. A. (1966). *Chem. Eng. Sci.* **21**, 397.

Baylor, E. R., Sutcliffe, W. H., Jr. and Hirschfeld, D. S. (1962). *Deep-Sea Res.* **9**, 120.

Beilke, S. and Lamb, D. (1974). *Tellus*, **26**, 268.

Berg, W. W. and Winchester, J. W. (1976). "Chlorine Chemistry in the Marine Atmosphere", Tech. Rept. **2–76**. The Florida State University, Tallahassee, U.S.A., 246 pp.

Berg, W. W. and Winchester, J. W. (1977). *J. Geophys. Res.* **82**, 5945.

Bezdek, H. F. and Carlucci, A. F. (1974) *Limnol. Oceanogr.* **19**, 126.

Blanchard, D. C. (1963). *In* "Progress in Oceanography" (M. Sears, ed.), Vol. 1, pp. 73–202. MacMillan, New York.

Blanchard, D. C. (1964). *Science, N.Y.* **146**, 396.

Blanchard, D. C. (1968). *In* "Proceedings of the International Conference on Cloud Physics, Toronto", pp. 25–29. American Meteorological Society, Boston, Massachusetts, U.S.A.

Blanchard, D. C. and Parker, B. C. (1977). *In* "The Aquatic Microbial Community" (J. Cairns, ed.), Vol. 15, pp. 625–653. Garland Pub., New York.

Blanchard, D. C. and Syzdek, L. D. (1972). *J. Geophys. Res.* **77**, 5087.

Blanchard, D. C. and Syzdek, L. D. (1974). *Limnol. Oceanogr.* **19**, 133.

Blanchard, D. C. and Woodcock, A. H. (1957). *Tellus*, **9**, 145.

Bloch, M. R. and Luecke, W. (1968). *Naturwissenschaften*, **91**, 441.

Bloch, M. R. and Luecke, W. (1970). *Israel J. Earth Sci.* **19**, 41.

Boyce, S. G. (1951). *Science, N.Y.* **113**, 620.

Buat-Menard, P. (1970). Thèse de Troisième Cycle. Faculté des Sciences, Paris

Buat-Menard, P., Morelli, J. and Chesselet, R. (1974). *J. Rech. Atmos.* **8**, 661.

Cadle, R. D. (1975). *J. Geophys. Res.* **80**, 1650.

Cadle, R. D. and Robbins, R. C. (1960). *Discuss. Faraday Soc.* **30**, 155.

Cauer, H. (1935). *Z. Analyt. Chem.* **103**, 321, 385.

Cauer, H. (1938). *Balneologe*, **5**, 409.

Cauer, H. (1951). *In* "Compendium of Meteorology", 1126–1136. American Meteorological Society, Boston, Massachusetts, U.S.A.

Cheng, H. H., Bremner, J. M. and Edwards, A. P. (1964). *Science, N.Y.* **146**, 1574.

Chesselet, R., Morelli, J. and Buat-Menard, P. (1972a). *J. Geophys. Res.* **77**, 5116.

Chesselet, R., Morelli, J. and Buat-Menard, P. (1972b) *In* "The Changing Chemistry of the Oceans" (D. Dyrssen and D. Jagner, eds). John Wiley and Sons, New York.

Conway, E. J. (1943). *Proc. R. Ir. Acad. B*, **48**, 161.

Cox, R. A. (1974). *Tellus*, **26**, 235.

Cuong, N. B., Bonsang, B. and Lambert, G. (1974). *Tellus*, **26**, 241.

Darzi, M. (1977). M.Sc. Thesis, The Florida State University, Tallahassee, U.S.A.

Day, J. A. (1964). *Q. Jl R. Met. Soc.* **90**, 72.

Dean, G. A. (1963). *N.Z. Jl Sci.* **6**, 208.

Delany, A. C., Parkin, D. W., Griffin, J. J., Goldberg, E. D. and Reimann, B. E. F. (1967). *Geochim. Cosmochim. Acta*, **31**, 885.

Delany, A. C., Pollock, W. H. and Shedlovsky, J. P. (1973). *J. Geophys. Res.* **78**, 6249.

Delany, A. C., Shedlovsky, J. P. and Pollock, W. H. (1974). *J. Geophys. Res.* **79**, 5646.

Delwiche, C. C. and Steyn P. L. (1970). *Environ. Sci. Tech.* **4**, 929.

Dequasie, H. L. and Grey, D. C. (1970). Am. Lab. **December**, 19.

Dilling, W. L., Tefertiller N. and Kallos G. (1975). *Environ. Sci. Tech.* **9**, 833.

Dinger, J. E., Howell, H. B. and Wojciechowski T. A. (1970). *J. Atmos. Sci.* **27**, 791.

Duce, R. A. (1969). *J. Geophys. Res.* **74**, 4597.

Duce, R. A. (1973). "The Chemistry of Chlorine Compounds in the Non-urban Atmosphere". University of Rhode Island, U.S.A., 35 pp.
Duce, R. A. and Hoffman E. J. (1976). *Ann. Rev. Earth Planet. Sci.* **3**, 187.
Duce, R. A. and Woodcock A. H. (1971) *Tellus*, **23**, 427.
Duce, R. A., Wasson J. T., Winchester J. W. and Burns F. (1963). *J. Geophys. Res.* **68**, 3943.
Duce, R. A., Winchester, J. W. and Van Nahl, T. W. (1965). *J. Geophys. Res.* **70**, 1175.
Duce, R. A., Woodcock A. H. and Moyers J. L. (1967) *Tellus*, **19**, 369.
Duce, R. A., Stumm, W. and Prospero, J. M. (1972). *J. Geophys. Res.* **77**, 5059.
Duce, R. A., Hoffman, G. L., Fasching, J. L. and Moyers J. L. (1973). *In* "Proceedings of the WMO/WHO Technical Conference on Observation and Measurement of Atmospheric Pollution", Helsinki.
Duce, R. A., Ray, B. J., Hoffman, G. L. and Walsh, P. R. (1976). *Geophys. Res. Lett.* **3**, 339.
Durbin, W. G. and White, G. D. (1961). *Tellus*, **13**, 260.
Eriksson, E. (1952). *Tellus*, **4**, 215, 280.
Eriksson, E. (1955). *Tellus*, **7**, 388.
Eriksson, E. (1959). *Tellus*, **11**, 375.
Eriksson, E. (1960). *Tellus*, **12**, 63.
Eriksson, E. (1963). *J. Geophys. Res.* **68**, 4001.
Environmental Protection Agency, United States. (1975). Pub. No. EPA-450/2-75-007 (September).
Fasching, J. L., Courant R. A., Duce R. A. and Piotrowicz S. R. (1974). *J. Rech. Atmos.* **8**, 649.
Folger, D. W. (1970). *Deep-Sea Res.* **17**, 337.
Foster, P. M. (1969). *Atmos. Environ.* **3**, 157.
Friedlander, S. K. (1973). *Environ. Sci. Tech.* **7**, 235.
Fuller, E. C. and Crist R. H. (1941). *J. Am. Chem. Soc.* **63**, 1644.
Gambell, A. W. J. (1962). *Tellus*, **14**, 91.
Garrett, W. D. (1967). *Deep-Sea Res.* **14**, 221.
Gast, J. A. and Thompson T. G. (1959). *Tellus*, **11**, 344.
Gatz, D. F. (1975). *Atmos. Environ.* **9**, 1.
Georgii, H. W. (1963). *J. Geophys. Res.* **68**, 3963.
Georgii, H. W. (1970). *J. Geophys. Res.* **75**, 2365.
Georgii, H. W. and Müller, W. J. (1974). *Tellus*, **26**, 180.
Gillette, D. A. (1974). *J. Rech. Atmos.* **8**, 735.
Goldberg, E. D., ed. (1976). "Strategies for Marine Pollution Monitoring". Wiley-Interscience, New York, 310 pp.
Gordon, C. M., Jones, E. C. and Larson, R. E. (1977). *J. Geophys. Res.* **82**, 988.
Gorham, E. (1955). *Geochim. Cosmochim. Acta*, **7**, 231.
Graedel, T. E. and Allara, D. L. (1976). *Atmos. Environ.* **10**, 385.
Green, W. D. (1972). *J. Geophys. Res.* **77**, 5152.
Grey, D. C. and Jensen, M. L. (1972). *Science, N.Y.* **177**, 1099.
Grimsrud, E. P. and Rasmussen, R. A. (1975a). *Atmos. Environ.* **9**, 1014.
Grimsrud, E. P. and Rasmussen, R. A. (1975b). *Atmos. Environ.* **9**, 1010.
Hahn, J. (1974). *Tellus*, **26**, 160.
Healy, T. V., McKay, H. A. C., Pilbeam, A. and Scargill, D. (1970). *J. Geophys. Res.* **75**, 2317.
Heard, M. J. and Wiffen R. D. (1969). *Atmos. Environ.* **3**, 337.

Hidy, G. M., Mueller, P. K., Wang, H. H., Karney, J., Twiss, S., Imada, M. and Alcocer, A. (1974). *J. Appl. Met.* **13**, 96.

Hoering, T. C. (1958). *Geochim. Cosmochim. Acta*, **15**, 154.

Hoffman, E. J. and Duce, R. A. (1974). *J. Geophys. Res.* **79**, 4474.

Hoffman, E. J. and Duce, R. A. (1976). *J. Geophys. Res.* **81**, 3667.

Houghton, H. G. (1932). *Physics*, **2**, 467.

Hsu, S. A. and Whelan, T. (1976). *Environ. Sci. Tech.* **10**, 281.

ISCSAPEP (collected papers) (1974). *J. Rech. Atmos.* **8**, No. 3–4, 501.

Jacobs, W. C. (1937). *Mon. Weath. Rev.* **65**, 147.

Johansson, T. B., Van Grieken, R. E. and Winchester, J. W. (1974). *J. Rech. Atmos.* **8**, 761.

Johansson, T. B., Van Grieken, R. E. and Winchester, J. W. (1976). *J. Geophys. Res.* **81**, 1039.

Junge, C. E. (1953). *Tellus*, **5**, 1.

Junge, C. E. (1956). *Tellus*, **8**, 127.

Junge, C. E. (1957). *Tellus*, **9**, 528.

Junge, C. E. (1963). "Air Chemistry and Radioactivity", Academic Press, New York and London, 382 pp.

Junge, C. E. (1972). *J. Geophys. Res.* **77**, 5183.

Junge, C. E. and Ryan, T. G. (1958). *Q. Jl R. Met. Soc.* **84**, 46.

Junge, C. E. and Werby, R. T. (1958). *J. Met.* **15**, 417.

Kellogg, W. W., Cadle, R. D., Allen, E. R., Lazrus, A. L. and Martell, E. A. (1972). *Science, N.Y.* **175**, 587.

Ketseridis, G., Hahn, J., Jaenicke, R. and Junge, C. (1976). *Atmos. Environ.* **10**, 603.

Kientzler, C. F., Arons, A., Blanchard, D. and Woodcock, A. (1954). *Tellus*, **6**, 1.

Köhler, H. (1922). "Zur Kondensation des Wasserdampfes in der Atmosphöre, Erste Mitteilung 1921, Zweite Mitteilung 1922." Geofysiske Pub. 2, Oslo.

Köhler, H. (1925). *Meddn St. Met.-Hydrogr. Anst.* **2**, 73.

Köhler, H. (1941). *Nova Acta R. Soc. Scient. Upsal.* **12**, 55.

Köhler, H. and Bath, M. (1952). *Nova Acta R. Soc. Scient. Upsal.* **15**, 24.

Komabayasi, M. (1962). *J. Met. Soc. Japan*, **40**, 25.

Kowalczyk, G. S., Choquette, C. E. and Gordon, G. E. (1978). *Amos. Environ.* **12**.

Lai, R. S. and Shemdin, O. H. (1974). *J. Geophys. Res.* **79**, 3055.

Lazrus, A. L., Baynton, H. and Lodge, J. P., Jr. (1970). *Tellus*, **22**, 106.

Liss, P. S. (1975). *In* "Chemical Oceanography" (J. P. Riley and G. Skirrow, eds), Vol. 2, 2nd ed., pp. 193–243. Academic Press, London and New York.

Lodge, J. P., Jr. (1960). *In* "Physics of Precipitation", pp. 252–256. Waverly Press, Baltimore, U.S.A.

Lodge, J. P., Jr., MacDonald, A. J. and Vihman, E. (1960). *Tellus*, **11**, 184.

Lovelock, J. E. (1971). *Nature, Lond.* **230**, 379.

Lovelock, J. E. (1972). *Atmos. Environ.* **6**, 917.

Lovelock, J. E. (1975). *Nature, Lond.* **256**, 193.

Lovelock, J. E., Maggs, R. J. and Rasmussen, R. A. (1972). *Nature, Lond.* **237**, 452.

Lovelock, J. E., Maggs, R. J. and Wade, R. J. (1973) *Nature, Lond.* **241**, 194.

Ludwig, F. L. (1976). *Tellus*, **28**, 427.

MacIntyre, F. (1968). *J. Phys. Chem.* **72**, 589.

MacIntyre, F. (1974). *In* "The Sea" (E. D. Goldberg, ed.), Vol. 5, pp. 245–299. John Wiley and Sons, New York.

MacIntyre, F. and Winchester, J. W. (1969). *J. Phys. Chem.* **72**, 2163.

McKay, H. A. C. (1969) *Chemy Ind.* 1162.
Manowitz, B., Smith, M. E. and Steinberg, M. (1970). *In* "Proceedings Second International Clean Air Congress, Washington D.C.".
Martell, E. A. and Moore, H. E. (1974). *J. Rech. Atmos.* **8**, 903.
Martens, C. S. (1973). *J. Geophys. Res.* **78**, 8867.
Martens, C. S. and Harriss, R. C. (1970). *In* "Precipitation Scavenging (1970)", AEC Symposium Series, Vol. 22, p. 319. U.S. Atomic Energy Commission.
Martens, C. S., Wesolowski, J. J., Harriss, R. C. and Kaifer, R. (1973). *J. Geophys. Res.* **78**, 8778.
Mason, B. J. (1957). *Geofis. Pura Appl.* **36**, 148.
Mason, B. (1966). "Principles of Geochemistry", 3rd ed. John Wiley and Sons, New York.
Meinert, D. L. and Winchester, J. W. (1977). *J. Geophys. Res.* **82**, 1778.
Metnieks, A. L. (1958). *Geophys. Bull., Dubl.* **15**, 1.
Miyake, Y. (1948). *Geophys. Mag.* **16**, 64.
Miyake, Y. and Tsunogai, S. (1963). *J. Geophys. Res.* **68**, 3989.
Molina, M. J. and Rowland, R. S. (1974). *Nature, Lond.* **249**, 810.
Moore, H. (1974). *Tellus*, **26**, 169.
Morelli, J., Buat-Menard, P. and Chesselet, R. (1974). *J. Rech. Atmos.* **8**, 961.
Moyers, J. L. and Duce, R. A. (1972a). *J. Geophys. Res.* **77**, 5330.
Moyers, J. L. and Duce, R. A. (1972b). *J. Geophys. Res.* **78**, 5229.
Nielson, H. (1974). *Tellus*, **26**, 213.
Okita, T., Kaneda, K., Kanaka, T. and Sugai, R. (1974). *Atmos. Environ.* **8**, 927.
Östlund, H. G. and Alexander, J. (1963). *J. Geophys. Res.* **68**, 3995.
Owens, J. S. (1940). *Q. Jl R. Met. Soc.* **66**, 2.
Parkin, D. W., Phillips, D. R., Sullivan, R. A. L. and Johnson, L. R. (1972). *Q. Jl. R. Met. Soc.* **98**, 798.
Paslawska, S. and Ostrowski, S. (1968a). *Acta Geophys. Pol.* **16**, 181.
Paslawska, S. and Ostrowski, S. (1968b). *Acta Geophys. Pol.* **16**, 329.
Petriconi, G. L. and Papee, H. M. (1972). *Water, Air, Soil Poll.* **1**, 117.
Pierson, D. H., Cawse, P. and Chambray, R. (1974). *Nature, Lond.* **251**, 675.
Prahm, L. P., Torp, U. and Stern, R. M. (1976). *Tellus*, **28**, 355.
Prospero, J. M. and Carlson, T. N. (1972) *J. Geophys. Res.* **77**, 5255.
Rahn, K. A. (1971). Ph.D. Thesis, University of Michigan, Ann Arbor, U.S.A., 309 pp.
Rahn, K. A. (1976a). "The Chemical Composition of the Marine Aerosol", Tech. Rept. July. University of Rhode Island, U.S.A.
Rahn, K. A. (1976b). *Atmos. Environ.* **10**, 597.
Rasmussen, R. A. (1974). *Tellus*, **26**, 254.
Riley, J. P. (1975). *In* "Chemical Oceanography" (J. P. Riley and G. Skirrow, eds), Vol. 1, 2nd ed., Appendix 2. Academic Press, London and New York.
Robbins, J. A. (1970). *In* "Precipitation Scavenging (1970)", AEC Symposium Series, Vol. 22, p. 325. U.S. Atomic Energy Commission.
Robbins, R. C., Cadle, R. D. and Eckhardt, D. L. (1959). *J. Met.* **16**, 53.
Robinson, E. and Robbins, R. C. (1968). "Sources, Abundance and Fate of Gaseous Atmospheric Pollutants", Final Rept. Proj. PR-6755, Stanford Research Inst., Menlo Park, California, U.S.A.
Robinson, E. and Robbins, R. C. (1970). *J. Air. Pollut. Control Ass.* **20**, 233.
Rodhe. H. (1972). *Tellus*, **24**, 128.
Rossby, C. G. and Egnér, H. (1955). *Tellus*, **7**, 118.
Rowland, F. S. and Molina, M. J. (1975). *Rev. Geophys. Space Phys.* **13**, 1.

Schroeder, W. H. and Urone, P. (1974). *Environ. Sci. Tech.* **8**, 756.

Schütz, K., Junge, C. E., Beck, R. and Albrecht, B. (1970). *J. Geophys. Res.* **75**, 2230.

Scott, W. D. and Hobbs, P. V. (1967). *J. Atmos. Sci.* **24**, 54.

Seto, Y. B. (1970). M.Sc. Thesis, University of Hawaii, U.S.A., 163 pp.

Seto, Y. B. and Duce, R. A. (1972). *J. Geophys. Res.* **77**, 5339.

Seto, Y. B., Duce, R. A. and Woodcock, A. H. (1969). *J. Geophys. Res.* **74**, 1101.

Spedding, D. J. (1972). *Atmos. Environ.* **64**. 583.

Stensland, G. J. and de Pena, R. G. (1975). *J. Geophys. Res.* **80**, 3410.

Su, C. W. and Goldberg, E. D. (1973). *Nature, Lond.* **245**, 27.

Toba, Y. (1965). *Tellus*, **17**, 131, 365.

Toba, Y. (1966). *Telus*, **18**, 132.

Tsunogai, S. (1974). *Tellus*, **27**, 51.

Tsunogai, S. and Fukuda, K. (1974). *Geochem. J.* **8**, 141.

Tsunogai, S., Saito, O., Yamada, K. and Nakaya, S. (1972). *J. Geophys. Res.* **77**, 5283.

TTPO (collected papers) (1975). "A Workshop on the Tropospheric Transport of Pollutants to the Ocean." National Research Council, Washington, D.C.

Turekian, K. K. (1969). *In* "Handbook of Geochemistry" (K. H. Wedepohl, ed.), Vol. 1, pp. 297–323. Springer–Verlag, Heidelberg.

Turekian, K. K. and Wedepohl, K. H. (1961). *Bull. Geol. Soc. Am.* **72**, 175.

Urone, P., Lutsep, H., Noyes, C. M. and Parcher, J. F. (1968). *Environ. Sci. Tech.* **2**, 611.

Valach, R. (1967). *Tellus*, **19**, 509.

Van den Heuvel, A. P. and Mason, B. J. (1963). *Q. Jl R. Met. Soc.* **89**, 271.

Van Grieken, R. E., Johansson, T. B. and Winchester, J. W. (1974). *J. Rech. Atmos.* **8**, 611.

Volynets, V. F., Zadorozhniy, I. L. and Florenskiy, K. P. (1967). *Geokhimiya*, **5**, 587.

Wada, S. and Kokubu, N. (1973). *Geochem. J.* **6**, 131.

Wedepohl, K. H., ed. (1969). *In* "Handbook of Geochemistry", Vol. 1, Ch. 7–8. Springer-Verlag, Heidelberg.

Wilkniss, P. E. and Bressan, D. J. (1971). *J. Geophys. Res.* **76**, 736.

Wilkniss, P. E. and Bressan, D. J. (1972). *J. Geophys. Res.* **77**, 5307.

Wilkniss, P. E., Swinnerton, J. W., Bressan, D. J., Lamontagne, R. A. and Larson, R. E. (1975). *J. Atmos. Sci.* **32**, 158.

Winchester, J. W. and Desaedeleer, G. G. (1977). *In* "Nondestructive Activation Analysis" (S. Amiel, ed.), (in press). Elsevier, Amsterdam.

Winchester, J. W. and Duce, R. A. (1977). *In* "Fate of Pollutants in the Air and Water Environments" (I. N. Suffet, ed.), pp. 24–47. Wiley-Interscience, New York.

Woodcock, A. H. (1948). *J. Mar. Res.* **7**, 56.

Woodcock, A. H. (1953). *J. Met.* **10**, 362.

Woodcock, A. H. (1972). *J. Geophys. Res.* **77**, 5316.

Woodcock, A. H., Kientzler, C. F., Arons, A. B. and Blanchard, D. C. (1953). *Nature, Lond.* **172**, 1144.

Woodcock, A. H., Duce, R. A. and Moyers, J. L. (1971). *J. Atmos. Sci.* **28**, 1252.

WSSAC (collected papers) (1972). *J. Geophys. Res.* **77**, No. 27, 5059.

Yeatts, L. R. B. and Taube, H. (1949). *J. Am. Chem. Soc.* **71**, 4100.

Zafiriou, O. C. (1974a). *J. Geophys. Res.* **79**, 2730.

Zafiriou, O. C. (1974b). *J. Geophys. Res.* **79**, 4491.

Zafiriou, O. C. (1975). *J. Mar. Res.* **33**, 75.

Zafonte, L., Nester, N. E., Stephens, E. R. and Taylor, O. C. (1975). *Atmos. Environ.* **9**, 1007.

Chapter 39

The Organic Chemistry of Marine Sediments*

BERND R. T. SIMONEIT

*Institute of Geophysics and Planetary Physics,
University of California, Los Angeles, California, U.S.A.*

* Contribution No. 1664: Institute of Geophysics and Planetary Physics, University of California, Los Angeles.

39.1. Introduction

About 71 % of the Earth's surface is covered by the oceans and seas, and the sediments beneath them represent a vast, still virtually unknown, realm for geological, palaeontological and geochemical exploration. In these environments organic geochemistry can serve a number of purposes and can cast light on a variety of problems: it can add further detail to our understanding of chemical evolution, hydrocarbon generation and accumulation, coal formation, post-depositional microbiological activity, diagenetic stability of organic compounds, molecular condensation and disproportionation reactions, organic–inorganic interactions, transport mechanisms and input sources, etc. Most of the deep-sea samples involved in such studies are collected by coring and drilling procedures (see e.g. Chapter 36).

Marine sediments usually occur in forms ranging from unconsolidated, virtually unaltered, shell oozes and clays to immature consolidated rock. They have been the principal depositories of posthumous organic debris throughout the history of the Earth; however, the existing oceanic sediment column only extends back to the Jurassic (see Chapter 24). The organic debris in such sediments contains both resistant autochthonous material and allochthonous organic residues mainly derived from continental weathering.

The biogenetic, and/or geogenetic, origin of marine sediments is reflected not only in their mineralogy and lithology, but also in the character of the soluble and insoluble organic matter which they contain. Compounds can often be identified in the solvent-soluble organic matter which can serve as specific biogenic markers yielding information on the origin of the organic matter. For example, such markers can be used as tags for terrestrial organic matter, and data acquired in this way can be supplemented by measurements of bulk chemical parameters of the insoluble organic matter (e.g. $^{13}C/^{12}C$ ratios (Sackett, 1964)). These organic markers can be employed in conjunction with other geological data to elucidate the mechanism by which such material is transported to the marine environment. Certain restricted geographical areas have well-defined drainage sinks and an examination of the sedimentary column in these sinks may enable conclusions to be drawn about palaeo-environments in the related catchment area. Thus, the microfossil and palynological records can be coupled with data on the composition of the terrigenous organic matter to give a greater insight into the palaeo-environmental conditions in the watershed area.

The distribution of the organic matter in marine sediments has been reviewed in Chapter 31, in which it was reported that generally the organic matter content lies between 0·1 % and 5%; usually, deep-sea sediments contain $\sim 0·2$%, geosynclinal sediments ~ 2% and shelf sediments 1–5%. Of the various marine sediments, the non-pelagic clays (see Chapters 24 and 31) are the best potential repositories for terrigenous allochthonous organic matter. In contrast, autochthonous organic matter is probably best preserved in the pelagic oozes, particularly those underlying areas of high primary production. Usually, the preservation of organic matter is better in anoxic regions than it is in oxic zones. The sediments best indicative of palaeoenvironmental conditions in the proto-oceans are those occurring furthest away from the respective mid-ocean ridge, at depths which can be sampled by deep-sea drilling (e.g. the Cretaceous shales of the early Atlantic found during the Deep-Sea Drilling Project).

The organic geochemistry of marine sediments is reviewed in the present chapter with particular reference to material recovered during the Deep-Sea Drilling Project (DSDP). Topics covered include the methods of analysis, and the relationship between variations in the lipid and non-soluble organic matter composition, and parameters such as transport mechanisms and alteration processes.

39.2. Determination of Organic Components of Marine Sediments

39.2.1. experimental precautions

Because the concentrations of individual organic compounds in sediments are very low, it is essential to take extreme precautions in the collection and analysis of samples in order to keep organic contamination to a minimum.

Contamination from the laboratory can be minimized by using clean environments, such as laminar flow hoods (Section 19.10.4.11.6). All solvents should be of either redistilled reagent grade or nanograde, distilled in glass, quality. Glassware should be cleaned with chromic acid solution and then washed with distilled water and an appropriate organic solvent. It can also be annealed before use. Small residual volumes of solvent should be evaporated in a stream of dry filtered nitrogen. Fractions should be stored in glass vials with polytetrafluoroethylene-lined screw caps, preferably while still dissolved in some solvent. In all work it is mandatory to carry blank determinations through the whole procedure to check for contamination.

39.2.2. PREPARATION OF SAMPLES

Marine sediments are usually wet and unconsolidated, and the removal of excess water is best accomplished by lyophilization (freeze drying). The dried sediment is then powdered with either a mortar and pestle or a shaker mill.

39.2.3. TECHNIQUES FOR ISOLATION, SEPARATION AND DETERMINATION

Many of the organic analytical isolation and determination techniques used in hard-rock and petroleum investigations are directly applicable to marine sediments and have been reviewed, for example, by Eglinton and Murphy (1969), Swain (1970) and Breger (1963). Those which are appropriate for such sediments will be described below.

39.2.3.1. *Lipids*

1. *Extraction.* Extraction of lipids from dry sediments is usually carried out by agitating for 15–30 minutes with a 3:1 toluene–methanol mixture using an ultrasonic vibrator. Extraction has also been carried out with a variety of other solvents, including benzene, which is undesirable because it is highly toxic. The sediment–solvent slurry is centrifuged and, after removal of the centrifugate, the extraction is repeated with further aliquots of the solvent until the extracts are colourless. The extracts are then combined and concentrated using a rotary evaporator. Dry sediments may also be extracted by Soxhlet extraction for ~300 cycles with the same solvent mixture.

Alternatively, wet samples (especially Recent sediments containing labile lipids) can be extracted directly by a modification (Brooks, 1974) of the method of Dole and Meinertz (1960) in which the sample is agitated with a 4:1 mixture of isopropanol/heptane using ultrasonic vibration. After centrifugation and removal of the centrifugate the extraction is repeated, either with the same solvent or with a 3:1 toluene–methanol mixture, until the extracts are colourless.

2. *Separation.* Various procedures are available for the separation of the lipid extracts into various classes; these are usually tailored to suit the particular compound series that is to be separated. Basically, three chromatographic techniques are employed: liquid (standard or high performance) (see e.g. Reed *et al.*, 1977), gel permeation, and thin-layer (TLC) (see e.g. Boon *et al.*, 1975a, 1977a; Simoneit, 1975). The last-named technique is preferable, since it is rapid and results in satisfactory separation of compound classes. The following compound series can be separated by these procedures: hydrocarbons, fatty acids (derivatized as methyl esters), fatty alcohols, steroids, triterpenoids, diterpenoids, tetraterpenoids, tetrapyrrole pigments, aromatic hydrocarbons and the "hump" compounds (some of these series may not be completely resolved from one another).

Separations by TLC, both analytical and preparative, are performed on glass plates coated with 0·3–1·0-mm layers of activated (120°C) silica gel which have been pre-eluted with ethyl acetate to remove impurities. Visualization of the separated lipids is carried out by spraying the plate with acetone solutions of either Rhodamine G or sodium fluorescein, or by exposing the plate to iodine vapour. Viewing of the treated plates is then carried out under u.v. light. Finally, the lipid bands are eluted with either hexane/diethyl ether (9:1) or methylene chloride.

For some classes of lipids it is advantageous to carry out the TLC separation on plates which have been coated with silica gel impregnated with 10% (by weight) of silver nitrate. Hydrocarbons and fatty acids (as their methyl esters) can be separated on such plates, according to their degree of unsaturation, by development with hexane and hexane/diethyl ether (95:5), respectively. After the separated bands have been rendered visible as described above, the various fractions are recovered from the silica gel by elution with diethyl ether or ethyl acetate. The eluate is then passed through a short (<0·5 cm) column of alumina (to retain the fluorescing dye), or through a fritted filter (for plates treated with iodine vapour).

Normal and branched/cyclic components can be separated by urea adduction or molecular sieve methods. The urea adduction technique has been described in detail by Gaskell (1974).

3. *Derivatization prior to gas–liquid chromatography.* For convenient gas chromatographic examination, fatty acid fractions can be methylated with diazomethane in ether (Fales *et al.*, 1973) or, alternatively, with boron trifluoride in methanol which is less likely to cause side reactions.

Alcohols can be converted to their trimethylsilyl derivatives by treatment with an excess of a solution of *N,O*-bis(trimethylsilyl)acetamide (BSA) in ethyl acetate.

4. *Analysis*. The most commonly used techniques for lipid analysis are gas chromatography (GC) and gas chromatography–mass spectrometry (GC–MS). The best GC separations, both in terms of peak resolution and throughput of labile and polar compounds, are obtained with wall-coated glass capillary columns. The GC–MS technique is well established for lipid analysis and is probably the most frequently applied. For a discussion of the GC–MS method see the review by McFadden (1973); for an account of mass spectrometry see Williams (1971, 1973), Johnstone (1975) and Burlingame *et al.* (1974, 1976), and for the basic interpretation of mass spectra McLafferty (1966) should be consulted. Not all lipid compounds are readily identifiable from a low-resolution mass spectrum, and frequently high-resolution mass spectrometry (HRMS) will prove of additional value. The potential of the interactive application of HRMS and GC–MS to environmental organic geochemistry has been illustrated by Simoneit *et al.* (1975), and the technique should prove of use in the analysis of marine sediments. High-resolution gas chromatography–HRGS for operation in real-time (using glass SCOT capillary columns) has been developed for the analysis of complex organic mixtures (Kimble *et al.*, 1975). This system should prove useful for determining the heteroatomic composition of unknown lipids from the geosphere.

Other ancillary techniques which may be applied to lipid analyses include u.v.-visible spectrophotometry, infra-red spectrophotometry and both proton and ^{13}C nuclear magnetic resonance spectroscopy (NMR).

39.2.3.2. *Humates and kerogen*

1. *Extraction*. Humic acids can be separated from exhaustively extracted sediments (from which carbonates have previously been removed) by shaking with aliquots of 0·1 M sodium hydroxide under nitrogen until the extracts are colourless (Stuermer and Simoneit, 1978). The combined extracts are centrifuged (17 000 g for 45 minutes) to remove suspended sediment and acidified with hydrochloric acid to pH 1 to precipitate the humic acid. The precipitate is then washed with 0·01 M hydrochloric acid to remove salts, and freeze-dried to yield dry humic acid in the protonated form. Sodium pyrophosphate has also been used to separate humic and fulvic material from sediments (Stevenson and Butler, 1969).

The technique used for the isolation of kerogen from sediments is essentially the same as that employed in hard-rock geochemistry (Burlingame and Simoneit, 1969; Djuricic *et al.*, 1972; Simoneit and Burlingame, 1973a). The sediment is first extracted to remove lipids and then demineralized by consecutive treatment with hydrochloric and 48% hydrofluoric acids which remove carbonates and silicates respectively. The solid remaining after these extractions is a kerogen concentrate from which kerogen with a very low ash content can be obtained by flotation with a dense liquid such as zinc bromide solution ($d_{20} = 2.0$).

2. *Characterization of humates and kerogen.* Humates are mixtures of complex macromolecules, and for this reason very little is known about the detailed structures of the compounds concerned. Indeed, the separation of the individual classes of compounds is a dauntingly difficult task, most workers being content to determine bulk properties of the humates. Functional group analysis has been extensively used to determine total acidity, and carboxyl, hydroxyl, phenolic hydroxyl, carbonyl and methoxyl groups (Martin et al., 1963; Stevenson and Butler, 1969; Fester and Robinson, 1964, 1966; Schnitzer and Khan, 1972). Physical methods of structure determination, e.g. infrared (Schnitzer and Khan, 1972) and ultra-violet spectrophotometry, NMR (Stuermer and Payne, 1976), ESR (Nissenbaum and Kaplan, 1972) and spectrofluorimetry, have also been applied to the study of humates, but with only very limited success. Humate molecular weight distributions have been investigated by the use of gel-permeation chromatography (Rashid and King, 1969).

The kerogens are complex mixtures of high molecular weight moieties of almost unknown structure. Some attempts have been made to characterize them by physical methods, such as i.r. and u.v. spectrophotometry, X-ray diffraction and $^{13}C/^{12}C$ ratio measurements (see e.g. Forsman and Hunt, 1958; Huc, 1973), but chemical methods of structure determination have yielded very little useful information. No detailed chemical degradative studies of marine kerogens have yet been carried out. However, data which are available for consolidated sediments of marine origin (Djuricic et al., 1972) and for a Recent lagoonal sediment (Philp and Calvin, 1976a) may be of value in interpreting degradation data for kerogen in marine samples when they become available (e.g. from the DSDP). Indeed, some qualitative evaluations of kerogens from the DSDP have already been published (Hunt, 1974a, b, c; Huc, 1973). Pyrolytic methods have recently been shown to be useful for characterizing kerogens and for assessing their potential for hydrocarbon generation (Philp et al., 1978).

39.2.3.3. *Amino acids and peptides*

Amino acids and peptides are very widespread in the geosphere and those from marine sedimentary environments have been extensively analysed. Free amino acids are usually extracted with water; following this, protein and peptide amino acids are then extracted by treatment with 6 M hydrochloric acid which hydrolyses them (e.g. Swain et al., 1959; Hare, 1969). The extracted amino acids are usually separated and determined by a combined ion exchange–spectrophotometric technique (see e.g. Hare, 1969). Alternatively, the amino acids may be determined gas chromatographically after conversion to volatile derivatives (see e.g. Kvenvolden *et al.*, 1970).

39.2.3.4. *Determination of lignin*

Lignin is identified and determined by treating the sample with either copper(II) oxide or nitrobenzene and characterizing and determining the resulting phenols (Leo and Barghoorn, 1970; Hedges, 1975).

39.2.3.5. *Determination and characterization of cutin and suberin*

The cutin and suberin residues in sediments are hydrolysed by refluxing with 3% potassium hydroxide solution and determining the resulting hydroxy-fatty acids by GC after their purification and derivatization (Cardoso *et al.*, 1977).

39.3. ORGANIC COMPONENTS OF MARINE SEDIMENTS

The increasing sophistication of GC and GC–MS techniques has vastly expanded reported analyses of marine sediment, mainly for paraffinic and aromatic hydrocarbons. In part, this springs from economic pressures consequent on the development of new petroleum resources on the continental shelves and has resulted in many "baseline" studies in potential lease areas. Two examples (both sponsored by the U.S. Bureau of Land Management) may be cited. These are: (i) the southern California Bight study (Reed *et al.*, 1977); and (ii) the north-eastern Gulf of Mexico investigation (Gearing *et al.*, 1976). Thus, potentially, much new data concerning the organic composition of marine sediments (predominantly those of the continental shelf and slope) are likely to become available. The deeper oceanic areas are currently being probed by the DSDP, now called DSDP/International Programme for Ocean Drilling (IPOD), and many new organic geochemical studies are associated with it. The following discussion includes only the most pertinent data, and the account given should not be considered to be an exhaustive compilation. Reviews dealing with particular aspects of the environmental organic chemistry of some marine realms have appeared (e.g. Eglinton, 1975). The organic chemistry of the sediments of oceans, fjords and anoxic basins has been reviewed by Morris and Culkin (1975) and that of marine and lacustrine sediments has been summarized by Yen (1977). The environmental chemistry of hydrocarbons in marine sediments has been briefly discussed by Farrington and Meyers (1975).

39.3.1. HYDROCARBONS

Only the alkanes, cycloalkanes and isoprenoids will be discussed below.

39.3.1.1. *Normal paraffinic hydrocarbons*

Normal alkanes (paraffins) are of ubiquitous occurrence in marine sediments; however, their concentrations and the distribution of the various members of the homologous series differ considerably from one sediment to another. Initial studies of Recent sediments from San Francisco Bay (Kvenvolden, 1962) and the Gulf of Mexico (Stevens *et al.*, 1965) indicated that the *n*-alkanes have terrigenous distribution patterns (i.e. they have a high carbon preference index (*CPI*); see e.g. Cooper and Bray (1963) and Kvenvolden (1966)). Deeper sediments extending to the mid-Miocene were obtained during the Mohole Project (Riedel *et al.*, 1961) and core sections have been examined for their physical properties, chemical characteristics and contents of the various lipid classes (Rittenberg *et al.*, 1963). The hydrocarbon examination was confined merely to a separation according to the degree of unsaturation and the results were reported as bulk fractions (Rittenberg *et al.*, 1963). The Mohole Project provided the impetus for the DSDP from which most of the following data are drawn. Some examples of normal alkanes derived from Recent sediments from many marine environments will be discussed below.

The lipid distributions in the sediments of the Black Sea reflect the terrigenous origin of the detritus forming the sediments in this basis (Simoneit, 1977a). The *n*-alkanes of most samples exhibit distributions in which odd carbon numbers predominate over even ones, and in which there are maxima at *n*-C_{29}, or less frequently *n*-$C_{29} \approx C_{31}$ (see Fig. 39.1A) (Simoneit, 1975, 1977a), which are typical of higher plant waxes (Simoneit, 1974a, 1977a; Peake *et al.*, 1974; Hunt, 1974d; Vuchev *et al.*, 1978; Geodekyan *et al.*, 1978). Sediments from the Persian Gulf contain *n*-alkanes in the carbon number range C_{23}–C_{31} (Fig. 39.1F), with a high *CPI* value* indicative of an origin from higher plants (Welte and Ebhardt, 1968a, b). Analyses for *n*-alkanes have also been reported for Recent sediments derived from:

(i) the deltas of the Euphrates, Orinoco, Amazon, Limpopo and Danube Rivers, the Straits of Japan, the Norwegian Sea, the Black Sea and the continental shelf and slope off Peru (Telkova *et al.*, 1976);

(ii) the Beaufort Sea (Peake *et al.*, 1972);

(iii) the north-western Atlantic (Farrington *et al.*, 1977a; Farrington and Tripp, 1977);

(iv) Narragansett Bay, Rhode Island (Farrington and Quinn, 1973a);

---

* For normal hydrocarbons the carbon preference index (*CPI*$_{\mathrm{H}}$) is usually expressed by:

$$CPI_{\mathrm{H}} = \frac{2\Sigma \text{ odd } C_{21}\text{-to-}C_{31}}{\Sigma \text{ even } C_{20}\text{-to-}C_{30} + \Sigma \text{ even } C_{22}\text{-to-}C_{32}}$$

(Cooper and Bray, 1963). However, other carbon number ranges may be used as long as they are specified.

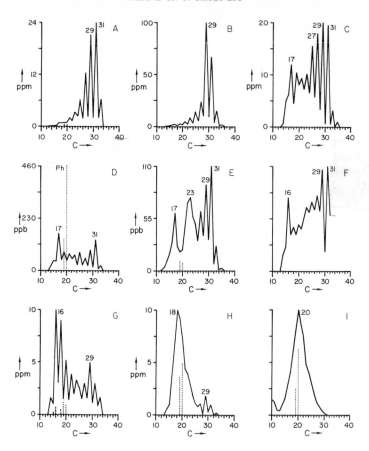

FIG. 39.1. Distribution diagrams of *n*-alkanes in lipids of sediments (·········· isoprenoids; ----- diterpenoids; heights of lines indicate approximate concentrations).
A. Black Sea (AII49-1462-5 m), Pleistocene (Simoneit, 1974a, 1977a).
B. North Atlantic Ocean (DSDP 12-114-5-5), Pliocene (Simoneit, 1975).
C. Gulf of Mexico (DSDP 10-92-5-4), Pleistocene (Simoneit *et al.*, 1973a).
D. Atlantic Ocean (DSDP 41-367-19-4), Cretaceous (Simoneit, 1977f).
E. Atlantic Ocean (DSDP 44-391C-52-2), Jurassic (Stuermer and Simoneit, 1978).
F. Persian Gulf (250–260 cm), Holocene (drawn from data of Welte and Ebhardt (1968a, b)).
G. North-eastern Pacific Ocean (DSDP 18-175-2-2), Pleistocene (Simoneit, 1975, 1977b).
H. Cariaco Trench (DSDP 15-147B-4-1), Pleistocene (Simoneit, 1975).
I. Norwegian–Greenland Sea (DSDP 38-345-4-1), Glacial (Simoneit, 1976).

(v) Choctawhatchee Bay, Florida (Palacas *et al.*, 1972);
(vi) the continental shelf off NW. Africa (Gaskell *et al.*, 1975), see Fig. 39.2D (p. 248);
(vii) the Laguna Mormona, Baja California (Cardoso *et al.*, 1976);

(viii) the San Nicolas and the Tanner Basins, California (Baedecker et al., 1977; Mitterer and Hoering, 1968);
(ix) the Baltic Sea (Zsolnay, 1971);
(x) the Dead Sea (Nissenbaum et al., 1973);
(xi) a lagoon on Surtsey Island (Sever and Haug, 1971).

Generally, typical higher plant wax distributions predominated.

Some examples of n-alkane distributions in older marine sediments are shown in Fig. 39.1. A glauconitic silty clay sediment of Pliocene age from the Reykjanes Ridge, North Atlantic Ocean, has been found to have an n-alkane distribution typical of a plant wax (Fig. 39.1B) (Simoneit, 1975, 1978c). Pleistocene–Pliocene sediments from the Gulf of Mexico exhibit bimodal n-alkane distributions, with one maximum which is typical of a higher plant wax and another (at n-C_{17}, and ranging to n-C_{23}) which is characteristic of the products of marine autochthonous production (Fig. 39.1C)(Aizenshtat et al., 1973; Simoneit et al., 1973a; Simoneit and Burlingame, 1974a).

The oldest sediments which have been examined to date are from the Atlantic Ocean and comprise black shales of Cretaceous age (Fig. 39.1D) and Jurassic claystones (Fig. 39.1E). The n-alkanes of the shales can be ascribed partly to a terrigenous and partly to a marine origin. The maximum at n-C_{17}, with the range from C_{13} to C_{23}, is indicative of autochthonous production, and the predominantly odd numbered homologues $>C_{23}$ indicate an allochthonous higher plant wax influx (Fig. 39.1D) (Simoneit, 1977e, f, 1978b). Similar distributions occur in Cretaceous sediments from both sides of the Mid-Atlantic Ridge (see e.g. Simoneit et al., 1973c; Boon et al., 1977b; Simoneit, 1977e, f, 1978b). The Jurassic claystones exhibit a slightly different distribution, in which the higher n-alkanes are characteristic of higher plant waxes and the maximum at n-C_{17} is typical of marine auto-chthonous production (Fig. 39.1E) (Stuermer and Simoneit, 1978). The "hump" with a maximum at n-C_{23} probably arises from the biodegradation products of algae; this origin is supported by the presence of a similar "hump" in the branched/cyclic fraction, with a maximum at the same GC retention time (Stuermer and Simoneit, 1978).

Distributions of n-alkanes having marked even carbon number pre-dominances have been observed for some sediments. This was first found for an oxic sediment from the Persian Gulf (Fig. 39.1F) in which n-hexadecane predominated in the region $<C_{23}$. Welte and Ebhardt (1968a, b) ascribed this predominance of even carbon numbers to the reduction of n-fatty acids under suitable micro-environmental conditions. Such distributions have also been observed in some light grey sediments (Pleistocene–Pliocene) of the north-eastern Pacific Ocean (Simoneit, 1975, 1977b). For these sediments, the even

carbon number predominance ranges from n-C_{14} to n-C_{22}, with a maximum at n-C_{16}, and a minor plant wax component is evident above C_{23}, with a maximum at n-C_{29} (Fig. 39.1G). The n-alkane concentration in such samples is usually less than that of the fatty acids (Simoneit, 1975).

A narrow n-alkane distribution envelope was found for some sediments from the Cariaco Trench (Fig. 39.1H); this appears to indicate the occurrence of the products of another type of autochthonous production (Simoneit et al., 1973b; Simoneit and Burlingame, 1974a; Simoneit, 1975). The maximum at n-C_{18} (C_{12}–C_{25}), and a *CPI* of $\sim 1 \cdot 0$, is attributable to marine production (Simoneit, 1975). A minor input of higher plant wax is also apparent ($> C_{25}$, with a maximum at n-C_{29}).

Petroleum has been shown to be present in marine sediments in certain areas (see Fig. 39.1I, which illustrates the n-alkane distribution of Pleistocene–Glacial Sequence sediments of the Norwegian–Greenland Sea (Simoneit, 1975, 1976)). The n-alkanes range from $< n$-C_{10} to n-C_{31}, with a maximum at n-C_{20}, and neither odd nor even carbon numbers predominate. This pattern, coupled with other parameters (e.g. the presence of steranes, triterpanes and a broad, branched/cyclic "hump") is conclusive evidence that this bitumen is petroleum, which has probably seeped into the lithological sequence (Simoneit, 1975, 1976).

n-Alkanes have also been determined in the older sediments from some other deep-sea areas: e.g. various areas in the Pacific Ocean (age range from Pleistocene to Palaeocene) (Simoneit and Burlingame, 1971a, b, 1972a, b); the Hatteras Abyssal Plain and the Bahama Banks (Cretaceous) (Simoneit et al., 1972); the Labrador and Mediterranean Seas (Simoneit and Burlingame, 1973b, 1974a; Deroo et al., 1978b); the Bengal Fan (Simoneit and Burlingame, 1974b); the North Atlantic (Deroo et al., 1978b; Cardoso et al., 1978; Stuermer and Simoneit, 1978; Simoneit, 1977e, f, 1978b); the south-eastern Atlantic (Forsman, 1978); and the Gulf of Aden (Cernock, 1974). In general, it has been found that sediments derived from near the continents contain n-alkanes attributable to both higher plant wax and marine autochthonous origins, whereas sediments from deep-sea depositional environments contain predominantly n-alkanes derived from marine sources.

39.3.1.2. Branched and cyclic hydrocarbons

Isoalkanes (e.g. I) and anteisoalkanes (e.g. II) are found as minor components in association with the normal alkanes. No reports of detailed studies of

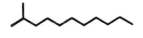

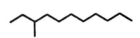

I.   isododecane (2-methylundecane)   II.   anteisododecane (3-methylundecane)

these compounds in marine sediments have appeared. A mixture of 7-
and 8-methylheptadecane has been identified in Recent lagoonal sediments
from the Virgin Islands (Philp *et al.*, 1976) and the Laguna Mormona, Baja
California (Cardoso *et al.*, 1976). These compounds (as 1:1 mixtures) have
been found to be major components of the lipids of blue-green algae (Han and
Calvin, 1970; Han *et al.*, 1968).

The isoprenoid hydrocarbons, mainly pristane (III) and phytane (IV), are

III. pristane   IV. phytane

ubiquitous in marine sediments. Recent sediments from the North Atlantic
contain predominantly pristane with only traces of phytane (Blumer *et al.*,
1964; Blumer and Snyder, 1965). However, in deeper sediments, and in closer
proximity to the continents, both homologues are found in varying proportions (Fig. 39.1; see also Simoneit, 1975; Gearing *et al.*, 1976; Telkova *et al.*,
1976; Reed *et al.*, 1977). Both pristane and phytane are ultimately derived
diagenetically from phytol (V) (Ikan *et al.*, 1975a; de Leeuw *et al.*, 1974, 1977;

V. phytol

Didyk *et al.*, 1978). The depths at which these transformations take place are
not certain and probably depend on the sediment composition and lithology.

Although cyclohexylalkanoic acids have been found in a thermophilic
bacterium (deRosa *et al.*, 1971), alkylcyclohexanes have not been detected
in marine sediments. The alkylcyclohexanes are, however, found as minor
components in petroleum, together with alkylpolycyclanes and alkylbenzenes, so they may be present in bituminous marine sediments from
greater lithological depths. A monocyclic hydrocarbon ($C_{20}H_{40}$) is present
in the branched/cyclic hydrocarbons of both the algal mat and the sediments
of the Laguna Mormona, Baja California (Cardoso *et al.*, 1976).

39.3.1.3. *Olefinic hydrocarbons*

Olefinic hydrocarbons (alkenes) have been found in Recent near-surface
sediments from continental shelf and slope areas. Heneicosahexene and
heptadecene have been reported to be present in polluted surface sediments
from Falmouth, Massachusetts (Ehrhardt and Blumer, 1972). Olefins have

also been found in sediments from the Hudson Canyon extending from the New York Bight to a water depth of $\sim 3500\,\mathrm{m}$ (Farrington et al., 1977a; Farringtom and Tripp, 1977). Reed et al. (1977) have identified three major olefinic hydrocarbons of formulae $C_{19}H_{38}$, $C_{21}H_{32}$ and $C_{25}H_{44}$ in sediments from the southern California Bight. The GC retention index and mass spectrum of the $C_{21}H_{32}$ compound match data presented by Lee and Loeblich (1971) for an unconjugated straight-chain heneicosahexaene isolated from algae. (Because of the poor resolving power of the GC column that was used, it is uncertain whether the 21:6 fraction was a single compound or a mixture of isomers.) The mass spectrum of the $C_{25}H_{44}$ compound matched that of a compound isolated by Farrington et al. (1977a) from Recent sediments of the Hudson Canyon. This compound, which contained four double bonds, yielded on hydrogenation a bicyclic hydrocarbon with the composition $C_{25}H_{48}$, and the minor component $C_{25}H_{48}$ (25:2), which is also present, was hydrogenated to the monocyclic hydrocarbon, $C_{25}H_{50}$ (Farrington et al., 1977a). Polyolefins of unknown structure having the compositions $C_{25}H_{50}$, $C_{25}H_{48}$, $C_{25}H_{46}$, $C_{25}H_{44}$ and $C_{23}H_{44}$ have been identified as major components of the lipids of continental shelf sediments of the Gulf of Mexico (Gearing et al., 1976). However, these olefins have not been detected in deeper sediments from the DSDP (Simoneit, 1975). Isoprenoidal olefins have been identified in some DSDP sediments (from e.g. the Cariaco Trench (Simoneit et al., 1973b)). Thus, phytenes, phytadienes and pristenes have been detected but decrease in concentration with increasing depth (Simoneit and Burlingame, 1974a); they appear to be secondary products derived from phytol (V) (de Leeuw et al., 1977). Normal alkenes have not been reported as constituents of older marine sediments.

39.3.1.4. Biogenic origin of hydrocarbons

Algae generally contain few hydrocarbons, and these are mainly low molecular weight n-alkanes and alkenes. Thus a wide variety of species of blue-green algae have been found to contain mainly $n\text{-}C_{15}$ and/or $n\text{-}C_{17}$ as the predominant alkane (Oró et al., 1967; Han et al., 1968; Gelpi et al., 1970; Blumer et al., 1971). Youngblood et al. (1971) have shown that $n\text{-}C_{15}$ and $n\text{-}C_{17}$ predominate in the brown marine algae and in the red algae, respectively. Gelpi et al. (1970) have observed that several algal species have a bimodal alkane distribution with maxima at $n\text{-}C_{17}$ (major) and $n\text{-}C_{27}$ (minor). In addition, a significant alkene content was noted, and other workers have commented on the presence of the mono- and diolefinic C_{15} and C_{17} components of marine algae (Youngblood et al., 1971; Blumer et al., 1971).

Two species of photosynthetic bacteria have been shown to have a hydrocarbon distribution with a maximum at $n\text{-}C_{17}$ (Han et al., 1968). The same workers have examined the alkanes of two non-photosynthetic bacteria, one aerobic and one anaerobic, and have shown that the former has n-C_{19} as its major alkane, whereas the latter has n-C_{18}. All these bacteria contained trace amounts of higher molecular weight alkanes. A species of bacterium from the Pacific Ocean has been found to contain n-C_{17} and two isomeric C_{17} alkenes as its major alkane constituents (Oró et al., 1967).

The alkanes of fungi have not been extensively studied. The surface wax of species grown on standardized media have alkane distribution patterns similar to those of higher plants (range, n-C_{18}–n-C_{35}; the most prominent components being n-C_{27}, n-C_{29} and n-C_{31} alkanes (Weete, 1972)).

The alkanes of higher plants have been thoroughly investigated (see e.g. Eglinton and Hamilton, 1963; Douglas and Eglinton, 1966; Caldicott and Eglinton, 1973; Kolattukudy and Walton, 1972). In plant waxes, the alkanes may range from n-C_7 to $\sim n$-C_{60}, but only rarely are there significant proportions of those lying outside the range C_{25}–C_{35}. The major components are odd carbon numbered homologues, usually the C_{27}, C_{29}, C_{31} and C_{33} n-alkanes (Eglinton and Hamilton, 1963; Eglinton et al., 1962). In contrast, the interiors of higher plants contain only low concentrations of alkanes (n-C_{16}–n-C_{28}, with no odd-to-even predominance (Weete, 1972)). It is apparent, therefore, that the characteristic alkanes of higher plants are found predominantly in the outer surface waxes.

39.3.2. FATTY ACIDS

39.3.2.1. n-*Saturated and* n-*unsaturated fatty acids*

The fatty acid (n-alkanoic acid) distributions in Recent marine sediments have been extensively studied. Recent investigations have included the following areas: The Tanner Basin, California (Baedecker et al., 1977); the San Nicolas Basin, California (Hoering, 1968); the North Atlantic (Cooper and Blumer, 1968); various estuaries (Farrington and Quinn, 1971a, b; Johnson and Calder, 1973); a lagoon on Surtsey Island (Sever and Haug, 1971); the Gulf of Mexico (Parker, 1967); the Dead Sea (Nissenbaum et al., 1973); the Black Sea (Simoneit, 1974a, 1977a); Buzzards Bay, Massachusetts (Farrington et al., 1977b); and Walvis Bay (Boon et al., 1975a).

Recent sediments from close to the continents usually have a predominance of even numbered fatty acids ranging from n-C_{10} to n-C_{36} and having a bimodal distribution with maxima at n-C_{16} and n-C_{24} or n-C_{26}. The fatty acids $< n$-C_{20} are believed to be of autochthonous origin, whereas those $> n$-C_{21} probably had a terrestrial origin. This distribution is typified by that in a Recent sediment from the Black Sea (Fig. 39.2A) in which the land plant origin of the $> n$-C_{21} acids is supported by the presence of higher n-alkanes typical of the same origin (Fig. 39.1A) (Simoneit, 1974a, 1975, 1977a). A similar bimodal pattern (Fig. 39.3F) is also found for the n-fatty acids of

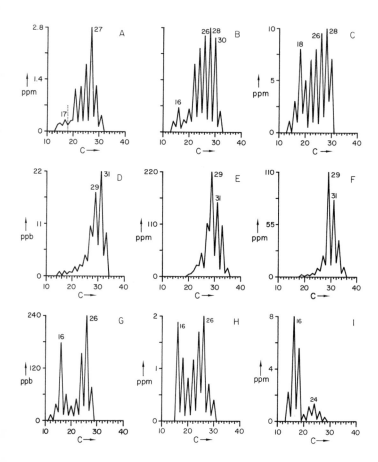

FIG. 39.2. Distribution diagrams of various homologues in lipids of sediments and aeolian dusts.
A. *n*-Methyl ketones, Black Sea (DSDP 42B-379A-30-3), Pleistocene (Simoneit, 1978a) (········ isoprenoid ketone).
B. *n*-Fatty alcohols, North Atlantic Ocean (DSDP 12-114-5-5), Pliocene (Simoneit, 1975).
C. *n*-Fatty alcohols, aeolian dust, off West Africa (E 2) (Simoneit and Eglinton, 1977).
D. *n*-Alkanes, Cap Blanc shelf, W. Africa (30–45 cm). Recent (drawn from data of Gaskell *et al.* (1975)).
E. *n*-Alkanes, aeolian dust, off equatorial Africa (DC 1) (Simoneit and Eglinton, 1977).
F. *n*-Alkanes, aeolian dust, off equatorial Africa (DC 2) (Simoneit and Eglinton, 1977).
G. *n*-Fatty acids, Cap Blanc shelf, W. Africa (30–45 cm). Recent (drawn from data of Gaskell *et al.* (1975)).
H. *n*-Fatty acids, aeolian dust, off equatorial Africa (DC 5) (Simoneit, 1977c).
I. *n*-Fatty acids, aeolian dust, off West Africa (E 2) (Simoneit and Eglinton, 1977).

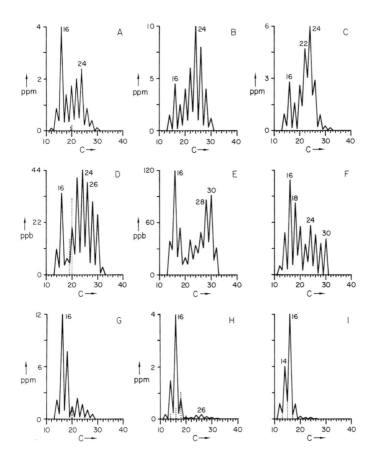

FIG. 39.3. Distribution diagrams of *n*-fatty acids of lipids of sediments.
A. Black Sea (AII49-1462-5 m), Pleistocene (Simoneit, 1974a, 1977a) (‒‒‒‒‒ diterpenoid; ········· phytanic acid).
B. North Atlantic Ocean (DSDP 12-114-5-5), Pliocene (Simoneit, 1975).
C. Gulf of Mexico (DSDP 10-92-5-4), Pleistocene (Simoneit *et al.*, 1973a).
D. Atlantic Ocean (DSDP 41-367-19-4), Cretaceous (Simoneit, 1977f) (........ isoprenoid acids).
E. Atlantic Ocean (DSDP 44-391C-52-2), Jurassic (Stuermer and Simoneit (1978)).
F. Persian Gulf (250–260 cm), Holocene (drawn from data of Welte and Ebhardt (1968a, b)).
G. North-eastern Pacific Ocean (DSDP 18-175-2-2), Pleistocene (Simoneit, 1975, 1977b) (‒‒‒‒ diterpenoid acid).
H. Cariaco Trench (DSDP 15-147B-4-1), Pleistocene (Simoneit, 1975) (......... monounsaturated acids).
I. Walvis Bay (KF6-B41-3-28 cm). Recent (drawn from data of Boon *et al.* (1975a)) (......... anteiso-acids).

Recent sediments from the Persian Gulf (Welte and Ebhardt, 1968a, b). Not all shelf sediments exhibit such a pronounced bimodal distribution. Thus, the fatty acids of Recent sediments from the continental shelf off western Africa are dominated by those of terrigenous origin (Fig. 39.2G) which probably have been brought to the sedimentary environment by aeolian transport of plant wax (Gaskell et al., 1975). In contrast, the sediments of Walvis Bay contain only those n-fatty acids which are characteristic of marine autochthonous production (Fig. 39.3I) (Boon et al., 1975a).

n-Fatty acid determinations have been made for many older marine sediments, including some which have been examined for hydrocarbons (see e.g. Section 39.3.1.1 and Fig. 39.3). Thus, a northern Atlantic Pliocene sediment had an n-fatty acid distribution indicative of a terrigenous higher plant origin (maximum at $n\text{-}C_{24}$, even-to-odd predominance; see Fig. 39.3B). A minor marine component (maximum at $n\text{-}C_{16}$, range $<n\text{-}C_{20}$) (Simoneit, 1975) was also present. The fatty acid distribution and the presumed origin of the acids are consistent with the n-alkane distribution in this sample (Fig. 39.1B). Pleistocene sediments from the Gulf of Mexico showed bimodal n-fatty acid distributions, the principal maximum being typical of a terrigeneous origin and the minor maximum, at $n\text{-}C_{16}$, being characteristic of autochthonous production (Fig. 39.3C) (see also Aizenshtat et al., 1973; Simoneit et al., 1973a; Simoneit and Burlingame, 1974a). Again, these distributions correlated well with those of the n-alkanes in the same samples (see Fig. 39.1C).

A Cretaceous black shale (Fig. 39.1D) from the Atlantic Ocean had a bimodal n-fatty acid distribution with maxima at $n\text{-}C_{16}$ and $n\text{-}C_{24}$ ($CPI^{*} = 5\cdot3$; see Fig. 39.3D), the former being indicative of autochthonous biogenic activity and the latter of an allochthonous higher plant influx (Simoneit, 1977e, f, 1978b). Similar distributions were found in Cretaceous sediments from both sides of the Mid-Atlantic Ridge. A Jurassic claystone from the same area had a bimodal n-fatty acid distribution, with a major maximum at $n\text{-}C_{16}$ and another at $n\text{-}C_{30}$ which is probably of terrigenous origin (Fig. 39.3E). The n-alkanes of this sample (Fig. 39.1E) probably originated in part from microbial production, and in part from degradation of algal material (maxima at $n\text{-}C_{17}$ and $n\text{-}C_{23}$). These origins are in accordance with the distributions of the $<C_{23}$ acids with a slight predominance of $n\text{-}C_{22}$ acid and a maximum at $n\text{-}C_{16}$ (Stuermer and Simoneit, 1978).

---

* For fatty acids the carbon preference index (CPI_{FA}) is usually expressed by:

$$CPI_{FA} = \frac{2\Sigma \text{ even } C_{14}\text{-to-}C_{26}}{\Sigma \text{ odd } C_{13}\text{-to-}C_{25} + \Sigma \text{ odd } C_{15}\text{-to-}C_{27}}$$

(Kvenvolden, 1966). However, other carbon number ranges may be used, as long as they are specified.

The n-fatty acid distributions in the north-eastern Pacific Ocean sediments together with the predominantly even carbon numbered n-alkanes (Fig. 39.1G) suggest that the organic matter is mainly of marine origin. This is illustrated in Fig. 39.3G which shows a maximum at n-C_{16}. However, the presence of minor amounts of homologues $>$$n$-$C_{20}$ and of dehydroabietic acid, a terrigenous marker for resinous plants, indicates a minor influx of higher plant matter (C_{22}–C_{28} acids) (Simoneit, 1975, 1977b).

The organic matter of the Pleistocene sediments of the Cariaco Trench appears to be mainly of marine origin, and this is reflected in its n-fatty acid distribution (Fig. 39.3H) (Simoneit, 1975; Hoering, 1973b; Simoneit et al., 1973b; Simoneit and Burlingame, 1974a). The maximum at n-C_{16} and the narrow range (n-C_{12}–n-C_{20}), are characteristic of autochthonous production. There is, however, evidence for a terrigenous higher plant contribution as n-C_{22}–n-C_{32} acids are present together with dehydroabietic acid (Simoneit, 1975).

The n-fatty acid distribution patterns of the older sediments of a number of other deep-sea areas have also been studied, including: the Pacific Ocean (Pleistocene–Palaeocene) (Simoneit and Burlingame, 1971a, b, 1972a, b); the Labrador and the Mediterranean Seas (Simoneit and Burlingame, 1973b, 1974a); and the Atlantic Ocean (Stuermer and Simoneit, 1978; Simoneit, 1977e, f, 1978b). These investigations have revealed that sediments from the vicinities of the continents exhibit bimodal n-fatty acid distributions and those from deep-sea environments have distributions typical of an autochthonous marine origin.

Unsaturated fatty acids have, up to the present, not been found in the pre-mid-Pleistocene sediments collected during the DSDP (Gaskell et al., 1975; Simoneit and Burlingame, 1972a, 1974a). They have, however, been detected in Recent sediments from a variety of locations viz., Puget Sound, Washington (Peterson, 1967), Narragansett Bay, Rhode Island (Farrington and Quinn, 1971a, b, 1973b), the continental slope off NW. Africa (Gaskell et al., 1975) and the Laguna Mormona, Baja California (Cardoso et al., 1976). In each instance the predominant unsaturated acid was 18:1. The ratio of unsaturated to saturated acids has been found to decrease with depth of burial. The geometrical isomers (i.e. cis and trans) of the unsaturated fatty acids in a sediment from an intertidal zone in Corner Inlet, Victoria, Australia have been examined by Volkman and Johns (1977).

39.3.2.2. Branched and cyclic fatty acids

Iso- and anteisoalkanoic acids have been reported to be present in a variety of Recent marine sediments (Leo and Parker, 1966; Cooper and Blumer, 1968; Hoering, 1969). In these sediments the anteiso-acids predominate over the iso-acids (Cooper and Blumer, 1968; Boon et al., 1975a). Only odd carbon numbered members (C_9–C_{17}) of the former are present, whereas both odd and even carbon numbered iso-acids from C_9–C_{19} are present. These compounds are believed to be formed by bacterial alteration of the lipids of diatoms (cf. Fig. 39.3I) as they are characteristic of bacteria (Boon et al., 1977a). Anteiso-fatty acids have also been found in older sediments, such as those from the Cariaco Trench (Simoneit, 1975).

Phytanic, pristanic and 4,8,12-trimethyltridecanoic acids have been identified in Recent marine sediments from various coastal areas (Blumer and Cooper, 1967) and phytanic acid has been found to be the dominant acid in the branched/cyclic fatty acid fraction of a Virgin Island lagoonal sediment (Philp et al., 1976). These compounds have also been identified in older marine sediments, such as Cretaceous shales from the Atlantic (Fig. 39.3D) (Simoneit, 1977f). Phytenic acid has been observed in diatomaceous ooze from Walvis Bay (Boon et al., 1975b).

Cyclopropanoid acids of bacteriogenic origin have been identified in a lacustrine sediment (Cranwell, 1973) and in the sediment from Laguna Mormona, Baja California (Cardoso et al., 1976), the major homologues being cis-9,10-methylenehexadecanoic and cis-9,10-, and cis-11,12-methylene-octadecanoic acids. Even carbon numbered α,ω-dicarboxylic acids (C_{16}–C_{24}) have been identified in Recent lagoonal, estuarine and coastal sediments (Johns and Onder, 1975). They appear to have originated either from mangrove detritus, or where this was not available, from in situ production by microbiota.

39.3.2.3. Biogenic origins of fatty acids

The saturated fatty acids of marine and freshwater algae have been shown to range mainly from n-C_{12} to n-C_{20}, and there is usually a single maximum at n-C_{16} and significant amounts of n-C_{14} and n-C_{18} are also present (Lee and Loeblich, 1971; Jamieson and Reid, 1972). The predominating unsaturated fatty acids are mono- and polyunsaturated unbranched acids of the C_{14}, C_{16}, C_{18} and C_{20} series (Chuecas and Riley, 1969; Klenk et al., 1963). The saturated carboxylic acids present in bacterial lipids range from n-C_{12} to n-C_{20}, and are dominated mainly by n-C_{16} and a significant amount of n-C_{14} (Kates, 1964; Hitchcock and Nichols, 1971). Homologues with chain lengths greater than n-C_{20} apparently do not have a wide occurrence in bacterial lipids. The saturated fatty acids from fungal lipids are dominated mainly by n-C_{16}, and in some cases shorter-chain homologues. However, homologues $> n$-C_{18} are rarely present (Brennan et al., 1974).

The iso-C_{15} and anteiso-C_{15} acids have been found to be abundant constituents of bacteria (Kates, 1964), and in Arthrobacter the total lipid fatty acids contain about 60–85% anteiso-C_{15} and about 5–30% anteiso-C_{17} (Shaw and Stead, 1971). In ten species of Bacillus, anteiso- and iso-acids together comprised more than 60% of the lipid fatty acids (Kaneda,

1967). The predominant homologues are anteiso-C_{15} and -C_{17}, but lesser amounts of iso-C_{14}, -C_{15}, -C_{16} and -C_{17} are also present. It appears, therefore, that bacteria are the main source of the anteiso- and iso-acids in sediments, and in the geosphere in general.

The saturated carboxylic acids of higher plant surface waxes have been both studied and reviewed extensively (see e.g. Hitchcock and Nichols, 1971). In general, such waxes contain normal carboxylic acids ranging in carbon number from C_{22} to C_{36}, with a strong even-to-odd predominance and a maximum at n-C_{24} or n-C_{26}.

39.3.3. FATTY ALCOHOLS, KETONES AND WAX ESTERS

39.3.3.1. n-*Alcohols and* n-*methyl ketones*

Normal fatty alcohols (n-alkanols) are probably ubiquitous in the marine sedimentary environment and have been identified in a few Recent marine sediments. For example, Sever and Parker (1969) have identified n-fatty alcohols ranging from n-C_{12} to n-C_{26} (even-to-odd carbon number predominance) in Recent sediments from Baffin Bay (Texas), the Gulf of Mexico and the San Nicolas Basin. Hoering (1969) has found n-fatty alcohols in the sediments of the Tanner Basin (range C_{16}–C_{28}, even carbon number predominance and a maximum at C_{22}). In another investigation, the n-fatty alcohols of Tanner Basin and Bandaras Bay (Gulf of California) sediments were observed to range from n-C_{14} to n-C_{24} (even carbon number predominance) and monounsaturated homologues from C_{22} to C_{24} were also detected (Ikan *et al.*, 1975b).

n-Fatty alcohols have also been detected in older marine sediments. Thus, a Pliocene sediment from the North Atlantic (DSDP sample 12-114-5-5) was found to have a n-alcohol distribution which ranged from n-C_{14} to n-C_{32}, with a strong even-to-odd carbon number predominance and a maximum at n-C_{28} (Fig. 39.2B). However, no unsaturated fatty alcohols were detected (Simoneit, 1978c). In Pleistocene samples from the Cariaco Trench (DSDP Site 15-147) the concentrations of the fatty alcohols were about equal to those of the fatty acids and were greatly in excess over those of the paraffins (Hoering, 1973a). Various DSDP sediments from the Gulf of Mexico and the western Atlantic Ocean have been found to contain n-fatty alcohols in the range n-C_{16}–n-C_{28}, with a strong even-to-odd carbon number predominance and with a maximum which is usually at n-C_{24} (Aizenshtat *et al.*, 1973).

Normal methyl ketones (e.g. undecan-2-one) have not been found in Recent marine sediments. The first data for their occurrence in older sedimentary rocks were obtained for the Bouxwiller lignite in which the series $C_nH_{2n}O$ was present with a strong odd-to-even carbon number predominance (Arpino, 1973). Pleistocene sediments from the Black Sea contained relatively large concentrations of *n*-methyl ketones which had a distribution range similar to that of the Bouxwiller lignite and again there was a strong odd-to-even carbon number predominance (Fig. 39.2A) (Simoneit, 1978a). The *n*-methyl ketones in these various sediments were probably derived microbiologically either from the β-oxidation of *n*-fatty acids or from the oxidation of *n*-alkanes (Arpino, 1973).

39.3.3.2. Branched alcohols and ketones

The predominant isoprenoidal alcohols of Recent marine sediments (from e.g. Baffin Bay (Gulf of Mexico), the San Nicolas Basin (California) and the Laguna Mormona) are phytol(V) and dihydrophytol (Sever and Parker, 1969; Ikan *et al.*, 1975b; Cardoso *et al.*, 1976; de Leeuw *et al.*, 1977). A number of sediments (e.g. those of the Tanner Basin) do not appear to contain free phytol or dihydrophytol, but these compounds were liberated by heating at ~100°C for extended periods (Ikan *et al.*, 1975b). The diagenesis of phytol in Recent sediments has been elucidated by Brooks (1974) and de Leeuw *et al.* (1977).

Isoprenoid ketones, which are probably products of the microbial oxidation of phytol, have been identified in both Recent and older sediments extending back to the Lower Cretaceous (see e.g. Ikan *et al.*, 1973; Simoneit, 1973). The major homologue present is 6,10,14-trimethylpentadecan-2-one (VI) and this is accompanied in some instances by minor amounts of 6,10-dimethylundecan-2-one (VII) (Simoneit, 1973, 1975, 1977a, 1978a, b).

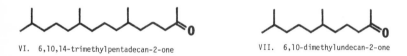

VI.   6,10,14-trimethylpentadecan-2-one   VII.   6,10-dimethylundecan-2-one

39.3.3.3. Wax esters

Wax esters are simple esters of long-chain fatty alcohols (usually even homologues C_{14}–C_{22}) with long-chain fatty acids (usually even homologues C_{14}–C_{24}); they serve as food reserves for many species of marine organisms and their structures have been investigated extensively (see the reviews by Nevenzel (1970) and Sargent *et al.* (1976)). Wax esters have also been observed in slicks (Lee and Williams, 1974) and clumps (Benson *et al.*, 1973) on the sea surface. In the marine sedimentary environment, these compounds have been identified in the uppermost 2–3 m of Recent sediments from both the Irish Sea and the North Sea (Sargent *et al.*, 1978; Sargent and Gatten, 1976). In these sediments the wax esters constituted ~50% of the total lipids and they may therefore be significant contributors to the organic matter in certain areas.

39.3.4. STEROIDS

The occurrence of sterols in all types of living organisms has been extensively documented (see the review by Gaskell (1974)). In general, primitive organisms (e.g. *Monera*) are the major contributors of sterols to marine sediments in highly productive environments; thus, blue-green algae contribute sterols with C_{27} and C_{29} 4-desmethyl skeletons (Reitz and Hamilton, 1968), and some bacteria provide C_{27}, C_{28} and C_{29} 4-desmethyl, C_{29} 4,4-dimethyl and C_{28} 4α-methyl skeletons (Schubert *et al.*, 1967, 1968; Bird *et al.*, 1971), but most bacteria contribute no steroids.

The earliest study of sterols in Recent sediments was made by Schwendinger and Erdman (1964) who found the total concentrations of sterols (as cholesterol) in a variety of freshwater and marine sediments to range from 60 to 300 μg g^{-1} of organic carbon. Subsequent investigations have concerned the identification of the individual steroids of the lipids of marine sediments. The limited data available show that steranes, stanols and sterenes are the principal steroid classes present, the principal homologues having C_{27}, C_{28} and C_{29} skeletons (Attaway and Parker, 1970; Simoneit and Burlingame, 1974a; Simoneit, 1974a; Ikan *et al.*, 1975b). Thus, Recent sediments from Baffin Bay, Texas, and from the San Pedro Basin, California, both contained cholesterol, campesterol, stigmasterol, and β-sitosterol as major steroids (Attaway and Parker, 1970). In Tanner Basin sediment, the major steroidal components are 22-dehydrocholesterol, brassicasterol, Δ^7-ergosterol, campesterol and β-sitosterol (Ikan *et al.*, 1975b). This array suggests a more marine origin for the steroids than do those in the Baffin Bay and San Pedro Basin sediments. Sediment from Laguna Mormona, Baja California, contained only mono- and diunsaturated sterenes (C_{27}–C_{29}) and no sterols (Cardoso, 1976). Various petrogenic steranes have been identified in Recent sediments from the continental shelf areas off southern California (Reed *et al.*, 1977). From this brief survey it is apparent that further work is required to evaluate the use of steroids as a diagnostic tool for determining the origins of the organic matter in Recent sediments.

The steroids in older marine sediments are usually reduced or altered homologues such as steranes (VIII, R = H.R' = H, CH_3, C_2H_5, A/B ring junction predominantly *trans*), sterenes (VIII, R = H, double bonds in ring A or B), stanols (VIII, R = OH) or stanones (VIII, R = O) (Peake *et al.*, 1974; Simoneit and Burlingame, 1974a; Simoneit, 1975, 1977a, b, e, f, 1978a, b). Various 4α-methylstanols (IX, R = OH) and 4α-methylsteranes (IX, R' = H) have been reported to be present in Messel shale, a lacustrine sediment (Mattern *et al.*, 1970). A 4-methyl C_{29} stanol (IX, R = OH, R = CH_3) has been identified as the major steroid in a Black Sea sapropelic sediment (Simoneit, 1974a) and has been shown to be 24-methyllophanol by

VIII. steroidal homologues   IX. 4(α)-methylsteroids

^{13}C NMR (Simoneit and Wilson, unpublished results; the stereochemistry of the side chain has not been established with certainty). Older sediments from the Black Sea also contain 4-methylstan-3-ones (IX, R = 0, probably 4αCH$_3$) (Simoneit, 1978a). Various other steroidal compounds have been identified in the lipids of DSDP sediments from the Black Sea (Simoneit, 1978a), the Atlantic (Cretaceous) (Simoneit, 1977e, f, 1978b), the Cariaco Trench (Pleistocene) (Simoneit, 1975), the North Atlantic (Pliocene) (Simoneit, 1975, 1978c) and the north-eastern Pacific (Pleistocene–Pliocene) (Simoneit, 1975, 1977b).

The backbone rearrangement of stanols to sterenes has been shown to occur both in the laboratory and under geological conditions (Rubinstein et al., 1975). Thus, the lipids of a marine shale (Jouy-aux-Arches, Toarcian Formation, Jurassic) contained (20R)- and (20S)-5β,14α-dimethyl-18,19-dinor-8α,9β,10α-ster-13(17)-enes, C_nH_{2n-8}, ranging from $n = 27$ to $n = 30$ (Rubinstein et al., 1975). These sterenes may well be useful as biological markers in the sedimentary environment.

39.3.5. TRITERPENOIDS

Recent work has shown that triterpenoids having hopane (X), the isomeric 17α(H)-hopane (XI) and moretane (XII) type skeletons are both ubiquitous and abundant in the geosphere (van Dorsselaer et al., 1974). These compounds exist mainly as the saturated and unsaturated hydrocarbons and carboxylic acids, but their alcohols and ketones also occur. The numbers of carbon atoms in these three skeletons range from C_{27} to C_{35}, according to the length of the side chain on ring E. Triterpenoids, especially those having a hopane skeleton, occur in a wide variety of biota (McCrindle and Overton, 1969; Kulshreshtha et al., 1972). However, recent evidence suggests that simple homologues are the sole triterpenoids of lower organisms. For example, hop-22(29)-ene (XIII-diploptene) has been identified in the bacteria *Methylococcus capsulatus* and *Bacillus acidocaldarius* (Bird et al., 1971; deRosa et al., 1971), the latter organism also containing hopane (X) and hop-17(21)-ene (XIV) (deRosa et al., 1973). Gelpi et al. (1970) have reported the presence

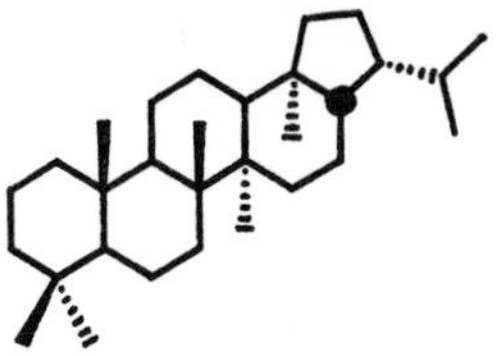

X. hopane

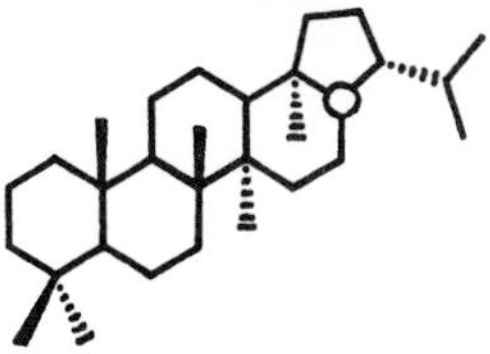

XI. 17α(H)-hopane

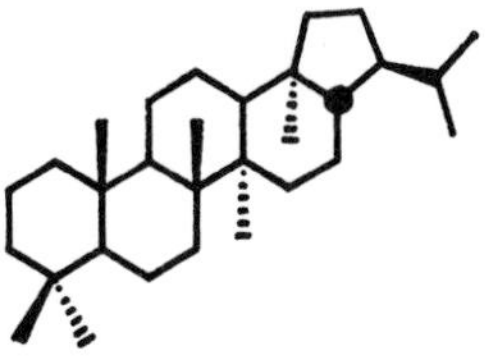

XII. moretane

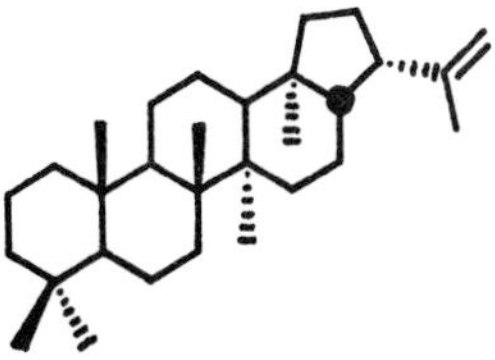

XIII. diploptene

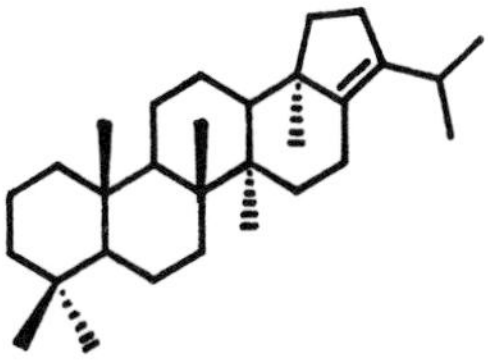

XIV. hop-17(21)-ene

in three blue-green algae of a triterpenoid which Bird *et al* (1971), on the basis of mass spectrometric correlations, have identified as also being diploptene (XIII). The extended hopanes, which were first identified in the geosphere, have now been reported to be present in the bacterium *Acetobacter xylinium* (Förster *et al.*, 1973); this organism also contains a tetrahydroxy C_{35} terpene of the hopane type (identified by Rohmer (1975) as 32,33,34,35-tetrahydroxy-bacteriohopane (XV)). Various polyhydroxy-bacteriohopanes and diplopterol have been identified as the predominant components of the alcohol fractions of the lipids of about 20 procaryotic species of bacteria (Rohmer,

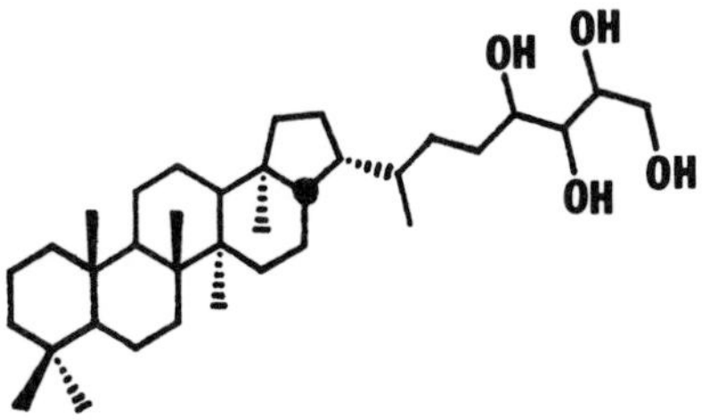

XV. 32,33,34,35-tetrahydroxybacteriohopane

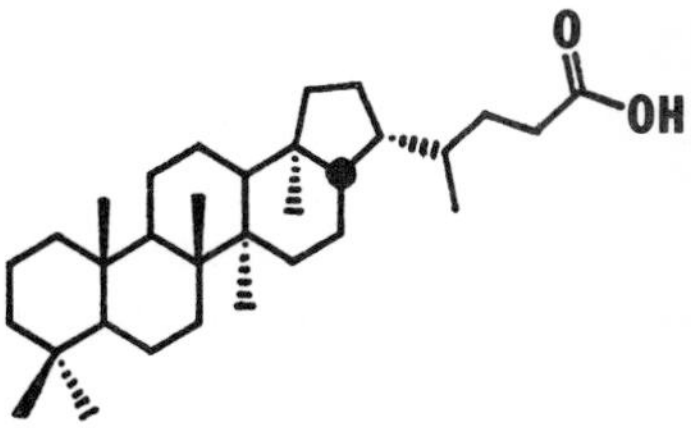

XVI. bishomohopanoic acid

1975). These polyols, which can be degraded to the extended hopanoic acids, (e.g. bishomohopanoic acid, (XVI)), have been identified in two lacustrine sediments and serve as markers for procaryotes (Rohmer, 1975).

The identification of triterpenoids involves the comparison of the GC retention index and the mass spectrum of the compound or its degradation product(s) with those of standards of known structure. Such data are available for the following compounds and/or homologous series: most of the basic triterpenoid skeletons of the hydrocarbons (Albrecht, 1969; Kimble, 1972; Gallegos, 1971; Whitehead, 1974; Kimble et al., 1974a, b); extended triterpenoid hydrocarbons $>C_{30}$ (Kimble et al., 1974b; van Dorsselaer, 1974; Ensminger et al., 1972, Ensminger, 1974); nortriterpenoidal hydrocarbons (Kimble et al., 1974b; Ensminger, 1974; van Dorsselaer, 1974); arborinone, trisnorhopan-21-one (Arpino, 1973); partially aromatized triterpenoids (Spyckerelle, 1975; Greiner et al., 1976); triterpenoidal alcohols (Brooks, 1974; Rohmer, 1975); and triterpenoidal acids (Arpino, 1973; Brooks, 1974; Rohmer, 1975).

Triterpenoids have been sought in a number of Recent marine sediments. For example, a variety of C_{27}, C_{29}, $C_{30}-C_{32}$ triterpanes, C_{30} triterpene, and C_{31} and C_{32} triterpenoid acids, mostly having hopane skeletons, have been found in lagoonal sediments from around the Virgin Islands (Eglinton et al., 1974). Triterpenoids have been found in all of a series of Recent sediment samples from the Black Sea (Simoneit, 1974a, 1977a). In these sediments the hydrocarbons identified were trisnorhopene (XVII), hopane (X), diploptene (XIII) and other C_{30} and C_{31}, triterpanes, together with another C_{30} triterpene. Oxygenated triterpenoids present were trisnorhopan-21-one

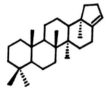

XVII. trisnorhopene   XVIII. trisnorhopan-21-one

(XVIII), two other ketones ($C_{28}H_{46}O$ and $C_{30}H_{50}O$) and bishomohopanoic acid (XVI). The sediments and the algal mat from Laguna Mormona, Baja California, contained only the triterpenoidal hydrocarbon diploptene (XIII) and 17β(H)-bishomohopanoic acid (XVI) (Cardoso, 1976; Cardoso et al., 1976). The major biogenic input to this region, as reflected by the lipid distributions, was from algae and bacteria (Cardoso et al., 1976). The triterpenoids in a surface sediment from the harbour of La Rochelle, France, have been found to contain hydrocarbons of the hopane series ranging from

C_{27}, C_{29} to C_{35} with 17α(H) stereochemistry, and extended hopanoic acids also with 17α(H) stereochemistry (Dastillung and Albrecht, 1976). The presence of these geologically older, thermodynamically more stable, triterpenoids, coupled with a "hump" of branched/cyclic hydrocarbon material, has been used as a marker for petroleum pollution in this Recent sediment.

Triterpenoids have been identified in many older marine sediments: e.g. those of the north-eastern Pacific Ocean (Simoneit, 1977b); the Atlantic Ocean, including the Blake–Bahama Basin, the Cape Verde Basin (Simoneit and Burlingame, 1974; Simoneit, 1977b, f; Deroo et al., 1977, 1978a; Stuermer and Simoneit, 1978) and the Angola Basin (Simoneit, 1977e); the Black Sea (Simoneit, 1978a); the Cariaco Trench (Simoneit and Burlingame, 1974a; Simoneit, 1975); and the Gulf of Mexico and the Mediterranean Sea (Simoneit and Burlingame, 1974a). Those sediments of Pleistocene–Pliocene age contained predominantly the 17β(H)-hopane series, with significant amounts of the 17α(H) homologues. The sediments of Miocene–Jurassic age (primarily Cretaceous) contained 17α(H)-hopanes > 17β(H)-hopanes. The major homologues that have been identified are: 17β(H)- and 17α(H)-trisnorhopane (XIX), 17α(H)- and 17β(H)-norhopane (XX), 17α(H)-hopane (XI), 17β(H)-hopane (X), some extended hopanes (XXI), diploptene (XIII), hop-17(21)-ene (XIV)

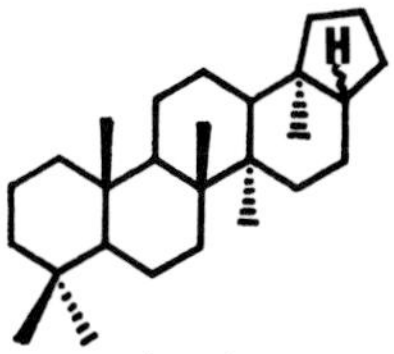

XIX. trisnorhopanes

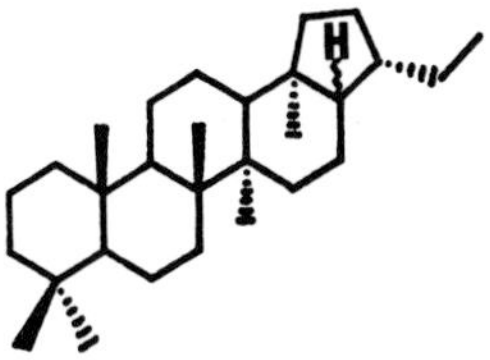

XX. norhopanes

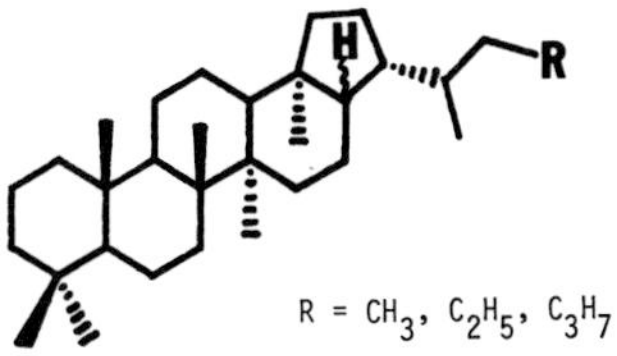

R = CH_3, C_2H_5, C_3H_7

XXI. extended hopanes

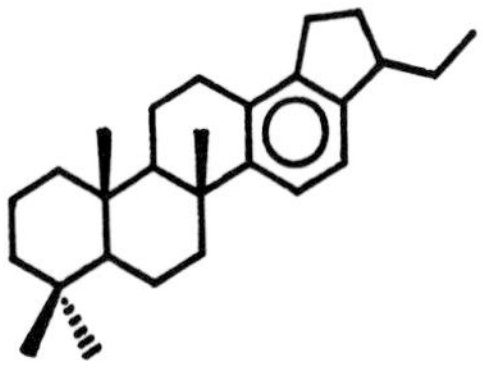

XXII. 14,18-bisnormethyladianta-13,15,17-triene

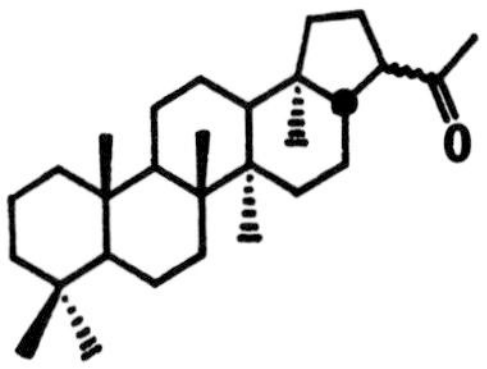

XXIII. adiantones

and 14,-18-bisnormethyladianta-13,15,17-triene (XXII). The last of these compounds was present as a major constituent in a Cretaceous black shale from the Atlantic Ocean (Simoneit 1977f). Because other aromatized triterpenoids have been identified in the Messel shale, they may also be present in marine sediments at low concentrations (Spyckerelle, 1975; Greiner et al., 1976). An example of the distribution of triterpenoid hydrocarbons in a North Atlantic sediment is shown in Fig. 39.4 (Simoneit, 1975, 1978c). In this, the major species are; $17\alpha(H)$- and $17\beta(H)$-trisnorhopane (XIX), $17\alpha(H)$- and $17\beta(H)$-norhopane (XX), hop-17(21)-ene (XIV), diploptene (XIII), $17\alpha(H)$-hopane (XI), $17\beta(H)$-hopane (X), and $17\alpha(H)$- and $17\beta(H)$-homohopane (XXI). The oxygenated triterpenoids of this sample are; trisnorhopan-21-one (XVIII), adiantones (XXIII), a C_{30} ketone ($C_{30}H_{50}O$) and triterpenoidal acids (e.g. XVI). More is known about the triterpenoidal ketones in Black Sea sediments (Simoneit, 1978a), and the extended hopanoic acids (major component C_{32}) have been found in a number of those sediments which have high lipid contents (e.g. Simoneit, 1974b, 1977b, 1978a). Triterpenoid alcohols have been identified in an algal ooze from Laguna Mormona, Baja California (Cardoso et al., 1976); these include a C_{30} pentacyclic triterpenoid alcohol (either 22- or 29-hydroxyhopane) and $17\beta(H)$-bishomohopanol (XXIV).

XXIV. $17\beta(H)$-bishomohopan-32-ol

39.3.6. DITERPENOIDS

The resins and supportive tissue of higher plants contain diterpenoids having abietane, pimarane or kaurane skeletons (see the review by Nakanishi et al. (1974)). As cyclic diterpenoids are not endogenous in the marine biota, and do occur in the geosphere (Thomas, 1969, 1970; Streibl and Herout, 1969), there is a possibility of using them as markers for the injection of terrigenous organic material into marine sediments.

The predominant diterpenoid which has been identified in marine sediments is dehydroabietic acid (XXV), concentrations of which probably lie in the range $0.2–20 \,\mu g \, g^{-1}$ (Simoneit, 1977b). Lesser amounts of its decarboxylation product, dehydroabietin (XXVI) and the aromatic hydrocarbon,

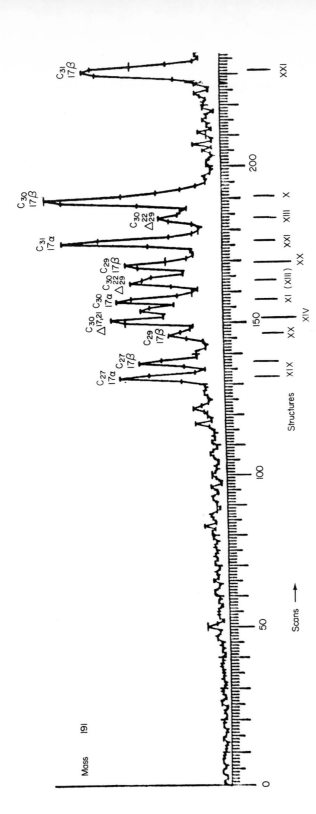

Fig. 39.4. Mass chromatogram of m/e 191, showing the triterpenoid distribution in a Pliocene sediment sample from the North Atlantic Ocean (DSDP 12-114-5-5) (Simoneit, 1975). The chemical structures of the various compounds are indicated by Roman numeral which refer to structures shown in the text.

retene (XXVII), are also often present (Simoneit, 1975). These compounds and other minor diterpenoidal constituents have been identified in sediment samples from the north-eastern Pacific Ocean, the North and the South Atlantic Ocean, the Cariaco Trench and the Black Sea (Simoneit, 1975, 1977a, b, d, e, f, 1978a, b; Stuermer and Simoneit, 1978). The occurrences of these diterpenoids in marine sediments are exemplified in Figs 39.2 and 39.3. The

XXV. dehydroabietic acid, $C_{20}H_{28}O_2$ XXVI. dehydroabietin, $C_{19}H_{28}$ XXVII. retene, $C_{18}H_{18}$

ages of the sediments ranged from Recent to Jurassic, and it would appear that these diterpenoids are excellent molecular markers for contributions to the sediments from terrigenous resinous plants (Simoneit, 1977b). These compounds have been found in sediments containing terrigenous clays, but not always in association with wax lipids of higher plants (Simoneit, 1975, 1977b).

39.3.7. TETRATERPENOIDS

Although early attempts were made to identify the tetraterpenoids of marine sediments using the comparatively crude techniques of spectrophotometry (see reviews by Schwendinger (1969) and Vallentyne (1957, 1960)), it was not until more sophisticated techniques became available that it was possible to characterize these compounds satisfactorily. β-Carotene (XXIX) has been found in Recent sediments from the Tanner Basin at a concentration of 80 ng g^{-1} dry sample (Ikan et al., 1975c). There is some evidence that marine sediments contain both partially and fully reduced carotenoids (e.g. XXVIII)

XXVIII. carotane, $C_{40}H_{78}$

XXIX. β-carotene, $C_{40}H_{56}$

and work is continuing on the mechanisms by which these compounds are produced (Simoneit and Burlingame, 1972a). In more recent studies, Watts (1975) has applied spectrophotometry, mass spectrometry and field desorption mass spectrometry to the identification of the tetraterpenoid pigments of marine sediments. Using these techniques, Watts (1975) and Watts and Maxwell (1977) determined the carotenoids in sediments from the Cariaco Trench and the Black Sea and were able to demonstrate that the progressive reduction of carotenoids commences in Recent sediments. The rate of reduction, however, differs for the various carotenoids. For example, β-carotene (XXIX) shows no reduction in a sediment $\sim 340\,000$ years old, whereas in the same sediment up to two double bonds had been reduced in the zeaxanthin (XXX) and up to five in the canthaxanthin (XXXI). Watts (1975)

XXX. zeaxanthin, $C_{40}H_{56}O_2$

XXXI. canthaxanthin, $C_{40}H_{52}O_2$

was also able to assign these sedimentary pigments to particular biogenic sources. For example, the major carotenoids in the upper level (~ 3200 years B.P.) of the Black Sea sediments were derived from a purely marine source. In contrast, those in the lower level ($\sim 14\,000$ years B.P.) had a lacustrine bacterial origin. These conclusions are borne out by other data, e.g. those for lipids (Simoneit, 1975).

39.3.8. TETRAPYRROLE PIGMENTS

The predominant biogenic precursors of the tetrapyrrole pigments (chlorins and porphyrins) in marine sediments are chlorophylls. Chlorophylls a and b are found in higher plants and green algae, and chlorophylls a and c, sometimes with d, occur in certain other marine plankton (Baker and Smith, 1974). General reviews of the tetrapyrrole pigments have been published by Vernon and Seely (1966), Hodgson et al. (1967, 1968), Baker et al. (1967), Baker (1969), Alturki (1972) and Didyk (1975). The initial conversion of chlorophyll to chlorins and their early diagenesis (essentially during the Pleistocene) has

264   BERND R. T. SIMONEIT

been reviewed by Baker and Smith (1974) whose diagenetic scheme is shown in Fig. 39.5. In this process of early diagenesis the conversion of chlorophyll to pheophytin (XXXII) is rapid, whereas the subsequent steps are slower.

Total chlorin contents have been determined in Recent marine sediments from the Tanner Basin, California (Ikan *et al.*, 1975c), the Beaufort Sea

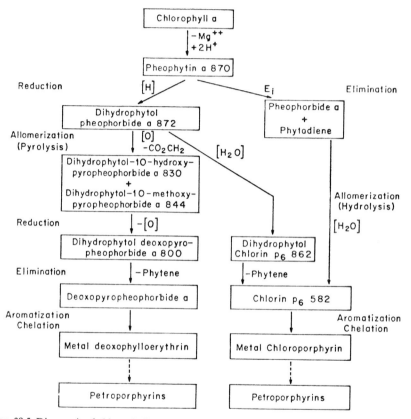

FIG. 39.5. Diagenesis of chlorophyll *a*; the numbers in the boxes are molecular weights. (Baker and Smith (1974), reproduced by permission, Éditions Technip, Paris.)

(Peake *et al.*, 1972), the continental shelf off Nova Scotia (Baker and Hodgson, 1968) and the Black Sea (Drozdova and Gursky, 1972; Peake *et al.*, 1974; Lorenzen, 1974).

The tetrapyrrole pigments in older marine sediments have been extensively elucidated (Baker *et al.*, 1976a). A series of dihydrophytol pheophorbides (in which the carboxyl group in XXXIII is esterified with dihydrophytol) have

XXXII. pheophytin a   XXXIII. pheophorbide a

been found in Pleistocene sediments from the Cariaco Trench (Baker and Smith, 1973). In successively older sediments, pheophorbides (e.g. XXXIII), chlorin p_6 and Ni porphyrins were observed at concentrations up to 120 ng g^{-1} (Baker and Smith, 1975b). The chlorins (e.g. pheophorbide) undergo further diagenesis by reduction, elimination and aromatization reactions yielding, for example, such compounds as XXXIV and XXXV petrophyrins

XXXIV. phylloerythrin   XXXV. deoxophylloerythroetioporphyrin

X = VO, Ni
R = H, CH$_3$, C$_2$H$_5$

XXXVI. metallated DPEP porphyrin   XXXVII. metallated Etio porphyrin

(Baker et al., 1976a). Chelation can occur with either nickel or vanadyl ions yielding, for example, the metallated DPEP series (XXXVI) which is transformed under thermal stress to the etioporphyrin series (XXXVII) (Didyk et al., 1975; Baker et al., 1978a).

Petroporphyrins have been identified in oil-bearing sediments from the Challenger Knoll, Gulf of Mexico (Baker, 1971), and lesser amounts of them ($<130 \,\mu g\, g^{-1}$) occur together with chlorins ($\leqslant 35 \,\mu g\, g^{-1}$) in older sediments from the North Atlantic, western Pacific and Indian Oceans (Baker and Smith, 1975b), the Sea of Japan and the Philippine Sea (Baker and Smith, 1975a; Baker et al., 1978b), the Norwegian–Greenland Sea (Baker et al., 1976b), the Angola and Cape Basins, the Walvis Ridge (Baker et al., 1977) and the Cape Verde Rise and Basin (Baker et al., 1978a). Marine petroporphyrins can also be formed in other ways. For example, they have been found to be present in an organic-rich shale which has been altered by the intrusion of a diabase sill (DSDP Leg 41, Site 368 at 957 m below the seabed) (Baker et al., 1976a, 1978a). Consideration of the various data given above suggest that porphyrins may be useful in sedimentary geochemical studies, such as those of geothermal history and maturation, source rock potential and organic diagenesis (Baker et al., 1976a).

39.3.9. AMINO ACIDS AND PEPTIDES

Amino acids, and to a lesser extent peptides and proteins, have a wide occurrence in the geosphere (see e.g. Hare, 1969), those which are found in marine sedimentary environments probably having a planktonic origin (Degens, 1970). In a study of the early diagenesis of amino acids a correlation has been found between the variation with depth of the total amino acid concentration and the C_{org}/N_{org} ratio in sediments from the Argentine Basin (Stevenson and Cheng, 1972). It was observed that, although the total amino acid concentration decreased with depth, there was little change in the N_{org} concentration, from which it was concluded that the amino acids became incorporated into humic material and thus became refractory (Stevenson, 1974). The geochemistry and racemization of amino acids have been reviewed by Hare (1969), Kvenvolden (1975) and Dungworth (1976).

There have been a number of studies of amino acids in Recent marine sediments: those from the Mohole project (Rittenberg et al., 1963), the Santa Barbara Basin, California (Degens et al., 1961, 1964), the San Diego Trough (Degens et al., 1963) the northern-east Pacific Ocean (Degens et al., 1964), the Beaufort Sea (Peake et al., 1972), the Black Sea (Starikova and Korzhikova, 1969, 1972a, b), the Indian and Atlantic Oceans (Aizenshtat et al., 1973; Degens, 1970; Whelan, 1977), the Gulf of Mexico (Aizenshtat et al., 1973) and off the NW. African coast (Morris, 1975). Older sediments have also been examined, originating mainly from the Caribbean Sea, the Cariaco Trench (Bada and Man, 1973; Hare, 1973), various sites in the Pacific Ocean (Wehmiller and Hare, 1972) and the Black Sea (Mopper et al., 1978). For example, the concentrations of total amino acids in older sediments from the north-eastern Pacific ranged from 0.16 nmol g^{-1} for an Eocene sample to a maximum 134 nmol g^{-1}. In general, the ratios of glycine/alanine and aspartic acid/glutamic acid were found to decrease with increasing age. Degens (1970) found that the contents of combined amino acids in sediments from the Atlantic and Indian Oceans decreased rapidly with depth in the upper 5 m. This author also reported that the abundances of the various amino acids were highly variable between different sediments and/or depositional environments. These variations probably reflect differences in the ease with which the various amino acids are degraded; thus, aromatic, hydroxy and neutral straight-chain amino acids appear to be the more readily degradable ones (Morris, 1975). Evidence for amino acid decarboxylation in sediments from the Gulf of Mexico and the Atlantic Ocean has been given by Aizenshtat et al. (1973). Marine fossils, which were separated from sedimentary rocks ranging in age from Pleistocene to Jurassic, were analysed for amino acids (Drozdova, 1974). The major analogues found were glycine, alanine, arginine and leucine.

The amino acids in living organisms have ($-$) stereoisomeric configurations, but in the geosphere they slowly epimerize yielding a ($\pm$) mixture. Glutamic acid and β-alanine have the fastest epimerization rates, followed by proline, phenylalanine, valine, leucine and isoleucine (Kvenvolden et al., 1970). The epimerization rate of ($-$)-isoleucine to ($-$)-alloisoleucine has been used to estimate the ages of sediments younger than $\sim$ 400 000 years B.P. (Wehmiller and Hare, 1971; Bada et al., 1970; King and Neville, 1977). The epimerization of ($-$)-isoleucine has also been used as a palaeotemperature indicator (McKenna et al., 1971; Bada et al., 1973).

39.3.10. PURINES AND PYRIMIDINES

Although purines and pyrimidines have not been detected in marine sediments they have been found in the sediments of those lakes exhibiting eutrophication (van der Velden and Schwartz, 1974, 1976a, b; van der Velden et al., 1974; Dungworth et al., 1977). The major purines which have been characterized are adenine (XXXVIII), guanine (XXXIX), hypoxanthine (XL) and xanthine (XLI). The major pyrimidines which have been found are cytosine (XLII), thymine (XLIII) and uracil (XLIV). The total concentrations reported for these classes of compounds in sediments from Lake Erie were 111 μg g^{-1} at the surface and 53 μg g^{-1} at a depth of 10–15 cm below the lake bed (van der Velden et al., 1974). These compounds represent the remains of the genetic material of organisms and may, if detectable, prove to be useful bio-organic markers in marine sediments.

Muramic acid, a proteinaceous cell-wall component found only in bacteria and blue-green algae, has been quantitatively determined in Black Sea sedi-

XXXVIII. adenine   XXXIX. guanine   XL. hypoxanthine

XLI. xanthine   XLII. cytosine   XLIII. thymine

XLIV. uracil

ments (King and White, 1978) and has been used to assess the microbial biomass at the time that the sediment was deposited.

39.3.11. CARBOHYDRATES

Carbohydrates (sugars) are ubiquitous in the marine environment and early studies of their geochemistry have been reviewed by Swain (1969). They occur as monosaccharides, oligosaccharides (di- or tri-), or polysaccharides. The common monosaccharides of sediments are D-glucose, D-mannose, D-fructose, D-galactose, D-xylose, L- and D-arabinose, L-rhamnose and D-glucuronic acid (Prashnowsky et al., 1961). The oligosaccharides of widest occurrence are sucrose, maltose and raffinose.

Recent sediments from the Santa Barbara Basin, California, have been found to contain roughly similar concentrations of galactose, mannose, glucose, rhamnose, ribose, xylose and arabinose, totalling 2.2 mg g^{-1} of dry sediment at the surface, and decreasing to very low concentrations 4 m below the seabed ($\sim$4000 years B.P.) (Prashnowsky et al., 1961). The same sugars were also detected in sediments from the San Diego Trough at a total concentration ranging up to 0.9 mg g^{-1} (Degens et al., 1963). The total sugars in continental shelf, slope and deep-ocean (Mohole drilling) sediments from off California have been found to decrease in concentration with depth of burial by a factor of ~20 in the upper few metres of sediment (Degens *et al.*, 1964). Abundant carbohydrates have also been detected in Black Sea sediments (Starikova and Yoblokova, 1972; Mopper *et al.*, 1978). Although Swain and Bratt (1978) were able to detect carbohydrate residues in north-western Atlantic sediments dating back to the Miocene, they observed no correlation of these with either depth or lithology.

Although amino sugars have been found in a sediment from Lake Ontario (Dungworth *et al.*, 1977), they do not appear to have been detected yet in marine sediments.

Various aspects of the biogeochemistry of carbohydrates in the marine environment have been reviewed by Mopper (1973) who concluded that the origin of the carbohydrates could not be established unambiguously. For example, the carbohydrates in coastal sediments from off California could have been derived from material having three different origins; viz. allochthonous (terrigenous), and autochthonous, either in the water column or in the sediments themselves (Prashnowsky *et al.*, 1961).

Inositols have been determined in various marine sediments (White and Miller, 1976) in which the major analogues identified were myo-, scyllo- and chiroinositol, with the highest concentrations ranging up to $87 \mu g g^{-1}$ for myo-inositol (XLV) in a dried Santa Barbara sediment. Low levels of

XLV. myo-inositol

inositols were still detectable in sediments 45×10^6 years old. Inositols are constituents of all higher organisms and of most micro-organisms, and they appear to be growth-promoting factors.

39.3.12. AROMATIC HYDROCARBONS

Alkylated aromatic hydrocarbons, such as alkylbenzenes, are common constituents of petroleum but have not yet been identified as major organic components of deep-sea sediments. Petroleum and petroleum-bearing shales have been encountered in the marine sedimentary column (Reed *et al.*, 1977; Simoneit, 1976) and it may be assumed, therefore, that alkylated aromatic hydrocarbons are probably present in them. More thorough analyses are necessary to assess the depth of burial and the lithological parameters necessary to generate these compounds.

Polynuclear aromatic hydrocarbons (PAH) have been identified in many marine sediments, in which they appear to have more than one source. Thus fluoranthene (XLVI), pyrene (XLVII), 1,2-benzpyrene (XLVIII) and 3,4-benzpyrene (XLIX), but not perylene (L) have been found to be the predominant PAH constituents of a polluted Severn Estuary sediment (Eglinton *et al.*, 1975) and most likely result from petroleum pollution. The

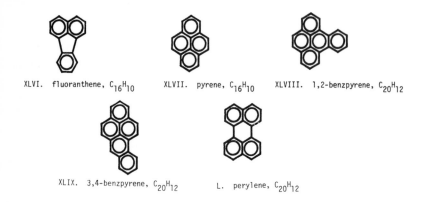

XLVI.   fluoranthene, $C_{16}H_{10}$ XLVII.   pyrene, $C_{16}H_{10}$ XLVIII.   1,2-benzpyrene, $C_{20}H_{12}$

XLIX.   3,4-benzpyrene, $C_{20}H_{12}$ L.   perylene, $C_{20}H_{12}$

presence of PAH is widespread in Recent marine sediments (Youngblood and Blumer, 1975; Blumer and Youngblood, 1975). However, in those sediments deposited prior to man's activities (excluding marine petroliferous seeps) the major PAH is perylene (Orr and Grady, 1967; Aizenshtat, 1973; Simoneit and Burlingame, 1974a; Simoneit, 1977a, f, 1978a). Because perylene has not been identified as a major component in PAH fractions derived from petroleum pollution or the combustion of fuels, it may perhaps be useful as a biogenic marker, although its precursor is at present not known for certain. However, this compound does appear to be associated with terrigenous lipids deposited in a reducing environment (Simoneit, 1975, 1978; Aizenshtat, 1973). (See also Addenda p. 311.)

Geogenetic aromatic hydrocarbons, e.g. retene (XXVII), have also been identified in marine sediments (see e.g. Simoneit, 1977b, e). Since partially aromatized homologues of such compounds have been identified in the geosphere (Spyckerelle, 1975; Greiner *et al.*, 1976) it appears that the fully aromatic homologues may have been generated by dehydrogenation reactions.

39.3.13. NATURAL POLYMERS

Some natural biopolymers that may be preserved in the geosphere include proteins (polypeptides), chitin, cellulose, lignin, cutin and suberin. However, the geological time periods over which these various polymers are stable are quite variable, but lignin, cutin and suberin, which can be readily identified in sediments, may be of value as complementary markers, with lipids, of terrigenous organic matter.

Proteins which are components of all living matter are easily hydrolysed to amino acids (see Section 39.3.9). Chitin is a common skeletal material of marine fauna and is a polymer of β-N-acetyl-D-glucosamine. Although no chemical data are available for its occurrence in marine sediments, it has been identified morphologically in fossils (Florkin, 1969). Cellulose, a major structural material of higher plants, has only been reported as discrete plant fragments in marine sediments. However, those carbohydrates which are formed by its hydrolysis have been identified in marine sediments (see Section 39.3.11). Lignin, another important skeletal material of higher plants, has been identified in marine sediments influenced by continental run-off (Hedges and Parker, 1976; Pocklington, 1976; Gardner and Menzel, 1974; Hedges, 1974; Leo and Barghoorn, 1970). For example, a rapid seaward decrease of lignin concentration was observed in sediments from several offshore transects in the western Gulf of Mexico (Hedges, 1975; Hedges and Parker, 1976). This lignin is believed to have originated mainly from flowering plants, grasses and tree leaves and to have been transported into the Gulf in the form of small fragments by the rivers. Similar data were observed for sediments from the U.S. Atlantic coast (Gardner and Menzel, 1974), the highest lignin concentrations being found in those sediments from the inner continental shelf.

The biopolymers cutin and suberin are structurally related and consist of interesterified and polymerized hydroxy fatty acids (Martin and Juniper, 1970; Kolattukudy and Walton, 1972; Holloway et al., 1972; Kolattukudy et al., 1975). The acids are generally C_{16} and C_{18}, and have a hydroxyl group in the terminal (ω) position and, in some cases, one-to-two additional hydroxyl substituents between positions 7 and 10 in the carbon chain. Cardoso (1976) has examined three marine sediments for cutins; a sapropelic sediment from the eastern Black Sea (∼3000 years B.P.) contained cutins, but these compounds could not be detected in DSDP sediments from either the Cariaco Trench (15-147-7-1, ∼340 000 years B.P.) or the Atlantic Ocean (44-391A-11-0, Miocene).

39.3.14. THE "HUMP"

The term "hump" has been applied to various types of mixtures which cannot be resolved using techniques such as GC and molecules sieving (see e.g. Eglinton et al., 1974; Farrington and Medeiros, 1975). These "humps" consist of a complex mixture of branched and cyclic hydrocarbons with various molecular weight distributions and maxima in which the cyclic hydrocarbons (naphthenes) predominate. Petroleum appears to be the major source of the broad "humps" in branched/cyclic hydrocarbon fractions of lipids of sediments (see e.g. Dastillung and Albrecht, 1976; Reed and Kaplan, 1977). An example of a capillary GC trace of such a broad "hump" is shown in Fig. 39.6. Broad "humps" (generally in the GC retention region of n-C_{12}–n-C_{33} alkanes with a maximum usually at $\sim n$-C_{25}–n-C_{27}) have been found for

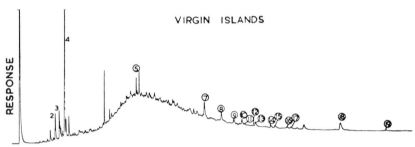

FIG. 39.6. An example of a "hump"; the branched/cyclic hydrocarbons from a Virgin Islands lagoonal sediment. (Eglinton *et al.* (1974), reproduced by permission, Éditions Technip, Paris.)

many Recent and older marine sediments (Blumer and Sass, 1972a, b; Cooper *et al.*, 1974; Zafirou, 1973; Farrington and Quinn, 1972a; Hunt, 1974c; Tissier and Oudin, 1974; Eglinton *et al.*, 1974, 1975; Reed and Kaplan, 1977), and most of them are thought to be derived from petroleum.

Some sediments, especially sapropels, exhibit narrower "humps" which extend over the GC retention region n-C_{17}–n-C_{26} alkanes and usually have a maximum at n-C_{22} or n-C_{23}. Such "humps" have been found for lipids extracted from various sediments including those from the California borderland (Reed *et al.*, 1977), the north-eastern Gulf of Mexico (Gearing *et al.*, 1976; Palacras *et al.*, 1976), coastal salt marshes and estuaries (Palacas *et al.*, 1972; Johnson and Calder, 1973) and Mangrove Lake, Bermuda (Hatcher *et al.*, 1977). Narrow "humps" have also been found for the anaerobic decomposition products of cultured soil micro-organisms (Jones, 1969), and they may, therefore, represent anaerobic decomposition products of algae and bacteria in general.

39.3.15. KEROGEN AND HUMATES OF MARINE SEDIMENTS

39.3.15.1. The nature of humates in marine sediments

Descriptions of the humic substances (as defined by Kononova (1966)) of marine sediments are sparse and are confined mainly to those from the continental shelf areas (see the review by Schnitzer and Khan (1972)). However, estimates of humic acids are available for some sediments from oceanic areas, e.g. the Bering Sea (Bordovskiy, 1965a, b) and the Indian Ocean (Kasatochkin *et al.*, 1968).

The molecular weight distributions of the humic and fulvic substances of the marine clays from the Scotian Shelf have been reported to lie in the range ∼700 to >2 000 000 (Rashid and King, 1969). Some data on bulk chemical properties (mainly C/H ratios) of marine sedimentary organic matter and humates have been published by Bordovskiy (1965a). This author has also investigated the capacity of humates to absorb ammonia. He concluded that part of the humate nitrogen may have been acquired by this process (Bordovskiy, 1965b).

The predominant functional groups of marine humates are carboxyl, phenolic and alcoholic hydroxyls, and carbonyl (Rashid and King, 1970). The principal functionality is carboxyl, and the total acidity ranges from 2 to 7 meq g^{-1} (Rashid and King, 1970). The oxygen-containing functional groups of marine humates are concentrated in the low molecular weight fractions, and the C/H ratio increases with increasing molecular weight (Rashid and King, 1971). The redox potential of the depositional environment does not appear to influence the chemical nature of the deposited marine humates (Rashid and King, 1970), which appear to resemble synthetic melanoidins made from amino acids and sugars (Hoering, 1973a). Marine humic substances have a significant cation exchange capacity (Picard and Felbeck, 1976) and are capable of strongly complexing metals, such as iron. They thus may serve as transporting agents for trace metals in the marine environment.

Work on marine humates includes a study of those extracted from the sediments of the San Pedro Basin, California, which were examined by a variety of instrumental techniques and also subjected to chemical derivatization and reduction (Hoering, 1971). The stable isotope ratios of carbon and sulphur in marine sedimentary humates and fulvates have been investigated by Nissenbaum and Kaplan (1972), who found that the $\delta^{13}C$ values were relatively constant (-20 to $-22\permil$) and that their $\delta^{34}S$ values indicated that the sulphur which is present had been incorporated from H_2S derived from sulphate reduction. It was concluded that these marine humates were formed *in situ* (Nissenbaum and Kaplan, 1972; Nissenbaum, 1974).

The humic substances from a DSDP core (Site 44-391) from the Blake–Bahama Basin spanning the Quaternary to the Upper Jurassic have been examined by Deroo *et al.* (1978a) and Stuermer and Simoneit (1978). The latter workers observed a 100-fold decrease in the humic acid carbon/total organic carbon ratio with depth, and concluded that humates may be converted to kerogen as the depth of burial increases. This process has also been demonstrated in the laboratory by heating Recent marine sediments (Ishiwatari *et al.*, 1976). Huc *et al.* (1977) and Huc and Durand (1977) have found that in immature sediments of various origins humic and fulvic acids decreased relative to organic carbon with increasing geological age.

A detailed review of marine humate chemistry and of the degradation of sea-water humates has been given by Stuermer (1975). The major differences between marine and terrestrial humic substances are summarized in Table 39.1.

39.3.15.2. *The nature of the kerogen in marine sediments*

Kerogen is an invaluable endogenous palaeoenvironmental marker for the origin of the bulk of the organic matter. Little detailed information is available about the nature of marine kerogen. However, some of the more important studies will be discussed below.

Djuricic *et al.* (1972) have degraded the kerogen from the Kimmeridge shale (marine-Jurassic) by oxidation. The oxidation products, which consisted predominantly of short-chain dicarboxylic acids and some aromatic acids, suggested that this kerogen was probably derived from marine biota. The $\delta^{13}C$ values and elemental compositions of kerogen from Recent sediments taken in the hot brine areas of the Red Sea have been determined by Saxby (1971). They ranged from $-20\permil$ to $-26\permil$, indicating that thermal alteration (diagenesis) of the organic matter had occurred. The major chromic acid oxidation products from a Cariaco Trench (DSDP Site 15-147B Pleistocene) core sample were found by Hoering (1973b) to be dicarboxylic acids, from succinic (C_4) to pimelic (C_7), and ketoacids with the same carbon number range. A volatile fraction was also produced; this consisted of predominantly low molecular weight *n*-fatty acids (C_2–C_{10}), and minor amounts of branched fatty acids and benzoic acid which is typical of marine kerogen.

39.4. Sources of Organic Matter and Use of Markers

39.4.1. lipids

The principal lipid constituents of marine sediments which are thought to have a terrigenous origin are *n*-alkanes (C_{23}–C_{33}, odd > even), *n*-alkanoic acids (C_{22}–C_{30}, even > odd), *n*-alkanols (C_{22}–C_{30}, even > odd) and diterpenoids. If the lipids of a sediment have a terrigeneous origin the distribution pattern of its *n*-alkanes, *n*-alkanoic acids and *n*-alkanols should resemble those in plant waxes (Eglinton and Hamilton, 1963; Hitchcock and Nichols, 1971; Kolattakudy and Walton, 1972).

The highest concentrations of terrigenous lipid matter occur in sediments from continental proximities (see e.g. Hunt, 1972), and these concentrations have been correlated with both the clay mineral (see e.g. Fan *et al.*, 1973; Müller and Stoffers, 1974) and the pollen (see e.g. Roman, 1974; Traverse,

TABLE 39.1

Summary of differences between marine (or lacustrine) and terrestrial humic substances

Character	Marine	Terrestrial	References
Aromaticity	Low (lake sediment, marine sediment)	High	Stuermer (1975), Ishiwatari (1969, 1971), Rashid and King (1970)
Phenol content	Low (marine sediment)	High	Huc (1973), Rashid and King (1970)
Nitrogen	High (marine sediment)	Low	Stuermer (1975), Nissenbaum and Kaplan (1972)
$\delta^{13}C$ (‰)	-22 to -24 (marine sediment)	-24 to -29	Stuermer (1975), Nissenbaum and Kaplan (1972)
$1540-1560$ cm^{-1} i.r. band	Present (lake sediment)	Absent	Ishiwatari (1967), Stuermer (1975)
Molecular weight	Low (sea-water) High (marine sediment)	High High	Stuermer (1975) Rashid and King (1969)

1974; Musich, 1973) content of the sediments. The proportions of terrigenous lipids are far greater than those of marine origin in sedimentary areas which receive a relatively large, solid input from the continents (e.g. the Black Sea) (Simoneit, 1974a, 1975, 1977a); conversely, in the sediments underlying those oceanic areas having a high primary productivity (e.g. the Cariaco Trench) the autochthonous marine lipids predominate (Simoneit, 1975).

Diterpenoids (predominantly of the abietane skeletal type) have been identified in a variety of sediments from various depositional environments (Simoneit, 1977b). These diterpenoids, which are largely derived from terrigenous resinous higher plants, can be detected at very low concentrations ($\sim 0.1\ \mu g\, g^{-1}$) and serve as a group of unique molecular markers since they are detectable even in sediments remote from the continents.

Biogenic markers having an exclusively marine autochthonous origin have not yet been unambiguously documented in sediments since the compounds which have been identified invariably also occur in biota of higher order. However, many natural products (e.g. sesterterpenoids, pentacyclic steroids, etc.) which have been characterized in marine biota probably do not occur in higher order biota (Faulkner and Andersen, 1974). Most of these compounds are present either at very low concentrations or are formed in organisms which have low population densities (i.e. which have a low overall biomass). Some lipids, pigments, etc. (e.g. anteisoalkanoic acids and chlorins) found in sediments remote from land must be of autochthonous origin, since they would not survive oxidation and metabolism during the times necessary for their transport to the area of deposition. Lipids, pigments and other types of compounds which have been found in marine sediments, and which may have originated from marine (phytoplankton and bacteria) and/or terrestrial (plant detritus) sources, are listed in Table 39.2, together with organic markers which are exclusively of terrigenous origin.

39.4.2. HUMIC AND FULVIC SUBSTANCES, AND KEROGEN

The source of marine humic and fulvic substances appears to be autochthonous material which has undergone the following reaction sequence: degraded cellular material → water-soluble complexes containing amino acids and carbohydrates → fulvic acids → humic acids → kerogen (Nissenbaum and Kaplan, 1972). During these reactions some lipids also become incorporated into the humates (Stuermer, 1975). The bulk of terrestial humates precipitate and deposit in, or around, estuaries where fresh-water and sea-water mix (see e.g. Beck et al., 1974), and this material is probably only brought to the deep-sea areas in small amounts by occasional rapid dispersal (e.g. by turbidite flows) down the continental slopes. Stable isotope ratios (of C, S and N) appear to provide the best key to the origin of the humate material in both Recent and ancient marine environments (see e.g. Nissenbaum and Kaplan, 1972; Nissenbaum, 1974; Stuermer and Simoneit, 1978).

The value of kerogen as an endogenous palaeoenvironmental marker could be enhanced if its bulk structure, and thus genetic origin, could be characterized. As with humates, stable isotope analyses and chemical degradation allow an evaluation of its origin to be made. The elemental composition and H/C and O/C ratios of the kerogen can be correlated using van Krevelen (1961) diagrams (see Section 39.7.3) to assess the magnitude of the allochthonous influx and the extent of the diagenesis of the kerogen (Durand et al., 1972; Deroo et al., 1978a). The potential of kerogen for the generation of petroleum in sediments from the Gulf of Aden has been investigated by Cernock (1974).

Kerogens have been shown to originate from either algal or higher plant debris, or from a combination of both (Tissot et al., 1974). On oxidation, algal kerogen yields aliphatic products. Thus, kerogen-like material, which upon degradation yielded products similar to those from degradations of ancient algal kerogen, has been isolated from Recent lagoonal sediments and algal mats (Philp and Calvin, 1976a, 1977), and cell-wall materials isolated from pure algal and bacterial cultures also yielded similar products on oxidation (Philp and Calvin, 1976b).

39.5. STABLE ISOTOPE GEOCHEMISTRY

Stable isotope geochemistry is a powerful ancillary tool for the elucidation of organic geochemical pathways and fluxes when used in conjunction with lipid analyses. The isotope geochemistry of the more important elements will be discussed below.*

39.5.1. CARBON-13

General reviews of the stable isotope geochemistry of carbon have been published by a number of authors (e.g. Degens, 1969; Garlick, 1974; Kaplan, 1975; Taylor, 1975). The $\delta^{13}C$ ranges of a variety of different materials are

---

* Stable isotope analysis data are normally expressed in the δ notation:

$$\delta^{13}C, \delta^{15}N, \delta^{34}S, \delta^{18}O, \delta D(\%_0) = \left[\frac{R_{sample} - R_{standard}}{R_{standard}} - 1 \right] \times 1000$$

where R is the $^{13}C/^{12}C$, $^{15}N/^{14}N$, $^{34}S/^{32}S$, $^{18}O/^{16}O$, or D/H ratio. The reference standards usually used are: Pee Dee belemnite for carbon, atmospheric N_2 for nitrogen, troilite from Canyon Diablo meteorite for sulphur and standard mean ocean water for oxygen and hydrogen.

TABLE 39.2

Summary of homologue distributions in lipids and allied compounds of marine sediments and their inferred origins

Constituents	Age	Significance of inferred biogenic source[c]			Relative abundance
		Terrigenous (e.g. higher plant wax)	Marine (e.g. bacteria, phytoplankton)	Diagenetic (maturation)	
n-Alkanes					
$n\text{-}C_{17}$ (C_{12}–C_{20})	Recent–Jurassic	–	+	+	Major
$n\text{-}C_{23}$ to $n\text{-}C_{33}$ (o/e)[a]	Recent–Jurassic	+	–	–	Major
Alkenes (C_{21}, C_{25} polyenes)	Recent	+	+	–	Intermediate
Iso- and anteisoalkanes	Recent	+	+	–	Minor
Isoprenoids (e.g. phytane)	Recent–Jurassic	–	–	+	Minor
"Hump"	Recent–Cretaceous	–	+	+	Minor
n-Fatty acids					
$n\text{-}C_{12}$ to $n\text{-}C_{18}$ (e/o)	Recent–Jurassic	+	+	–	Major
$n\text{-}C_{22}$ to $n\text{-}C_{32}$ (e/o)	Recent–Jurassic	+	? (some polyenoic to C_{24})	–	Minor
$C_{18:1}$, $C_{16:1}$	Recent	+	+	–	Minor
Iso- and anteiso-C_{13}–C_{19} acids	Recent–Pliocene	–	+	–	Minor
Isoprenoid acids (e.g. pristanic)	Recent–Cretaceous	?	?	+	Trace
Hydroxy acids	Recent	+	?	–	Minor
α,ω-Dicarboxylic acids	Recent	+	?	+	Minor
n-Fatty alcohols (C_{12}–C_{30}, e/o)	Recent–Pliocene	+	+	–	Minor
Isoprenols	Recent	+	+	+	Intermediate
n-Methyl ketones (C_{15}–C_{31}, o/e)	Recent	–	+	+	Intermediate
Isoprenoid ketones	Recent–Cretaceous		–	+	Minor
Wax esters	Recent	(+)[b]	+	–	Minor
Steroids (e.g. sterenes, steranes, stanols)	Recent–Cretaceous	+	+	+	Minor

TABLE 39.2—continued

Constituents	Age	Significance of inferred biogenic source			Relative abundance
		Terrigenous (e.g. higher plant wax)	Marine (e.g. bacteria, phyto-plankton)	Diagenetic (maturation)	
Triterpenoids (e.g. hopanes, 17(21)-hopene)	Recent–Cretaceous	+	+	+	Minor
Triterpenoidal acids (e.g. bishomohopanoic acid)	Recent–Cretaceous	?	+	+	Minor
Triterpenoidal ketones (e.g. adiantones)	Recent–Pliocene	?	+	+	Minor
Diterpenoids (e.g. dehydroabietic acid)	Recent–Cretaceous	+	–	+	Minor
Tetraterpenoids (e.g. carotenoids)	Recent–Cretaceous	(+)	+	+	Minor
Tetrapyrrole pigments (e.g. chlorins, porphyrins)	Recent–Cretaceous	(+)	+	+	Minor
Amino acids and peptides	Recent–Miocene	(+)	+	+	Intermediate
Carbohydrates	Recent–Eocene	(+)	+	–	Intermediate
Aromatic hydrocarbons	Recent–Cretaceous	–	–	+	Minor
Natural polymers (e.g. cutin, lignin, chitin)	Recent–Pliocene	+	+	+	Minor
Humates (also fulvates)	Recent–Cretaceous	(+)	+	+	Major
Kerogens	Recent–Cretaceous	(+)	+	+	Major

[a] Carbon number predominance (o/e = odd/even; e/o = even/odd).
[b] Parentheses indicate possible, but unlikely, source to marine sediments.
[c] + = has a biogenic source; – = does not have a biogenic source.

shown in Fig. 39.7, from which it is apparent that the δ^{13}C value of the organic matter in Recent marine sediments ($\sim -20\%_0$) is about the same as that of the organisms living in the environment of deposition, except in those areas which receive a large allochthonous influx. It is, however, $\sim 5\%_0$ greater than that of the organic matter in lacustrine sediments.

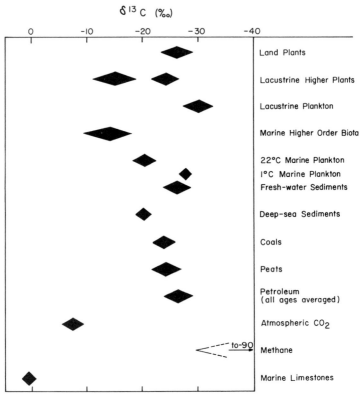

FIG. 39.7. δ^{13}C in various biological and geological materials (redrawn from data from Degens (1969) and the author's laboratory).

There have been many studies of the ^{13}C/^{12}C ratio of the total organic carbon of marine sediments, e.g. Recent sediments from the Antarctic (Sackett *et al.*, 1974a), the Gulf of Mexico (Sackett, 1964; Newman *et al.*, 1973), the Indian Ocean (Calder *et al.*, 1974; Erdman *et al.*, 1974) and the Tasman Sea (Erdman *et al.*, 1975). In some instances the organic material was divided into classes prior to analysis. Thus, δ^{13}C values have been determined for lipids and kerogens from DSDP sediments from the Atlantic Ocean and Mediterranean and Black Seas (Erdman and Schorno, 1977, 1978a, b, c).

The same components have also been separated from sediments from the Norwegian–Greenland Sea, for which $\delta^{13}C$ values ranging from $-25.8\%_0$ to $-28.8\%_0$ and $-22.8\%_0$ to $-28.2\%_0$ were found for the lipids and kerogens respectively (Erdman and Schorno, 1976). Methane, aromatic hydrocarbons and asphaltenes were separated from samples from the same area, and the reported $\delta^{13}C$ values ranged from $-71.2\%_0$ to $-87.3\%_0$ for methane and from $-24.4\%_0$ to $-27.2\%_0$ for both the asphaltenes and the aromatic hydrocarbons (Morris, 1976). Similarly, the total fatty acids and humates of the sediments from an offshore transect in the Gulf of Mexico had $\delta^{13}C$ values ranging from $-24.4\%_0$ to $-30.4\%_0$ and from $-19.6\%_0$ to $-24.0\%_0$ respectively (Hoering, 1974). Isotopic data for humic acids and kerogens have been determined for sediments from the Gulf of Mexico and the western Atlantic Ocean by Aizenshtat et al. (1973) who were able to detect the effects of terrigenous influxes into these deep-sea sediments. The humic acids isolated from Recent sediments from the Pacific and Atlantic Oceans, and from off California, had $\delta^{13}C$ values ranging from $-17.2\%_0$ to $-27.4\%_0$, reflecting both marine and terrigenous influences (Nissenbaum and Kaplan, 1972). Work has also been carried out on carbon isotope ratios of the interstitial gases (mainly CH_4) of sediments. Thus, the $\delta^{13}C$ values of the methane of samples from the north-eastern Atlantic were typical of organic matter having a biogenic origin ($-74\%_0$ to $-65\%_0$). However, the methane in the interstitial gases in a section between two diabase sills was enriched with respect to carbon-13 ($\delta^{13}C = -51\%_0$), which was thermally generated (Doose et al., 1977). Faber et al. (1978) have reported $\delta^{13}C$ values for headspace methane from Black Sea sediments.

There is some overlap between the $\delta^{13}C$ values of organic materials of different origins (Fig. 39.7). This makes it difficult to distinguish unambiguously between the sources (e.g. marine or terrigenous) of the organic material in sediments. In order to overcome this problem it is necessary to use data for other stable isotopes (e.g. those of nitrogen or hydrogen) in conjunction with those for carbon-13, and it is preferable to make the measurement on a variety of different types of organic material (e.g. fatty acids, humates, kerogen) isolated from the same sample. Data obtained in this manner should be correlated with information obtained from lipid markers, pollen and fossils, etc.

39.5.2. NITROGEN-15

Stable nitrogen isotope analysis has only been used to a limited extent for the examination of the organic matter in marine sediments and has been reviewed by Kaplan (1975). The $\delta^{15}N$ ranges for some biological and geological materials are shown in Fig. 39.8 (Hoering, 1955; Hoering and Moore, 1958; Miyaki and Wada, 1967; Kaplan, 1975), from which it can be seen that there is considerable overlap between the values for materials of different origins. This, together with difficulties arising from atmospheric contamination, reduces the usefulness of $\delta^{15}N$. However, if the latter problem can be overcome, the use of $\delta^{15}N$ data in conjunction with those for other stable isotopes (e.g. ^{13}C and ^{34}S) may be of value in marine geochemistry. In fact, $\delta^{15}N$ values have been successfully employed to indicate that terrigenous organic nitrogen constitutes a significant proportion of the organic nitrogen of the

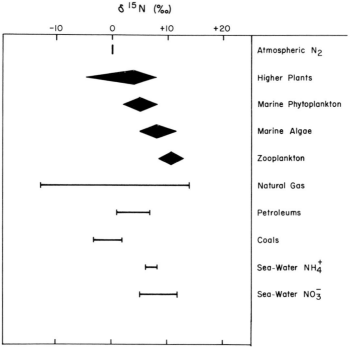

FIG. 39.8. $\delta^{15}N$ in various biological and geological materials (redrawn from data from Kaplan (1975) and the author's laboratory).

sediments from the Santa Barbara Basin, California (Sweeney et al., 1978; Sweeney and Kaplan, 1978). This nitrogen fraction was evidently not readily biodegradable, and the ammonium nitrogen present in the interstitial water was found to be derived from autochthonous planktonic debris.

39.5.3. SULPHUR-34

Stable isotope fractionation in the sulphur cycle has been reviewed by Goldhaber and Kaplan (1974), Garlick (1974) and Taylor (1975). Most of the available data for marine sediments relate to the anaerobic reduction of sulphate by bacteria (see e.g. Goldhaber and Kaplan, 1975; Sweeney, 1972). The range of $\delta^{34}S$ values found for the sulphate of sea-water is narrow ($\delta^{34}S = +20‰$ to $+21‰$), whereas that for sedimentary sulphides is much wider, ranging from $\sim +9‰$ to $-40‰$ (Kaplan, 1975). In contrast, magmatic sulphide ores have a much more restricted range, from $\sim +2‰$ to $-2‰$ (Kaplan, 1975).

Kaplan et al. (1963) have found that ^{32}S is enriched in the sulphides and organic sulphur in the sediments on the continental shelf off southern California. The sulphate of the interstitial water and the proteinaceous (i.e. cellular and skeletal) sulphur had a $\delta^{34}S$ value of $+20‰$, identical with that of sea-water sulphate. The non-proteinaceous organic sulphur ranged from $\delta^{34}S = +8‰$ to $-22‰$. Typical $\delta^{34}S$ values for marine algae, marine animals and shells are $\sim +19\cdot6‰$, $\sim +15\cdot3‰$ and $\sim +22‰$, respectively (Kaplan et al., 1963) (see also Addendum p. 311).

39.5.4. DEUTERIUM

Hydrogen isotopes have only been used to a limited extent for the study of the origin of the organic matter in marine sediments. Lipid and humate fractions from marine sediments collected on a transect off the coast in the Gulf of Mexico were examined for both δD and $\delta^{13}C$ by Hoering (1974), who also determined the δD value of the water overlying the sediment samples. The δD values of the fatty acids and humates ranged from $-123\cdot2‰$ to $-177\cdot9‰$ and from $-90\cdot1‰$ to $-114.4‰$, respectively. In contrast, the δD value of the water ranged from $+5\cdot1‰$ to $-53\cdot2‰$. Both the $\delta^{13}C$ and δD values of the fatty acids increased with distance from the shore. The $\delta^{13}C$ value of the humates also increased away from shore, but the δD values only changed slightly, No explanation was offered for these variations. Isotope ratios for both carbon and hydrogen have been determined for salt marsh biota (Smith and Epstein, 1970) and for various materials from other biogenic sources (Hoering, 1975). The δD values for plants were, in general, $50‰$ lower than those of water.

39.6. MECHANISMS FOR THE TRANSPORT OF TERRIGENOUS ORGANIC MATTER TO THE MARINE ENVIRONMENT

39.6.1. AUTOCHTHONOUS ORGANIC MATTER

The genesis of autochthonous organic matter in the water column overlying marine sediments depends on the biogenic productivity. Most of the organic matter produced in the photic zone passes through the food web and the balance, plus the waste products from the predators, undergoes vertical and lateral transport to the sea floor (Menzel, 1974). There it can undergo further modification by the bottom-feeding population (Heezen and Hollister, 1971) and by the sedimentary microbiota (Jannasch and Wirsen, 1977). The residual autochthonous lipids, and other organic material, are thus end-products from the food web.

39.6.2. ALLOCHTHONOUS ORGANIC MATTER

The amount of allochthonous material reaching the deep sea is small. However, over geological time such material makes a significant contribution to near-shore and hemipelagic sediments to which it is transported from the land by rivers, wind and ice-rafting mechanisms (see Chapter 24). These and other less important transport processes will be discussed briefly below.

39.6.2.1. *River transport*

Although river transport is obviously the major input mechanism for terrigenous organic matter to the marine environment, very little is known about the quantity and general nature of the organic material introduced in this way. The influence of rivers is evidenced in the Black Sea (Degens and Ross, 1974; Simoneit, 1974a, 1977a), as well as in the Atlantic into which the Amazon River system transports terrigenous organic detritus as far out as the Mid-Atlantic Ridge (Aizenshtat *et al.*, 1973; Gibbs, 1976).

39.6.2.2. *Aeolian transport*

The wind transports fine material far out to sea, and in certain areas wind-blown dust constitutes a significant proportion of the abyssal sediments (Heezen and Hollister, 1971). An appreciable quantity of dust is transported by the strong NE. trade-winds blowing across the Sahara and savannah of North Africa; this dust contains, together with the inorganic material, fresh-water diatoms (Delany *et al.*, 1967) and the siliceous remains of land plants (Folger, 1970; Parmenter and Folger, 1974).

The organic content of some aeolian dusts (e.g. those collected on Bermuda; Hoffman and Duce, 1974) is significant. Lepple and Brine (1976) have determined some volatile lipids and polar constituents in aeolian dusts collected off the north-western coast of Africa. Simoneit and Eglinton (1977) and Simoneit (1977c) have examined the organic components of atmospheric dusts collected from the NE. trade-winds off West Africa. Analyses of these dusts indicated that the major lipid component is plant wax (see Fig. 39.2, and Simoneit, 1975, 1977c). The *n*-alkane (dust DC1) and *n*-fatty acid (dust DC5) distributions in the figure are similar to those of a Recent sediment from the Cap Blanc Shelf, which is under the dust fall-out area (compare Figs 39.2E and H with Figs 39.2D and G). Similarly, the distributions of the *n*-alkanes and n-fatty acids in a Pliocene sediment sample from the North Atlantic (DSDP 12-114-5-5) strongly resemble those of trade-wind aeolian dusts (compare Fig. 39.1B with Fig. 39.2F, and Fig. 39.3B with Fig. 39.2H). An analogous relationship was also found between the distributions of the n-alcohols (Simoneit, 1975). The distributions of the n-alkanes, n-alkanoic acids and n-alcohols in the dusts also resembled those found by Kolattakudy and Walton (1972) for plant waxes (see Table 39.3). The presence of quartz grains coated with iron oxide (Wüstenquarz) in the older deep-sea sediments from this area (von Rad and Rösch, 1972) is an unambiguous indication that a proportion of these sediments was formed from terrestrial material by aeolian activity. This conclusion is also supported by the similarities in the distributions of the organic marker compounds in the sediments to those in the aeolian dusts (Simoneit and Eglinton, 1977; Simoneit, 1977f; Simoneit et al., 1977).

39.6.2.3. Rafting

Rafting of terrigenous material out to the oceans occurs mainly by ice and, to a much lesser degree, by biological processes. Sackett et al. (1974b) have demonstrated that in some polar areas (e.g. the Ross Sea) as much as 90% of the organic matter has a terrigenous origin and consists of kerogen. However, ice-rafting of material appears to be a localized and erratic process. Simoneit (1976) was able to show that only a small proportion of the terrigenous organic matter in sediments from the Norwegian–Greenland Sea consisted of lipids which had been transported to the sediment by this process.

39.6.2.4. Natural seepage

Natural seepage of both petroleum and gas into the marine environment is a significant transport process in some areas, one of the best documented instances of this being that associated with the Santa Barbara Coal Oil Point (Landes, 1973). Wilson et al. (1974) have given a summary of the known seeps (about 190) and have estimated that on a world-wide scale oil seepage amounts to $\sim 0.6 \times 10^9$ kg year^{-1}. Reed and Kaplan (1977) have studied the chemistry of marine petroleum seeps with particular reference to those of the California borderland and the Gulf of Alaska. According to Reed et al. (1977), petroleum seeps may contribute bitumen to marine sediments in certain areas, and indeed oil drops have been identified dispersed throughout a sediment of glacial flour at a depth of 28.3 m below the seabed at DSDP Site 345 (Simoneit, 1976). Natural gas seeps have been detected by shipboard instrumentation, and bubbles emanating from the seabed have been photographed.

   To summarize, the principal sources of the organic material in marine

TABLE 39.3

Major constituents of leaf waxes and their occurrence in the geosphere

Type	Range	Frequency	Occurrence in marine sediments	Occurrence in aeolian dusts[a,b]
Alkanes	Normal: odd C_{21}–C_{37}	Common (especially C_{29} and C_{31})	Present as major components[a]	Present as major components[a,b]
	Normal: even C_{20}–C_{34}	Common minor constituents	Present[a]	Present[a,b]
Alcohols (usually as esters)	Branched: C_{27}–C_{33}	Infrequent	Present[a]	Absent (not detected)[a,b]
	Primary: even C_{22}–C_{32}	Common	Present[c]	Present as major components[a,b]
	Primary: odd C_{25}–C_{31}	Infrequent	Present[c]	Present[a,b]
	Secondary: odd C_{21}–C_{33}	Common	Absent (not detected)[c]	Present[a,b]
	Diols and ketones	Rare	Not determined	Not determined
	Terpene alcohols	Infrequent	Present[b]	Present[b]
Aldehydes (as polymers)	Normal: C_{24}–C_{34}	Rare	Not determined	Not determined
Ketones	Di-n-alkyl ketones	Rare	Absent (not detected)[b]	Absent (not detected)[a,b]
Acids (usually as esters)	Normal: even C_{14}–C_{34}	Common	Present[a,b]	Present[a,b]
	Normal: odd C_{15}–C_{33}	?	Absent (not detected)[b]	Absent (not detected)[a,b]
	Ketoacids	Rare	Absent (not detected)[b]	Absent (not detected)[a,b]
	Dibasic acids	Rare	Present[d]	Absent (not detected)[a,b]
Esters	Between n-acids and primary and secondary alcohols	Common	Absent (not detected)	Absent (not detected)[a,b]
	Estolides of hydroxyacids	Infrequent ?	Not determined	Not determined

[a] Simoneit (1975).
[b] Simoneit (1977c).
[c] Sever and Parker (1969).
[d] Boon et al. (1975a).

sediments are autochthonous production and allochthonous material transported from the land mainly by rivers and wind.

39.6.2.5. *Redistribution by turbidite and lutite flows*

Once material has been deposited in the marine environment it may be redistributed. The primary dynamic processes involved in this are slumping and sliding, gravity flows (turbidite flows) and contour currents (lutite flows), or some combination of these (see also Chapter 24, and Laporte (1968)). Turbidite flows and slumping and sliding are particularly important on the continental shelf and slope. Sediments in pelagic environments can be scoured by bottom currents which can remove whole sequences, resulting in a hiatus in the sedimentary record (see Chapter 36).

39.7. DIAGENESIS AND LITHIFICATION

39.7.1. DIAGENESIS OF ORGANIC MATTER

Diagenesis, in the context of organic matter, refers to chemical and/or biochemical alteration of the organic compounds in sediments. The general reactions involved in this diagenesis include reduction, decarboxylation, deamination, cyclization, demethylation, disproportionation and aromatization (Hunt, 1974a). These processes appear to take place at sediment depths of ~ 1 m to ~ 1000 m, at temperature ranging between 25°C and 50°C (Hunt, 1974a). Little is known about the diagenesis of many types of naturally occurring organic compounds; however, the diagenesis of a few compounds has been studied. For example, it is known that diterpenoids in fossil wood undergo rearrangement reactions, such as the conversion of the pimarane skeleton to the abietane skeleton (Schuller and Conrad, 1966; Joyce and Lawrence, 1961). It has also been observed that disproportionation reactions occur in the diagenesis of monoterpenes (Skrigan, 1951) and diterpenes (Skrigan, 1964).

Another diagenetic process which has been studied is the stereoisomeric conversion of the hopanes (C_{29} to C_{35}) from the 17β(H) form (X), which appears to be the form present in all living organisms, to the thermodynamically more stable 17α(H) isomer (XI) over geological time (Dastillung and Albrecht, 1976). Although this conversion is more likely to be the result of a maturation process, there is a possibility that it may be brought about by diagenesis (thermal and catalytic) (Dastillung and Albrecht, 1976). The progressive aromatization of triterpenoids to, for example, substituted chrysenes and phenanthrenes, which occurs at great depths, is also thought to be the result of diagenesis (Greiner *et al.*, 1976). The backbone rearrangement of stanols and sterenes to (20RS)-5β,14β-dimethyl-18,19-dinor-

$8\alpha,9\beta,10\alpha$-ster-13(17)-enes may prove to be another example of a diagenetic process occurring in marine sediments (Rubinstein *et al.*, 1975).

The racemization of amino acid stereoisomers has been extensively investigated and some aspects of it have been reviewed in Chapter 31.

There have been a number of attempts to study diagenetic processes *in situ*. Thus, Rhead *et al.* (1971) injected ^{14}C-labelled oleic acid into a Severn Estuary sediment and found it to be rapidly reduced to stearic acid. Similar experiments with labelled sterols (mainly cholesterol) were carried out in Severn Estuary sediments by Gaskell and Eglinton (1974, 1975) who observed that the reduction of sterols to stanols proceeded much more slowly than did that of oleic acid; however, over geological time the former conversion is probably a significant diagenetic process. Analogous results have also been obtained for Recent sediments from Buzzards Bay, Massachusetts (Lee *et al.*, 1977).

39.7.2. THE GENERATION OF METHANE AND THE GASOLINE-RANGE HYDROCARBONS

Thermal alteration of organic material takes place at depths of burial ranging from 1000 to 6000 m, i.e. below the zone at which the diagenesis described above occurs. This type of alteration, which causes the cracking of large molecules to form small ones, is believed to occur in the temperature window 50–170°C (Hunt, 1974a), i.e. in the main zone of petroleum genesis. During the formation of petroleum, large amounts of C_4–C_7 hydrocarbons (gasoline-range hydrocarbons) are generated, and this makes these compounds excellent markers for oil genesis (Zarrella *et al.*, 1967).

Many deep-sea sediment samples have been analysed for the gasoline-range hydrocarbons. The most detailed analyses were carried out on the oil and sediments from the Challenger Knoll in the Gulf of Mexico (Deep-Sea Drilling Project, Vol. 1, 1969). Subsequent analyses have been carried out on sediments from the Mediterranean Sea (McIver, 1973c), the Bengal Basin (Hunt, 1974b), the Red Sea and the Gulf of Aden (Hunt, 1974c; McIver, 1974b), the Shatsky Rise (McIver, 1974a), the Tasman Sea (Hunt, 1975a), the Norwegian–Greenland Sea (Hunt, 1976), the south-eastern Atlantic Ocean (Hunt, 1978) and the Black Sea (Whelan and Hunt, 1978; Hunt and Whelan, 1978). Up to 25 compounds in the C_4–C_7 range have been identified at concentrations varying between $\sim$0·5 and 6400 ng g^{-1}. The total concentrations of gasoline hydrocarbons are variable, but are generally high in sediments with high contents of organic carbon, or in those near igneous intrusions (Hunt, 1975b). These compounds have also been found in oceanic sediments from much shallower sediment depths ($\leqslant$ 1000 m), indicating that they may also be derived from diagenetic reactions which had occurred much earlier in the sedimentary history, and at lower temperatures, than had previously been thought (Hunt, 1975b).

Metamorphism occurs at depths in excess of ~ 6000 m and at temperatures higher than $\sim 175°C$ (Hunt, 1974a), and ultimately converts the organic matter to methane, graphite and carbon dioxide. Methane and other short-chain hydrocarbons have been determined in many DSDP sediments, including those from the Cariaco Trench (McIver, 1973d), the Gulf of Aden (McIver, 1974b), the Arabian Sea (McIver, 1974c), the Gulf of Mexico, the Atlantic Ocean, the Mediterranean Sea, the Bering Sea and the north-eastern Pacific Ocean (Claypool et al., 1973; McIver, 1973a, b, c), the Black Sea (Hunt, 1974d; McIver, 1978), the Timor Trough (McIver, 1974d), the Ross Sea (McIver, 1975a), the Tasman Sea (McIver, 1975b) and the South Atlantic Ocean (Doose et al., 1977). The concentrations of methane found in these sediments are highly variable, and lie within the range ~ 0.5–$170\,000$ ppm (vol.).

By taking account of the above three different types of diagenesis and lithification zones, it is possible to estimate the temperature history of a sediment column from the degree of carbonization of its organic matter. In this context it is interesting to note that most of the DSDP sediments which have been examined are inferred to have had relatively low temperature regimes, and have therefore been classified as immature (an example of this is the sediments of the Shikoku Basin) (see Ames and Littlejohn, 1975).

Methane is also found in shallow-water marine sediments, sometimes together with trace amounts of carbon dioxide and ethane (Claypool and Kaplan, 1974). The methane in such sediments originates predominantly from bacterial reduction of carbon dioxide, and its origin and distribution have been reviewed by Claypool and Kaplan (1974). Barnes and Goldberg (1976) have made a mass balance calculation for the production and consumption of methane in the sediments from the Santa Barbara Basin, California, and concluded that the primary sink for anaerobically generated methane is bacterial sulphate reduction.

39.7.3. LITHIFICATION

Lithification is the process of turning sediments into rock by both consolidation and, particularly, induration. This process basically involves the removal of water by the compaction resulting from the overburden. Its effect on organic matter is not entirely clear. However, it is known that during lithification some of the functional groups are lost and the organic matter becomes more condensed, and, at the same time, the low molecular weight by-products $(H_2O, H_2S, NH_3, CO_2$, etc.) are removed or mineralized (Degens, 1967). This process results in the loss of hydrogen and oxygen relative to carbon. This fact led to the application by Tissot et al. (1974) to sedimentary kerogens of a technique first applied by van Krevelen (1961) to the study of coals. In this approach, Tissot *et al.* (1974) used H/C and O/C diagrams to assess the degree of coalification or maturation of the kerogen, both of which are a consequence of lithification.

39.8. SEDIMENTARY REDOX ENVIRONMENTS

A detailed treatment of sedimentary redox environments is obviously beyond the scope of this chapter. However, a brief summary will be presented below; for further details Laporte (1968), Selley (1970) and Chapter 30 should be consulted.

39.8.1. OXIC ENVIRONMENTS

The most common environments of present-day oceans are oxic. Conditions in a typical oxic environment are illustrated in Fig. 39.9A and apply to most deep-sea sediments and to all those from continental shelf areas in which circulation is not restricted. In oxic environments, the rate of organic decay is high, and it is more rapid in sands than in clays (DeSitter, 1947). However, in certain environments in which the water column is oxic, very high production in the euphotic zone can lead to the accumulation of relatively large amounts of organic matter in the underlying sediments (e.g. in Tanner Basin, California, and Walvis Bay). The rate of microbial degradation of the organic matter in deep-sea environments appears to be much slower than it is in shallow waters (Jannasch and Wirsen, 1977). Thus, when the submersible R.V. *Alvin* sank in 1540m of water it was found on recovery ten months later that food was well preserved (Jannasch *et al.*, 1971). A subsequent examination indicated that microbial activity was at least 10–1000 times slower at that depth.

Predation and sediment ingestion by bottom-feeders occurs in both shallow and deep oxic environments (Bruun and Wolff, 1961; Heezen and Hollister, 1971).

39.8.2. ANOXIC BASINS

Some typical sedimentary redox environments are shown diagrammatically in Fig. 39.9B and C in which three general types of anoxic basins are distinguished. These are:

(i) basins in which the oxygen in the overlying water is depleted and the sediment is anoxic (e.g. the Santa Barbara Basin) or those in which the water becomes anoxic and euxinic for part of the year and in which the sediments are anoxic (e.g. Saanich Inlet, British Columbia; see Fig. 39.9B); (ii) basins in which both the bottom waters and the sediments are anoxic

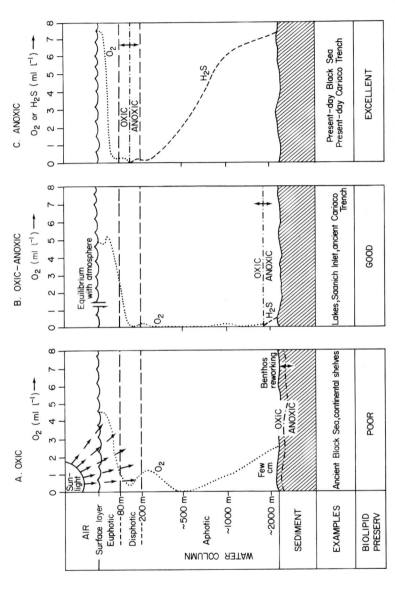

Fig. 39.9 Various sedimentary redox environments (Didyk *et al.* (1978); reproduced with permission from MacMillan Journals Ltd.). See text for details.

and the oxic–anoxic interface lies below the compensation point (e.g. the Cariaco Trench, see Fig. 39.9C);
(iii) basins in which both the bottom waters and the sediments are anoxic, but in which the oxic–anoxic interface lies closer to the water surface than in (ii), resulting in anaerobic biosynthesis in the euphotic layer (e.g. the Black Sea, see Fig. 39.9C).

The anoxic water column prevents oxidative alteration of the organic biomass before sedimentation and, as a consequence, the highest degree of preservation of organic matter is observed in the more anoxic basins. Both bacterial and higher order predation is much reduced under such anoxic conditions, and this results in a higher input of organic material to the sediments. In addition, the sulphate available for anaerobic reduction is limited, and this in turn limits the microbial consumption of organic matter (Orr and Gaines, 1974), as indicated by the reaction:

$$SO_4^{2-} + \text{organic matter} \xrightarrow[\text{reduction}]{\text{microbiological}} S^{2-} + CO_2\uparrow$$

$$+ \text{altered organic matter.}$$

This enhanced preservation of organic matter is directly reflected by the presence of labile markers, such as chlorins, carotenoids and phytol or its reduction products (dihydrophytol or phytane), in the sediments (Didyk *et al.*, 1978). The total concentration of organic matter preserved in the sediments of anoxic basins far exceeds that in those of oxic environments (Lijmbach, 1975).

39.8.3. ORGANIC MARKERS AND PALAEOENVIRONMENTAL CORRELATIONS

In oxic environments, labile compounds (e.g. chlorins, carotenoids, etc.) will be rapidly destroyed or will show changes reflecting the oxidative conditions to which they have been exposed. In anoxic environments these labile compounds tend to be preserved, resulting in the sediments having higher contents of these biolipids, or their direct degradation products, which still show the characteristic structural features imprinted by the biosynthetic processes.

Significant proportions of carotenoids have been detected in sediments from the Black Sea and the Cariaco Trench (Watts, 1975; Watts and Maxwell, 1977). Similarly, relatively large amounts of chlorins have been identified in sediments from these basins (Peake *et al.*, 1974; Baker and Smith, 1973), phytane being present in excess of pristane (Pr/Ph < 1). Because phytol undergoes geochemical alteration by two different pathways to either phytane (via reduction) or pristane (via oxidation) according to the prevailing redox conditions, the ratio of these compounds, together with the abundance of chlorins and sulphur, has been used as an indicator of palaeoenvironmental redox conditions (Didyk *et al.*, 1978). This approach has been applied to ancient sediments (e.g. Cretaceous shales from the Atlantic Ocean) (Simoneit, 1977e, f, 1978b) for which the abundance of porphyrins and sulphur, together with a Pr/Ph ratio of < 1, has been interpreted as indicating deposition in anoxic conditions; corroborative evidence was provided by the absence of bioturbation and the presence of plant detritus (Thiede and van Andel, 1977).

39.9. PETROLEUM GENESIS

The general concepts of petroleum genesis have been reviewed and elaborated upon in a series of papers by Philippi (1965, 1974, 1975, 1977). Many of these concepts are applicable to the present-day marine realm and will be summarized below.

39.9.1. METHANE AND THE GASOLINE-RANGE HYDROCARBONS VERSUS ORGANIC MATTER

Many studies have demonstrated that hydrocarbons are generated from kerogen as the depth of burial, and consequently the temperature, increases (see e.g. Tissot *et al.*, 1971; Vassoyevich *et al.*, 1969; Albrecht and Ourisson, 1969). A general scheme of hydrocarbon genesis is summarized in Fig. 39.10 which shows the relative amounts of hydrocarbons generated from sedimentary organic matter as a function of depth of burial. In Recent sediments, the hydrocarbons are biogenic methane and lipids (including the "geochemical fossil" markers). Beyond a certain depth (which may depend upon the particular environment), geogenetic hydrocarbons are formed by the thermal breakdown of kerogen and dilute the endogenous lipid fossils. This is the principal zone of petroleum formation (Vassoyevich *et al.*, 1969; Tissot *et al.*, 1974). At still greater depths there is increased cracking of carbon–carbon bonds in both the remaining kerogen and the previously formed oil (if it remains trapped), leading to the generation of the gasoline-range hydrocarbons and ultimately methane (Tissot *et al.*, 1974).

The organic matter of the source sediment ultimately determines the nature of the geogenetic products which can be formed from the kerogen. There are three types of kerogen and their respective paths of evolution are shown in Fig. 39.11, from which it is evident that kerogens having a low hydrogen content (Path III) give rise mainly to methane at depth.

It was mentioned above that the lipid content of a source sediment becomes diluted by the hydrocarbons formed at depth from kerogen. However, sediments with high lipid contents are usually also rich in kerogen which is normally of Types I or II (Fig. 39.11), i.e. rich in hydrogen and relatively poor in oxygen. Source sediments such as these have the potential to generate abundant oil (Tissot *et al.*, 1974). Many workers have investigated the petroleum generating potentials of sediments from various DSDP sites. Areas which have been investigated include, for example, the Angola Basin and the Walvis Ridge (Kendrick *et al.*, 1978a), the Cape Verde Rise and Basin (Dow, 1977; Kendrick *et al.*, 1977), the eastern Mediterranean and Black Seas (Kendrick *et al.*, 1978b) and the Blake–Bahama Basin (Dow, 1978; Kendrick *et al.*, 1978c).

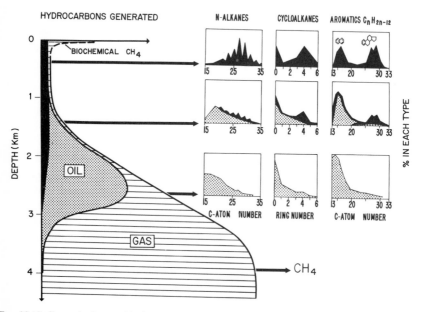

FIG. 39.10. General scheme of hydrocarbon generation. Depth scale shown is based on examples of Mesozoic and Palaeozoic source rocks. This is only a generalized profile and may vary according to the nature of the original organic matter, the burial history and the geothermal gradient. (Tissot *et al.* (1974), reproduced by permission of the American Association of Petroleum Geologists, Tulsa.)

39.9.2. IMPLICATIONS OF SEA-FLOOR SPREADING AND PLATE TECTONICS FOR PETROLEUM GENERATION

The implications of the concepts of sea-floor spreading and plate tectonics (see Chapter 36) are just beginning to be applied to the exploration for potential mineral (see e.g. Tarling, 1973a; Rona, 1973) and petroleum and gas (see e.g. Tarling, 1973b) deposits. Because gradual generation of petroleum hydrocarbons in sediments or sedimentary rocks is accelerated by small temperature increases, plate tectonics has a definite bearing on it. Thus, although heating during tectonic collision between two continental blocks, for example, is locally probably too severe, it may be very conducive to oil and gas generation at those portions of the margins where the disturbance is less (Tarling, 1973b). This process, coupled with burial heating, may be responsible for the accumulations of oil and gas at the continental extensions of ocean ridges (viz. the Arctic Ridge which is in proximity to the gas fields of

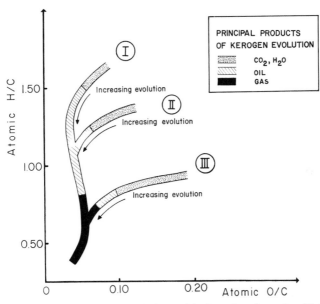

FIG. 39.11. Principal steps of kerogen evolution and hydrocarbon formation. Flatter parts of curves indicate mostly oxygen loss as CO_2 and H_2O. Steepest parts show loss of hydrogen and hydrocarbon generation. Paths I and II correspond to kerogens able to generate abundant oil, whereas Path III is representative of hydrogen-poor kerogens, producing mostly methane at depth. (Tissot *et al.* (1974), reproduced by permission of the American Association of Petroleum Geologists, Tulsa.)

NW. Siberia, and the East Pacific Rise which underlies the Los Angeles oil fields) (Tarling, 1973b). It may be concluded, therefore, that a knowledge of plate tectonics may prove of considerable value in selecting as yet unsuspected sites for oil and gas exploration.

39.10. CONCLUSIONS

A knowledge of organic geochemistry is of great value in investigating the history of the Earth's carbon cycle. For example, characterization of the lipid, humate (and fulvate) and kerogen fractions of the organic components of marine sediments may provide information about the origin of the organic matter, the palaeoenvironmental conditions prevailing at the time of sedimentation and the diagenetic and thermal history of the sediments. Information gained from such studies may also be useful in interpreting geological phenomena and may provide an insight into palaeogeography, palaeoclimate and palaeo-oceanography. In addition, it may be of economic value in oil exploration.

It is not yet possible to define, in detail, the input budgets of organic material to marine sediments; however, it is possible to distinguish between those components which are allochthonous and those which are autochthonous. Most of the former appear to be brought into the oceans by rivers, but aeolian fall-out is also significant. The majority of the organic material will be mineralized before it reaches the sea floor, but that which escapes this process will be incorporated into the sediment, from which some of it may be removed by recycling. Sediments deposited on the continental shelf may be subsequently redistributed to the deep-sea floor by turbidite flows and slides, which although intermittent, are important in the organic carbon cycle over a geological time scale.

ACKNOWLEDGEMENTS

Partial financial assistance from the National Science Foundation, the Bureau of Land Management and the Energy Research Development Administration is gratefully acknowledged.

REFERENCES

Aizenshtat, Z. (1973). *Geochim. Cosmochim. Acta*, **37**, 559.
Aizenshtat, Z., Baedecker, M. J. and Kaplan, I. R. (1973). *Geochim. Cosmochim. Acta*, **37**, 1881.
Albrecht, P. (1969). Ph.D. Thesis, University of Strasbourg, France.
Albrecht, P. and Ourisson, G. (1969). *Geochim. Cosmochim. Acta*, **33**, 138.
Alturki, Y. I. A. (1972). Ph.D. Thesis, University of Bristol, England.
Ames, R. L. and Littlejohn, R. (1975). *In* "Initial Reports of the Deep Sea Drilling Project", Vol. 31, pp. 621–627. U.S. Government Printing Office, Washington D.C.
Arpino, P. (1973). Ph.D. Thesis, University of Strasbourg, France.
Attaway, D. and Parker, P. L. (1970). *Science, N.Y.*, **169**, 674.
Bada, J. L. and Man, E. H. (1973). *In* "Initial Reports of the Deep Sea Drilling Project", Vol. 20, pp. 946–950. U.S. Government Printing Office, Washington D.C.
Bada, J. L., Luyendyk, B. P. and Maynard, J. B. (1970). *Science, N.Y.* **170**, 730.
Bada, J. L., Protsch, R. and Schroeder, R. A. (1973). *Nature, Lond.* **241**, 394.
Baedecker, M. J., Ikan, R., Ishiwatari, R. and Kaplan, I. R. (1977). *In* "Chemistry of Marine Sediments" (T. F. Yen, ed.), pp. 55–72. Ann Arbor Science Publishers, Ann Arbor, U.S.A.
Baker, B. L. and Hodgson, G. W. (1968). *Chem. Geol.* **3**, 119.

Baker, E. W. (1969). *In* "Organic Geochemistry—Methods and Results" (G. Eglinton and M. T. J. Murphy, eds), pp. 464–497. Springer-Verlag, Berlin and New York.

Baker, E. W. (1971). *Chem. Geol.* **7**, 45.

Baker, E. W. and Smith, G. D. (1973). *In* "Initial Reports of the Deep Sea Drilling Project", Vol. 20, pp. 942–945. U.S. Government Printing Office, Washington D.C.

Baker, E. W. and Smith, G. D. (1974). *In* "Advances in Organic Geochemistry 1973" (B. Tissot and F. Bienner, eds.), pp. 649–660. Éditions Technip, Paris.

Baker, E. W. and Smith, G. D. (1975a). *In* "Initial Reports of the Deep Sea Drilling Project", Vol. 31, pp. 629–632. U.S. Government Printing Office, Washington D.C.

Baker, E. W. and Smith, G. D. (1975b). *In* "Initial Reports of the Deep Sea Drilling Project", Vol. 31, pp. 905–909. U.S. Government Printing Office, Washington D.C.

Baker, E. W., Yen, T. F., Dickie, J. P., Rhodes, R. E. and Clark, L. F. (1967). *J. Am. Chem. Soc.* **89**, 3631.

Baker, E. W., Dereppe, C., Billig, M., Smith, G. D., Rankin, J. G., Palmer, S. E. and Huang, W. Y. (1976a). *In* "Geol. Soc. Amer., Symp. Org. Geochem. of DSDP Cores", Denver (abstract), p. 763.

Baker, E. W., Palmer, S. E. and Parrish, K. L. (1976b). *In* "Initial Reports of the Deep Sea Drilling Project", Vol. 38, pp. 785–789. U.S. Government Printing Office, Washington D.C.

Baker, E. W., Palmer, S. E. and Huang, W. Y. (1977). *In* "Initial Reports of the Deep Sea Drilling Project", Vol. 41, pp. 825–837. U.S. Government Printing Office, Washington D.C.

Baker, E. W., Palmer, S. E. and Huang, W. Y. (1978a). *In* "Initial Reports of the Deep Sea Drilling Project", Vol. 40, pp. 639–647. U.S. Government Printing Office, Washington D.C.

Baker, E. W., Palmer, S. E. and Huang, W. Y. (1978b). *In* "Initial Reports of the Deep Sea Drilling Project", Vol. 44, pp. 639–643. U.S. Government Printing Office, Washington D.C.

Barnes, R. O. and Goldberg, E. D. (1976). *Geology*, **4**, 297.

Beck, K. C., Reuter, J. H. and Perdue, E. M. (1974). *Geochim. Cosmochim. Acta*, **38**, 341.

Benson, A. A., Lee, R. F. and Nevenzel, J. C. (1973). *Biochem. Soc. Symp.* **35**, 175.

Bird, C. W., Lynch, J. M., Pirt, F. J., Reid, W. W., Brook, C. J. W. and Middleditch, B. S. (1971). *Nature, Lond.* **230**, 473.

Blumer, M. and Cooper, W. J. (1967). *Science, N.Y.* **158**, 1463.

Blumer, M. and Sass, J. (1972a). *Science, N.Y.* **176**, 1120.

Blumer, M. and Sass, J. (1972b). *Mar. Poll. Bull.* **3**, 92.

Blumer, M. and Snyder, W. D. (1965). *Science, N.Y.* **150**, 1588.

Blumer, M. and Youngblood, W. W. (1975). *Science, N.Y.* **188**, 53.

Blumer, M., Mullin, M. M. and Thomas, D. W. (1964). *Helgoländer Wiss. Meeresunters.* **10**, 187.

Blumer, M., Guillard, R. R. and Chase, T. (1971). *Mar. Biol.* **8**, 183.

Blumer, M., Dorsey, T. and Sass, J. (1977). *Science, N.Y.* **195**, 283.

Boon, J. J., de Leeuw, J. W. and Schenck, P. A. (1975a). *Geochim. Cosmochim. Acta*, **39**, 1559.

Boon, J. J., Rijpstra, W. I. C., de Leeuw, J. W. and Schenck, P. A. (1975b). *Nature, Lond.* **258**, 414.

Boon, J. J., de Lange, F., Schuyl, P. J. W., de Leeuw, J. W. and Schenck, P. A. (1977a). *In* "Advances in Organic Geochemistry 1975" (R. Campos and J. Goni, eds), pp. 255–272. Revista Española de Micropaleontologia, Madrid.

Boon, J. J., van der Meer, F. W., Schuyl, P. J. W., de Leeuw, J. W., Schenck, P. A. and Burlingame, A. L. (1977b). *In* "Initial Reports of the Deep Sea Drilling Project", Vol. 40. U.S. Government Printing Office, Washington D.C.

Bordovskiy, O. K. (1965a). *Mar. Geol.* **3**, 33.

Bordovskiy, O. K. (1965b). *Mar. Geol.* **3**, 83.

Breger, I. A., ed. (1963). "Organic Geochemistry". Pergamon Press, New York, 658 pp.

Brennan, P. J., Griffin, P. F. S., Lösel, D. M. and Tyrell, D. (1974). *In* "Progress in the Chemistry of Fats and Other Lipids" (R. T. Holman, ed.), Vol. 14, Pt. 2, pp. 51–89. Pergamon Press, New York.

Brisou, J. (1969). *C.R. Séanc. Soc. Biol.* **163**, 722.

Brooks, P. W. (1974). Ph.D. Thesis, University of Bristol, England.

Bruun, A. F. and Wolff, T. (1961). In "Oceanography" (M. Sears, ed.), pp. 391–397. American Association for the Advancement of Science, Washington.

Burlingame, A. L. and Simoneit, B. R. T. (1969). *Nature, Lond.* **222**, 741.

Burlingame, A. L., Cox, R. E. and Derrick, P. J. (1974). *A. Rev. Analyt. Chem.* **46**, 248R.

Burlingame, A. L., Kimble, B. J. and Derrick, P. J. (1976). *A. Rev. Analyt. Chem.* **48**, 368R.

Calder, J. A., Horvath, G. J., Shultz, D. J. and Newman, J. W. (1974). *In* "Initial Reports of the Deep Sea Drilling Project", Vol. 26, pp. 613–617. U.S. Government Printing Office, Washington D.C.

Caldicott, A. B. and Eglinton, G. (1973). *In* "Phytochemistry" (L. P. Miller, ed.), Vol. 3, pp. 162–194. Van Nostrand Reinhold, New York.

Cardoso, J. N. (1976). Ph.D. Thesis, University of Bristol, England.

Cardoso, J., Brooks, P. W., Eglinton, G., Goodfellow, R., Maxwell, J. R. and Philp, R. P. (1976). *In* "Environmental Biogeochemistry" (J. O. Nriagu, ed.), pp. 149–174. Ann Arbor Science Publishers, Ann Arbor, U.S.A.

Cardoso, J. N., Eglinton, G. and Holloway, P. J. (1977). *In* "Advances in Organic Geochemistry 1975" (R. Campos and J. Goni, eds), pp. 273–287. Revista Española de Micropaleontologia, Madrid.

Cardoso, J. N., Wardroper, A. M. K., Watts, C. D., Barnes, P. J., Maxwell, J. R., Eglinton, G., Mound, D. G. and Speers, G. C. (1978). *In* "Initial Reports of the Deep Sea Drilling Project", Vol. 44, pp. 617–623. U.S. Government Printing Office, Washington D.C.

Cernock, P. J. (1974). *In* "Initial Reports of the Deep Sea Drilling Project", Vol. 24, pp. 791–797. U.S. Government Printing Office, Washington D.C.

Chucecas, L. and Riley, J. P. (1969). *J. Mar. Biol. Ass. U.K.* **49**, 117.

Claypool, G. E. and Kaplan, I. R. (1974). *In* "Natural Gases in Marine Sediments" (I. R. Kaplan, ed.), pp. 99–139. Plenum Press, New York and London.

Claypool, G. E., Presley, B. J. and Kaplan, I. R. (1973). *In* "Initial Reports of the Deep Sea Drilling Project", Vol. 19, pp. 879–884. U.S. Government Printing Office, Washington D.C.

Cooper, B. S., Harris, R. C. and Thompson, S. (1974). *Mar. Poll. Bull.* **5**, 15.

Cooper, J. E. and Bray, E. E. (1963). *Geochim. Cosmochim. Acta*, **27**, 1113.

Cooper, W. J. and Blumer, M. (1968). *Deep-Sea Res.* **15**, 535.

Cranwell, P. A. (1973). *Chem. Geol.* **11**, 307.

Dastillung, M. and Albrecht, P. (1976). *Mar. Poll. Bull.* **7**, 13.

Degens, E. T. (1967). *In* "Diagenesis in Sediments" (G. Larsen and G. V. Chilingar, eds), pp. 343–390. Elsevier, Amsterdam.

Degens, E. T. (1969). *In* "Organic Geochemistry—Methods and Results" (G. Eglinton and M. T. J. Murphy, eds), pp. 304–329. Springer-Verlag, Berlin and New York.

Degens, E. T. (1970). *In* "Symposium on Organic Matter in Natural Waters" (D. W. Hood, ed.), pp. 77–106. University of Alaska Press, College.

Degens, E. T. and Ross, D. A., eds (1974). "The Black Sea—Geology, Chemistry, and Biology". Memoir 20, American Association of Petroleum Geologists, Tulsa, U.S.A., 633 pp.

Degens, E. T., Prashnowsky, A., Emery, K. O. and Pimenta, J. (1961). *Neues Jb. Geol. Paläont. Mh.* 1961, 413.

Degens, E. T., Emery, K. O. and Reuter, J. H. (1963). *Neues Jb. Geol. Paläont. Mh.* 1963, 231.

Degens, E. T., Reuter, J. H. and Shaw, K. N. F. (1964). *Geochim. Cosmochim. Acta,* **28**, 45.

Delany, A. C., Delany, A. C., Parkin, D. W., Griffin, J. J., Goldberg, E. D. and Reimann, B. E. F. (1967). *Geochim. Cosmochim. Acta,* **31**, 885.

de Leeuw, J. W., Correia, V. A. and Schenck, P. A. (1974). *In* "Advances in Organic Geochemistry 1973" (B. Tissot and F. Bienner, eds), pp. 993–1004. Éditions Technip, Paris.

de Leeuw, J. W., Simoneit, B. R. T., Boon, J. J., Rijpstra, W. I. C., de Lange, F., van der Leeden, J. C. W., Correia, V. A., Burlingame, A. L. and Schenck, P. A. (1977). *In* "Advances in Organic Geochemistry 1975" (R. Campos and J. Goni, eds), pp. 61–79. Revista Española de Micropaleontologia, Madrid.

Deroo, G., Herbin, J. P., Roucaché, J., Tissot, B., Albrecht, P. and Schaeffle, J.'(1977) *In* "Initial Reports of the Deep Sea Drilling Project", Vol. 41, pp. 865–873. U.S. Government Printing Office, Washington D.C.

Deroo, G., Herbin, J. P., Roucaché, J., Tissot, B., Albrecht, P and Dastillung, H. (1978a). *In* "Initial Reports of the Deep Sea Drilling Project", Vol. 44, pp. 593–598. U.S. Government Printing Office, Washington D.C.

Deroo, G., Herbin, J. P. and Roucaché, J. (1978b). *In* "Initial Reports of the Deep Sea Drilling Project", Vol. 42A, pp. 465–472. U.S. Government Printing Office, Washington D.C.

deRosa, M., Gambacorta, A., Minale, L. and Bu'Lock, J. D. (1971). *Chem. Comm.* 619.

deRosa, M., Gambacorta, A., Minale, L. and Bu'Lock, J. D. (1973). *Phytochemistry,* **12**, 1117.

DeSitter, L. U. (1947). *Bull. Am. Assoc. Pet. Geol.* **31**, 2030.

Didyk, B. M. (1975). Ph.D. Thesis, University of Bristol, England.

Didyk, B. M., Alturki, Y. I. A., Pillinger, C. T. and Eglinton, G. (1975). *Nature, Lond.* **256**, 563.

Didyk, B. M., Simoneit, B. R. T., Brassell, S. C. and Eglinton, G. (1978). *Nature, Lond.* **272**, 216.

Djuricic, M. V., Vitorovic, D., Andresen, B. D., Hertz, H. S., Murphy, R. C., Preti, G. and Biemann, K. (1972). *In* "Advances in Organic Geochemistry 1971" (H. R. von Gaertner, and H. Wehner, eds), pp. 305–321. Pergamon Press, Oxford, England.

Dole, V. P. and Meinertz, H. (1960). *J. Biol. Chem.* **235**, 2595.

Doose, P. R., Sandstrom, M., Jodele, R. Z. and Kaplan, I. R. (1977). *In* "Initial Reports of the Deep Sea Drilling Project", Vol. 41, pp. 861–863. U.S. Government Printing Office, Washington D.C.

Douglas, A. G. and Eglinton, G. (1966). *In* "Comparative Phytochemistry" (T. Swain, ed.), pp. 57–77. Academic Press, New York and London.

Dow, W. G. (1977). *In* "Initial Reports of the Deep Sea Drilling Project", Vol. 41, pp. 821–824. U.S. Government Printing Office, Washington D.C.

Dow, W. G. (1978). *In* "Initial Reports of the Deep Sea Drilling Project", Vol. 44, pp. 625–634. U.S. Government Printing Office, Washington D.C.

Drozdova, T. V. (1974). *In* "Advances in Organic Geochemistry 1973" (B. Tissot and F. Bienner, eds), pp. 285–292. Éditions Technip, Paris.

Drozdova, T. V. and Gursky, J. N. (1972). *Geokhimiya*, **3**, 323.

Dungworth, G. (1976). *Chem. Geol.* **17**, 135.

Dungworth, G., Thijssen, M., Zuurveld, J., van der Velden, W. and Schwartz, A. W. (1977). *Chem. Geol.* **19**, 295.

Durand, B., Espitalie, J., Nicaise, G. and Combaz, A. (1972). *Revue Inst. Fr. Pétrole* **27**, 865.

Eglinton, G., sr. rptr. (1975). "Environmental Chemistry", Vol. 1. The Chemical Society, London, 199 pp.

Eglinton, G. and Hamilton, R. J. (1963). *In* "Chemical Plant Taxonomy" (T. Swain, ed.), pp. 187–217. Academic Press, New York and London.

Eglinton, G. and Murphy, M. T. J., eds (1969). "Organic Geochemistry—Methods and Results". Springer-Verlag, Berlin and New York, 828 pp.

Eglinton, G., Hamilton, R. J., Raphael, R. A. and Gonzales, A. G. (1962). *Nature, Lond.* **193**, 739.

Eglinton, G., Maxwell, J. R. and Philp, R. P. (1974). *In* "Advances in Organic Geochemistry 1973" (B. Tissot and F. Bienner, eds), pp. 941–961. Éditions Technip, Paris.

Eglinton, G., Simoneit, B. R. T. and Zoro, J. A. (1975). *Proc. R. Soc. Lond. B.* **189**, 415.

Ehrhardt, M. and Blumer, M. (1972). *Environ. Poll.* **3**, 179.

Ensminger, A. (1974). M.Sc. Thesis, University of Strasbourg, France.

Ensminger, A., Albrecht, P., Ourisson, G., Kimble, B. J., Maxwell, J. R. and Eglinton, G. (1972). *Tetrahedron Lett.* **36**, 3861.

Erdman, J. G. and Schorno, K. S. (1976). *In* "Initial Reports of the Deep Sea Drilling Project", Vol. 38, pp. 791–799. U.S. Government Printing Office, Washington D.C.

Erdman, J. G. and Schorno, K. S. (1977). *In* "Initial Reports of the Deep Sea Drilling Project", Vol. 41, pp. 849–853. U.S. Government Printing Office, Washington D.C.

Erdman, J. G. and Schorno, K. S. (1978a). *In* "Initial Reports of the Deep Sea Drilling Project", Vol. 40, pp. 651–658. U.S. Government Printing Office, Washington D.C.

Erdman, J. G. and Schorno, K. S. (1978b). *In* "Initial Reports of the Deep Sea Drilling Project", Vol 42B, pp. 717–721. U.S. Government Printing Office, Washington D.C.

Erdman, J. R. and Schorno, K. S. (1978c). *In* "Initial Reports of the Deep Sea Drilling Project", Vol. 44, pp. 605–616. U.S. Government Printing Office, Washington D.C.

Erdman, J. G., Schorno, K. S. and Scanlan, R. S. (1974). *In* "Initial Reports of the Deep Sea Drilling Project", Vol. 24, pp. 1169–1176. U.S. Government Printing Office, Washington D.C.

Erdman, J. G., Schorno, K. S. and Scanlan, R. S. (1975). *In* "Initial Reports of the Deep Sea Drilling Project", Vol. 31, pp. 911–916. U.S. Government Printing Office, Washington D.C.

Faber, E., Schmitt, M. and Stahl, W. (1978). *In* "Initial Reports of the Deep Sea Drilling Project", Vol. 42B, pp. 667–672. U.S. Government Printing Office, Washington D.C.

Fales, H. M., Jaouni, T. M. and Babashak, J. F. (1973). *Analyt, Chem.* **45**, 2302.

Fan, P.-F., Rex, R. W., Cook, H. E. and Zemmels, I. (1973). *In* "Initial Reports of the Deep Sea Drilling Project", Vol. 15, pp. 847–921. U.S. Government Printing Office, Washington D.C.

Farrington, J. W. and Medeiros, G. C. (1975). *In* "Proceedings of the 1975 Conference on Prevention and Control of Oil Pollution", pp. 115–122, EPA/API/USCG, San Francisco.

Farrington, J. W. and Meyers, P. A. (1975). *In* "Environmental Chemistry" (G. Eglinton, sr. rptr.), Vol. 1, pp. 109–136. The Chemical Society, London.

Farrington, J. W. and Quinn, J. G. (1971a). *Nature, Phys. Sci.* **230**, 67.

Farrington, J. W. and Quinn, J. G. (1971b). *Geochim. Cosmochim. Acta*, **35**, 735.

Farrington, J. W. and Quinn, J. G. (1973a). *Estuar. Coastal Mar. Sci.* **1**, 71.

Farrington, J. W. and Quinn, J. G. (1973b). *Geochim. Cosmochim. Acta*, **37**, 259.

Farrington, J. W. and Tripp, B. W. (1977). *Geochim. Cosmochim. Acta*, **41**, 1627.

Farrington, J. W., Frew, N. M., Gschwend, P. M. and Tripp, B. W. (1977a). *Estuar. Coastal Mar. Sci.* **5**, 793.

Farrington, J. W., Henrichs, S. M. and Anderson, R. (1977b). *Geochim. Cosmochim. Acta*, **41**, 289.

Faulkner, D. J. and Andersen, R. J. (1974). *In* "The Sea" (E. D. Goldberg, ed.), Vol. 5, pp. 679–714. John Wiley and Sons, New York.

Fester, J. I. and Robinson, W. E. (1964). *Analyt. Chem.* **36**, 1392.

Fester, J. I. and Robinson, W. E. (1966). *In* "Coal Science" (R. F. Gould, ed.), pp. 22–31. American Chemical Society, Washington.

Florkin, M. (1969). *In* "Organic Geochemistry—Methods and Results" (G. Eglinton and M. T. J. Murphy, eds), pp. 498–520. Springer-Verlag, Berlin and New York.

Folger, D. W. (1970). *Deep-Sea Res.* **17**, 337.

Fontes, J.-Ch., Letolle, R., Nesteroff, W. D. and Ryan, W. B. F. (1973). *In* "Initial Reports of the Deep Sea Drilling Project", Vol. 13, pp. 788–796. U.S. Government Printing Office, Washington D.C.

Forsman, J. B. (1978). *In* "Initial Reports of the Deep Sea Drilling Project", Vol. 40. pp. 557–567. U.S. Government Printing Office, Washington D.C.

Forsman, J. B. and Hunt, J. M. (1958). *In* "Habitat of Oil" (L. G. Weeks, ed.), pp. 747–778. American Association of Petroleum Geologists, Tulsa, U.S.A.

Förster, H. J., Biemann, K., Haigh, G., Tattrie, N. H. and Colvin, J. R. (1973). *Biochem. J.* **135**, 133.

Gallegos, E. J. (1971). *Analyt. Chem.* **43**, 1151.

Gardner, W. S. and Menzel, D. W. (1974). *Geochim. Cosmochim. Acta*, **38**, 813.

Garlick, G. D. (1974). *In* "The Sea" (E. D. Goldberg, ed.), Vol. 5, pp. 393–425. John Wiley and Sons, New York.

Gaskell, S. J. (1974). Ph.D. Thesis, University of Bristol, England.

Gaskell, S. J. and Eglinton, G. (1974). *In* "Advances in Organic Geochemistry 1973" (B. Tissot and F. Bienner, eds), pp. 963–976. Éditions Technip, Paris.

Gaskell, S. J. and Eglinton, G. (1975). *Nature, Lond.* **254**, 209.

Gaskell, S. J., Morris, R. J., Eglinton, G. and Calvert, S. E. (1975). *Deep-Sea Res.* **22**, 777.

Gearing, P., Gearing, J. N., Lytle, T. F. and Lytle, J. S. (1976). *Geochim. Cosmochim. Acta*, **40**, 1005.

Gelpi, E., Schneider, H., Mann, J. and Oró, J. (1970). *Phytochemistry*, **9**, 603.

Geodekyan, A. A., Ul'mishek, G. F., Tchernova, T. G., Avilov, V. I., Bokovoy, A. P., Verkhovskaya, Z. I. and Federova, M. S. (1978). *In* "Initial Reports of the Deep Sea Drilling Project", Vol. 42B, pp. 683–696. U.S. Government Printing Office, Washington D.C.

Gibbs, R. J. (1976). *Geology*, **4**, 45.

Goldhaber, M. B. and Kaplan, I. R. (1974). *In* "The Sea" (E. D. Goldberg, ed.), Vol. 5, pp. 569–655. John Wiley and Sons, New York.

Goldhaber, M. B. and Kaplan, I. R. (1975). *Soil Sci.* **119**, 42.

Greiner, A. C., Spyckerelle, C. and Albrecht, P. (1976). *Tetrahedron Lett.* **32**, 257.

Han, J. and Calvin, M. (1970). *Chem. Comm.* 1970, 1490.

Han, J., McCarthy, E. D., Calvin, M. and Benn, M. H. (1968). *J. Chem. Soc.*, 1968, 2785.

Hare, P. E. (1969). *In* "Organic Geochemistry—Methods and Results" (G. Eglinton and M. T. J. Murphy, eds), pp. 438–463. Springer-Verlag, Berlin and New York.

Hare, P. E. (1973). *In* "Initial Reports of the Deep Sea Drilling Project", Vol. 20, pp. 940–941. U.S. Government Printing Office, Washington D.C.

Hase, A. and Hites, R. A. (1976). *Geochim. Cosmochim. Acta,* **40**, 1141.

Hatcher, P. G., Simoneit, B. R. T. and Gerchakov, S. M. (1977). *In* "Advances in Organic Geochemistry 1975" (R. Campos and J. Goni, eds), pp. 469–484. Revista Española de Micropaleontologia, Madrid.

Hedges, J. I. (1974). *Carnegie Inst. Yb.* **73**, 581.

Hedges, J. I. (1975). Ph.D. Thesis, University of Texas, Austin.

Hedges, J. I. and Parker, P. L. (1976). *Geochim. Cosmochim. Acta,* **40**, 1019.

Heezen, B. C. and Hollister, C. D. (1971). "The Face of the Deep". Oxford University Press, New York, 659 pp.

Hitchcock, C. and Nichols, B. W. (1971). "Plant Lipid Biochemistry". Academic Press, London and New York, 387 pp.

Hodgson, G. W., Baker, B. L. and Peake, E. (1967). *In* "Fundamental Aspects of Petroleum Geochemistry" (B. Nagy and U. Colombo, eds), pp. 177–259. Elsevier, London.

Hodgson, G. W., Hitchon, B., Taguchi, K., Baker, B. L. and Peake, E. (1968). *Geochim. Cosmochim Acta,* **32**, 737.

Hoering, T. C. (1955). *Science, N.Y.* **122**, 1233.

Hoering, T. C. (1968). *Carnegie Inst. Yb.* **66**, 515.

Hoering, T. C. (1969). *Carnegie Inst. Yb.* **67**, 199.

Hoering, T. C. (1971). *Carnegie Inst. Yb.* **69**, 334.

Hoering, T. C. (1973a). *Carnegie Inst. Yb.* **72**, 682.

Hoering, T. C. (1973b). *In* "Initial Reports of the Deep Sea Drilling Project", Vol. 20, pp. 936–938. U.S. Government Printing Office, Washington D.C.

Hoering, T. C. (1974). *Carnegie Inst. Yb.* **73**, 590.

Hoering, T. C. (1975). *Carnegie Inst. Yb.* **74**, 598.

Hoering, T. C. and Moore, H. E. (1958). *Geochim. Cosmochim. Acta,* **13**, 225.

Hoffman, E. J. and Duce, R. A. (1974). *J. Geophys. Res.* **79**, 4474.

Holloway, P. J., Baker, E. A. and Martin, J. T. (1972). *An. Quim. R. Soc. Esp. Fis. Quim.* **68**, 905.

Huc, A. Y. (1973). Ph.D. Thesis, University of Nancy, France.

Huc, A. Y. and Durand, B. M. (1977). *Fuel,* **56**, 73.

Huc, A. Y., Durand, B. and Monin, J. C. (1978). *In* "Initial Reports of the Deep Sea Drilling Project", Vol. 42B, pp. 737–748. U.S. Government Printing Office, Washington D.C.

Hunt, J. M. (1972). *Bull. Am. Assoc. Pet. Geol.* **56**, 2273.

Hunt, J. M. (1974a). *In* "Advances in Organic Geochemistry 1973" (B. Tissot and F. Bienner, eds), pp. 593–605. Éditions Technip, Paris.

Hunt, J. M. (1974b). *In* "Initial Reports of the Deep Sea Drilling Project", Vol. 22, pp. 673–675. U.S. Government Printing Office, Washington D.C.

Hunt, J. M. (1974c). *In* "Initial Reports of the Deep Sea Drilling Project", Vol. 24, pp. 1165–1167. U.S. Government Printing Office, Washington D.C.

Hunt, J. M. (1974d). *In* "The Black Sea—Geology, Chemistry, and Biology" (E. T. Degens and D. A. Ross, eds), pp. 499–504. Memoir 20, American Association of Petroleum Geologists, Tulsa, U.S.A.

Hunt, J. M. (1975a). *In* "Initial Reports of the Deep Sea Drilling Project", Vol. 31, pp. 901–903. U.S. Government Printing Office, Washington D.C.

Hunt, J. M. (1975b). *Nature, Lond.* **254**, 411.

Hunt, J. M. (1976). *In* "Initial Reports of the Deep Sea Drilling Project", Vol. 38, pp. 807–808. U.S. Government Printing Office, Washington D.C.

Hunt, J. M. (1978). *In* "Initial Reports of the Deep Sea Drilling Project", Vol. 40, pp. 649–650. U.S. Government Printing Office, Washington D.C.

Hunt, J. M. and Whelan, J. K. (1978). *In* "Initial Reports of the Deep Sea Drilling Project", Vol. 42B, pp. 661–665. U.S. Government Printing Office, Washington D.C.

Ikan, R., Baedecker, M. J. and Kaplan, I. R. (1973). *Nature, Lond.* **244**, 154.

Ikan, R., Baedecker, M. J. and Kaplan, I. R. (1975a). *Geochim. Cosmochim. Acta*, **39**, 187.

Ikan, R., Baedecker, M. J. and Kaplan, I. R. (1975b). *Geochim. Cosmochim. Acta*, **39**, 195.

Ikan, R., Aizenshtat, Z., Baedecker, M. J. and Kaplan, I. R. (1975c). *Geochim. Cosmochim. Acta*, **39**, 173.

Ishiwatari, R. (1967). *Geochem. J.* **1**, 61.

Ishiwatari, R. (1969). *Soil Sci.* **107**, 53.

Ishiwatari, R. (1971). Ph.D. Thesis, Tokyo Metropolitan University, Tokyo.

Ishiwatari, R., Ishiwatari, M., Kaplan, I. R. and Rohrback, B. G. (1976). *Nature, Lond.* **264**, 347.

Jamieson, G. R. and Reid, E. H. (1972). *Phytochemistry*, **11**, 1423.

Jannasch, H. W. and Wirsen, C. O. (1977). *Scient. Am.* **236**, 42.

Jannasch, H. W., Eimhjellen, K., Wirsen, C. O. and Farmanfarmaian, A. (1971). *Science, N.Y.* **171**, 672.

Johns, R. B. and Onder, O. M. (1975). *Geochim. Cosmochim. Acta*, **39**, 129.

Johnson, R. W. and Calder, J. A. (1973). *Geochim. Cosmochim. Acta*, **37**, 1943.

Johnstone, R. A. W., sr. rptr. (1975). "Mass Spectrometry", Vol. 3. The Chemical Society, London, 402 pp.

Jones, J. G. (1969). *J. Gen. Microbiol.* **59**, 145.

Joyce, N. M. and Lawrence, R. V. (1961). *J. Org. Chem.* **26**, 1024.

Kaneda, T. (1967). *J. Bacteriol.* **93**, 894.

Kaplan, I. R. (1975). *Proc. R. Soc. Lond.* B **189**, 183.

Kaplan, I. R., Emery, K. O. and Rittenberg, S. C. (1963). *Geochim. Cosmochim. Acta*, **27**, 297.

Kasatochkin, V. K., Bordovskii, I. K., Larina, N. K. and Cherkinskaya, K. (1968). *Dokl. Akad. Nauk. SSSR*, **179**, 690.

Kates, M. (1964). *Adv. Lipid Res.* **2**, 17.

Kendrick, J. W., Hood, A. and Castaño, J. R. (1977). *In* "Initial Reports of the Deep Sea Drilling Project", Vol. 41, pp. 817–819. U.S. Government Printing Office, Washington D.C.

Kendrick, J. W., Hood, A. and Castaño (1978a). *In* "Initial Reports of the Deep Sea Drilling Project", Vol. 40, pp. 671–676. U.S. Government Printing Office, Washington D.C.

Kendrick, J. W., Hood, A. and Castaño, J. R. (1978b). *In* "Initial Reports of the Deep Sea Drilling Project", Vol. 42B, pp. 729–735. U.S. Government Printing Office, Washington D.C.

Kendrick, J. W., Hood, A. and Castaño, J. R. (1978c). *In* "Initial Reports of the Deep Sea Drilling Project", Vol. 44, pp. 599–603. U.S. Government Printing Office, Washington D.C.

Kimble, B. J. (1972). Ph.D. Thesis, University of Bristol, England.

Kimble, B. J., Maxwell, J. R., Philp, R. P. and Eglinton, G. (1974a). *Chem. Geol.* **14**, 173.

Kimble, B. J., Maxwell, J. R., Philp, R. P., Eglinton, G., Albrecht, P., Ensminger, A., Arpino, P. and Ourisson, G. (1974b). *Geochim. Cosmochim. Acta*, **38**, 1165.

Kimble, B. J., Walls, F. C., Olsen, R. W. and Burlingame, A. L. (1975). *In* "Proceedings of the 23rd Annual Conference on Mass Spectrometry and Allied Topics", pp. 503–505. American Society for Mass Spectrometry.

King, J. D. and White, D. C. (1978). *In* "Initial Reports of the Deep Sea Drilling Project", Vol. 42B, pp. 765–770. U.S. Government Printing Office, Washington D.C.

King, K., Jr. and Neville, C. (1977). *Science, N.Y.* **195**, 1333.

Klenk, E., Knipprath, W., Eberhagen, D. and Koof, H. P. (1963). *Hoppe-Seyler's Z. Physiol. Chem.* **334**, 44.

Knorr, M. and Schenk, D. (1968). *Arch. Hyg. Bakt.* **152**, 282.

Kolattakudy, P. E. and Walton, T. J. (1972). *In* "Progress in the Chemistry of Fats and Other Lipids" (R. T. Holman, ed.),Vol. 13, Pt. 3, pp. 121–175. Pergamon Press, New York.

Kolattukudy, P. E., Kronman, K. and Poulose, A. J. (1975). *Pl. Physiol.* **55**, 567.

Kononova, M. M. (1966). "Soil Organic Matter", 2nd ed. Pergamon Press, Oxford, England, 544 pp.

Kulshreshtha, M. J., Kulshreshtha, D. K. and Rastogi, R. P. (1972). *Phytochemistry*, **11**, 2369.

Kvenvolden, K. A. (1962). *Bull. Am. Assoc. Pet. Geol.* **46**, 1643.

Kvenvolden, K. A. (1966). *Nature, Lond.* **209**, 573.

Kvenvolden, K. A. (1975). *Ann. Rev. Earth Planet. Sci.* **3**, 183.

Kvenvolden, K. A., Peterson, E. and Brown, F. S. (1970). *Science, N.Y.* **69**, 1079.

Landes, K. K. (1973). *Bull. Am. Assoc. Pet. Geol.* **57**, 637.

Laporte, L. F. (1968). "Ancient Environments". Prentice-Hall, Englewood Cliffs, 116 pp.

Lee, C., Gagosian, R. B. and Farrington, J. W. (1977). *Geochim. Cosmochim. Acta*, **41**, 985.

Lee, R. F. and Loeblich, A. R., III (1971). *Phytochemistry*, **10**, 593.

Lee, R. F. and Williams, P. M. (1974). *Naturwissenschaften*, **61**, 505.

Lee, R. F., Hirota, J. and Barnett, A. M. (1971). *Deep-Sea Res.* **18**, 1147.

Leo, R. F. and Barghoorn, E. S. (1970). *Science, N.Y.* **168**, 582.

Leo, R. F. and Parker, P. L. (1966). *Science, N.Y.* **152**, 649.

Lepple, F. K. and Brine, C. J. (1976). *J. Geophys. Res.* **81**, 1141.

Lijmbach, G. W. M. (1975). *In* "Proceedings of the 9th World Petroleum Congress", Vol. 2, pp. 357–369. Applied Science Publishers, London.

Lorenzen, C. J. (1974). *In* "The Black Sea—Geology, Chemistry, and Biology" (E. T. Degens and D. A. Ross, eds), pp. 426–428. Memoir 20, American Association of Petroleum Geologists, Tulsa, U.S.A.

Martin, F., Dubach, P., Mehta, N. C. and Deuel, H. (1963). *Z. PflErnähr. Düng. Bodenk.* **103**, 27.

Martin, J. T. and Juniper, B. E. (1970). "The Cuticles of Plants". Edward Arnold, London, 347 pp.

Mattern, G., Albrecht, P. and Ourisson, G. (1970). *Chem. Comm.* 1970, 1570.

McCrindle, R. and Overton, K. H. (1969). *In* "Rodd's Chemistry of Carbon Compounds", Vol. IIc, pp. 369–482. Elsevier, Amsterdam.

McFadden, W. H. (1973). "Techniques of Combined Gas Chromatography/Mass Spectrometry: Applications in Organic Analysis". John Wiley and Sons, New York, 463 pp.

McIver, R. D. (1973a). *In* "Initial Reports of the Deep Sea Drilling Project", Vol. 18, pp. 1013–1014. U.S. Government Printing Office, Washington D.C.

McIver, R. D. (1973b). *In* "Initial Reports of the Deep Sea Drilling Project", Vol 19, pp. 875–877. U.S. Government Printing Office, Washington D.C.

McIver, R. D. (1973c). *In* "Initial Reports of the Deep Sea Drilling Project", Vol. 13, pp. 813–816. U.S. Government Printing Office, Washington D.C.

McIver, R. D. (1973d). *In* "Initial Reports of the Deep Sea Drilling Project", Vol. 20, pp. 934–935. U.S. Government Printing Office, Washington D.C.

McIver, R. D. (1974a). *Bull. Am. Assoc. Pet. Geol.* **58**, 163.

McIver, R. D. (1974b). *In* "Initial Reports of the Deep Sea Drilling Project", Vol. 24, pp. 1157–1158. U.S. Government Printing Office, Washington D.C.

McIver, R. D. (1974c). *In* "Initial Reports of the Deep Sea Drilling Project", Vol. 23, pp. 971–973. U.S. Government Printing Office, Washington D.C.

McIver, R. D. (1974d). *In* "Initial Reports of the Deep Sea Drilling Project", Vol. 27, pp. 453–454. U.S. Government Printing Office, Washington D.C.

McIver, R. D. (1975a). *In* "Initial Reports of the Deep Sea Drilling Project", Vol. 28, pp. 815–817. U.S. Government Printing Office, Washington D.C.

McIver, R. D. (1975b). *In* "Initial Reports of the Deep Sea Drilling Project", Vol. 31, pp. 899–900. U.S. Government Printing Office, Washington D.C.

McIver, R. D. (1978). *In* "Initial Reports of the Deep Sea Drilling Project", Vol. 42B, pp. 679–681. U.S. Government Printing Office, Washington D.C.

McKenna, M. C., Bada, J. L. and Luyendyk, B. P. (1971). *Science, N.Y.* **172**, 503.

McLafferty, F. W. (1966). "Interpretation of Mass Spectra". Benjamin, New York, 229 pp.

Menzel, D. W. (1974). *In* "The Sea" (E. D. Goldberg, ed.), Vol. 5, pp. 659–678. John Wiley and Sons, New York.

Mitterer, R. M. and Hoering, T. C. (1968). *Carnegie Inst. Yb.* **66**, 510.

Miyaki, Y. and Wada, E. (1967). *Rec. Oceanogr. Wks Japan,* **9**, 37.

Mopper, K. (1973). Ph.D. Thesis, Massachusetts Institute of Technology, Cambridge, and Woods Hole Oceanographic Institution, Woods Hole.

Mopper, K., Michaelis, W., Garrasi, C. and Degens, E. T. (1978). *In* "Initial Reports of the Deep Sea Drilling Project", Vol. 42B, pp. 697–705. U.S. Government Printing Office, Washington D.C.

Morris, D. A. (1976). *In* "Initial Reports of the Deep Sea Drilling Project", Vol. 38, pp. 809–814. U.S. Government Printing Office, Washington D.C.

Morris, R. J. (1975). *Geochim. Cosmochim. Acta,* **39**, 381.

Morris, R. J. and Culkin, F. (1975). *In* "Environmental Chemistry" (G. Eglinton, sr. rptr.), Vol. 1, pp. 81–108. The Chemical Society, London.

Müller, G. and Stoffers, P. (1974). *In* "The Black Sea—Geology, Chemistry, and Biology" (E. T. Degens and D. A. Ross, eds), pp. 200–248. Memoir 20, American Association of Petroleum Geologists, Tulsa, U.S.A.

Musich, L. F. (1973). *In* "Initial Reports of the Deep Sea Drilling Project", Vol. 18, pp. 799–815. U.S. Government Printing Office, Washington D.C.

Nakanishi, K. and others (1974). *In* "Natural Products Chemistry" (K. Nakanishi, T. Goto, S. Itô, S. Natori and S. Nozoa, eds), Vol. 1, pp. 185–312. Academic Press, New York and London.

Nevenzel, J. C. (1970). *Lipids*, **5**, 308.

Newman, J. W., Parker, P. L. and Behrens, E. W. (1973). *Geochim. Cosmochim. Acta*, **37**, 225.

Niaussat, P., Mallet, L. and Ottenwaelder, J. (1969). *C.R. Hebd. Séanc. Acad. Sci., Paris*, **268D**, 1189.

Niaussat, P., Auger, C. and Mallet, L. (1970). *C.R. Hebd. Séanc. Acad. Sci., Paris*, **270D**, 1042.

Nissenbaum, A. (1974). *In* "Advances in Organic Geochemistry 1973" (B. Tissot and F. Bienner, eds), pp. 39–52. Éditions Technip, Paris.

Nissenbaum, A. and Kaplan, I. R. (1972). *Limnol. Oceanogr.* **17**, 570.

Nissenbaum, A., Baedecker, M. J. and Kaplan, I. R. (1973). *Geochim. Cosmochim. Acta*, **36**, 709.

Oró, J., Tornabene, T. G., Nooner, D. W. and Gelpi, E. (1967). *J. Bacteriol.* **93**, 1811.

Orr, W. L. and Gaines, A. G., Jr. (1974). *In* "Advances in Organic Geochemistry 1973" (B. Tissot and F. Bienner, eds), pp. 791–812. Éditions Technip, Paris.

Orr, W. L. and Grady, J. R. (1967). *Geochim. Cosmochim. Acta*, **31**, 1201.

Palacas, J. G., Love, A. H. and Gerrild, P. M. (1972). *Bull. Am. Assoc. Pet. Geol.* **56**, 1402.

Palacas, J. G., Gerrild, P. M., Love, A. H. and Roberts, A. A. (1976). *Geology*, **4**, 81.

Parker, P. L. (1967). *Contrib. Mar. Sci.* **4**, 135.

Parmenter, C. and Folger, D. W. (1974). *Science, N.Y.* **184**, 695.

Peake, E., Baker, B. L. and Hodgson, G. W. (1972). *Geochim. Cosmochim. Acta*, **36**, 867.

Peake, E., Casagrande, D. J. and Hodgson, G. W. (1974). *In* "The Black Sea—Geology, Chemistry, and Biology" (E. T. Degens and D. A. Ross, eds), pp. 505–523. Memoir 20, American Association of Petroleum Geologists, Tulsa, U.S.A.

Peterson, D. H. (1967). Ph.D. Thesis, University of Washington, Seattle.

Philippi, G. T. (1965). *Geochim. Cosmochim. Acta*, **29**, 1021.

Philippi, G. T. (1974). *Geochim. Cosmochim. Acta*, **38**, 947.

Philippi, G. T. (1975). *Geochim. Cosmochim. Acta*, **39**, 1353.

Philippi, G. T. (1977). *Geochim. Cosmochim. Acta*, **41**, 33.

Philp, R. P. and Calvin, M. (1976a). *In* "Environmental Biogeochemistry" (J. O. Nriagu, ed.), Vol. 1, pp. 131–148. Ann Arbor Science Publishers, Ann Arbor, U.S.A.

Philp, R. P. and Calvin, M. (1976b). *Nature, Lond.* **262**, 134

Philp, R. P. and Calvin, M. (1977). *In* "Advances in Organic Geochemistry 1975" (R. Campos and J. Goni, eds), pp. 735–752. Revista Española de Micropaleontologia, Madrid.

Philp, R. P., Maxwell, J. R. and Eglinton, G. (1976). *Sci. Prog., Oxford*, **63**, 521.

Philp, R. P., Calvin, M., Brown, S. and Yang, E. (1978). *Chem. Geol.* **22**, 207.

Picard, G. L. and Felbeck, G. T., Jr. (1976). *Geochim. Cosmochim. Acta*, **40**, 1347.

Pocklington, R. (1976). *J. Fish. Res. Bd Can.* **33**, 93.

Prashnowsky, A., Degens, E. T., Emery, K. O. and Pimenta, J. (1961). *Neues Jb. Geol. Paläont. Mh.* 1961, 400.

Rashid, M. A. and King, L. H. (1969). *Geochim. Cosmochim. Acta*, **33**, 147.

Rashid, M. A. and King, L. H. (1970). *Geochim. Cosmochim. Acta*, **34**, 193.

Rashid, M. A. and King, L. H. (1971). *Chem. Geol.* **7**, 37.

Reed, W. E. and Kaplan, I. R. (1977). *J. Geochem. Explor.* **7**, 255.

Reed, W. E., Kaplan, I. R., Sandstrom, M. and Mankiewicz, P. (1977). "Proceedings of the 1977 Oil Spill Conference", pp. 183–188, EPA/API/ISCG.

Reitz, R. C. and Hamilton, J. G. (1968). *Comp. Biochem. Physiol.* **25**, 401.

Rhead, M. M., Eglinton, G., Draffan, G. H. and England, P. J. (1971). *Nature, Lond.* **232**, 327.

Riedel, W. R., Ladd, H. S., Tracey, J. I., Jr. and Bramlette, M. N. (1961). *Bull. Am. Assoc. Pet. Geol.* **45**, 1793.

Rittenberg, S. C., Emery, K. O., Hülsemann, J., Degens, E. T., Fay, R. C., Reuter, J. H., Grady, J. R., Richardson, S. H. and Bray, E. E. (1963). *J. Sediment. Pet.* **33**, 140.

Rohmer, M. (1975). Ph.D. Thesis, University of Strasbourg, France.

Roman, S. (1974). *In* "The Black Sea—Geology, Chemistry, and Biology" (E. T. Degens and D. A. Ross, eds), pp. 396–410. Memoir 20, American Association of Petroleum Geologists, Tulsa, U.S.A.

Rona, P. A. (1973). *Scient. Am.* **229**, 86.

Rubinstein, I., Sieskind, O. and Albrecht, P. (1975). *J. Chem. Soc., Perkin Trans. 1*, 1833.

Sackett, W. M. (1964). *Mar. Geol.* **2**, 173.

Sackett, W. M., Eadie, B. J. and Exner, M. E. (1974a). *In* "Advances in Organic Geochemistry 1973" (B. Tissot and F. Bienner, eds), pp. 661–671. Éditions Technip, Paris.

Sackett, W. M., Poag, C. W. and Eadie, B. J. (1974b). *Science, N.Y.* **185**, 1045.

Sargent, J. R. and Gatten, R. R. (1976). "Abstracts, Joint Oceanographic Assembly", Session C5, Edinburgh.

Sargent, J. R., Lee, R. F. and Nevenzel, J. C. (1976). *In* "The Chemistry and Biochemistry of Natural Waxes" (P. Kolattukudy, ed.), pp. 49–89. Elsevier, Amsterdam.

Sargent, J. R., Gatten, R. R. and McIntosh, R. (1978). *Mar. Chem.* (in press).

Saxby, J. D. (1971). *Chem. Geol.* **9**, 233.

Schnitzer, M. and Khan, S. U. (1972). "Humic Substances in the Environment". Marcel Dekker, New York, 327 pp.

Schubert, K., Rose, G. and Hörhold, C. (1967). *Biochim. Biophys. Acta,* **137**, 168.

Schubert, K., Rose, G., Wachtel, H., Hörhold, H. and Ikekawa, N. (1968). *Eur. J. Biochem.* **5**, 246.

Schuller, W. H. and Conrad, C. M. (1966). *J. Chem. Engng Data,* **11**, 89.

Schwendinger, R. B. (1969). *In* "Organic Geochemistry—Methods and Results" (G. Eglinton and M. T. J. Murphy, eds), pp. 425–437. Springer-Verlag, Berlin and New York.

Schwendinger, R. B. and Erdman, J. G. (1964). *Science, N.Y.* **144**, 1575.

Selley, R. C. (1970). "Ancient Sedimentary Environments". Chapman and Hall, London, 237 pp.

Sever, J. R. and Haug, P. (1971). *Nature, Lond.* **234**, 447.

Sever, J. R. and Parker, P. L. (1969). *Science, N.Y.* **164**, 1052.

Shanks, W. C., Bischoff, J. L. and Kaplan, I. R. (1974). *In* "Initial Reports of the Deep Sea Drilling Project", Vol. 23, pp. 947–950. U.S. Government Printing Office, Washington D.C.

Shaw, N. and Stead, D. (1971). *J. Bacteriol.* **107**, 130.

Simoneit, B. R. T. (1973). *In* "Initial Reports of the Deep Sea Drilling Project", Vol. 21, pp. 909–923. U.S. Government Printing Office, Washington D.C.

Simoneit, B. R. T. (1974a). *In* "The Black Sea—Geology, Chemistry, and Biology" (E. T. Degens and D. A. Ross, eds), pp. 477–498. Memoir 20, American Association of Petroleum Geologists, Tulsa, U.S.A.

Simoneit, B. R. T. (1974b). *In* "Initial Reports of the Deep Sea Drilling Project", Vol. 24, pp. 1159–1163. U.S. Government Printing Office, Washington D.C.

Simoneit, B. R. T. (1975). Ph.D. Thesis, University of Bristol, England.

308 BERND R. T. SIMONEIT

Simoneit, B. R. T. (1976). In "Initial Reports of the Deep Sea Drilling Project", Vol. 38, pp. 805–806. U.S. Government Printing Office, Washington D.C.

Simoneit, B. R. T. (1977a). Deep-Sea Res. 24, 813.

Simoneit, B. R. T. (1977b). Geochim. Cosmochim. Acta, 41, 463.

Simoneit, B. R. T. (1977c). Mar. Chem. 5, 443.

Simoneit, B. R. T. (1977d). In "Initial Reports of the Deep Sea Drilling Project", Vol. 39, pp. 497–500. U.S. Government Printing Office, Washington D.C.

Simoneit, B. R. T. (1977e). In "Initial Reports of the Deep Sea Drilling Project", Vol. 40, pp. 459–662. U.S. Government Printing Office, Washington D.C.

Simoneit, B. R. T. (1977f). In "Initial Reports of the Deep Sea Drilling Project", Vol. 41, pp. 855–858. U.S. Government Printing Office, Washington D.C.

Simoneit, B. R. T. (1978a). In "Initial Reports of the Deep Sea Drilling Project", Vol. 42B, pp. 749–753. U.S. Government Printing Office, Washington D.C.

Simoneit, B. R. T. (1978b). In "Initial Reports of the Deep Sea Drilling Project", Vol. 43 (in press). U.S. Government Printing Office, Washington D.C.

Simoneit, B. R. T. (1978c). In "Organic Geochemistry of Deep Sea Drilling Project Sediments" (E. W. Baker, ed.) (in press). Science Press, Princeton, U.S.A.

Simoneit, B. R. T. and Burlingame, A. L. (1971a). In "Initial Reports of the Deep Sea Drilling Project", Vol. 7, pp. 889–912. U.S. Government Printing Office, Washington D.C.

Simoneit, B. R. T. and Burlingame, A. L. (1971b). In "Initial Reports of the Deep Sea Drilling Project", Vol. 8, pp. 873–900. U.S. Government Printing Office, Washington D.C.

Simoneit, B. R. T. and Burlingame, A. L. (1972a). In "Advances in Organic Geochemistry 1971" (H. R. von Gaertner and H. Wehner, eds), pp. 189–228. Pergamon Press, Oxford, England.

Simoneit, B. R. T. and Burlingame, A. L. (1972b). In "Initial Reports of the Deep Sea Drilling Project", Vol. 9, pp. 859–901. U.S. Government Printing Office, Washington D.C.

Simoneit, B. R. T. and Burlingame, A. L. (1973a). Geochim. Cosmochim. Acta, 37, 595.

Simoneit, B. R. T. and Burlingame, A. L. (1973b). In "Initial Reports of the Deep Sea Drilling Project", Vol. 17, pp. 561–590. U.S. Government Printing Office, Washington D.C.

Simoneit, B. R. T. and Burlingame, A. L. (1974a). In "Advances in Organic Geochemistry 1973" (B. Tissot and F. Bienner, eds), pp. 629–648. Éditions Technip, Paris.

Simoneit, B. R. T. and Burlingame. A. L. (1974b). In "Initial Reports of the Deep Sea Drilling Project", Vol. 22, pp. 681–692. U.S. Government Printing Office, Washington D.C.

Simoneit, B. R. T. and Eglinton, G. (1977). In "Advances in Organic Geochemistry 1975" (R. Campos and J. Goni, eds), pp. 415–430. Revista Española de Micropaleontologia, Madrid.

Simoneit, B. R. T., Scott, E. S., Howells, W. G. and Burlingame, A. L. (1972). In "Initial Reports of the Deep Sea Drilling Project", Vol. 11, pp. 1013–1045. U.S. Government Printing Office, Washington D.C.

Simoneit, B. R. T., Scott, E. S. and Burlingame, A. L. (1973a). In "Initial Reports of the Deep Sea Drilling Project", Vol. 10, pp. 625–636. U.S. Government Printing Office, Washington D.C.

Simoneit, B. R. T., Howells, W. G. and Burlingame, A. L. (1973b). In "Initial Reports of the Deep Sea Drilling Project", Vol. 20, pp. 907–933..U.S. Government Printing Office, Washington D.C.

Simoneit, B. R. T., Scott, E. S. and Burlingame, A. L. (1973c). *In* "Initial Reports of the Deep Sea Drilling Project", Vol. 16, pp. 575–600. U.S. Government Printing Office, Washington D.C.

Simoneit, B. R. T., Smith, D. H. and Eglinton, G. (1975). *Archs Environ. Contam. Toxicol.* **3**, 385.

Simoneit, B. R. T., Chester, R. and Eglinton, G. (1977). *Nature, Lond.* **267**, 682.

Skrigan, A. I. (1951). *Dokl. Akad. Nauk. SSSR*, **80**, 607.

Skrigan, A. I. (1963). *Trudy Vses. Nauchno-tekhn. Soveshch. Gorki*, 1963, 108.

Smith, B. N. and Epstein, S. (1970). *Pl. Physiol.* **46**, 738.

Spyckerelle, C. (1975). Ph.D. Thesis, University of Strasbourg, France.

Starikova, N. D. and Korzhikova, L. I. (1969). *Oceanology*, **9**, 509.

Starikova, N. D. and Korzhikova, L. I. (1972a). *Dokl. Akad. Nauk. SSSR*, **204**, 203.

Starikova, N. D. and Korzhikova, L. I. (1972b). *Geokhimiya*, **2**, 230.

Starikova, N. D. and Yoblokova, O. G. (1972). *Oceanology*, **12**, 363.

Stevens, N. P., Bray, E. E. and Evans, E. D. (1965). *Bull. Am. Assoc. Pet. Geol.* **40**, 975.

Stevenson, F. J. (1974). *In* "Advances in Organic Geochemistry 1973" (B. Tissot and F. Bienner, eds). pp. 701–714. Éditions Technip, Paris.

Stevenson, F. J. and Butler, J. H. A. (1969). *In* "Organic Geochemistry—Methods and Results" (G. Eglinton and M. T. J. Murphy, eds), pp. 534–557. Springer-Verlag, Berlin and New York.

Stevenson, F. J. and Cheng, C. N. (1972). *Geochim. Cosmochim. Acta*, **36**, 653.

Streibl, M. and Herout, V. (1969). *In* "Organic Geochemistry—Methods and Results" (G. Eglinton and M. T. J. Murphy, eds), pp. 401–424. Springer-Verlag, Berlin and New York.

Stuermer, D. H. (1975). Ph.D. Thesis, Massachusetts Institute of Technology, Cambridge, and Woods Hole Oceanographic Institution, Woods Hole.

Stuermer, D. H. and Payne, J. R. (1976). *Geochim. Cosmochim. Acta*, **10**, 1109.

Stuermer, D. H. and Simoneit, B. R. T. (1978). *In* "Initial Reports of the Deep Sea Drilling Project", Vol. 44, pp. 587–591. U.S. Government Printing Office, Washington D.C.

Swain, F. M. (1969). *In* "Organic Geochemistry—Methods and Results" (G. Eglinton and M. T. J. Murphy, eds), pp. 374–400. Springer-Verlag, Berlin and New York.

Swain, F. M. (1970). "Non-Marine Organic Geochemistry". Cambridge University Press, England, 445 pp.

Swain, F. M. and Bratt, J. M. (1978). *In* "Initial Reports of the Deep Sea Drilling Project", Vol. 44, pp. 653–654. U.S. Government Printing Office, Washington D.C.

Swain, F. M., Blumentals, A. and Millers, R. (1959). *Limnol. Oceanogr.* **4**, 119.

Sweeney, R. E. (1972). Ph.D. Thesis, University of California, Los Angeles.

Sweeney, R. E. and Kaplan, I. R. (1978). *Mar. Chem.* (in press).

Sweeney, R. E., Liu, K. K. and Kaplan, I. R. (1978). *In* "Stable Isotopes in the Earth Sciences" (B. W. Robinson, ed.), *DSIR Bulletin*, **220**, pp. 9–26. DSIR, Wellington, New Zealand.

Swift, D. J. P., Duane, D. B. and Pilkey, O. H. (1972). "Shelf Sediment Transport: Process and Pattern". Dowden, Hutchinson and Ross, Stroudsburg, 656 pp.

Tarling, D. H. (1973a). *Nature, Lond.* **243**, 193.

Tarling, D. H. (1973b). *Nature, Lond.* **243**, 277.

Taylor, H. P., Jr. (1975). *Rev. Geophys. Space Phys.* **13**, 102, 159.

Telkova, M. S., Rodionova, K. F., Shlyakhov, A. F. and Dyuzhikova, T. N. (1976). *Geokhimiya*, **7**, 1084.

Thiede, J. and van Andel, T. H. (1977). *Earth Planet. Sci. Lett.* **33**, 301.

Thomas, B. R. (1969). *In* "Organic Geochemistry—Methods and Results" (G. Eglinton and M. T. J. Murphy, eds), pp. 599–618. Springer-Verlag, Berlin and New York.

Thomas, B. R. (1970). *In* "Phytochemical Phylogeny" (J. B. Harborne, ed.), pp. 59–79. Academic Press, London and New York.

Tissier, M. J. and Oudin, J. L. (1974). *In* "Advances in Organic Geochemistry 1973" (B. Tissot and F. Bienner, eds), pp. 1029–1041. Éditions Technip, Paris.

Tissot, B., Califet-Debyser, Y., Deroo, G. and Oudin, J. L. (1971). *Bull. Am. Assoc. Pet. Geol.* **55**, 2177.

Tissot, B., Durand, B., Espitalié, J. and Combaz, A. (1974). *Bull. Am. Assoc. Pet. Geol.* **58**, 499.

Traverse, A. (1974). *In* "The Black Sea—Geology, Chemistry, and Biology" (E. T. Degens and D. A. Ross, eds), pp. 381–388. Memoir 20, American Association of Petroleum Geologists, Tulsa, U. S. A.

Vallentyne, J. R. (1957). *J. Fish. Res. Bd Can.* **14**, 33.

Vallentyne, J. R. (1960). *In* "Comparative Biochemistry of Photoreactive Systems" (M. A. Allen, ed.), pp. 96–103. Academic Press, New York and London.

van der Velden, W. and Schwartz, A. W. (1974). *Science, N.Y.* **185**, 691.

van der Velden, W. and Schwartz, A. W. (1976a). *Chem. Geol.* **18**, 273.

van der Velden, W. and Schwartz, A. W. (1976b). *In* "Environmental Biogeochemistry" (J. O. Nriagu, ed.), pp. 175–183. Ann Arbor Science Publishers, Ann Arbor, U.S.A.

van der Velden, W., Chittenden, G. J. F. and Schwartz, A. W. (1974). *In* "Advances in Organic Geochemistry 1973" (B. Tissot and F. Bienner, eds), pp. 293–304. Éditions Technip, Paris.

van Dorsselaer, A. (1974). M.Sc. Thesis, University of Strasbourg, France.

van Dorsselaer, A., Ensminger, A., Spyckerelle, C., Dastillung, M., Sieskind, D., Arpino, P., Albrecht, P., Ourisson, G., Brooks, P. W., Gaskell, S. J., Kimble, B. J., Philp, R. P., Maxwell, J. R. and Eglinton, G. (1974). *Tetrahedron Lett.* 1974, 1349.

van Krevelen, D. W. (1961). "Coal". Elsevier, Amsterdam, 514 pp.

Vassoyevich, N. B., Korchagina, Yu. I., Lopatin, N. V. and Chernyshev, V. V. (1969). *Moskov. Univ. Vestnik* **6**, 3; *Internat. Geol. Rev.* (in English), **12**, 1276 (1970).

Vernon, L. P. and Seely, G. R., eds (1966). "The Chlorophylls". Academic Press, New York and London, 679 pp.

Vinogradov, A. P., Gruenko, V. A. and Ustinov, V. I. (1962). *Geokhimiya*, **10**, 973.

Volkman, J. K. and Johns, R. B. (1977). *Nature, Lond.* **267**, 693.

von Rad, U. and Rösch, H. (1972). *In* "Initial Reports of the Deep Sea Drilling Project", Vol. 14, pp. 727–751. U.S. Government Printing Office, Washington D.C.

Vuchev, V. T., Ivanov, C. P., Kabakchieva, M. St., Petrov, L. P., Stojanova, R. Zh., Stephanov, D. D., Djakova, D. N. and Petrova, L. K. (1978). *In* "Initial Reports of the Deep Sea Drilling Project", Vol. 42B, pp. 723–728. U.S. Government Printing Office, Washington D.C.

Watts, C. D. (1975). Ph.D. Thesis, University of Bristol, England.

Watts, C. D. and Maxwell, J. R. (1977). *Geochim. Cosmochim. Acta*, **41**, 493.

Weete, J. D. (1972). *Phytochemistry*, **11**, 1201.

Wehmiller, J. and Hare, P. E. (1971). *Science, N.Y.* **173**, 907.

Wehmiller, J. and Hare, P. E. (1972). *In* "Initial Reports of the Deep Sea Drilling Project", Vol. 9, pp. 903–905. U.S. Government Printing Office, Washington D.C.

Welte, D. H. and Ebhardt, G. (1968a). *Geochim. Cosmochim. Acta*, **32**, 465.

Welte, D. H. and Ebhardt, G. (1968b). "Meteor" Forschungsergebnisse, Reihe C, Heft 1, Gebr, Borntraeger, Berlin-Stuttgart, pp. 43–52.

Whelan, J. K. (1977). *Geochim. Cosmochim. Acta*, **41**, 803.

Whelan, J. K. and Hunt, J. M. (1978). *In* "Initial Reports of the Deep Sea Drilling Project", Vol. 42B, pp. 673–677. U.S. Government Printing Office, Washington D.C.

White, R. H. and Miller, S. L. (1976). *Science, N.Y.* **193**, 885.

Whitehead, E. V. (1974). *In* "Advances in Organic Geochemistry 1973" (B. Tissot and F. Bienner, eds), pp. 225–243. Éditions Technip, Paris.

Williams, D. H., sr. rptr. (1971). "Mass Spectrometry", Vol. 1. The Chemical Society, London, 323 pp.

Williams, D. H., sr. rptr. (1973). "Mass Spectrometry", Vol. 2, The Chemical Society London, 356 pp.

Wilson, R. D., Monaghan, P. H., Osanik, A., Price, L. C. and Rogers, M. A. (1974). *Science, N.Y.* **184**, 857.

Yen, T. F., ed. (1977). "Chemistry of Marine Sediments". Ann Arbor Science Publishers, Ann Arbor, U.S.A., 265 pp.

Youngblood, W. W. and Blumer, M. (1975). *Geochim. Cosmochim. Acta,* **39**, 1303.

Youngblood, W. W., Blumer, M., Guillard, R. R. L. and Fiore, F. (1971). *Mar. Biol.* **8**, 190.

Zafiriou, O. C. (1973). *Estuar. Coastal Mar. Sci.* **1**, 81.

Zarrella, W. M., Mousseau, R. J., Coggeshall, N. D., Norris, M. S. and Schrayer, G. J. (1967). *Geochim. Cosmochim. Acta,* **31**, 1155.

Zsolnay, A. (1971). *Kieler Meeresforsch.* **27**, 135.

ADDENDA

p. 270

It has been reported that PAH (especially 3,4-benzpyrene, XLIX) are synthesized by anaerobic bacteria (Knorr and Schenk, 1968; Brisou, 1969; Niaussat *et al.*, 1969, 1970), but mixed anaerobic bacterial cultures from recent marine sediments only bio-accumulated PAH and *de novo* synthesis could not be detected (Hase and Hites, 1976). Nitrogen heterocyclic PAH (aza-arenes) have been identified in Recent marine sediments and are probably derived from combustion processes (Blumer *et al.*, 1977).

p. 283

The isotopic compositions of sulphur compounds have been determined for Black Sea sediments (Vinogradov *et al.*, 1962), for evaporites and shales from various DSDP sites in the Red Sea (Shanks *et al.*, 1974), and for sediments and interstitial waters in the Mediterranean Sea (Fontes *et al.*, 1973).

Chapter 40

Determination of Marine Chronologies Using Natural Radionuclides

KARL K. TUREKIAN and J. KIRK COCHRAN

Department of Geology and Geophysics,
Yale University, New Haven, Connecticut, U.S.A.

40.1. INTRODUCTION

The radioactive geochronometry of marine deposits has been established as an important aid to the understanding of the history of ocean basins. So great have the triumphs been, that measurement of radionuclides in the pursuit of a date, or a rate, appears now to have become a routine approach. This is an error, and is one frequently found in the attitudes of the most obvious beneficiaries of such information—the sedimentologist, stratigrapher and palaeontologist. Indeed, radionuclides are chronometers, but they respond to a variety of time-dependent processes of which the rate of sediment accumulation is only one. The natural radioactive clock factory has time-pieces that have uses that range from the determination of fluxes of various components through the oceans to the measurement of the reworking rates of sediments.

The modern approach to radioactive geochronometry clearly cannot be the simple one of the past, but must rely on information gained from several different clock systems. These systems are divisible into two groups: the members of the uranium and thorium decay series, and the cosmogenic nuclides. This division will be followed in the treatment presented below, although the various nuclides do share some common behaviour. Since several detailed general reviews of the use of natural and man-made radio-nuclides in marine problems have recently appeared (Goldberg and Bruland, 1974; Ku, 1976; see also Chapter 18), this chapter will focus on those aspects which provide novel, or different, points of view.

40.2. THE GROUND RULES

There are certain assumptions which must be made before the data from radioactive clocks can be interpreted. These assumptions involve:

(i) the manner in which the supply or production of each natural radio-active nuclide is controlled;
(ii) the best way in which the behaviour of each radionuclide in the ocean water column can be described;
(iii) the mechanisms which control the distribution of each radionuclide in the sediment column.

These will be discussed below.

40.2.1. PRODUCTION AND SUPPLY

Radionuclides supplied to the sea are produced on land, in the atmosphere or within the ocean system. Those that are produced in the atmosphere (or have an important fraction that is produced in the atmosphere) include the cosmogenic nuclides ^{7}Be (half-life 53 days), ^{3}H (12 years), ^{32}Si (~ 300 years), ^{14}C (5730 years), ^{26}Al (0.75×10^6 year), ^{10}Be (1.5×10^6 year) and ^{210}Pb (22 years) which is produced from atmospheric ^{222}Rn (3.8 days). Among the cosmogenic nuclides, ^{7}Be shows the greatest variability in its supply pattern in both time and space, and this may seriously affect its general usefulness in oceanic studies. Tritium (^{3}H), ^{32}Si and ^{14}C all have long lives and, like their stable analogues, they have long residence times in the

oceans; this tends to damp out the effects on their distribution due to fluctuations in production rates with time and place. Although the production
rate of ^{14}C has been shown to vary over long time scales, its long residence
time at depth in the ocean, and the resolving power of its variations in deep-
sea deposits, allows it to be treated as virtually invariant in supply with time.
In contrast, the short residence times of ^{10}Be and ^{26}Al in both the atmosphere
and the ocean, and the long half-lives of these nuclides, suggest that the persistent sites of accelerated leakage from the stratosphere should influence their
delivery to the ocean in the same way as they influence the supply of bomb-
produced nuclides from the stratosphere.

It is generally assumed that the major mechanism for the transport of
continentally derived natural radionuclides to the oceans is surface run-off.
The world-wide flux of river-water has been estimated to lie between
3.23×10^{16} litre year^{-1} (Livingstone, 1963) and 3.65×10^{16} litre year^{-1}
(Alekin and Brazhnikova, 1961). If the area of the oceans (both deep and
shallow) is considered to be 3.6×10^{18} cm^2, a river flux value of 3.6×10^{16}
litre year^{-1} implies that 10 cm^3 of river-water is delivered each year to every
square centimetre of the oceans (this is of course balanced by evaporation).
Thus, the average concentration of any component in river-water can be
directly translated into a flux. This assumption ignores transport by ground-
waters directly into the coastal ocean, transport as aerosols and reactions
occurring in the estuarine zone.

The production of natural radionuclides in the ocean system by *in situ*
decay of radioactive parents and their delivery by release from bottom
sediments will be discussed below.

40.2.2. BEHAVIOUR AND TRANSPORT IN THE WATER COLUMN

The mechanisms of delivery of the radionuclides to the ocean floor are
diverse and include sinking both after incorporation in tests (e.g. carbon-14
and silicon-32) and adsorption on particles (e.g. ^{230}Th and ^{210}Pb). The
natural radionuclides provide a means of determining the mechanisms and
rates of processes such as these, because they vary from place to place in the
ocean system.

There are two ways of treating data on radionuclide distributions in the
water column to arrive at these transfer rates. In the first of these, each
specified part of the ocean is regarded as a box into which substances are
delivered and processed and from which they are subsequently transported.
Equations describing the box model have the form:

$$\text{Flux-into-box} = \text{decay-in-box} + \text{flux-out-of-box}.$$

The decay term is expressed as λN, where N is the atomic concentration of the radionuclide and λ is its characteristic decay constant. It is common practice to assume that the "flux-out-of-box" term obeys first order kinetics. This is purely a device to enable us to express the "flux-out-of-box" term as kN, where N, as above, is the atomic concentration and k is the first order removal constant. Because the "flux-into-box" term itself is a production term for members of the uranium (or thorium) decay series, it also can be represented by a λN product, where λ and N refer to the parent of the radionuclide under examination. Thus, for the coupled members of a decay chain, the material balance in the box is:

$$\lambda_P N_P = \lambda_D N_D + k_D N_D \tag{1}$$

where the subscripts refer to the parent (P) and daughter (D), respectively. The activity (A) is identically equivalent to λN, leading to the following equation:

$$A_P = A_D \left(1 + \frac{k_D}{\lambda_D} \right). \tag{2}$$

If the activity ratio A_D/A_P in the water parcel or box can be measured, the value of k_D can be calculated directly. The reciprocal of k_D which is in units of time, is called the mean residence time of the nuclide relative to its physical removal from the box, and should be distinguished from its mean radioactive life ($= 1/\lambda$).

The second way to characterize the transport of radionuclides in the water column is by the use of an advection–diffusion equation of the form:

$$K_z \frac{\partial^2 N}{\partial z^2} + w \frac{\partial N}{\partial z} - \lambda N + J = 0 \tag{3}$$

where K_z is the vertical eddy diffusion coefficient in the oceanic water column, w is the advective velocity, z is the depth and J is a production (or removal) term (see also Chapter 1). In the discussion of radioactive decay chains it is commonly assumed, purely for convenience, that J obeys the same first order law discussed above, thus taking the form $J = kN$. The solution of this equation depends on curve-fitting using at least one independently known parameter.

40.2.3. DISTRIBUTION IN SEDIMENTS

One of the most valuable ways in which natural radionuclides can be used to study marine deposits is as chronometers. For such deposits this has generally involved the dating of layers in sediments or of growth bands in such chemical accumulations as manganese nodules and tests of organisms. Radionuclides can also be used as chronometers for determining the rates of reworking of marine sediments by organisms. If each of these two processes (i.e. accumulation and reworking) can be isolated, then the use of a radioactive chronometer is fairly simple. However, this is not always possible especially within several centimetres of the sediment–water interface. In oversimplified terms, the three limiting equations describing all of the possible processes, assuming steady-state conditions, are:

(i) sediment accumulation only,

$$S \frac{\partial N}{\partial z} - \lambda N = 0; \tag{4}$$

(ii) reworking only (radionuclide supplied continuously without sediment carrier),

$$K_b \frac{\partial^2 N}{\partial z^2} - \lambda N = 0; \tag{5}$$

(iii) both sediment accumulation and reworking,

$$K_b \frac{\partial^2 N}{\partial z^2} - S \frac{\partial N}{\partial z} - \lambda N = 0; \tag{6}$$

where N is the concentration of the radionuclide which varies with the depth (z) in the sediment, K_b is a mixing coefficient representing the reworking process, S is the sediment accumulation rate and λ is the decay constant of the radioactive nuclide.

The general solution to each of these equations involves an exponential function. Thus, the form of the concentration decrease with depth in a core does not alone provide information about which of the three equations is valid. Only the fact that reworking is commonly restricted primarily to the top of the sediment pile allows these different processes to be resolved.

40.3. URANIUM AND THORIUM DECAY SERIES NUCLIDES

Thorium and the two naturally occurring isotopes of uranium have half-lives within a factor of ten of the age of the Earth. These radionuclides, which were produced under conditions in the cosmos that are as yet not well understood, decay step-wise to stable isotopes of lead. The intermediate nuclides in the decay chains all have different half-lives, and most have different chemical properties. It is these two facts that have made the uranium and thorium decay series attractive for the study of rates and mechanisms of earth surface processes. Figure 40.1 shows the decay schemes for ^{232}Th, ^{235}U and ^{238}U which are distributed in the earth according to the general geochemical laws that govern the distribution of the elements. However, the

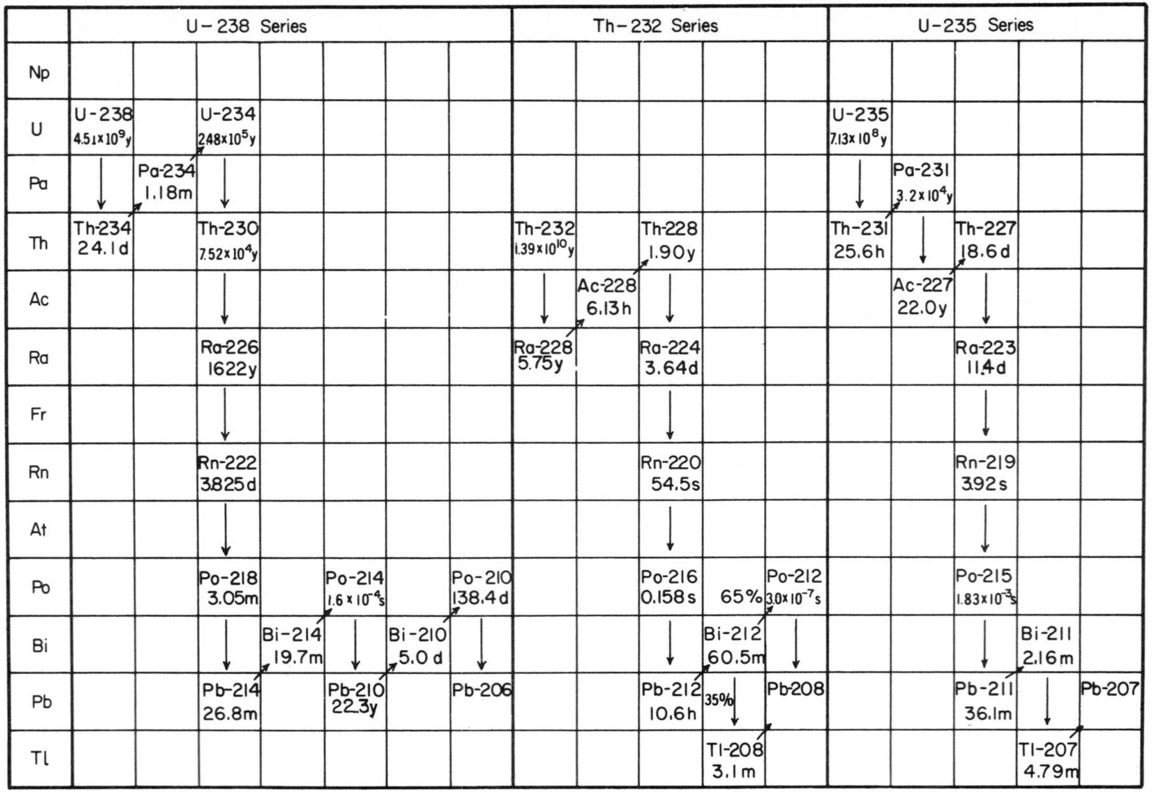

FIG. 40.1. ^{238}U, ^{232}Th and ^{235}U decay series.

behaviour of the radioactive daughters of these nuclides is complicated by the fact that their formation is accompanied by a release of energy, a transmutation of the element and a change in electrical charge. These forces lead to mobilization over and above that ascribable to lattice solution. Indeed, the nascent atoms produced by radioactive decay may have abnormal chemical properties, at least until adjustment to the environment occurs.

The chemical behaviour of radioactive nuclides produced by neutron bombardment has been studied in great detail. This field has come to be known as "hot atom" chemistry and many of its principles, if not the details, are applicable to the uranium and thorium decay chains. The radioactive transformation involving α-decay results in the recoil of the transformed atom in order to conserve momentum. If the decay occurs in a solid, then in most cases the energy of recoil is sufficiently large to break the restraining chemical bonds. Movement of the recoiled atom is then a function of the steric and charge constraints of the solid in which the transformation occurs. The greatest possibility for mobilization occurs when the recoil takes place at, or on, the surfaces of a solid phase in contact with a fluid, such as air or water. A number of processes serve to influence the distribution of the newly formed nuclide in the fluid. These include reactions such as adsorption, precipitation, and ion exchange with the solid surfaces in contact with the fluid, complexation and oxidation–reduction effects based on the chemistry of the fluid phase, radioactive decay, diffusion, and bulk transport by advection of the fluid.

After delivery to the aqueous domain, a radionuclide will eventually take on the chemical properties characteristic of that element. Its subsequent history will then become chemically, rather than radiochemically, controlled. In the following discussion all the isotopes of an element will therefore be treated under one heading and their coupled and disparate behaviours will be noted when they occur.

40.3.1. URANIUM

40.3.1.1. The separation and transport of the uranium isotopes

Of the isotopes of uranium, ^{238}U and ^{235}U are primordial and ^{234}U is produced by ^{238}U decay. The primary long-lived isotopes occur in nature in a constant $^{238}U/^{235}U$ ratio of 137·88/1. During weathering, the release of these uranium isotopes is primarily in response to the general chemical destruction of the rock. Indeed, as Fig. 40.2 shows, there is a correlation between the concentration of total dissolved ions in a stream or groundwater and the uranium concentration. Despite the fact that in oxidizing aqueous systems uranium forms stable uranyl carbonate complexes, e.g. $[UO_2(CO_3)_3^{4-}]$, there does not seem to be any evidence that the total concentration of species of the carbonate system exerts a control on the uranium concentration in rivers, as has been suggested by Broecker (1974). However, this total concentration may control the kinetics of dissolution.

The dissolved uranium concentration in river-water varies considerably from one river to another. Thus, Bertine, *et al.* (1970) found a range of $< 0.01 \, \mu g \, l^{-1} – 1.22 \, \mu g \, l^{-1}$ for a large number of rivers, the average concentration being $0.27 \, \mu g \, l^{-1}$. Turekian and Chan (1971) decided on the

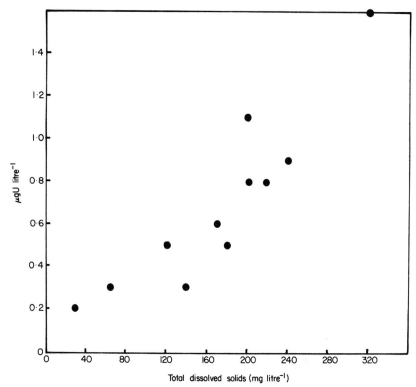

FIG. 40.2. Uranium concentration versus total dissolved solids for the MacKenzie River (Turekian and Chan, 1971).

basis of these, and data for the MacKenzie River, that the average for world rivers is $\sim 0.3 \, \mu g \, l^{-1}$. This average, with the world average annual river flux given above, yields a dissolved uranium flux to the oceans of $3 \, \mu g \, cm^{-2}$ $(10^3 \, \text{year})^{-1}$. The concentration of uranium in the oceans is virtually constant at $3.3 \, \mu g \, l^{-1}$ for a salinity of $35\%_{0}$ (Turekian and Chan, 1971).

Uranium is adsorbed to some degree on the particulate matter in rivers. The exact adsorption mechanism is, as yet, uncertain, but Fig. 40.3 shows the data from which its existence has been inferred. The release of uranium from these adsorption, or other superficial sequestering, sites when sea-water is encountered will increase the uranium flux to the oceans. A knowledge of the supply rate of uranium (mainly ^{238}U) to the oceans is the keystone to the interpretation of data for the fluxes of a variety of nuclides.

Uranium-234 is produced by the radioactive decay of ^{234}Th and, for the reasons given above, is preferentially mobilized during weathering relative to ^{238}U. This has the effect of producing a separation of ^{234}U from ^{238}U and provides the basis of its use as a tracer and chronometer.

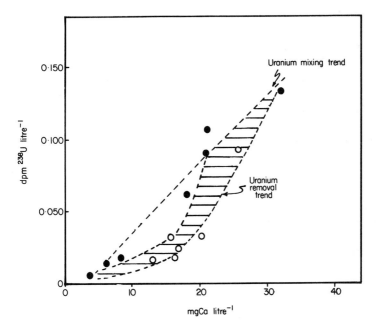

FIG. 40.3. Dissolved uranium versus dissolved calcium in the Susquehanna River and its tributaries (after Lewis, 1976). Solid circles = source waters from small tributaries; open circles = main stream waters downstream from tributaries.

The mechanism of release of ^{234}U to aqueous systems is not a direct one. Kigoshi (1971) has shown that a major factor in enriching continental ground-waters in ^{234}U is the recoil of the ^{234}Th produced during ^{238}U decay which detaches it from the mineral lattice. It seems likely that the recoiled ^{234}Th is quickly adsorbed onto mineral surfaces in the ground-water regime, since the average distance to a surface is much smaller than it is in coastal ocean water in which, as will be seen later, the residence time is about a day. This ^{234}Th subsequently undergoes radioactive decay and its daughter ^{234}U is subsequently released by selective chemical leaching. The evidence for the recoil enrichment of ^{234}U in continental ground-waters is extensive and has been reviewed by several authors (see e.g. Cherdyntsev, 1969; Osmond *et al.*, 1974). The activity ratio of ^{234}U/^{238}U in river-waters is highly variable and an average value is difficult to ascertain but, according to a number of workers (see e.g. Turekian and Chan, 1971), it appears to be $\sim 1 \cdot 20$. In all ocean waters this ratio is $\sim 1 \cdot 15$ (see Chapter 18). The maintenance of a constant ^{234}U/^{238}U activity ratio is basic to several types of chronometry.

40.3.1.2. *Chronologies based on the uranium isotopes*

All of the uranium isotopes have long half-lives and can properly be thought of as providing the framework for all the shorter-term chronologies. Although the stable lead isotopes resulting from the ultimate decay of ^{238}U and ^{235}U can, in principle, be used for dating deposits which are old enough for secular equilibrium to have been established, no deep-sea deposits have ever been dated by this technique. The possibility of using this approach has been explored, but the uranium concentration of most marine materials is too low, and the background lead concentrations too high, to warrant serious efforts in measurement. However, there are two types of chronologies that are possible in marine systems. These are based on the observed ^{234}U/^{238}U activity ratio of $1 \cdot 15$ in sea-water. One of these is to assess the extent to which this ratio approaches equilibrium in those calcareous tests (corals) and calcareous precipitates (oolites) which contain indigenous uranium. The other is to determine the relative fluxes of unsupported ^{234}U to the ocean from the continents and deep-sea deposits.

The curve for the change in the ^{234}U/^{238}U activity ratio as a function of time in a closed system having an initial ratio of $1 \cdot 15$ is shown in Fig. 40.4. In this figure the error bar represents the one-sigma error commonly obtained under normal analytical conditions. It is evident that because of this large error the accuracy of the ages obtained solely by following changes in the ^{234}U/^{238}U activity ratio with time is very poor. Indeed, the most satisfactory way to use measurements of this ratio is with systems free of initial ^{230}Th (such as the corals and oolites) and to plot the change with time of the ^{230}Th/^{234}U activity ratio against that of the ^{234}U/^{238}U activity ratio (Fig. 40.5). Each value of the ratios ^{230}Th/^{234}U/^{238}U (as activities) is unique for a given age, and is constrained by the initial ^{234}U/^{238}U activity ratio of $1 \cdot 15$ typical of sea-water. Any departure from this concordance implies that the system has been open to uranium or ^{230}Th, and so the determined age is in doubt. Thus, in initially ^{230}Th-free systems, the major use of the ^{234}U/^{238}U system is to monitor whether or not the system is the closed one required for accurate dating.

If all the uranium in deep-sea sediments is derived from sea-water, or if the authigenic fraction can be efficiently separated, it should be possible to

use the decay of the 15 % excess of ^{234}U in the sediment column as a chronological indicator. In fact, the distribution of the ^{234}U/^{238}U activity ratio with
depth in clay-rich cores usually resembles the profile shown in Fig. 40.6.
Ku (1965) found that such a pattern was best explained in terms of ^{234}U
recoil, followed by diffusion in the sediment column via the pore waters. These

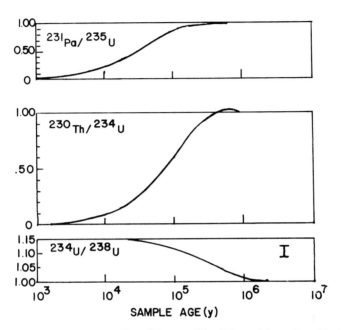

FIG. 40.4. Variation of ^{231}Pa/^{235}U, ^{230}Th/^{234}U and ^{234}U/^{238}U activity ratios with time in a
closed system with an initial ^{234}U/^{238}U ratio of 1·15 (after Broecker *et al.*, 1968). Error bar on
^{234}U/^{238}U plot shows typical counting uncertainty for this ratio.

processes, as well as the previously mentioned constraints involved in dating
calcareous material, make excess ^{234}U less useful for dating. However, if the
recoiled ^{234}U is trapped by authigenic phases such as manganese micro-
nodules forming in the sediment, its build-up as recorded by these phases
is itself a chronological indicator (Immel and Osmond, 1976).

Profiles such as that shown in Fig. 40.6 provide constraints on one parameter of importance in the balance of uranium in the World Ocean, the
diffusive flux of ^{234}U from sediments. The mass balance for ^{234}U and ^{238}U
in the oceans may be written as (Veeh, 1967; Turekian and Chan, 1971):

$$\frac{d\,^{234}U_o}{dt} = J^{234}_{sed} + J^{234}_s - J^{234}_r - (\lambda_{234}\,^{234}U_o) + (\lambda_{238}\,^{238}U_o) \qquad (7)$$

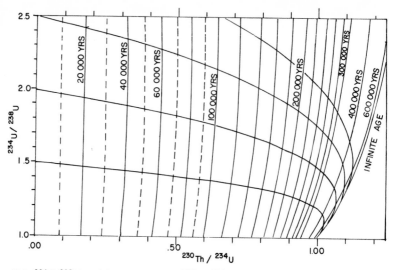

FIG. 40.5. $^{234}U/^{238}U$ activity ratio versus $^{230}Th/^{234}U$ activity ratio (after Kaufman, 1964). Vertical lines represent isochrons. Horizontal curves correspond to the variation in the activity ratios of a sample with the indicated initial $^{234}U/^{238}U$ ratio and no ^{230}Th.

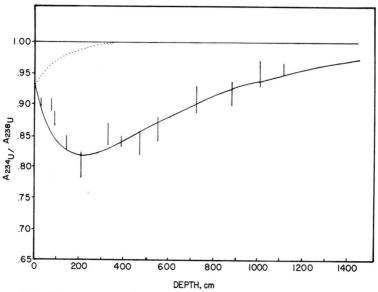

FIG. 40.6. $^{234}U/^{238}U$ activity ratio versus depth in North Atlantic red clay core V10-95 (after Ku, 1965). The horizontal line represents a closed system with a $^{234}U/^{238}U$ ratio of 1·00. Dashed curve represents theoretical profile for a closed system (no ^{234}U migration) and an initial $^{234}U/^{238}U$ ratio of 0·93 (Scott, 1968). The difference between these closed system curves and that observed gives a steady-state ^{234}U loss from the sediment of 0·6 dpm cm^{-2} year^{-1}.

and

$$\frac{d\,^{238}U_o}{dt} = J_s^{238} - J_r^{238} - (\lambda_{238}\,^{238}U_o) \tag{8}$$

where $^{234}U_o$, $^{238}U_o$ = amount of ^{234}U and ^{238}U in the ocean (atoms cm^{-2}), J_s^{234}, J_s^{238} = rate of river-borne supply of ^{234}U and ^{238}U (atoms cm^{-2} year^{-1}), J_r^{234}, J_r^{238} = rate of removal of ^{234}U and ^{238}U from the ocean (atoms cm^{-2} year^{-1}), J_{sed}^{234} = rate of diffusion of ^{234}U out of deep-sea sediments (atoms cm^{-2} year^{-1}).

Assuming the oceans are in steady-state for the uranium isotopes and

$$\frac{\lambda_{234}J_s^{234}}{\lambda_{238}J_s^{238}} = \, ^{234}U/^{238}U \text{ activity ratio of rivers } (R_{river})$$

and

$$\frac{\lambda_{234}J_r^{234}}{\lambda_{238}J_r^{238}} = \, ^{234}U/^{238}U \text{ activity ratio of the ocean}$$

$$= 1·15.$$

Equations (7) and (8) can be combined to give

$$J_{sed}^{234} = \left(0·15 - 1·15\frac{\lambda_{238}}{\lambda_{234}}\right)\lambda_{238}\, ^{238}U_o + \frac{\lambda_{238}}{\lambda_{234}}J_s^{238}(1·15 - R_{river}). \tag{9}$$

Or, in terms of activities, and approximating the term $(0·15 - 1·15\frac{\lambda_{238}}{\lambda_{234}})$ by $0·15$:

$$F_{sed}^{234} = 0·15\lambda_{234}A_0^{238} + F_s^{238}(1·15 - R_{river}) \tag{10}$$

where A_0^{238} is the activity of $^{238}U_o$ and F = flux of the indicated isotope from sediments or rivers in dpm cm^{-2} year^{-1}.

Figure 40.7 shows plots of the relationship of Equation (10) for different values of F_{sed}^{234}, F_s^{238} and R_{river} (from Turekian and Chan, 1971). It can be seen that for $F_{sed}^{234} \simeq 0·6$ dpm cm^{-2} year^{-1} (see Fig. 40.6), a concentration of uranium in rivers of $0·3 - 1$ µg U l^{-1} gives a $^{234}U/^{238}U$ activity ratio in rivers of $1·06 - 1·13$. This is in reasonable agreement with data observed for these parameters.

40.3.2. THORIUM (AND PROTACTINIUM)

The fundamental property of thorium in natural aqueous systems is that it is strongly sequestered by surfaces. The concentration of dissolved ^{232}Th in the oceans is certainly less than $0·000017$ dpm kg^{-1} ($0·00007$ µg kg^{-1}) (Kaufman, 1969). Most of the higher ^{232}Th values which have been reported for sea-water are probably due to particulate thorium, the source and concentration of which may vary from place to place in the oceans. The other thorium isotopes, ^{234}Th, ^{230}Th and ^{228}Th, are produced in the oceans from the decay of dissolved ^{238}U, ^{234}U and ^{228}Ra, respectively. ^{231}Pa, a nuclide behaving somewhat like thorium, is produced by the decay of ^{235}U.

40.3.2.1. Residence times of thorium in the oceans

Departures of the concentrations of each of the thorium isotopes from those expected on the assumption of equilibrium with their parents enables the residence time of the isotopes to be calculated on the basis of their removal by

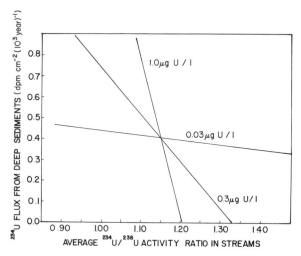

FIG. 40.7. Relationship between ^{234}U flux from deep-sea sediments and the ^{234}U/^{238}U activity activity ratio in rivers for several average ^{238}U concentrations of rivers (Equation (10) in text). (After Turekian and Chan, 1971.)

particulate matter as discussed above. If it is assumed, as a first approximation, that the concentrations of ^{230}Th, ^{234}Th and ^{228}Th in the deep and surface waters of the oceans are controlled by adsorption onto the surfaces of particles, and that this is strictly diffusion-controlled, the mean residence time of any one of these isotopes can be used for all the others or, indeed, for any other chemical species behaving in the same way. On the basis of observations in the open-ocean environment, a mean residence time for these nuclides of one year in surface waters and 100 years in deep waters is a useful generalization (Table 40.1). Closer to the continental margins, the mean residence time of thorium in the water column decreases since the increased biological activity and physical supply of suspended material leads to an increase in the total concentration of particulate matter. The data of Broecker *et al.* (1973) show that the ^{228}Th/^{228}Ra activity ratios in surface sea-water decrease by almost a factor of ten as the coasts are approached, even before the continental

TABLE 40.1

Residence times for U/Th series nuclides in the oceans

Nuclide	$\tau_{surface}$ (years)[a]	τ_{deep} (years)[b]
[234]Th	0·38	>0·25
	Matsumoto (1975)	Amin et al. (1974)
[228]Th	0·5	—
	Broecker et al. (1973)	
[210]Pb	1·4–2·3 (N. Atlantic)	~50 (Pacific)
	Bacon et al. (1976)	Craig et al. (1973)
	1·7 (N. Pacific)	~100 (S. Pacific)
	Nozaki et al. (1976)	Thomson and Turekian (1976)
	0·5–1·0 (NW. Pacific)	<96 (NW. Pacific)
	Nozaki and Tsunogai (1976)	Nozaki and Tsunogai (1976)
[210]Po	0·3–0·6 (N. Atlantic)	4 (N. Atlantic)
	Bacon et al. (1976)	Bacon et al. (1976)
	0·6 (N. Pacific)	2 (S. Pacific)
	Nozaki et al. (1976)	Thomson and Turekian (1976)
	Nozaki and Tsunogai (1976)	

[a] Surface, 0– ~400 m.
[b] Deep, 1000–4000 m.

shelf, with its generally high particle concentration, is reached. In this context, Bhat et al. (1969) found similar results for [234]Th relative to [238]U. An example of the ultimate efficiency of thorium scavenging in coastal systems is the observation by Aller and Cochran (1976) that virtually all of the [234]Th produced in the waters of Long Island Sound is present in the sediments; this corresponds to a mean residence time of about one day for this nuclide in the water column.

40.3.2.2. *The chronology of sediment mixing by organisms*

If no significant quantity of sediment has been accumulating during several half-lives of a nuclide supplied to the ocean floor at a constant rate, the distribution of the radionuclide in the sediment would be due to mixing processes alone and would provide an estimate of the mixing coefficient by the use of Equation (5).

The half-life of [234]Th is sufficiently short (24 days) that, below the immediate sediment surface, the sea-water-derived (i.e. excess) [234]Th should show no variation with depth in the sediment on the basis of sediment accumulation alone. Thus, in near-shore sediments accumulating at a rate of ~3 mm year^{-1}, excess [234]Th should be present only in the top ~1 mm. This feature makes [234]Th an excellent tracer for studying physical and biological particle mixing in sediments (Aller and Cochran, 1976). As mentioned above, the high particle density in near-shore waters leads to the rapid removal of ^{234}Th from the water to the sediments. Figures 40.8A and 40.8B show comparative ^{234}Th profiles and related X-radiographs of cores taken in Long Island Sound and the New York Bight. In both cores the mixing is dominated by deposit-feeding bivalves, i.e. by *Yoldia limatula* in Long Island Sound and *Nucula* sp. in the New York Bight. In these two cores the ^{234}Th mixing depths were a few centimetres, and when the data for them was fitted to Equation (5), which applies to steady-state patterns due to reworking alone, they yielded a value of $K \simeq 5 \times 10^{-7}$ cm^{2}s^{-1}. It should be noted, however, that the mode and depth of sediment reworking in near-shore locations are very much dependent upon the particular benthic environment and on the nature of the primary bioturbating fauna (see also Chapter 29).

40.3.2.3. *Deep-sea deposit-dating using excess* ^{230}*Th and* ^{231}*Pa*

Deep-sea sediments are also subject to mixing by bioturbation which is limited to the uppermost 5–10 cm of the deposit (Arrhenius, 1963; Berger and Heath, 1968; Glass, 1969; Hanor and Marshall, 1971; Guinasso and Schink, 1975). Excess ^{230}Th is plentiful in such sediments because of the large reservoir of uranium available in the oceanic water column. The relatively long life of this nuclide, coupled with the slow sediment accumulation rate on the deep-sea floor, implies that below the mixing zone (in which the excess ^{230}Th concentration is homogeneous) the exponential decrease of ^{230}Th is due to decay alone (see Goldberg and Koide, 1962). Thus, ^{230}Th can be used as a dating tool below the mixed layer; however, its use is subject to a number of constraints which are discussed below.

Initial attempts to use ^{230}Th for such dating involved the measurement of its daughter, ^{226}Ra (see e.g. Piggott and Urry, 1942). Later work (see e.g. Pettersson, 1951; Kröll, 1953; Koczy, 1958) indicated that ^{226}Ra is mobilized in deep-sea sediments and migrates out of them, thus suggesting that it might only be of limited use as an index of the ^{230}Th activity. This criticism is actually valid only for the top part of the core because the molecular diffusion coefficient of ^{226}Ra is sufficiently small to attenuate the loss of this nuclide at greater depths. Thus, with increasing depth in the sediment column, ^{226}Ra approaches secular equilibrium with ^{230}Th. Under such conditions, it is possible to estimate rates of accumulation of deep-sea sediments from determinations of ^{230}Th based on either direct measurement of ^{226}Ra (or its daughter ^{222}Rn) or non-destructive measurements of the γ-emitting daughters (Osmond and Pollard, 1967; Yokoyama *et al.*, 1968; Bhandari *et al.*, 1971; Cochran and Osmond, 1974, 1976). Recently, α-track measurements of deep-sea cores (Fisher, 1977; Andersen and Macdougall, 1977) have been successfully made, and this technique may offer a further approach to the problem.

The fundamental requirements for ^{230}Th-dating are that the sediment accumulation rate has been constant, that the production and removal rates of ^{230}Th per unit area of the ocean bottom have been constant over time, and that only ^{230}Th measurements made below the zone of mixing are used in calculating accumulation rates (Equation 4). In fact, it is only rarely that all these requirements are satisfied. The production rate of ^{230}Th is dependent solely on the uranium concentration in the oceans, which, on the basis of coral compositions, can be assumed to have been constant over the past several hundred thousand years (see e.g. Broecker, 1974). The short residence time of ^{230}Th in sea-water indicates that this nuclide is effectively removed from the water column on a very rapid time scale. The flux of ^{230}Th to the sediments is not, however, a simple function of its production rate in the overlying water column. At a given location, sediment accumulating by particle-by-particle settling is commonly augmented, or diminished, by bottom-transported material on either a continuous or an episodic basis. Depending on the extent to which this process operates, the ^{230}Th inventory in the sediment at a specific site may be greater or less than that which would arise solely from production in the water column (Ku, 1966; Scott et al., 1972; Cochran and Osmond, 1976).

On an areal basis, the ^{230}Th production in the water column must be balanced by its accumulation in the sediments. The larger the area of the ocean basin dimension used in the material balance, the greater will be the possibility of approaching a true balance with respect to the production of ^{230}Th and its removal to the sediments (Cochran and Osmond, 1976). Constancy of production of the nuclide in the water column and its removal from it, implies that a change in the rate of sediment accumulation will cause the specific activities of excess ^{230}Th to vary as a function of sedimentation rate. This was the major argument used by Ku et al. (1968) to explain why the accumulation rates yielded by the ^{230}Th distribution in Atlantic deep-sea cores were lower than those determined by radiocarbon dating.

The ^{230}Th flux to a specific site at the ocean bottom, however, can change with time when the sediment accumulation rate remains constant. The reason for this is not clear, but it may be that the ^{230}Th is associated with a particular phase, such as manganese oxide, rather than with the bulk sediment. This correlation with manganese is demonstrated by the high excess of ^{230}Th in the surface layer of manganese nodules. Indeed, this association has been used to determine the growth rate of manganese nodules. The decrease in ^{230}Th with depth in a manganese nodule has been interpreted as being the result of radioactive decay. The history of using ^{230}Th, as well as ^{231}Pa, to date manganese nodules has recently been reviewed with elegance by Ku (1977) and will not be discussed again here (see Chapter 18). The conclusion reached from the most direct interpretation of such data is that most

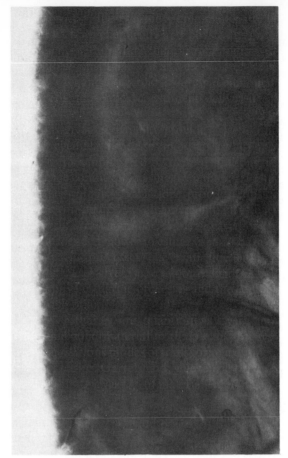

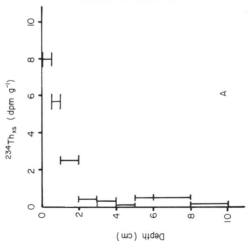

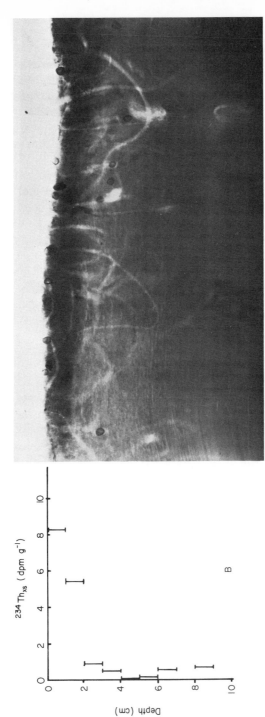

Fig. 40.8. Excess ^{234}Th versus depth in sediment column and X-radiographs of box cores. A. Long Island Sound (Aller and Cochran, 1976). X-ray shows protobranch bivalve in feeding position (left) and relict burrow of *Cerianthus* sp. (centre). B. New York Bight (X-ray shows core analysed. Cochran and Aller, (unpublished)). X-ray shows numerous small bivalves (*Nucula* sp.), burrows of polychaetes and anemones. Particle mixing coefficients, calculated using Equation (6), are 2×10^{-7} cm^2 s^{-1} for the Long Island core and 3×10^{-7} cm^2 s^{-1} for the New York Bight core.

manganese nodules grow at the rate of ~ 4 mm $(10^6 \text{ year})^{-1}$ and this is supported by independent calculations based on ^{10}Be measurements. Lalou and Brichet (1972) have, however, argued that in some nodules, at least, growth may actually be considerably faster than this commonly accepted rate, and that the radionuclide distributions found in them are due primarily to diffusion or to sampling problems.

The use of ^{231}Pa for dating will not be discussed in detail but it should be noted that it is subject to the same constraints as ^{230}Th. Rosholt et al. (1961) have suggested that ^{230}Th and ^{231}Pa, which are both produced from dissolved uranium isotopes in sea-water, are equally refractory and are therefore deposited in sediments in the same ratio as they are produced. The changing ratios of these nuclides down a core would therefore be expected to be directly related to age; however, this has not been found to be generally the case. Ku (1966) has shown that there is a marked separation of ^{230}Th and ^{231}Pa in deep-sea sediments and Sackett (1966) has demonstrated that manganese nodules have a preference for ^{231}Pa over ^{230}Th. Turekian and Chan (1971) used these observations to assess, by indirect methods, the accumulation rate of manganese nodules which was found to be compatible with the slow-growth model (i.e. ~ 4 mm $(10^6 \text{ year})^{-1}$).

40.3.2.4. Dating of marine deposits on the basis of the ingrowth of ^{230}Th and ^{231}Pa

Certain marine deposits, such as corals, oolites and some types of manganese deposits, are formed so rapidly that the uranium which they incorporate from sea-water has not produced any significant amounts of its refractory daughters ^{230}Th (from ^{234}U) and ^{231}Pa (from ^{235}U). However, with time, ^{230}Th and ^{231}Pa will grow in until secular equilibrium with their respective parents is achieved (Fig. 40.4). These daughters are primarily used for dating those corals (and, more rarely, other suitable carbonate materials) which occur above sea-level as fossil outcrops. Using these nuclides, the existence of a world-wide high sea stand ~ 120000 years ago was established (Veeh, 1966). This was several metres higher than the present sea-level and has been correlated with the general global warming which occurred soon after the last major glacial cycle. Broecker et al. (1968) and Mesolella et al. (1969) have used coral terraces on Barbados to establish the existence of three separate high sea stands which occurred 120000, 105000 and 80000 years ago, and these dates have been confirmed by Ku (1968) using ^{231}Pa. These stands are thought to have originated because Barbados has been rising virtually continuously from the sea.

The same nuclides have been used in dating certain manganese deposits which are growing at faster rates than are the more normal manganese nodules discussed above. Ku and Glasby (1972) used a technique based on these nuclides to demonstrate that each layer can yield a growing-in age, and the rate of increase of these ages with depth in the manganese deposit was then used to calculate a rate of growth. The rapidly growing manganese deposit which they examined was from a near-shore environment, and clearly consisted of manganese that was mobilized from the sediment during diagenesis and subsequently accumulated on an active growth site. Scott *et al.* (1974) have shown similar rapid growth rates for manganese deposits along the Mid-Atlantic Ridge. They ascribe this fast accretion to a supply of hydrothermal manganese, although a diagenetic source from the sediments cannot be ruled out.

40.3.2.5. *Thorium-228: chronometer for sediment accumulation or bioturbation?*

After its formation in sediments, ^{228}Ra, like ^{226}Ra, is rapidly mobilized and diffuses out into the overlying water. Subsequently, its daughter, ^{228}Th, is adsorbed by particulate matter and removed from sea-water. Thus, ^{228}Th excesses are observed in both near-shore and deep-sea sediments (Koide *et al.*, 1973; Thomson *et al.*, 1975; Cochran *et al.*, 1976). The accumulation and decay of excess ^{228}Th in sediments is largely restricted to the top few centimetres of the sediment–water interface, i.e. the zone in which bioturbation is most important. The general differential equation (Equation 6), which includes sedimentation, mixing and decay terms, should be used together with independent information to assess the relative importance of sediment accumulation and bioturbation in any particular locality.

The role of biological mixing is minimal in near-shore sediments accumulating under anoxic (or nearly so) waters, and under these conditions the decay of excess ^{228}Th can be used to establish the sediment geochronology for the previous decade (Koide *et al.*, 1973). In other near-shore environments, such as Long Island Sound, the ^{228}Th profile is complicated by sediment mixing and by ^{228}Ra loss (Thomson *et al.*, 1975). In such instances the profile obviously cannot be used to determine sediment accumulation rates unless it is possible to independently resolve the rates of the other processes with sufficient accuracy. In the uppermost few centimetres of deep-sea sediments, the role of bioturbation (and ^{228}Ra loss) is dominant in controlling the ^{228}Th profile (Fig. 40.9).

40.3.3. RADIUM

Both ^{226}Ra (1620 years) and ^{228}Ra (5·75 years (Reid, 1976)) are produced directly by the radioactive decay of thorium isotopes (^{230}Th and ^{232}Th, respectively). It has been shown above that thorium is very efficiently scavenged from sea-water and it is therefore obvious that there should not be any direct *in situ* production of radium in the oceans. However, both of these radium nuclides are present in the water column at unexpectedly high concentrations. In the deep Pacific, for example, the activity ratio of

^{226}Ra/^{234}U is about 0·1—a significantly high value considering the virtual absence from the water column of the parent thorium nuclide. The identification of the sources of the radium isotopes in the sea and of the controls on their distribution and ultimate disposition provides the framework for their use as clocks. The fact that the source of most of the radium in the oceans is the sediment makes the cycles of these nuclides through the oceans different from those of all of the others discussed in this chapter, with the exception of ^{234}U.

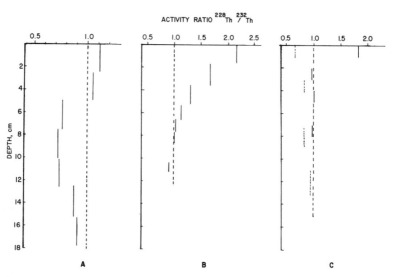

FIG. 40.9. ^{228}Th/^{232}Th activity ratios with depth in near-shore and deep-sea sediments. A. Long Island Sound (Thomson *et al.*, 1975). B. Santa Barbara Basin (Koide *et al.*, 1973). C. Deep-sea sediments. Solid bars = Mid-Atlantic Ridge (Cochran *et al.*, 1976); dashed bars = North equatorial Pacific (Cochran and Krishnaswami, 1977).

40.3.3.1. *Radium loss from sediments*

Early interest in the distribution of ^{226}Ra in deep-sea sediments centred on its use as an index for ^{230}Th, the decay of which could be used to determine the sediment accumulation rate. The pattern of ^{226}Ra distribution in a deep-sea core should follow that of a typical short-lived daughter coming to equilibrium with its longer-lived parent. That is, there should be low ^{226}Ra concentrations near the core top, and these should increase with depth to a maximum and then decrease as a result of decrease of ^{230}Th with depth. However, recoil of ^{226}Ra (and ^{228}Ra) atoms as they are produced from the α-decay of ^{230}Th (and ^{232}Th) provides a mechanism for the introduction of radium into sediment pore waters. Thus mobilized, radium is able to migrate out of the sediment column, although other processes, such as ion exchange, adsorption and authigenic mineral formation, will affect the ultimate flux of radium to the bottom water. Available pore water data (Somayajulu and Church, 1973; Cochran and Krishnaswami, 1977) indicate that both ^{226}Ra and ^{228}Ra concentrations are about an order of magnitude greater in the pore water than in the overlying bottom water. Irregular patterns of ^{226}Ra with depth in the solid phases of the sediments were observed by Pettersson (1951) and Kröll (1954), and led to suggestions that radium diffusion was indeed occurring.

Direct measurements of the ^{226}Ra flux out of sediments have not been made. However, the total flux from the ocean bottom can be assessed from the standing-crop in the world oceans and this is being fairly accurately measured as part of the GEOSECS programme. It is evident, however, that the flux will vary within the various ocean basins because of differences in both the accumulation rates and mineralogies of the sediments.

As the major source of the ^{226}Ra flux from sediments is ^{230}Th adsorbed from the sea-water onto sediment grains, the highest production rate is at the top of the sediment pile because ^{230}Th decreases with depth as a result of radioactive decay. The concentration of ^{226}Ra per unit mass or volume of sediment will be inversely proportional to the rate of sediment accumulation if the same amount of ^{230}Th is rapidly removed from the water column at each equivalent depth in the ocean independently of the number of particles sedimenting. Thus, in an area of slowly accumulating red clay there should be a higher flux of ^{226}Ra than there would be from a more rapidly accumu-lating calcareous area.

Using measured sediment accumulation rates and calculated ^{226}Ra deficiencies, Cochran et al. (1976) and Cochran and Krishnaswami (1977) tested this theory using data from a North Atlantic carbonate core and a clay-rich core taken north of the Equator in the Central Pacific (see Table 40.2). The red clay which had accumulation rates of 0.14–0.30 cm $(10^3$ year$)^{-1}$ had a ^{226}Ra flux of ~ 0.2 dpm cm^{-2} year^{-1}, whereas the calcareous sediment which had an accumulation rate of 2.9 cm $(10^3$ year$)^{-1}$ yielded a flux of 0.006 dpm cm^{-2} year^{-1}. Because its half-life is very much shorter than that of ^{226}Ra it is difficult to make similar measurements for ^{228}Ra. However, evidence for its release from both deep-sea and near-shore sediments does exist (Moore, 1969a, b; Thomson et al., 1975; Cochran et al., 1976), but only the crudest calculations of flux have been made from such information.

In principle it should be possible to estimate the flux of ^{226}Ra to the oceans from sediments if the sediment accumulation rates are coupled with the radium fluxes. The application of this approach is complicated by the fact that there may be radium "traps" in the sediment which would tend to inhibit its loss to the overlying water. On the basis of similarities of their chemical behaviour,

TABLE 40.2

Radium fluxes from deep-sea sediments

Core	Location	Sediment type	Sed. rate (cm $(10^3$ year)$^{-1}$)	ρ (g cm^{-3})	Ra flux (dpm cm^{-2} year^{-1})	Reference
Monsoon 49G	Indian Ocean	Siliceous ooze	0·28	0·3 (assumed)	0·07	Goldberg and Koide (1963)
Monsoon 57G	Indian Ocean	Calcareous ooze	0·30	0·7 (assumed)	0·17	Goldberg and Koide (1963)
DOMES A47-16	N. eq. Pacific	Siliceous ooze–red clay	0·15	0·30	0·12	Cochran (1978)
DOMES B52-39	N. eq. Pacific	Siliceous ooze–red clay	0·31	0·34	0·23	Cochran (1978)
DOMES C57-58	N. eq. Pacific	Siliceous ooze–red clay	0·14	0·46	0·25	Cochran (1978)
FAMOUS 527-3	Atlantic (Mid-Atlantic Ridge)	Calcareous ooze	2·9	0·76	0·006	Cochran et al. (1976)

it is probable that the most likely radium traps will be phases containing barium. The two minerals which have been best studied in connection with radium migration are phillipsite (Bernat and Goldberg, 1969; Bernat et al., 1970) and barite (Church and Bernat, 1972; Borole and Somayajulu, 1977), both of which occur commonly in the South Pacific: the phillipsite in areas of intraplate volcanism and the barite in areas of high productivity and along the East Pacific Rise (see Cronan (1974) for a review). High ^{228}Th/^{232}Th ratios have been observed in both of these phases, and the simplest explanation of this is that migrating ^{228}Ra became incorporated into the mineral as it formed and later decayed to its granddaughter ^{228}Th. In fact, mass balance calculations based on the ^{228}Th/^{232}Th data of Bernat et al. (1970) and Church and Bernat (1972) suggest that 40–70% of the ^{228}Ra produced in the sediment from the dominant ^{232}Th-bearing clay minerals is associated with authigenic barite or phillipsite. Borole and Somayajulu (1978) have found 300–800 dpm ^{226}Ra g^{-1} and ^{226}Ra/^{230}Th activity ratios of ~ 20 in barite separated from a Pacific deep-sea core, which demonstrates that ^{226}Ra is also incorporated in this mineral. A consequence of the existence of these radium traps in sediments is that the radium flux may be substantially lowered in regions in which such traps occur, even though the sediment accumulation rate is low.

40.3.3.2. *Time scales derivable from the distribution of the radium isotopes in the oceans*

Once it was realized that there is a significant flux of both radium isotopes from the ocean floor into the water column, attention was given to the horizontal and vertical oceanic distributions of these nuclides, particularly ^{226}Ra, in the oceans. The broad features of the distribution of ^{228}Ra are also known but, because of its short half-life and the difficulty of sampling, this nuclide has not been extensively used in large-scale studies of ocean circulation. However, it is possible that its usefulness may come to rival that of ^{226}Ra for such purposes.

The use of ^{226}Ra as a chronometer for oceanic circulation is based on the assumption that the ocean bottom is the primary source of this nuclide. Thus, if the radium flux from the ocean bottom is assumed to be the same everywhere, then the concentration would decrease from the bottom of the ocean upwards. The distribution profile could then, in principle, be analysed using an advection–diffusion model (Equation 3), or even a box model, to arrive at the time-dependent mixing parameters. This analysis is made possible by the fact that ^{226}Ra is a clock having a half-life comparable to the mixing times of the oceans. It has, indeed, been found that ^{226}Ra concentrations do increase in the oceans with depth (Fig. 40.10), but then so do those of many other elements, e.g. the micronutrients. For phosphate, nitrate and silicate the distribution patterns are clearly ascribable to their extraction from surface waters by organisms and their transfer and regeneration at depth; this effectively yields steady-state profiles in which the concentration increases from low values at the surface to much higher ones at depth. A similar pattern has also been demonstrated for barium, first by Chow and Goldberg (1960), and then with ever increasing accuracy by others. The

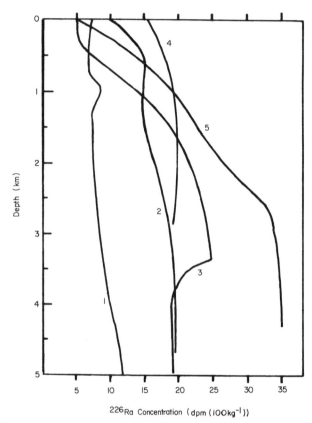

^{226}Ra Concentration (dpm (100kg^{-1}))

FIG. 40.10. ^{226}Ra concentrations (dpm(100 kg)$^{-1}$) versus depth in the ocean (after Chung, 1974). 1 = NW. Atlantic; 2 = Antarctic (59°S.); 3 = SW. Pacific; 4 = Antarctic (65°S.); 5 = NE. Pacific.

culmination of this was the extensive measurements made together with those of ^{226}Ra as part of the GEOSECS programme (for a summary see Chan et al., 1976). The GEOSECS data show that throughout most of the ocean basins the ^{226}Ra distribution, because of its evident biological involvement, cannot be used as a simple tracer of circulation processes. This means that once again information about sources and sinks in the oceans has been partially obscured by the great circulation and biological cycles.

A great deal of careful analysis of the abundant data is therefore required before the distribution of ^{226}Ra can be related to the large-scale oceanic circulation; however, it is possible to make two generalizations.

First, although Ra and Ba are strongly correlated throughout the oceans, there is a marked departure from this in the bottom waters of the eastern North Pacific, implying that in this part of the ocean the radium flux from the bottom is sufficiently high to overcome the effect of homogenization. This phenomenon again raises the question of whether different types of sediment yield different radium fluxes. The first order control on the radium flux for a given depth of water (and thus the ^{230}Th supply) is the rate of sedimentation. The second order control must be the releasability of the radium from the sediment. On the basis of the ^{228}Th distribution patterns in barite- and phillipsite-bearing deposits (see above) it is evident that at least half of the radium produced by decay in a deep-sea sediment must move through the pore waters. However, the concentrations of ^{226}Ra in pore waters are low relative to ^{226}Ra produced from *in situ* ^{230}Th decay (Somayajulu and Church, 1973; Cochran and Krishnaswami, 1977), probably because its standing-crop in the pore waters will depend on the efficiency with which it is sequestered by the mineral phases. Since the rate of diffusion of ^{226}Ra depends on the concentration difference between the pore water and the overlying water, this difference acts as a constraint on the flux.

Second, Atlantic surface waters have ~ 7.4 dpm ^{226}Ra $(100\ \text{kg})^{-1}$ (Broecker *et al.*, 1976), whereas the Pacific surface waters have ~ 6.3 dpm ^{226}Ra $(100\ \text{kg})^{-1}$ (Chung, 1976). This difference is remarkable when it is borne in mind that the deep waters of the Pacific have a radium concentration at least twice as great as those of the Atlantic. There are several possible explanations for these differences.

(i) The Atlantic receives more than half of the world's river run-off, and this could supply an incremental flux of radium to the Atlantic Ocean surface. The size of this river contribution can be estimated as follows: if half the world river discharge is into the Atlantic, then 0.02 litre cm^{-2} year^{-1} would be delivered to its surface. Assuming that the maximum river concentration of ^{226}Ra is 0.2 dpm litre^{-1}, then 0.004 dpm ^{226}Ra cm^{-2} year^{-1} would be delivered to this ocean. If the deep ocean is upwelling at a rate of 4 m year^{-1} and the deep water is assumed to have a concentration of 10 dpm $(100\ \text{kg})^{-1}$, then 0.04 dpm ^{226}Ra cm^{-2} year^{-1} would be supplied to the surface from this source. Thus, on this basis, rivers would increase the supply of ^{226}Ra to the Atlantic by 10% over that to the Pacific if no radium were supplied to the Pacific by streams, and the upward flux of radium was the same as in the Atlantic. However the actual excess in the Atlantic is $\sim 20\%$.

(ii) The average productivity in the Pacific Ocean is $\sim 20\%$ greater than that in the Atlantic and this is likely to lead to more efficient removal of radium from the surface. However, it is very difficult to assess the significance of this factor at the present time.

(iii) Because the surface waters in both oceans circulate through the ^{226}Ra-rich water around Antarctica, the journey time for the Atlantic surface waters is about 20% less per unit volume of water than it is for the Pacific surface waters.

The nuclide ^{228}Ra (half-life 5·75 years), like ^{226}Ra, is produced in sediments by the decay of a thorium parent, in this instance ^{232}Th; this is the most abundant of the thorium isotopes and is supplied to the oceans in the lattices of clay minerals in which it occurs at an average concentration of ~ 15 ppm. On the basis of an argument similar to that used for ^{226}Ra, Moore, (1969a) established that ^{228}Ra also migrates out of bottom sediments. However, because of its shorter half-life it is affected by oceanic mixing to a more limited extent than is ^{226}Ra. As a consequence, ^{228}Ra is most useful (i) for tracing the rapid surface circulation of the oceans, and (ii) as an extension of the use of ^{222}Rn in determining bottom-ocean circulation parameters (see below).

In the oceans the concentration of ^{228}Ra (Trier et al., 1972; Sarmiento et al., 1976) is high in both the surface and near-bottom waters, with near-zero values at mid-depths (Fig. 40.11). The high values near the bottom are a direct result of the flux of the nuclide from the sediments, and the decrease away from the bottom is a function of upwelling, vertical eddy diffusion and radioactive decay. The high surface values can be explained in terms of the addition of ^{228}Ra to shallow coastal waters, primarily from sediments, but also in part from continental run-off. The influence of this run-off decreases away from the continental landmasses (see Fig. 40.12) and the ^{228}Ra distribution profile can be used to derive the horizontal eddy diffusion coefficient for surface waters (Kaufman et al., 1973).

40.3.3.3. Geochronometric uses of the radium isotopes in marine deposits

Although ^{226}Ra has been used as an index for ^{230}Th in the dating of deep-sea deposits, this cannot be considered to be its primary use in chronology. The involvement of radium in biological cycling in the water column suggests that if the phases which carry it reach the bottom in sufficient quantity, there will be an excess of this nuclide in the sediment. Koide et al. (1976) have suggested that the excess ^{226}Ra which they found in a sediment core taken in the San Clemente Basin off California resulted from this mechanism.

Decay of excess ^{226}Ra is useful for determining the sediment accumulation rates for sediments deposited during the last 6000 years, thus making it applicable to sediments which deposit at rates intermediate between those of the near-shore and deep sea. Essential requirements for the method are that the ^{226}Ra/^{230}Th ratio in the accumulating sediment should be substantially greater than 1·0, that the sediment accumulation rate and ^{226}Ra/ ^{230}Th activity ratio of freshly deposited sediment be constant with time and

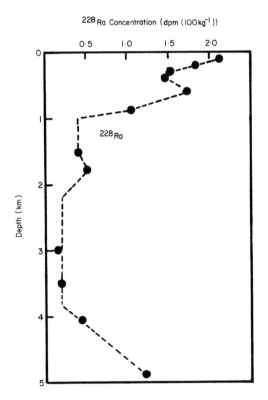

FIG. 40.11. ^{228}Ra concentration (dpm(100 kg)$^{-1}$) versus depth in the North Atlantic (after Trier *et al.*, 1972).

that the excess ^{226}Ra is immobile. This latter assumption is reasonable if the mobility of radium depends primarily on the α-recoil during its production. This process would not affect the excess ^{226}Ra brought to the sediment from the overlying water. The technique appears to be a promising one for the dating of sediments from coastal areas of high productivity (e.g. upwelling areas of the west coast of the United States and South America) and possibly for sediments from some lakes.

Radium isotopes, in particular ^{228}Ra, in the calcium carbonate skeletons of certain marine organisms are also useful for obtaining skeletal growth rates and in evaluating the magnitude of the radium flux from the bottom sediments. The underlying assumption in this dating application is that

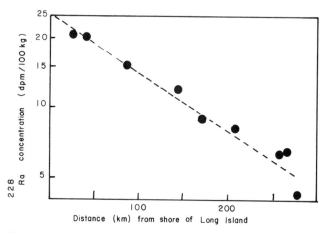

FIG. 40.12. ^{228}Ra concentrations (dpm(100 kg)$^{-1}$) in surface waters away from the coast. The line is based on a horizontal eddy diffusion coefficient of 10^6 cm^2 s^{-1} (after Kaufman et al., 1973).

radium is incorporated into the $CaCO_3$ lattice in the same ratio to Ca as that in the ambient sea-water (see Moore, 1969b; Blanchard and Oakes, 1965). The concentrations of radium in a number of $CaCO_3$-secreting organisms from both shallow waters and the deep sea are shown in Table 40.3. The values for ^{228}Ra shown are those in excess of that which would be in secular equilibrium with the ^{232}Th content, although the latter fraction can be assumed to be small. The radium/calcium ratio in calcareous tests does not depart significantly from that in sea-water. Suspension-feeding bivalves, such as *Mercenaria*, the wood-borer *Xylophaga* and epifauna such as the deep-sea ahermatypic corals contain ^{228}Ra concentrations similar to those of near-shore and near-bottom deep-sea waters (Figs 40.11, 40.12). Deposit-feeding molluscs (such as *Tindaria*) appear to derive the cations for their shells from pore water, which has a higher ^{228}Ra concentration than the overlying bottom water (Somayajulu and Church, 1973; Cochran and Krishnaswami, 1977).

If the radium concentrations shown in Table 40.3 are representative of those in the water from which the animal is secreting its shell, then analysis of the epifauna will serve to determine the concentration of radium at the

TABLE 40.3

Radium isotopes in molluscs and corals

Sample	Location	Water depth (m)	$^{228}Ra_{ex}$ (dpm(gCa)$^{-1}$)	^{228}Ra ^{226}Ra	Sea-water equivalent ^{228}Ra (dpm(100 kg)$^{-1}$)	^{226}Ra	Reference
Atlantic Coast							
Bivalve (growing edge) *Mercenaria* sp.	Cape Cod, Mass.	<10	0·20	—	8·4	—	Turekian *et al.* (1975)
Mollusc	Montauk Pt., Long Island	<10	0·08	0·08	3·3	3·3	Moore (1969b)
Mollusc	Miami, Fla.	<10	0·05	0·10	2·1	4·2	Moore (1969b)
Mollusc	Biloxi, Miss.	<10	0·28	0·16	12	6·8	Moore (1969b)
Pacific Coast							
Mollusc	La Jolla, Calif.	<10	0·08	0·11	3·1	4·5	Moore (1969b)
Central Atlantic							
Hermatypic coral *Montastrea annularis*	Jamaica	<10	0·10	0·15	4·3	6·4	Dodge and Thomson (1974)
Hermatypic coral *Diploria labrynthiformis*	Bermuda	<10	0·07	0·15	2·8	6·4	Dodge and Thomson (1974)
Central Pacific							
Mollusc	Tahiti	<10	0·003	0·10	0·10	4·2	Moore (1969b)
Deep Sea							
Ahermatypic coral (1·2 cm)	S. Atlantic	1000	0·14	0·48	5·9	20	Cochran (unpublished)
Bivalve (wood-boring) *Xylophaga* sp.	N. Atlantic	2800	0·09	—	4·0	—	Cochran (unpublished)
Bivalve (1·7 mm size class) *Tindaria callistiformis*	N. Atlantic	3800	0·59	—	25	—	Turekian *et al.* (1975)

sediment–water interface. This is particularly important for ^{228}Ra, which shows a strong concentration gradient away from the bottom in the deep sea. This interfacial ^{228}Ra concentration, which is not readily obtainable by conventional water sampling, may be used together with water column data to place limits on the ^{228}Ra flux from the bottom.

An important application of ^{228}Ra to $CaCO_3$-secreting organisms, which is not obvious from Table 40.3, is its use as a chronological indicator for the determination of the growth rates of calcareous marine organisms. Two approaches to this are possible. One, which was followed by Moore and co-workers (1972, 1973, 1974) and by Dodge and Thomson (1974), involves the growth increment sectioning of an individual specimen (such as coral) and the analysis of the sections for ^{228}Ra (or its daughter ^{228}Th). The resultant decay curve with distance from the surface in the coral is then used to derive the growth rate. The alternative approach, which can be used in those instances in which the individual samples are too small to be sectioned effectively (e.g. those of deep-sea assemblages), is to analyse several size fractions of a single population (Turekian *et al.*, 1975). The variation in the activity between the size fractions is then assumed to be a function of the rate of addition of new material and as well as the rate of radioactive decay.

40.3.4. LEAD-210

The decay of ^{222}Rn leads, via a sequence of short-lived daughters, to the production of ^{210}Pb which has a half-life of 22 years. This nuclide has become of particular interest, both as a clock for measuring time over the past century in lake and marine deposits and as a tracer for the heavy metals in earth surface reservoirs.

40.3.4.1. *The supply of ^{210}Pb to the ocean from the continents*

When ^{210}Pb is produced in ground-water by the decay of ^{222}Rn, it is soon removed by adsorption onto surfaces and very little of it makes it way into river-water (Benninger *et al.*, 1975). Under unusual conditions of intense acid attack on rocks, such as those which occur during the strip mining of coal or sulphide-rich deposits, ^{210}Pb is solubilized, together with manganese and iron, and for as long as the stream retains a low pH (<4) both the ^{210}Pb and the manganese will remain in solution. If, however, the stream is neutralized, both manganese and ^{210}Pb precipitate rapidly. The existence of this pathway for ^{210}Pb was first demonstrated by Rama *et al.* (1961) for the Colorado River in Arizona, and it has been more recently observed by Lewis (1976) for the west branch of the Susquehanna River in Pennsylvania. Both of these studies confirm the observation that ^{210}Pb does not remain in solution in river-water. Because very little ^{210}Pb is supplied to streams, and because that which is, is easily sequestered by particles, it is evident that rivers are not a major pathway of this nuclide to coastal oceanic reservoirs.

The major source of ^{210}Pb to the surface veneer of the continents and oceans is the atmosphere in which radon emanating from the continents decays to ^{210}Pb which is precipitated out to the Earth's surface. Most of the ^{210}Pb which falls on land is retained by the top soil because it is strongly chelated by the organic compounds produced by the decay of plant remains, and virtually none of it is transported to rivers. A small portion of it is, however, transported in association with particles liberated during bank erosion by the rivers. Benninger et al. (1975) and Lewis (1976) have calculated that the mean residence time of lead in soils of the north-eastern United States is of the order of thousands of years. It is apparent, therefore, that there is no significant flux of either dissolved or particulate ^{210}Pb to the oceans from the continents. In this context, Benninger (1976) has shown that virtually the entire ^{210}Pb budget in a shallow (~ 20 m) coastal area, such as Long Island Sound, is dependent on the atmospheric flux.

40.3.4.2. The scavenging of ^{210}Pb in coastal waters

The ^{210}Pb concentration in coastal waters is lower than that in the surface waters of the open ocean. This can be ascribed to the efficient scavenging of ^{210}Pb by the sustained high concentrations of particles arising from the high biological productivity in the water column and the physical disruption of the sediments. For these reasons, the ^{210}Pb concentrations in the waters of areas such as Long Island Sound (Benninger et al., 1975; Beninger, 1976) the Santa Barbara Basin (Krishnaswami et al., 1975) and the Gulf of California (Bruland et al., 1974) are virtually zero.

40.3.4.3. Lead-210-dating of coastal sediments

The intense scavenging of primarily atmospherically-derived ^{210}Pb in coastal regions has encouraged the investigation of the use of this nuclide as a tool for dating the rapidly depositing sediments of such areas. The first attempt to use this approach was made by Koide et al. (1972) who examined the annually varved sediments of the Santa Barbara Basin and showed that, when sediments accumulate in a relatively undisturbed fashion, without strong physical or biological post-depositional mixing, a true ^{210}Pb radioactive decay curve can be obtained with depth. This is generally true only for sediments which have been deposited in contact with bottom waters which are either anoxic or have an oxygen content sufficiently low to inhibit bioturbation. However, bioturbation of the sediment may occur in areas in which the overlying waters have higher oxygen concentrations and this will perturb the time record. If the sediment accumulation rate is constant, and the rate of bioturbation is also constant, then the distribution of ^{210}Pb will show an exponential decrease with depth and yield an apparent sediment accumulation rate that is faster than the true one. If mixing is limited to a certain depth, and both the sedimentation and ^{210}Pb supply rates are fast enough, then the measurable excess ^{210}Pb concentration below the mixing layer will be subject only to radioactive decay and the sediment accumulation rate can be determined from the depth distribution of ^{210}Pb in the core below the mixing zone (Bruland, 1974). Shokes (1976) has shown that this is attained in deeper water around the Mississippi delta. If, however, the mixing zone itself is not sharply defined on a long time scale, even the deep record can be perturbed and will yield an incorrect sediment accumulation rate.

Benninger et al. (1976) have given examples of each of these limiting conditions. These indicate that even where there is no physical disturbance of the record a range of different types of decay curves can be obtained for coastal sediment overlain by oxygenated water. In such cases it is clear that the ^{210}Pb is of more value as a tracer of the disturbance of the record than as a chronometer for sediment accumulation.

The one type of coastal deposit which appears to be free of this constraint is the salt marsh. McCaffrey and Thomson (1978) have shown that the ^{210}Pb chronology of a Connecticut salt marsh is compatible with that inferred from regional sea-level curves derived from tide gauge data (Fig. 40.13). The importance of this observation may lie not so much in the use of this nuclide in chronology as in the possibility of using the salt marsh record as a detector of changes in the atmospheric flux of metals (McCaffrey, 1977).

40.3.4.4. *The scavenging of ^{210}Pb in open-ocean waters*

In the open oceans ^{210}Pb is supplied to the surface primarily from atmospheric precipitation, but a significant amount is also produced directly from the radium dissolved in the water column. Lead-210 is scavenged from both the surface waters (Nozaki et al., 1976; Bacon et al., 1976) and the deep-water column (Craig et al., 1973; Bacon et al., 1976; Nozaki and Tsunogai, 1976; Somayajulu and Craig, 1976; Thomson and Turekian, 1976). The calculated mean residence time of ^{210}Pb in the surface waters of most of the North Pacific is about 1·6 years, but it decreases sharply in the high productivity boundary regions (Nozaki et al., 1976). In the deep ocean there is a deficiency of ^{210}Pb relative to ^{226}Ra (Fig. 40.14) which is interpreted as reflecting the scavenging of the former by particles. Mean residence times for ^{210}Pb relative to particulate removal, calculated either from one dimensional advection–diffusion models or from box models, range from 50 to 100 years (Table 40.1).

The exact mode by which ^{210}Pb is removed is as yet unknown. However, recent work by Somayajulu and Craig (1976) suggests that the active scavenging agents are particles having a diameter of $\sim 4\,\mu m$. In contrast,

Bacon *et al.* (1976) believe that ^{210}Pb is scavenged primarily at the sediment–water interface, suggesting that horizontal transport of newly formed ^{210}Pb is of more importance than is its vertical scavenging by particles. It should be borne in mind that scavenging may also be effected by particles having high Stokes' law settling velocities which would often escape sampling.

The recent finding of ^{210}Pb in the tops of some deep-sea cores retrieved

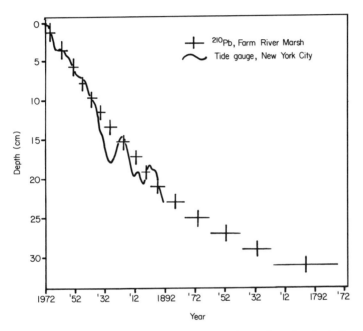

FIG. 40.13. ^{210}Pb age as a function of depth in a salt marsh core from Branford, Connecticut (after McCaffrey and Thomson, 1978). The tide gauge data show the rise in sea-level. The agreement with the ^{210}Pb data suggests that the surface of the marsh rises with sea-level.

by a submersible (Cochran *et al.*, 1976; Nozaki *et al.*, 1977) shows a possible way of resolving whether the horizontal or the vertical mechanism is the more important method for the removal of ^{210}Pb from the oceans. The distribution of ^{210}Pb in one such core from the FAMOUS site on the Mid-Atlantic Ridge is shown in Fig. 40.15. The standing-crop of ^{210}Pb deduced from this study is slightly greater than that predicted from the vertical supply of ^{210}Pb to that site from both atmospheric and deep-water sources. Similar measurements are at present being made in the Pacific Ocean, and preliminary results indicate that ^{210}Pb scavenging is also vertically effected. The effect of bioturbation in deep-sea environments is also shown in

Fig. 40.15. The calculated diffusion-modelled mixing coefficient is $0.6 \times 10^{-8} \, cm^2 \, s^{-1}$, with a definite mixing horizon of $\sim 8 \, cm$ (inferred from radiocarbon data, see Fig. 40.16). This provides the first accurate estimate of the rate of bioturbation in the deep sea, although similar, but less accurate, values have been obtained from geological (tektites) or man-made (^{244}Pu) spikes (Guinasso and Schink, 1975).

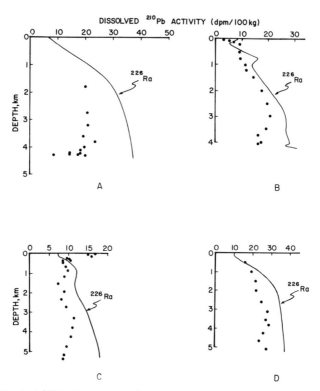

FIG. 40.14. Dissolved ^{210}Pb (dpm$(100 \, kg)^{-1}$) versus depth in water column. Solid curve represents ^{226}Ra distribution. A. NE. Pacific (Craig et al., 1973). B.S. Pacific (Thomson and Turekian, 1976). C. N. Atlantic (Bacon, 1976). D. NW. Pacific (Nozaki and Tsunogai, 1976).

40.4. THE COSMOGENIC NUCLIDES

The possibility of dating marine deposits using radionuclides produced in the atmosphere by the action of cosmic rays was exploited soon after the identification of the production and pathways of ^{14}C by W. F. Libby. Since then, ^{14}C has been extensively used for dating deep-sea sediments and the results have focused on two important features of the deposition of such sediments. These are: (i) the changes which have occurred in the accumulation rates of the various components of deep-sea deposits since the start of the last ice-age, and (ii) the role of biological and physical reworking of deep-sea sediments in influencing the reliability of ages obtained by measurements of both ^{14}C and other radionuclides. Once the success of ^{14}C had been

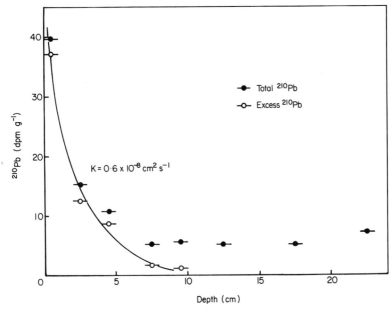

FIG. 40.15. Total and excess ^{210}Pb (dpm g^{-1}) versus depth in a calcium carbonate rich core from the Mid-Atlantic Ridge (FAMOUS 527-3) (after Nozaki *et al.*, 1977). The data are fitted by the solution to Equation (6) which yields a biological mixing coefficient of 0.6×10^{-8} cm^2 s^{-1}. See text for discussion and further information on this core.

established, other cosmogenic nuclides were also examined to assess their value in marine studies. The cosmogenic nuclides most often used in marine studies are listed in Table 40.4. Their use has encompassed not only sediments but also, where possible, the water column (see also Chapter 18).

40.4.1. SILICON-32

This cosmogenic nuclide was first identified in the marine system by Lal *et al.* (1960) and since that time there have been a large number of measurements for siliceous sponges (Lal *et al.*, 1976) and some for sea-water samples

(Somayajulu, 1969; Krishnaswami *et al.*, 1972a; Somayajulu *et al.*, 1973). From these data it is apparent that ^{32}Si may be useful not only as a chronometer for tracing oceanic mixing, but also for studying processes involving the emplacement of biogenic siliceous sediments. Such applications have been hindered somewhat by an uncertainty about the half-life of ^{32}Si, estimates of which range from 200 to 700 years (Lederer *et al.*, 1967), with the best current value at 300 years (Jantsch, 1967; Clausen, 1973).

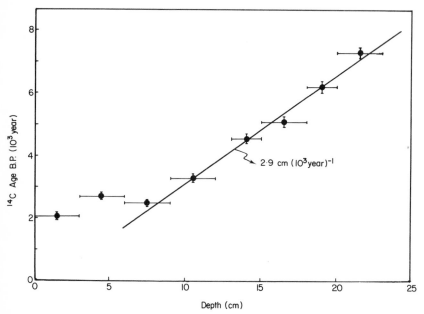

FIG. 40.16. ^{14}C age versus depth for FAMOUS core 527-3 (Mid-Atlantic Ridge) (after Nozaki *et al.*, 1977).

There have been only a few attempts to apply ^{32}Si to the dating of marine sediments. The depth and intensity of bioturbation are limited in those sediments accumulating under oxygen-depleted water, and the decrease in ^{32}Si with depth can then reasonably be assigned to radioactive decay alone. Certain sediments from the Gulf of California fulfil this condition, and data from them, coupled with some from siliceous sponges from the same region, yield accumulation rates within about an order of magnitude of those obtained from varve counting and radiocarbon techniques (Kharkar *et al.*, 1963; Calvert, 1966).

For deep-sea sediments, Kharkar *et al.* (1969) compared data for ^{32}Si and ^{230}Th in some Antarctic siliceous cores. They reported a ^{32}Si accumulation rate of ~ 47 cm $(10^3 \, \text{year})^{-1}$ for the cores for which the average ^{230}Th accumulation rate was 0·6 cm $(10^3 \, \text{year})^{-1}$. Their interpretation was that if, indeed, both of these were true sedimentation rates, then the rapid rates shown by the ^{32}Si data for the last 2000 years were short-term ones, and sufficient episodic losses had occurred on the longer time scales involved in the ^{230}Th measurements. These losses decreased the long-term average to 1/80 of the rapid rate. However, bioturbation can seriously affect the interpretation of an exponential decrease with depth in a core of a constantly

TABLE 40.4

Cosmogenic nuclides

Nuclide	Half-life
^{7}Be	52·9 days
^{3}H	12·3 years
^{32}Si	300 years
^{14}C	5730 years
^{10}Be	$1·5 \times 10^6$ years
^{53}Mn	$3·7 \times 10^6$ years
^{26}Al	$0·75 \times 10^6$ years

supplied radionuclide and, according to D. J. De Master (Yale University, personal communication), this may explain the high ^{32}Si data of Kharkar *et al.* The mixing coefficient (Equation 6) required to produce the observed ^{32}Si profiles is $\sim 10^{-6} \, \text{cm}^2 \, \text{s}^{-1}$ (for a true sedimentation rate of 0·6 cm $(10^3 \, \text{year})^{-1}$). The difficulties in accepting such an explanation are that mixing depths of 20–45 cm are required and that the calculated specific activity (dpm ^{32}Si $(\text{kg SiO}_2)^{-1}$) of the silicon supplied to the sediment–water interface is greater than that actually observed in sponges in the area and in deep water of Antarctic origin. Clearly, further work is required to resolve these uncertainties. However, it does appear that ^{32}Si may be a powerful tool for investigating biogenic mixing processes in siliceous deep-sea sediments.

Water column data for ^{32}Si have been obtained both using *in situ* extraction methods (Schink, 1962; Somayajulu, 1969; Krishnaswami *et al.*, 1972a), and from analyses of siliceous sponges (Kharkar *et al.*, 1966; Lal *et al.*, 1970, 1976). Open-ocean profiles of ^{32}Si (Somayajulu *et al.*, 1973) show that the absolute activities are low in surface waters and increase with depth. This distribution is a consequence of the biological cycle which affects both ^{32}Si and stable silicon. However, ^{32}Si specific activity decreases with depth. Detailed modelling of such profiles must await more intensive sampling, although residence times in deep waters have been determined from the existing data using a box model (Kharkar *et al.*, 1966; Lal, 1969) or a one-dimensional advection–diffusion model (Somayajulu, 1969).

40.4.2. CARBON-14

The use of radiocarbon for dating marine deposits has, by itself, been one of the most important developments in the study of the events of the last glacial age. The half-life of ^{14}C is long enough (5730 years) to allow accurate dates to be resolved even when the accumulation rates vary in a core. This was first shown by Broecker *et al.* (1958) and carbon-14 data have subsequently been used either directly, or in conjunction with palaeontological and oxygen isotope stratigraphy, to assist in the interpretation of the sedimentary events of the last 35 000 years.

A particular problem inherent in the use of radiocarbon in the study of deep-sea cores is that the top of the core is frequently found to have too old an age. However, two recent investigations have shed some light on this, and both Peng *et al.* (1977) and Nozaki *et al.* (1977) have shown that the apparently anomalously old age of the tops of some cores is the result of bioturbation rather than of sediment loss. Peng and his co-workers studied a piston core from the Indian Ocean, whereas Nozaki *et al.* used a short core obtained from the Mid-Atlantic Ridge by a research submersible under visually controlled conditions. Both cores showed a constant ^{14}C age in the top few centimetres, and the Mid-Atlantic Ridge data are illustrated in Fig. 40.16. If this constancy is due to a continuous mixing by organisms down to some particular depth, then a consideration of the ^{14}C balance in the core yields a relationship between the depth of mixing, the rate of accumulation and the ^{14}C age for the mixed layer. This relationship is shown graphically in Fig. 40.17, and a knowledge of two of these parameters suffices to determine the third. When this approach is applied to the data for the Mid-Atlantic Ridge core (Fig. 40.17), using the sedimentation rate determined from ^{14}C data below the mixed layer in conjunction with the observed mixing depth, a mixed layer age of 2400 years is obtained which is in agreement with the observed age. The mixing coefficient (Equation 6) cannot be determined on the basis of the ^{14}C data, but shorter-lived nuclides such as ^{210}Pb (see above) are useful for this purpose. Indeed, the use of ^{14}C and ^{210}Pb data for a core, together with Fig. 40.17, would enable any loss of core top during sampling to be assessed. It would also enable it to be established whether or not the constant ^{14}C age of the surface layers is the result of artificial homogenization during sampling or of a natural process, such as bioturbation. In some instances, the surface layers of sediment piles are lost naturally due to bottom transport processes, in contrast to loss during coring. Unless the natural event occurred long enough ago (>100 years) for the ^{210}Pb profile to be re-established, it is not possible to distinguish between these two possibilities.

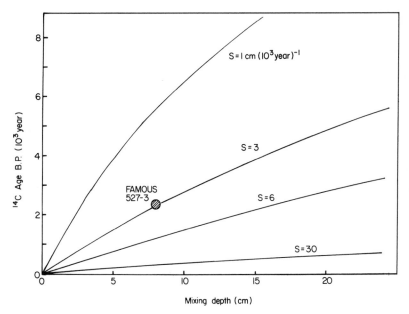

FIG. 40.17. ^{14}C age in the mixed zone versus the mixing depth for different sediment accumulation rates (after Nozaki *et al.*, 1977).

40.4.3. BERYLLIUM-10 AND ALUMINIUM-26

These two cosmogenic nuclides, ^{10}Be (half-life 1.5×10^6 year) and ^{26}Al (0.75×10^6 year), are supplied primarily by production in the atmosphere. Although it was thought at one time that their supply might be dominated by meteoritic dust, measurements of the ^{10}Be/^{26}Al ratios in a Greenland ice-cap core (McCorkell *et al.*, 1967) and in deep-sea sediments (Reyss *et al.*, 1976) corresponded with those expected for production in the atmosphere rather than those found for meteorites. Indeed, the only radionuclide in surface deposits which is known to be uniquely associated with meteoritic (or "cosmic") dust is the ^{53}Mn which is found in Antarctic ice (Bibron *et al.*, 1974).

Aluminium-26 has a very low abundance and because it is difficult to measure it has not been exploited as a dating tool. In contrast, ^{10}Be is about 50 times more abundant, and can be accurately measured under appropriate conditions. The earliest estimations of ^{10}Be in deep-sea sediments were made by Arnold (1956), Goel *et al.* (1957) and Merrill *et al.* (1960) who showed that it was feasible to make accurate determinations of this nuclide in marine deposits. However, it became evident that its decay is too slow to permit its use for the relatively short period dating of normal deep-sea cores. Following this work ^{10}Be has been used in two important ways.

(i) In the determination of the growth rates of manganese nodules. It was shown above that the growth rates of manganese nodules, derived from ^{230}Th measurements, are ~ 4 mm $(10^6 \text{ year})^{-1}$. Recently, doubt has been cast on the validity of this very slow accumulation rate since it has been suggested that the ^{230}Th distribution in nodules is the result of diffusion rather than of radioactive decay (Lalou and Brichet, 1972). However, accumulation rates based on ^{10}Be measurements in nodules (Fig. 40.18) con-

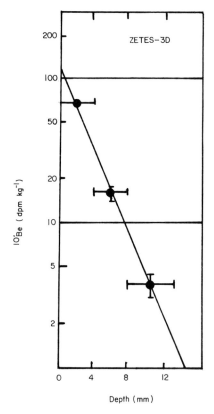

FIG. 40.18. ^{10}Be concentration (dpm kg^{-1}) versus depth for a N. Pacific manganese nodule. The data yield a growth rate of 0·8 mm $(10^6 \text{ year})^{-1}$ (Krishnaswami *et al.*, 1972b).

firm this slow rate (Krishnaswami *et al.*, 1972b; Bhat *et al.*, 1973) suggesting that diffusion of thorium is of minor importance in determining the distribution of ^{230}Th in many nodules. Clearly, additional work is required utilizing both ^{10}Be and ^{230}Th measurements on the same nodules to identify normal growth and anomalous growth patterns.

(ii) In the assessment of long-term sediment accumulation rates. Amin *et al.* (1966, 1975) have dated deep-sea cores using essentially single measurements on the assumption that ^{10}Be can be treated as a virtually stable isotopic tracer. If the supply rate to the bottom of the ocean is known, then the ^{10}Be concentration will be directly related to the sediment accumulation rate. Inoue and Tanaka (1976) have related changes in the ^{10}Be concentration down a core in the western Pacific to different degrees of dilution by volcanic debris and Somayajulu (1977) has related it to climatic variations.

Recently, Finkel *et al.* (1977) have tried to relate the ^{10}Be flux delivered to the bottom of the ocean to the latitude-dependent tropospheric penetration of ^{10}Be from the stratosphere (Fig. 40.19); they were able to use this approach since the fall-out patterns of nuclides such as ^{90}Sr have been fairly well established. The actual ^{10}Be flux to the ocean bottom must be attenuated in the Arctic Ocean at the present time because of the virtually continuous ice cover. Because ^{10}Be accumulates on the sea-ice it is in part transferred from

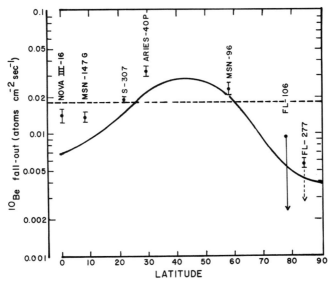

FIG. 40.19. ^{10}Be flux to the ocean determined from ^{10}Be concentrations in deep-sea cores (after Finkel *et al.* 1977). The sold curve represents the atmospheric fall-out pattern.

the Arctic Ocean southwards, thus reducing the flux to the Arctic Ocean floor and increasing it in the region where the ice melts. The mean life of the Arctic sea-ice relative to southward transport has been estimated to be only about ten years (K. Hunkins, personal communication), thus the transport of ^{10}Be is going on continuously. When the Arctic was ice-free, all the ^{10}Be falling on the ocean surface would have been expected to reach the sediments in which the higher flux would then be recorded; this could be used as a criterion for the existence of ice-free conditions.

40.4.4. BERYLLIUM-7

Unlike ^{10}Be, ^{7}Be has a much shorter half-life (53 days). After precipitating on the ocean surface, it is transported to depth by mixing of the surface waters and also by particles. In the open ocean the distribution of ^{7}Be with depth has been used to obtain vertical eddy diffusion coefficients through the thermocline (Silker, 1972). In near-shore waters, in which particle concentrations are high, the rapid removal of ^{7}Be to sediments makes it useful as a tracer for particle mixing, in much the same way as ^{234}Th (see Section 40.3.2.2). Krishnaswami et al. (1978 and in preparation) have tested this application of ^{7}Be in several cores collected from Long Island Sound (as well as in a lake core). They observed exponentially decreasing concentrations of ^{7}Be in the top 2–3 cm, and calculated mixing coefficients of $\sim 3 \times 10^{-7}$ cm^2 s^{-1}. These values agree with those obtained using ^{234}Th (see Section 40.3.2.2).

REFERENCES

Alekin, O. A. and Brazhnikova, L. V. (1961). Gidrokhim. Mater. 32, 12.
Aller, R. C. and Cochran, J. K. (1976). Earth Planet. Sci. Lett. 29, 37.
Amin, B. S., Kharkar, D. P. and Lal, D. (1966). Deep-Sea Res. 13, 805.
Amin, B. S., Krishnaswami, S. and Somayajulu, B. L. K. (1974). Earth Planet. Sci. Lett. 21, 343.
Amin, B. S., Lal, D. and Somayajulu, B. L. K. (1975). Geochim. Cosmochim. Acta, 39, 1187.
Andersen, M. E. and Macdougall, J. D. (1977). EOS, Trans. Am. Geophys. Un. 58, 420.
Arnold, J. R. (1956). Science, N.Y. 124, 584.
Arrhenius, G. (1963). In "The Sea" (M. N. Hill, ed.), Vol. 3, p. 655. John Wiley and Son, New York.
Bacon, M. P. (1976). "Applications of Pb-210/Ra-226 and Po-210/Pb-210 Disequilibria in the Study of Marine Geochemical Processes". Ph.D. Thesis, Woods Hole Oceanographic Institute, Massachusetts.
Bacon, M. P., Spencer, D. W. and Brewer, P. G. (1976). Earth Planet. Sci. Lett. 32, 277.
Benninger, L. K. (1976). "The Uranium-series Radionuclides as Tracers of Geochemical Processes in Long Island Sound". Ph.D. Thesis, Yale University, Connecticut.

Benninger, L. K., Lewis, D. M. and Turekian, K. K. (1975). In "Marine Chemistry in the Coastal Environment" (T. M. Church, ed.), Am. Chem. Soc. Symp. Ser. 18, 202.
Benninger, L. K., Aller, R. C., Cochran, J. K. and Turekian, K. K. (1976). EOS, Trans. Am. Geophys. Un. 57, 931.
Berger, W. H. and Heath, G. R. (1968). J. Mar. Res. 26, 134.
Bernat, M. and Goldberg, E. D. (1969). Earth Planet. Sci. Lett. 5, 308.
Bernat, M., Bieri, R. H., Koide, M., Griffin, J. J. and Goldberg, E. D. (1970). Geochim. Cosmochim. Acta, 34, 1053.
Bertine, K. K., Chan, L. H. and Turekian, K. K. (1970). Geochim. Cosmochim. Acta, 34, 641.
Bhandari, N., Bhat, S. G., Krishnaswami, S. and Lal, D. (1971). Earth Planet. Sci. Lett. 11, 121.
Bhat, S. G., Krishnaswami, S., Lal, D., Rama and Moore, W. S. (1969). Earth Planet. Sci. Lett. 5, 483.
Bhat, S. G., Krishnaswami, S., Lal, D., Rama and Somayajulu, B. L. K. (1973). In "Proceedings of the Symposium on Hydrogeochemistry and Biogeochemistry", Vol. 1, p. 443. Clarke Co., Washington.
Bibron, R., Chesselet, R., Crozaz, G., Leger, G., Mennessier, J. P. and Picciotto, E. (1974). Earth Planet. Sci. Lett. 21, 109.
Blanchard, R. L. and Oakes, D. (1965). J. Geophys. Res. 70, 2911.
Borole, D. V. and Somayajulu, B. L. K. (1978). Mar. Chem. (in press).
Broecker, W. S. (1974). "Chemical Oceanography". Harcourt Brace Jovanovich, New York, 214 pp.
Broecker, W. S., Turekian, K. K. and Heezen, B. C. (1958). Am. J. Sci. 256, 503.
Broecker, W. S., Thurber, D. L., Goddard, J., Ku, T.-L., Matthews, R. K. and Meso-lella, K. J. (1968). Science, N.Y. 159, 297.
Broecker, W. S., Kaufman, A. and Trier, R. M. (1973). Earth Planet. Sci. Lett. 20, 35.
Broecker, W. S., Goddard, J. and Sarmiento, J. L. (1976). Earth Planet. Sci. Lett. 32, 220.
Bruland, K. W. (1974) "Pb-210 Geochronology in the Coastal Marine Environment". Ph.D. Thesis, University of California San Diego, 106 pp.
Bruland, K. W., Koide, M. and Goldberg, E. D. (1974). J. Geophys. Res. 79, 3083.
Burton, J. D. (1975). In "Chemical Oceanography" (J. P. Riley and G. Skirrow, eds), Vol. 3, 2nd Edn, p. 91. Academic Press, London and New York.
Calvert, S. E. (1966). Bull. Geol. Soc. Am. 77, 569.
Chan, L. H., Edmond, J. M., Stallard, R. F., Broecker, W. S., Chung, Y. C., Weiss, R. F. and Ku, T.-L. (1976). Earth Planet. Sci. Lett. 32, 258.
Cherdyntsev, V. V. (1969). "Uranium-234". Atmoizdat, Moscow, 234 pp. (Translated 1971, Israeli Programme for Scientific Translations, Jersualem.)
Chow, T. J. and Goldberg, E. D. (1960). Geochim. Cosmochim. Acta, 20, 192.
Chung, Y. C. (1974). Earth Planet. Sci. Lett. 23, 125.
Chung, Y. C. (1976). Earth Planet. Sci. Lett. 32, 249.
Church, T. M. and Bernat, M. (1972). Earth Planet. Sci. Lett. 14, 139.
Clausen, H. B. (1973). J. Glaciol. 12, 411.
Cochran, J. K. (1978). "The Geochemistry of ^{226}Ra and ^{228}Ra in Marine Deposits". Ph.D. Thesis, Yale University, Connecticut.
Cochran, J. K. and Krishnaswami, S. (1977) EOS, Trans. Am. Geophys. Un. 58, 420.
Cochran, J. K. and Osmond, J. K. (1974). Deep-Sea Res. 21, 721.
Cochran, J. K. and Osmond, J. K. (1976). Deep-Sea Res. 23, 193.
Cochran, J. K., Nozaki, Y., Turekian, K. K. and Keller, G. (1976). EOS, Trans. Am. Geophys. Un. 57, 936.

Craig, H., Krishnaswami, S. and Somayajulu, B. L. K. (1973). *Earth Planet. Sci. Lett.* **17**, 295.
Cronan, D. S. (1974). *In* "The Sea" (E. D. Goldberg, ed.), Vol. 5, p. 491. Wiley–Inter-Science, New York.
Dodge, R. E. and Thomson, J. (1974). *Earth Planet. Sci. Lett.* **23**, 313.
Finkel, R., Krishnaswami, S. and Clark, L. (1977). *Earth Planet. Sci. Lett.* **35**, 199.
Fisher, D. E. (1977). *Nature, Lond.* **265**, 227.
Glass, B. P. (1969) *Earth Planet. Sci. Lett.* **6**, 409.
Goel, P. S., Kharkar, D. P., Lal, D., Narsappaya, N., Peters, B. and Yatirajam, V. (1957). *Deep-Sea Res.* **4**, 202.
Goldberg E. D. and Bruland, K. (1974). *In* "The Sea" (E. D. Goldberg, ed.), Vol. 5, p. 451. Wiley–Interscience, New York.
Goldberg, E. D. and Koide, M. (1962). *Geochim. Cosmochim. Acta*, **26**, 417.
Goldberg, E. D. and Koide, M. (1963). *In* "Earth Science and Meteoritics" (J. Geiss and E. D. Goldberg, eds), p. 90. Wiley–Interscience, New York.
Guinasso, N. L. and Schink, D. R. (1975). *J. Geophys. Res.* **80**, 3032.
Hanor, J. S. and Marshall, N. F. (1971). *In* "Trace Fossils" (B. F. Perkins, ed.), p. 127. Louisiana State University, Baton Rouge.
Immel, R. L. and Osmond, J. K. (1976). *Chem. Geol.* **18**, 263.
Inoue, T. and Tanaka, S. (1976). *Earth Planet. Sci. Lett.* **29**, 155.
Jantsch, K. (1967). *Kernenergie*, **10**, Jahrg. Ht. **3**, 89.
Kaufman, A. (1964). "Th-230/U-234 Dating of Carbonates from Lakes Lahontan and Bonneville". Ph.D. Thesis, Columbia University, New York.
Kaufman, A. (1969). *Geochim. Cosmochim. Acta*, **33**, 717.
Kaufman, A., Trier, R. M., Broecker, W. S. and Feely, H. W. (1973). *J. Geophys. Res.* **78**, 8827.
Kharkar, D. P., Lal, D. and Somayajulu, B. L. K. (1963). *In* "Proceedings of the Symposium on Radioactive Dating", p. 175. International Atomic Energy Agency, Vienna.
Kharkar, D. P., Lal, D., Nijampurkar, V. N., Goldberg, E. D. and Koide, M. (1966). *In* "Abstracts of the 2nd International Oceanographic Congress, Moscow", p. 192.
Kharkar, D. P., Turekian, K. K. and Scott, M. R. (1969). *Earth Planet. Sci. Lett.* **6**, 61.
Kigoshi, K. (1971). *Science, N.Y.* **173**, 47.
Koczy, F. F. (1958). *In* "Proceedings of the 2nd U.N. International Conference on the Peaceful Uses of Atomic Energy", Vol. 18, p. 351.
Koide, M., Soutar, A. and Goldberg, E. D. (1972). *Earth Planet. Sci. Lett.* **14**, 442.
Koide, M., Bruland, K. W. and Goldberg, E. D. (1973). *Geochim. Cosmochim. Acta*, **37**, 1171.
Koide, M., Griffin, J. J. and Goldberg, E. D. (1975). *J. Geophys. Res.* **80**, 4153.
Koide, M., Bruland, K. W. and Goldberg, E. D. (1976). *Earth Planet. Sci. Lett.* **31**, 31.
Krishnaswami, S., Lal, D., Somayajulu, B. L. K., Dixon, F. S., Stonecipher, S. A. and Craig, H. (1972a). *Earth Planet. Sci. Lett.* **16**, 84.
Krishnaswami, S., Somayajulu, B. L. K. and Moore, W. S. (1972b). *In* "Ferroman-ganese Deposits in the Ocean Floor" (D. R. Horn, ed.), p. 117. IDOE, National Science Foundation, Washington D.C.
Krishnaswami, S., Somayajulu, B. L. K. and Chung, Y. (1975). *Earth Planet. Sci. Lett.* **27**, 388.
Krishnaswami, S., Benninger, L. K., Allen, R. C. and Van Damm, K. L. (1978). *EOS, Trans. Am. Geophys. Un.* **59**, 296.
Kröll, V. S. (1953). *Nature, Lond.* **171**, 742.

Kröll, V. S. (1954). *Deep-Sea Res.* **1**, 211.

Ku, T.-L. (1965). *J. Geophys. Res.* **70**, 3457.

Ku, T.-L. (1966). "Uranium Series Disequilibrium in Deep-sea Sediments". Ph.D. Thesis, Columbia University, New York.

Ku, T.-L. (1968). *J. Geophys. Res.* **73**, 2271.

Ku, T.-L. (1976). *Ann. Rev. Earth Planet. Sci.* **4**, 347.

Ku, T.-L. (1977). *In* "Marine Manganese Deposits" (G. P. Glasby, ed.), p. 249. Elsevier, Amsterdam.

Ku, T.-L. and Glasby, G. P. (1972). *Geochim. Cosmochim. Acta,* **36**, 699.

Ku, T.-L., Broecker, W. S. and Opdyke, N. (1968). *Earth Planet. Sci. Lett.* **4**, 1.

Lal, D. (1969). *In* "Morning Review Lectures of the 2nd International Oceanographic Congress", p. 29. UNESCO, Paris.

Lal, D., Goldberg, E. D. and Koide, M. (1960). *Science, N.Y.* **132**, 332.

Lal, D., Nijampurkar, V. N. and Somayajulu, B. L. K. (1970). *Galathea Reports,* **11**, 247.

Lal, D., Nijampurkar, V. N., Somayajulu, B. L. K., Koide, M. and Goldberg, E. D. (1976). *Limnol. Oceanogr.* **21**, 285.

Lalou, C. and Brichet, E. (1972). *C.R. Hebd. Seanc. Acad. Sci., Paris,* **275D**, 815.

Lederer, C. M., Hollander, J. M. and Perlman, I. (1967). "Table of Isotopes". John Wiley and Sons, New York, 594 pp.

Lewis, D. M. (1976). "The Geochemistry of Manganese, Iron, Uranium, Pb-210 and Major Ions in the Susquehanna River". Ph.D. Thesis, Yale University, Connecticut.

Livingstone, D. A. (1963). *Prof. Pap. U.S. Geol. Surv.* **440G**, 64 pp.

Matsumoto, E. (1975). *Geochim. Cosmochim. Acta,* **39**, 205.

McCaffrey, R. J. (1977). "A Record of the Accumulation of Sediment and Trace Metals in a Connecticut, U.S.A. Salt Marsh". Ph.D. Thesis, Yale University, Connecticut.

McCaffrey, R. J. and Thomson, J. (1978). *Adv. Geophys.* (in press).

McCorkell, R., Fireman, E. K. and Langway, C. C., Jr. (1967). *Science, N.Y.* **158**, 1690.

Merrill, J. R., Lyden, E. F. X., Honda, M. and Arnold, J. R. (1960). *Geochim. Cosmochim. Acta,* **18**, 108.

Mesolella, K. J., Matthews, R. K., Broecker, W. S. and Thurber, D. K. (1969). *J. Geol.* **77**, 250.

Moore, W. S. (1969a). "Oceanic Concentration of Radium and a Model for its Supply". Ph.D. Thesis, SUNY-Stony Brook, New York.

Moore, W. S. (1969b). *J. Geophys. Res.* **74**, 694.

Moore, W. S. and Krishnaswami, S. (1972). *Earth Planet. Sci. Lett.* **15**, 187.

Moore, W. S., Krishnaswami, S. and Bhat, S. G. (1973). *Bull. Mar. Sci.* **23**, 157.

Moore, W. S. and Krishnaswami, S. (1974). *In* "Proceedings of the 2nd International Coral Reef Symposium", Vol. 2, p. 269.

Nozaki, Y. and Tsunogai, S. (1976). *Earth Planet. Sci. Lett.* **32**, 313.

Nozaki, Y., Thomson, J. and Turekian, K. K. (1976). *Earth Planet. Sci. Lett.* **32**, 304.

Nozaki, Y., Cochran, J. K., Turekian, K. K. and Keller, G. (1977). *Earth Planet. Sci. Lett.* **34**, 167.

Osmond, J. K. and Pollard, L. D. (1967). *Earth Planet. Sci. Lett.* **3**, 476.

Osmond, J. K., Kaufman, M. I. and Cowart, J. B. (1974). *Geochim. Cosmochim. Acta,* **38**, 1083.

Peng, T.-H., Broecker, W. S., Kipphut, G. and Shackleton, N. (1977). *In* "The Fate of Fossil Fuel CO_2 in the Oceans" (N. H. Andersen and A. Malakoff, eds), p. 355. Plenum Press, New York.

Pettersson, H. (1951). *Nature, Lond.* **167**, 942.
Piggot, C. S. and Urry, W. D. (1942). *Bull. Geol. Soc. Am.* **53**, 1187.
Rama, Koide, M. and Goldberg, E. D. (1961). *Science, N.Y.* **134**, 98.
Reid, D. F. (1976). *Geophys. Res. Lett.* **3**, 253.
Reyss, J. L., Yokohama, Y. and Tanaka, S. (1976). *Science, N.Y.* **193**, 1119.
Rosholt, J. N., Emiliani, C., Geiss, J., Koczy, F. F. and Wangersky, P. J. (1961). *J. Geol.* **69**, 162.
Sackett, W. M. (1966). *Science, N.Y.* **154**, 646.
Sarmiento, J. L., Feely, H. W., Moore, W. S., Bainbridge, A. E. and Broecker, W. S. (1976). *Earth Planet. Sci. Lett.* **32**, 357.
Schink, D. R. (1962). "The Measurement of Dissolved Si-32 in Sea Water". Ph.D. Thesis, University of California, San Diego.
Scott, M. R. (1968). *Earth Planet. Sci. Lett.* **4**, 245.
Scott, M. R., Osmond, J. K. and Cochran, J. K. (1972). *In* "Antarctic Oceanology II: The Australian–New Zealand Sector" (D. E. Hayes, ed.), p. 317. American Geophysical Union, Washington D.C.
Scott, M. R., Scott, R. B., Rona, P. A., Butler, L. W. and Nalwalk, A. J. (1974). *Geophys. Res. Lett.* **1**, 355.
Shokes, R. F. (1976). "Rate Dependent Distributions of ^{210}Pb and Interstitial Sulfate in Sediments of the Mississippi River Delta". Ph.D. Thesis, Texas A and M University College Station, 122 pp.
Silker, W. B. (1972). *Earth Planet. Sci. Lett.* **16**, 131.
Somayajulu, B. L. K. (1969). "Study of Marine Processes Using Naturally Occurring Radioactive Nuclides". Ph.D. Thesis, Tata Institute, Bombay.
Somayajulu, B. L. K. (1977). *Geochim. Cosmochim. Acta,* **41**, 909.
Somayajulu, B. L. K. and Church, T. M. (1973). *J. Geophys. Res.* **78**, 4529.
Somayajulu, B. L. K. and Craig, H. (1976). *Earth Planet. Sci. Lett.* **32**, 268.
Somayajulu, B. L. K., Lal, D. and Craig, H. (1973). *Earth Planet. Sci. Lett.* **18**, 181.
Thomson, J. and Turekian, K. K. (1976). *Earth Planet. Sci. Lett.* **32**, 297.
Thomson, J., Turekian, K. K. and McCaffrey, R. J. (1975). *In* "Estuarine Research" (L. E. Cronin, ed.), Vol. 1, p. 28. Academic Press, New York and London.
Trier, R. M., Broecker, W. S. and Feely, H. W. (1972). *Earth Planet. Sci. Lett.* **16**, 141.
Turekian, K. K. and Chan, L. H. (1971). *In* "Activation Analysis in Geochemistry and Cosmochemistry" (A. O. Brunfelt and E. Steinnes, eds), p. 311. Universitetsforlaget, Oslo.
Turekian, K. K., Cochran, J. K., Kharkar, D. P., Cerrato, R. M., Vaisnys, J. R., Sanders, H. L., Grassle, J. F. and Allen, J. A. (1975). *Proc. Natn. Acad. Sci. U.S.A.* **72**, 2829.
Veeh, H. H. (1966). *J. Geophys. Res.* **71**, 3379.
Veeh, H. H. (1967). *Earth Planet. Sci. Lett.* **3**, 145.
Veeh, H. H. (1968). *Geochim. Cosmochim. Acta,* **32**, 117.
Yokoyama, Y., Tobailem, J., Grjebine, T. and Labeyrie, J. (1968). *Geochim. Cosmochim. Acta,* **32**, 347.

Chapter 41

Estuarine Chemistry

S. R. ASTON

Department of Environmental Sciences, University of Lancaster, England

41.1. INTRODUCTION

An estuary may be defined in several ways. One of the most suitable and widely adopted definitions is that offered by Cameron and Pritchard (1963):

> An estuary is a semi-enclosed coastal body of water which has a free connexion with the open sea and within which sea water is measurably diluted with fresh water derived from land drainage.

More recently, an alternative definition has been suggested by Hedgpeth (1967):

> The estuarine system is a mixing region between sea and inland waters of such shape and depth that the net residence time of suspended matter exceeds the flushing time.

In combination, these two quite different approaches to defining an estuary bring out some important concepts.

Firstly, it is within estuaries that sea-water and land-derived fresh waters are mixed, producing a wide range of brackish waters of intermediate salinities. In this mixing process, the two most dominant natural waters on the Earth's surface interact both chemically and physically. Secondly, estuaries are aquatic environments within which there are interactions of the suspended matter with the soluble constituents of a succession of fresh, brackish and marine waters. The exchange of solid–solution components in estuaries determines the nature of the chemical constituents delivered to the oceans from continental drainage.

The two definitions given above may provide a useful framework within which to discuss the chemical processes in estuaries, but one very pertinent feature of estuaries is not emphasized by these definitions, i.e. the temporal fluctuations in estuarine conditions arising from the variability of run-off and tidal conditions. Meteorological variations produce considerable changes in the amount of river run-off from a catchment into an estuary. These fluctuations often occur in an erratic manner, and any variations in the supply of dissolved and solid suspended matter to estuaries must be taken into account when considering estuarine processes. The erratic fluctuations in river run-off and the rhythmic tidal cycles give rise to major difficulties in the sampling of estuaries, the modelling of estuarine processes and the comparison of individual estuaries with each other.

On a global basis, estuaries are quantitatively very important to the supply of dissolved and solid material to the oceans. Figure 41.1 diagrammatically illustrates the various modes of transport of continental weathering products to the ocean basins. Garrels and Mackenzie (1971) have estimated that a total of $\sim 250 \times 10^{14}$ g year^{-1} of material enters the oceans from the continents, of which $\sim 210 \times 10^{14}$ g year^{-1} is transported via rivers and estuaries. Thus, almost 85% of all soluble and particulate weathering products pass through the estuarine environment before entering coastal waters and $\sim 170 \times 10^{14}$ g year^{-1} of this river-transported material reaches estuaries in solution, while the remaining $\sim 40 \times 10^{14}$ g year^{-1} is present in the form of suspended solids. The other important transport paths for continental material to the oceans involve atmospheric and glacial processes (see Chapter 24).

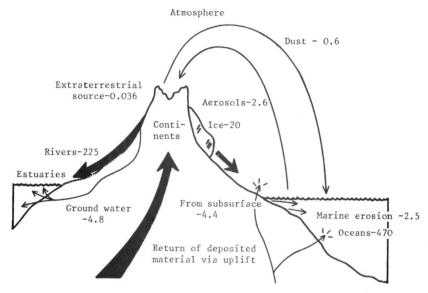

FIG. 41.1. Total masses of materials transported to and leaving the oceans by various agencies (from Garrels and Mackenzie, 1971). Mass values are (per year) 10^{14} g, except for the total dissolved solids in the oceans which are in units of 10^{20} g.

Although estuaries dominate the transport of natural weathering products to the oceans in the hydrological cycle, they are also involved in the transport of pollutants. In recent years pollution of estuaries has received considerable attention from marine scientists whose studies have provided much of our present knowledge of the physical, chemical and biological nature of estuarine systems. Pollution of estuaries can occur in two ways: firstly, by the direct introduction of sewage and industrial wastes into them; and secondly, by the downstream transport of pollutants already present in rivers. Estuarine pollution is particularly relevant because of the fact that estuaries and their hinterlands are often sites of human settlement, commercial enterprise and recreation.

Before dealing with the chemical processes in estuaries, it is necessary to briefly outline their physical nature and the physical processes occurring in them.

41.2. The Development and Topographies of Estuaries

The geological development and the observed topography of estuaries are often used as a basis for their classification. These topographical effects influence both the physical circulation and those chemical processes which are related to circulation patterns, e.g. the restricted flushing and circulation of fjords which can lead to the formation of anoxic waters. The chemistry of marine reducing environments, land-locked basins and fjords has been reviewed in Chapters 15 and 16 and will not be discussed here.

Pritchard (1952) has adopted a simple system of topographical classification of estuaries, dividing them into the four types described below.

41.2.1. COASTAL PLAIN ESTUARIES

These are the most common topographical type of estuary and are often referred to as drowned river valley estuaries, as they have resulted from the Flandrian transgression (c. 3000 BC) when previously incised valleys were flooded by ice melt-water. Their topography remains very similar to that of a river valley because in these estuaries sedimentation has not exceeded inundation. Drowned river valley estuaries are commonly found in temperate latitudes where the sediment transport is low despite relatively high river flow.

Other characteristics of such estuaries are the frequent occurrence of extensive mudflats and saltings, and an exponential increase in cross-sectional area towards the mouth of the estuary.

41.2.2. BAR BUILT ESTUARIES

Although they are similar in most respects to coastal plain estuaries, bar built estuaries have characteristic bars across their mouths. Bar built estuaries were also formed during the Flandrian transgression, the rates of subsequent sedimentation being greater than those for simple coastal plain estuaries, and as a result of this a bar has been deposited, usually at the break-point of waves. These estuaries are normally associated with tropical regimes and coasts of active deposition. Rarely more than a few metres deep, bar built estuaries often have extensive shallow lagoons associated with them (see Chapter 42).

41.2.3. FJORDS

During the Pleistocene era, the pressure of overlying ice sheets led to the deepening and widening of river valleys. Frequently, rock sills or bars were left as residual features, and these have produced the characteristically restricted circulation associated with fjords. Thus, exchange of water between a fjord and the open sea is limited by the sills which are very shallow (1–10 m) in most fjords.

The deepening and widening of pre-existing river valleys has led to the fjords having rectangular cross-sections and small width-to-depth ratios. River discharges into fjords, which only occur in high-latitude mountainous regions, are small, and the floors of fjords have thin layers of sediment in comparison to flood plain and bar built estuaries as a consequence of the meagre detrital inputs.

41.2.4. OTHER TOPOGRAPHICAL TYPES

Some estuaries do not conveniently fit into the three topographical types described above. Most of these other estuaries have been formed by local geological phenomena, e.g. faulting, landslips or volcanism. These tectonic events may lead to the drowning of the lower reaches and mouths of rivers by vertical displacement or the removal of material.

41.3. PHYSICAL PROCESSES IN ESTUARIES

The chemistry of estuaries should be considered in the context of the physical processes of water circulation which occur in them, since the distributions of dissolved and particulate substances are controlled by the circulation and mixing of their waters. The major influences on water circulation in estuaries are river discharge and tidal movements, and it is apparent that there will be considerable variations in circulation patterns, mixing and stratification between individual estuaries.

It is convenient to group the various observed salinity distributions found in estuaries into four characteristic types (Fig. 41.2), thus providing an alternative basis for estuarine classification.

41.3.1. VERTICALLY MIXED ESTUARIES

In a vertically mixed estuary the salinity–depth profile at any given site does not vary with depth (Fig. 41.2A). Such estuaries are usually shallow, with the salinity increasing along the estuary from the head to the mouth at all depths. The saline water progresses towards the head of the estuary by eddy diffusion

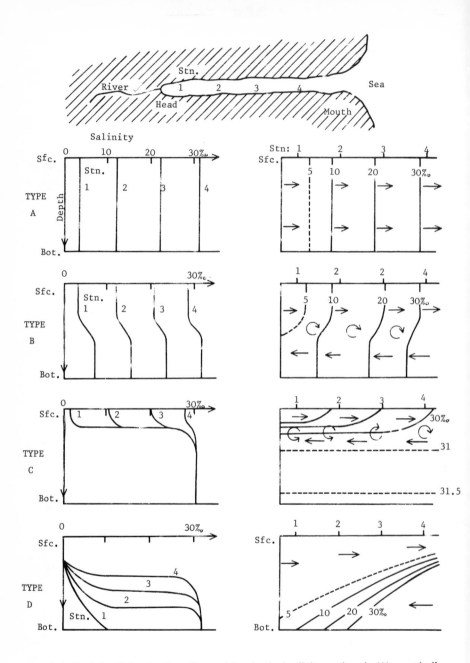

FIG. 41.2. Typical salinity–depth profiles and longitudinal salinity sections in (A) a vertically mixed estuary, (B) a slightly stratified estuary, (C) a highly stratified estuary and (D) a salt wedge estuary (after Pickard, 1975).

at all depths. The net direction of flow of fresh water is shown in the right-hand portion of Fig. 41.2A and is from head to mouth at all depths.

41.3.2. SLIGHTLY STRATIFIED ESTUARIES

Some shallow estuaries exhibit a slight degree of stratification in which a layer of fresh water at the surface overlays deeper saline water of greater density (Fig. 41.2B). As a result of a more dominant fresh water discharge the stratification appears as a net flow of river-water in the upper layer with a corresponding net inflow of saline water in the lower layer and an intermediate layer of mixing by eddy diffusion. There is a longitudinal variation in salinity in both the upper and lower water layers.

41.3.3. HIGHLY STRATIFIED ESTUARIES

In a highly stratified estuary, the salinity of the upper water layer increases from fresh water at the head of the estuary to almost that of sea-water at the mouth. The lower water layer is, on the other hand, of nearly constant salinity from head to mouth and always close to that of sea-water (Fig. 41.2C).

Highly stratified estuaries have a dominant halocline between their upper and lower water layers, with some degree of mixing in which the underlying saline water is incorporated into the upper layer by the action of internal waves (Pickard, 1975). The depth of the halocline remains essentially uniform from the head to the mouth of a highly stratified estuary. Hence, the cross-sectional area of the upper layer remains the same while its volume transport increases due to the upward mixing of saline water from the deeper layers. Consequently, the velocity of the outflowing upper layer may increase substantially from head to mouth, and this has some importance when the disposal of wastes into estuaries is considered.

41.3.4. SALT WEDGE ESTUARIES

Figure 41.2D illustrates the salinity distribution of a salt wedge estuary; the reason for this name being obvious from the longitudinal section in which a wedge of dense saline water is shown to underlie the seaward flow of river-water. In common with stratified estuaries, salt wedge types have a horizontal salinity gradient at the bottom and a dominant vertical salinity gradient. However, there is no salinity gradient at the surface of a salt wedge estuary, the outflowing water remaining fresh until it reaches the mouth of the estuary. The salt wedge phenomenon only occurs in systems in which the volume of river run-off is large.

41.3.5. DISPERSION AND FLUSHING IN ESTUARIES

The detailed dynamics of estuarine mixing and circulation are beyond the scope of the present chapter, but it is worthwhile considering some simple concepts of dispersion and flushing in estuaries. This is particularly important in the context of estuarine pollution by contaminated run-off or by the direct disposal of effluents into estuaries.

The time spent by a pollutant, or other chemical constituent, in an estuary is pertinent to its biological effects and to its availability for interaction with sediments within the estuary. The most commonly used concept for discussing the movement of a pollutant, or other dissolved constituent, in an estuary is the flushing time (T). This is defined as the time required to replace the existing fresh water in the estuary at a rate equal to the river discharge. Hence:

$$T = Q/R$$

where Q is the total amount of fresh water accumulated in the whole, or a section of, the estuary and R is the rate of river discharge into the estuary. The fraction of fresh water (f) at any location is given in terms of the salinity at that location (S) and the salinity of undiluted sea-water (σ) by:

$$f = \frac{\sigma - S}{\sigma}.$$

The total volume of fresh water in the estuary (V_f) is:

$$V_f = \int f \, dv = \bar{f} V$$

where $\bar{f}$ is the averaged fraction of fresh water present and V is the total volume of the estuary. Hence, the flushing time T is easily found from the salinity distribution in the estuary and the river discharge using the expression:

$$T = \frac{\bar{f} V}{R}.$$

This simple approach has been applied to many estuaries; thus, Hughes (1958) has calculated a flushing time of 5·3 days for the Mersey Narrows (U.K.) at a typical river discharge of 25·7 m^3 s^{-1}.

More realistic estimates of flushing times can be gained from a consideration of the tidal prism. In this method the water entering the estuary on the flood time is assumed to mix completely with the water already present in the estuary. On the ebb tide the same volume of water is removed. If V_L is the low-water volume and V_p is the intertidal volume (tidal prism), the flushing time is given by:

$$T = \frac{V_L + V_p}{V_p}.$$

The assumption that complete mixing occurs during the tidal cycle is inadequate since the fresh water at the head of the estuary does not necessarily reach the mouth during the ebb, and some of the water which has been removed may possibly return on the next flood. The tidal prism method has been modified by Ketchum (1951) who divided the estuary into segments within which complete mixing could reasonably be assumed to occur during the tidal cycle. The total flushing time of the complete estuary is the sum of the flushing times for the separate segments. Ketchum (1955) has extended the tidal prism concept to the modelling of effluent residence in an estuary under various conditions.

Further information on mixing and flushing in estuaries is provided by Stommel and Farmer (1952), Pritchard (1957), Dyer (1973) and Officer (1976).

41.4. Major Elements and Nutrients

The mixing of river-water and sea-water in estuaries gives rise to a continuous spectrum of brackish waters of intermediate salinities and compositions. Table 41.1 shows the major element composition of sea-water and, for

TABLE 41.1

Average abundances of nutrient and major ions in river-water and sea-water[a]

Element	Concentration in river-water ($\mu g\, l^{-1}$)	Concentration in sea-water[b] ($\mu g\, l^{-1}$)
Cl	8×10^3	1.987×10^7
S	3.7×10^3	9.28×10^5
Br	20	6.8×10^4
F	100	1.4×10^3
B	10	4.5×10^3
Na	9×10^3	11.05×10^6
Mg	4.1×10^3	1.326×10^6
Ca	1.5×10^3	4.22×10^5
K	2.3×10^3	4.16×10^5
Sr	50	8.5×10^3
N	2.5×10^2	500
P	20	70
Si	$6.1 \times 10^{3\,c}$	1000

[a] Data from Riley and Chester (1971).
[b] Salinity = 35‰.
[c] Data from Livingstone (1963).

comparison, the average composition of river-water. River-waters contain, on average, only $\sim 0.3\%$ of the concentration of dissolved solids which is found in sea-water. Although the salinity of sea-water is always high compared to that of river-water, and the major elements of sea-water are present at much lower concentrations in river-water, this is not the case for some of the nutrient elements, e.g. nitrogen, phosphorus, and especially silicon. The concentrations of these substances and some trace elements (Section 41.6) are often greater in fresh water than in sea-water. These observations imply that processes must exist by which some substances, e.g. the major ions of sea-water, are enriched in the oceans relative to their supply by river-water run-off, whereas other substances undergo a depletion in the sequence of transport from river to estuary to ocean. This is not to suggest that the processes of enrichment or depletion necessarily occur exclusively during estuarine mixing. Enrichment and depletion of the dissolved constituents of sea-water may take place as a result of various geochemical and biological processes within the oceans themselves. The question of whether or not various dissolved major and minor constituents of river-water behave in a conservative manner during estuarine mixing has, however, aroused considerable interest in recent years.

41.4.1. CONSERVATIVE AND NON-CONSERVATIVE BEHAVIOUR DURING ESTUARINE MIXING

In general terms, it is possible for a dissolved substance to remain in solution during mixing, or to react with pre-existing particulate matter in suspension or to form a new particulate phase by precipitation. The obvious changes in ionic strength and other physico-chemical factors, e.g. pH and Eh, during estuarine mixing, and the abundance of suspended solids with active adsorption sites, have resulted in speculation about the conservative or non-conservative behaviour of the major ions, nutrients and other dissolved solids.

Liss (1976) has discussed the necessity for specifying what is meant by the term dissolved, as applied to the composition of natural waters, when considering conservative and non-conservative behaviour. The speciation of many elements in natural waters is not well known. They exist in true ionic solution and as ion pairs, complexes and/or polymers. For many elements there will be a continuous spectrum of chemical forms extending from true solution, through the sub-micrometre colloidal range, and into the regions of well-defined particles. The practical aspects of natural water filtration for particulate and dissolved components have been summarized in Chapter 19 in which the widespread application of a 0.45 µm average diameter for filter pores was noted. When considering the conservative or non-conservative behaviour of elements during estuarine mixing, the arbitary nature of this frequently adopted filter pore size should be borne in mind.

The principles of the theoretical dilution line, as applied by several workers to conservative and non-conservative behaviour of major elements and nutrients, have been discussed by Liss (1976). Briefly, a suite of water samples is collected from an estuary and the dissolved components, whose behaviour is to be studied, are measured together with a conservative parameter such as salinity, chlorinity or chlorosity. Figure 41.3 illustrates the way in which

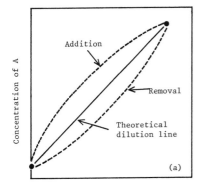
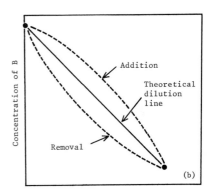

Conservative index of mixing

FIG. 41.3. Idealized representation of the relationship between the concentration of a dissolved component and a conservative index of estuarine mixing, e.g. chlorinity (from Liss, 1976); (a) for a component A whose concentration is greater in sea-water than in river-water, (b) for a component B whose concentration is greater in river-water than in sea-water.

the index of conservative behaviour, which is subject only to physical mixing, produces a theoretical dilution line. In conservative mixing the data points will lie on, or very close to, the straight line joining the end-members, pure river-water and sea-water. When a component is not behaving conservatively, i.e. it undergoes losses from solution or additions to solution, the data points will not lie on the theoretical dilution line but will show deviations above or below the line indicating addition to, or removal from, solution respectively. The dilution curves for a component A, whose concentration is greater in sea-water than in river-water, and for a component B, whose concentration is greater in river-water than in sea-water, are illustrated in Fig. 41.3.

Although the concept of dilution curves is simple, there are a considerable number of pragmatic problems involved in their use for the study of major element, nutrient and other components during estuarine mixing. Several of these problems have been reviewed by Liss (1976), and it is pertinent to consider them here in some detail since they are vital in the assessment of the removal from, or addition to, solution of any chemical constituent during mixing. There are three principal areas of difficulty in the application of the dilution curve: (i) the choice of a conservation index; (ii) sampling and analysis; and (iii) interpretation of the dilution curve. These will be discussed below.

(i) The choice of a conservative index for the construction of a theoretical dilution curve lies between chlorinity, chlorosity and salinity (for a discussion of these terms see Chapter 6). Chlorinity and chlorosity are probably more satisfactory than salinity. These two indices are easily determined with a high degree of accuracy by titration, and there is little or no evidence that the halide ions show any nonconservative behaviour during estuarine mixing (see Section 41.4.4). Chlorinity has been adopted by some workers (see e.g. Hosokawa et al, 1970; Carpenter, cited in Warner, 1972). However, other investigators, some specific examples of whose work are discussed below, have used salinity as an index. Salinity is easily determined accurately, but the drawback in adopting it as an index lies in possible variations in the concentrations of individual components contributing to the salinity of river-waters. Because the relative composition of sea-water is constant, it is satisfactory to adopt salinity as an index of the sea-water contribution during mixing. Hosokawa et al. (1970) have investigated the conservative behaviour of some of the major ions of sea-water and have found that sulphate, magnesium and calcium are all conservative in the Chikugogawa River estuary. Warner (1972) has cited the results of Carpenter who found that sodium, potassium and calcium, but not magnesium, behaved conservatively in the Potomac Estuary. Burton (1976) has suggested that the dominance of sea-salts early in estuarine mixing makes the problem of the variations of the relative proportions of the dissolved ions in river-waters of little significance when using salinity as an index. He further suggested that measurements of most of the constituents studied by this approach are subject to analytical errors which are substantially greater than the uncertainty inherent in the estimation of salinity from a parameter such as conductivity. Recently, the alternative use of $^{18}O/^{16}O$ isotopic ratios for the estimation of the contribution of fresh water during estuarine mixing has been suggested by Boyle et al. (1974).

(ii) In unstratified estuaries, the collection of water samples for a study of the conservative behaviour of dissolved constituents is straightforward since surface samples should be representative of the entire water column at that location. In stratified estuaries, care must be taken to provide samples of estuarine waters derived from both the upper (outflowing) and lower (inflowing) water bodies. Other difficulties of sampling arise from the direct input of fresh water into the estuary by tributaries not associated with the main river input and, of even greater importance, the collection of water samples which truly represent the end-members of the mixing series, i.e.

river-water and sea-water. Care in the spacing of samples to gain representation of the whole of the salinity range is an obvious precaution for estuarine surveys.

The difficulties associated with the definition of dissolved and particulate forms of a constituent have already been noted above. The problem has no simple solution, but may be alleviated, in part, by the consistent use of an arbitarily chosen filter pore diameter, e.g. 0·45 μm. The analysis of estuarine waters for their dissolved constituents is not easy, even for the major components; such waters usually contain a wide range of concentrations of individual elements, and the methods of analysis developed for ocean waters, or for river-waters, are often not directly applicable to brackish waters (Aston, 1978).

(iii) The interpretation of data on the behaviour of a dissolved constituent during estuarine mixing and the fitting of field data to the theoretical dilution curve lead to several problems. If a close fit of data points to the theoretical dilution line is obtained for the whole salinity range, then conservative behaviour may be assumed within the precision and accuracy of the analysis. Deviations from the theoretical line may reflect non-conservative behaviour due to chemical reaction, e.g. ion-exchange, precipitation and adsorption. However, incorrect sampling of end-members, or undetected dilution by direct tributary run-off into the estuary, or biological activity may all give rise to misinterpretations.

Boyle et al. (1974) have developed an analytical approach to the problem of the interpretation of theoretical dilution curves and the fitting of data to them. Using salinity as the index of mixing, they have derived the following expression for the variation of the flux of a given constituent with salinity during estuarine mixing:

$$\frac{dQ_c}{dS} = -Q_w(S - S_r)\frac{d^2C}{dS^2}$$

where Q_w = flux of river-water, Q_c = flux of a dissolved constituent transported by river-water, S_r = salinity of river-water, S = salinity at a given isohaline surface and C = concentration of the constituent at the isohaline surface.

For conservative behaviour there is no loss or gain of the constituent in solution during the estuarine mixing, hence:

$$\frac{dQ_c}{dS} = 0 = \frac{d^2C}{dS^2}$$

and a plot of the concentration of the dissolved constituent against salinity will be a straight line, If there is non-conservative behaviour, the second derivative d^2C/dS^2 will not be equal to zero, and the graph of concentration against salinity will be a curve described by the equation given above.

Before dealing with the conservative and non-conservative behaviour of some major ion and nutrient species in detail, it is worthwhile noting the three alternative approaches which have been made to the problems of estuarine mixing. Firstly, a total budget approach has been applied by Turekian (1971) to the behaviour of silicon in estuaries, and by Head (1970) to the behaviour of nutrient discharges into estuarine waters. The main problems in the total budget approach to conservative behaviour are the collection of sufficient suitable data to allow a realistic budget of inputs and outputs of the dissolved constituents to be formed, and the shortcoming of this method with regard to the zones over which any additions or removal from solution may occur. Secondly, the theoretical dilution curve approach, based on the choice of an index of conservative behaviour, which has been widely adopted by various workers. The third, and most recent, investigation of the problem of estuarine mixing has been made by Sholkovitz (1976) who has adopted a product approach. Rather than examining the behaviour of constituents by measuring their dissolved concentrations during mixing *in situ* (the reactant approach), he has investigated the conservative/non-conservative behaviour of selected constituents by examination of particulate materials produced during the experimental mixing of river-water and sea-water. Sholkovitz has criticised the reactant approach, i.e. that based on the deviations from a theoretical dilution curve, for several reasons. (i) The reactant approach is very dependent on the choice of the river-water end-member concentrations; these can show large variations during the period of an isolated estuarine study. (ii) The possibility of there being three or more end-members (Boyle *et al.*, 1974) in the mixing complicates the situation further. (iii) Estuarine studies have shown the difficulties of demonstrating that removal products exist as either suspended matter or sediments. This is apparent from the studies by Coonley *et al.* (1971), Hair and Bassett (1973) and Bewers *et al.* (1974). The problem lies in distinguishing between river-borne suspended matter, resuspended estuarine sediments, and the recently formed products of non-conservative behaviour. (iv) The reactant approach does not give any idea of the mechanisms involved in mixing processes, although a study of the nature of the products formed may do so. It must be noted, however, that the product approach adopted by Sholkovitz is based on laboratory experimental data and, as such, is open to the criticisms frequently levelled at laboratory simulations of the chemical processes occurring in natural waters.

In the following sections the behaviour of some selected major ion and nutrient constituents of natural waters during estuarine mixing are considered.

41.4.2. THE BEHAVIOUR OF SILICON DURING ESTUARINE MIXING

The behaviour of only a few of the major ion and nutrient constituents of natural waters have been studied in estuarine mixing processes in any detail. Of these few components, silicon has received the greatest attention from estuarine researchers.

Silicon is present in river run-off in three principal forms: dissolved silicon, alumino-silicate weathering products (e.g. clay minerals) and detrital quartz. Here, only the behaviour of dissolved silicon will be considered and the reader should refer to Chapters 26 and 27 for discussions of the chemistry of alumino-silicates and quartz in the marine environment, including their transport to the oceans. Dissolved silicon in river-waters is derived from the incongruent weathering of silicate and alumino-silicate rocks at the Earth's surface. Some typical weathering reactions leading to the formation of dissolved silicon in natural waters are shown in Table 41.2. These, and other similar weathering reactions, may be summarized by a general weathering reaction of the type:

$$XAl\text{-silicate(s)} + H_2CO_3 + H_2O \rightarrow X^+ + HCO_3^- + H_4SiO_4$$
$$+ Al\text{-silicate(s)}.$$

For waters of pH < 9, Siever (1971) has shown that dissolved silicon will be present almost exclusively as silicic acid (H_4SiO_4) and, since the pH values of both sea-water and river-waters are less than 9, it is this form of dissolved silicon which must be considered in estuarine mixing processes. Early workers considered that because sea-water contains, on average, much lower concentrations of dissolved silicon than do river-waters (see Table 41.1), the removal of silicon from solution in estuaries is the most probable mechanism to account for oceanic depletion. The assumption has frequently been made that most of the dissolved silicon in river-water is present in a polymeric, colloidal form which could undergo flocculation as the electrolyte concentration increases during estuarine mixing (Krauskopf, 1956). The more recent work of Siever (1971) and Burton et al. (1970a) suggests that there is no evidence to support this idea. Most workers have defined dissolved silicon as that which will react with molybdate ions to form silico-molybdic acid. Burton et al. (1970b) have used methods for measuring both molybdate-reactive and total dissolved silicon, the former fraction consisting mainly of monomers, but also perhaps including some low molecular weight polymers (Alexander, 1953). These workers reported no evidence for the presence of polymeric forms of silicic acid in river- or sea-waters, and also showed that when polymeric forms were added to either sea- or river-waters there was a rapid depolymerization of silicic acid to monomers. These results suggest

TABLE 41.2

Some typical weathering reactions leading to the presence of dissolved silicon in natural waters

(1) $SiO_{2(s)} + 2H_2O \rightarrow H_4SiO_4$
Quartz

(2) $NaAlSi_3O_{8(s)} + H_2CO_3 + 4 \cdot 5\,H_2O \rightarrow Na^+ + HCO_3^- + 2H_4SiO_4 + 0 \cdot 5\,Al_2Si_2O_5(OH)_{4(s)}$
Albite $\qquad\qquad\qquad\qquad\qquad\qquad\qquad\qquad\qquad\qquad\qquad\qquad$ Kaolinite

(3) $3KAlSi_3O_{8(s)} + 2H_2CO_3 + 12H_2O \rightarrow 2K^+ + 2HCO_3^- + 6H_4SiO_4 + KAl_3Si_3O_{10}(OH)_{2(s)}$
Orthoclase $\qquad\qquad\qquad\qquad\qquad\qquad\qquad\qquad\qquad\qquad\qquad\qquad$ Mica

(4) $KMg_3AlSi_3O_{10}(OH)_{2(s)} + 7H_2CO_3 + 0 \cdot 5H_2O \rightarrow K^+ + 3Mg^{2+} + 7HCO_3^- + 2H_4SiO_4 + 0 \cdot 5Al_2Si_2O_5(OH)_{4(s)}$
Biotite $\qquad\qquad\qquad\qquad\qquad\qquad\qquad\qquad\qquad\qquad\qquad\qquad\qquad$ Kaolinite

(5) $Al_2Si_2O_5(OH)_4 + 5H_2O \rightarrow 2H_4SiO_4 + Al_2O_3 \cdot 3H_2O(s)$
Kaolinite $\qquad\qquad\qquad\qquad\qquad\qquad$ Gibbsite

Non-stoichiometric representations have been adopted for some reactions for the sake of simplicity. The use of H_2CO_3 in the left-hand side of the reactions is intended to indicate the source of H^+ ions for weathering reactions brought about by dissolved CO_2 in the weathering solutions.

that the direct coagulation and flocculation of polymeric silicic acid during estuarine mixing is unlikely, and support the much earlier criticisms of the work of Krauskopf (1956).

The fact that it is unlikely that silicon is removed from solution by coagulation does not, however, imply that estuaries may not be instrumental in partly controlling the dissolved silicon content of the oceans. Over the last twenty years or so, considerable effort has been put into the application of the idealized dilution curve to the problem of silicon behaviour in estuaries. The principal investigations have been collated by Liss (1976) and these, together with the most recent publications, are summarized in Table 41.3. There has been considerable disagreement between the results of the various investigations of silicon behaviour. Early results, e.g. those obtained by Maéda (1952, 1953). Makimoto et al. (1955), Maéda and Tsukamoto (1959) and Maéda and Takesue (1961) for Japanese estuaries, are in contrast to the findings of the early investigations by Bien et al. (1958) for the Mississippi Estuary. The Japanese workers found a linear relationship between dissolved silicon and chlorinity, but their data are of limited value for the investigation of conservative or non-conservative behaviour because of the narrow range of chlorinities sampled.

The work of Bien et al. (1958) on the Mississippi led these authors to conclude that extensive removal of silicon from solution occurred during the mixing of sea-water and river-water. A typical plot of dissolved silicon concentration against chlorosity for the Mississippi Delta is shown in Fig. 41.4. The points corresponding to waters of intermediate salinity all lie below the theoretical dilution curve, suggesting silicon removal from solution. These results are quite typical of those obtained by other workers (see Table 41.3), but the concept of the almost complete removal of dissolved silicon suggested by Bien et al. (1958) has been criticized by later workers. For example, Schink (1967) has re-interpreted the data of Bien et al. (1958) and has estimated that only a ~ 10–20% removal is indicated. Fanning and Pilson (1973) have further investigated the Mississippi system, and have concluded that $\sim 7\%$ removal of dissolved silicon occurs.

Burton and co-workers (e.g. Burton et al., 1970b; Burton, 1970; Burton and Liss, 1973) have investigated silicon behaviour in several British estuaries and in the Vellar Estuary, India. Except for Southampton Water, all the estuaries investigated showed removal of dissolved silicon during mixing, typically to an extent of ~ 10–20%. These studies, together with the other recent investigations by Fanning and Pilson (1973), Liss and Spencer (1970) and Liss and Pointon (1973), all suffer from the problems of sampling and interpretation discussed above. These limitations on recent behaviour studies of silicon have been summarized by Liss (1976), who concluded that where silicon removal has been reasonably proven it is always $< 30\%$. The

TABLE 41.3

Principal investigations of the behaviour of dissolved silicon during estuarine mixing

Estuary	Finding	Reference	River concentration of SiO_2 (mg l^{-1})
Various, Japan	Linear relationship between Si and chlorinity; numerous surveys generally over a limited chlorinity range	Maéda (1952, 1953), Makimoto et al. (1955), Maéda and Tsukamoto (1959), Maéda and Takesue (1961)	—
Mississippi, U.S.A.	10–20% removal	Data of Bien et al. (1958) interpreted by Schink (1967)	6
Columbia, U.S.A.	No significant removal	Stefánsson and Richards (1963), Park et al. (1970)	12
Jiu-Long, China	Considerable removal (20–30%[a])	Fa-Si et al. (1964)	16
Kiso and Nagara, Japan	Linear relationship between Si and chlorinity; surveys in both estuaries over a limited chlorinity range	Kobayashi (1967)	—
Southampton Water, U.K.	No significant removal	Burton et al. (1970b)	10
Hamble, U.K.	10% removal	Burton et al. (1970b)	12
Vellar, India	10% minimum removal (see text)	Burton (1970)	39
Chikugogawa, Japan	Considerable removal (20–30%[a])	Hosokawa et al. (1970)	30
Conway, U.K.	10–20% removal	Liss and Spencer (1970)	3
St. Marks, U.S.A.	Considerable removal	Stephens and Oppenheimer (1972)	12
Alde, U.K.	25–30% removal	Liss and Pointon (1973)	7
Orinoco, Venezuela	Inconclusive, lack of data on fresh water end-member	Fanning and Pilson (1973)	—
Savannah, U.S.A.	Probably no significant removal	Fanning and Pilson (1973)	9
Mississippi, U.S.A.	7% removal	Fanning and Pilson (1973)	6
Merrimack, U.S.A.	No significant removal	Boyle et al. (1974)	5
Various, Scotland	No significant removal	Sholkovitz (1976)	0·3–1·3

[a] Extent of removal not estimated quantitatively. Figure given is from interpretation by Burton and Liss (1973) (after Liss, 1976).

fact that the removal is observed at times of low biological production (e.g. in winter in temperate latitudes) suggests that it occurs by a non-biological mechanism.

Indeed, the mechanism of silicon removal from solution in estuaries is not clearly understood. The work of Krauskopf (1956) on silicon polymerization and precipitation showed that removal under natural conditions in fresh

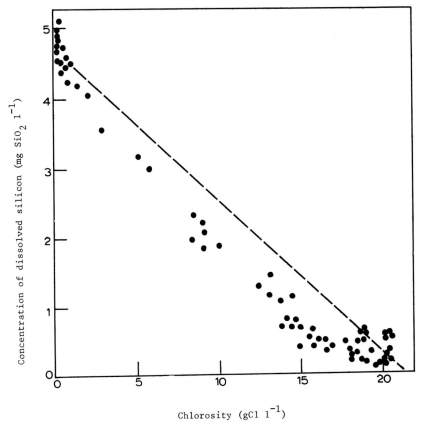

FIG. 41.4. The relationship between dissolved silicon concentration and chlorosity in the Mississippi Delta in June 1953, (after Bien *et al.*, 1958). The broken line represents the theoretical dilution curve.

or saline waters is only significant over periods of several months. The mechanism of polymerization and precipitation involves the loss of a water molecule from monomeric silicic acid and the formation of a siloxane bond, —Si—O—Si—:

$$2H_4SiO_4 \rightarrow H_6Si_2O_7 + H_2O.$$
disilicic
acid

Further, three-dimensional linking through siloxane bonds converts the disilicic acid into a polymer of similar geometry to cristobalite (Iler, 1955). The studies by Burton et al. (1970a, b) cited above indicate that polymeric silicic acid is not present in river-waters and this, together with the slow kinetics of precipitation, suggests that polymerization and flocculation are not important in the removal of silicon from solution in estuaries.

The first laboratory studies of the mechanism of inorganic removal of dissolved silicon were those of Bien et al. (1958). By simple experiments, these workers demonstrated that in order to remove dissolved silicon from river-water both suspended sedimentary particles and sea-water electrolytes must be present. Natural sedimentary suspended matter from the Mississippi River, bentonite and alumina were all found to produce a rapid removal of silicon from solution in the presence of sea-water. The natural river-borne detritus was observed to be most effective, and because silicon removal was found to be a function of the concentration of suspended matter Bien et al. concluded that it was removed by adsorption onto the detritus. These results are in agreement with those of the later investigations by Liss and Spencer (1970), who also demonstrated that there is no significant desorption of silicon from suspended detritus. However, Fanning and Pilson (1973) found no evidence of silicon removal in experiments performed with Mississippi river-water. The disagreement between their findings and those of both Liss and Spencer (1970) and Bien et al. (1958) has been attributed by Fanning and Pilson to differences in experimental conditions, particularly temperature, which may have affected the solubility of silicon sufficiently to give rise to the contrasting results.

Silicon flocculation during estuarine mixing has been investigated by the product approach (see above) by Sholkovitz (1976); the results of his experiments with four river-waters from Scotland are shown in Fig. 41.5. Although large random fluctuations occurred, the silica contents of the floccu-lants formed by mixing the river-waters with sea-water were found to increase as the salinity rose from 0‰ to 30‰. His results showed that removal of silica amounted to 3–6%; however these results do not help to elucidate the mechanism by which removal occurs, since adsorption onto pre-existing flocculants as well as direct flocculation of dissolved silicic acid may be taking place.

Liss (1976) has attempted to summarize the reasons why the evidence on the behaviour of silicon during estuarine mixing is so conflicting, and he has related the discrepancies to some of the apparent controlling factors. The following important points may be drawn from Liss's summary.

(i) The role of suspended particulates in silicon removal from solution in estuaries has been well established by laboratory experiments. The sources of suspended matter in estuaries are various, e.g. river-borne detritus, resuspended sediments, biological detritus and flocculants. Differences in

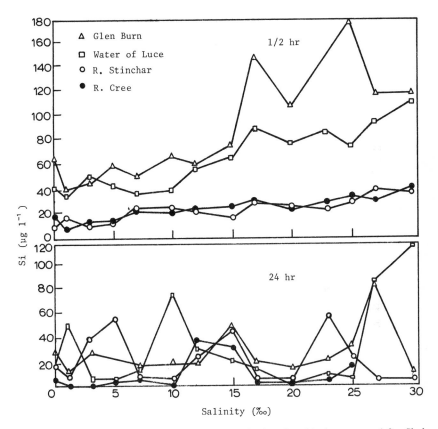

FIG. 41.5. Silicon flocculation as a function of salinity for four Scottish river systems (after Sholkovitz, 1976). Results for $\frac{1}{2}$ hour and 24 hour experimental runs are illustrated.

the nature of the suspended loads between individual estuaries may lead to removal, or lack of removal, of silicon, but this has not been established.

(ii) Sea-water electrolytes seem to be important factors in silicon removal, but this removal, when observed, appears to take place during early mixing at low salinities, i.e. 0–5‰ Harder (1965) has shown, however, that silicon removal can take place onto amorphous solids in both fresh and saline waters. Since all estuaries have some sea-water electrolytes present, the variations in salinity distributions observed between individual estuaries should not control the presence or absence of the silicon removal process.

(iii) Variations in the concentration of dissolved silicon in river run-off from different catchments, under a variety of weathering conditions, may control the rate and extent of silicon adsorption during estuarine mixing. The relationship between silicon concentration in water and the amount removed has been investigated by Liss and Spencer (1970) whose results show that the proportion removed tends to increase with increasing silicon concentration. Wollast (1973) has shown experimentally that dissolved silicon concentrations tend to equilibrate to ~ 14 mg $SiO_2 \, l^{-1}$ during simulated estuarine mixing; this is in close agreement with the experimental results of both Mackenzie *et al.* (1967) and Siever and Woodford (1973). Liss (1976) has concluded that the close agreement between these results and the similarity of the typical equilibrium value of ~ 14 mg$SiO_2 \, l^{-1}$ to the mean dissolved silicon concentration of World river-waters (Livingstone, 1963) suggest the existence of a silicon buffering mechanism. Furthermore, in all estuaries investigated to date for which the river-water silicon concentration exceeds 14 mg$SiO_2 \, l^{-1}$, removal is observed, whereas in those estuaries which exhibit a linear dissolved silicon–salinity (or silicon–chlorinity) relationship, the dissolved silicon concentration in the inflowing river-water is less than ~ 14 mg$SiO_2 \, l^{-1}$. Liss (1976) has pointed out that the concentration of dissolved silicon cannot be the only controlling factor, as removal has been found in some estuaries in which the river-water silicon concentrations are less than ~ 14 mg$SiO_2 \, l^{-1}$. Removal would, however, seem most likely to occur during the earlier stages of mixing before the diluting effect of sea-water reduces the dissolved silicon concentration to well below the adsorption equilibrium value.

41.4.3. THE BEHAVIOUR OF ALUMINIUM DURING ESTUARINE MIXING

Most of the aluminium derived from the weathering of continental crustal rocks reaches the oceans in the form of alumino-silicate minerals, and the concentrations of dissolved aluminium in both river-water and sea-water are extremely low for such an abundant element. Sea-water contains, on average, considerably less dissolved aluminium than do river-waters. Thus, Hydes (1974) has reported a range of 5–800 $\mu g \, l^{-1}$ for dissolved aluminium in river-waters, whereas sea-water typically contains < 1–10 $\mu g \, l^{-1}$ (Hydes, 1974; Sackett and Arrhenius, 1962). The wide range of aluminium concentrations reported for river-waters may, in part, be due to the difficulty of defining dissolved and particulate species. Alumino-silicate weathering products, of which the clay minerals are the most abundant, occur in natural waters as particulates of sub-micrometric diameter with a particle size range extending over most of the colloidal range. This, together with the ability of aluminium in solution to form polymeric hydroxy species by hydrolysis, leads to severe difficulties in defining the dissolved–particulate boundary for the element in natural waters. In effect, it is not realistic to try to clearly separate dissolved and particulate species when a continuous size spectrum of colloidal polymeric species exists. The pragmatic solution to this problem has been the adoption of a particular filter pore size (see Section 41.4.1).

The ease with which aluminium present in solution in river-water can undergo hydrolysis, polymerization and precipitation suggests that the increases of pH caused by the mixing of river-water with sea-water in estuaries might well cause the behaviour of this element to be non-conservative. The estuarine chemistry of aluminium has been investigated both in the field by Hosokawa *et al.* (1970) and Hydes (1974), and experimentally by Sholkovitz (1976) and Eckert and Sholkovitz (1976).

The results found by Hosokawa *et al.* (1970) for the Chikugogawa Estuary, Japan, suggest that a large-scale removal of aluminium from solution occurs during early mixing. The relationship between dissolved aluminium concentration and chlorinity for this estuary is shown in Fig. 41.6. The extent of aluminium loss from solution during mixing is such that waters of intermediate salinity have lower dissolved aluminium concentrations than does sea-water. Hosokawa *et al.* (1970) have suggested a two-stage process to explain this effect. Line A–B in Fig. 41.6 represents the initial rapid removal of aluminium from solution which appears to be complete when a chlorinity of $\sim 2\%_{0}$ is reached. Line B–C represents the subsequent mixing of this very low-aluminium water with normal sea-water which is relatively richer in dissolved aluminium.

The studies by Hydes (1974) of aluminium behaviour in two British estuaries have also demonstrated that aluminium is removed from solution during early mixing. In the Conway Estuary, the magnitude of removal was not as great as that reported by Hosokawa *et al.* (1974), but did represent a $\sim 30\%$ loss of the dissolved aluminium entering the estuary in the water of the Conway River. Hydes (1974) has postulated that this $\sim 30\%$ removal from solution is a result of the coagulation of aluminium particulates or polymers in the size range 0.1–$0.45\,\mu m$, producing particulates of an average size greater than $0.45\,\mu m$. This interpretation is based on the observation that $\sim 30\%$ of the dissolved aluminium in Conway river-water would pass through a $0.45\,\mu m$ membrane filter, but would not pass through a $0.10\,\mu m$ filter. In contrast, Conway Bay sea-water showed the same dissolved aluminium concentration when passed through either a $0.45\,\mu m$ or a $0.10\,\mu m$ filter. Hydes (1974) suggested that the coagulation process is a consequence of the increases in pH and ionic strength resulting from the mixing of the Conway river-water with sea-water.

Using a product approach (see above), Sholkovitz (1976) has recently examined the flocculation of aluminium during the experimental mixing of river- and sea-waters. He found that the removal of dissolved aluminium during mixing increased with increasing salinity until a salinity of ~20‰ was reached, above which no further removal occurred. The extent of aluminium flocculation was found to lie in the range 10–70% of the initial

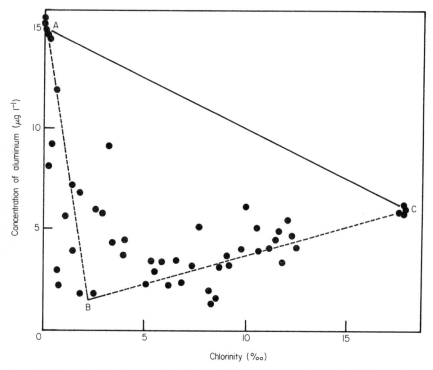

FIG. 41.6. Relationship between dissolved aluminium concentration and chlorinity in the Chikugogawa Estuary (from Hosokawa *et al.*, 1970). Line A–C is the theoretical dilution curve. Lines A–B and A–C represent alternative mixing patterns.

dissolved concentration, the degree of flocculation depending on the initial concentrations present in the individual river-waters studied. For example, a river-water containing $70 \, \mu g \, l^{-1}$ of dissolved aluminium gave rise to 70% flocculation on mixing with sea-water, whereas under the same experimental conditions a river-water containing $40 \, \mu g \, l^{-1}$ gave rise to only 10% flocculation.

In order to investigate the possible role of organic matter, including humic materials, on the flocculation of aluminium and other elements,

Sholkovitz (1976) has studied the composition of flocculants resulting from the experimental raising and lowering of pH. A striking association between aluminium and river-water humics was found in the flocculation experiments, and Sholkovitz concluded that dissolved organic matter, particularly humic acids, plays a major role in controlling the concentrations of dissolved and particulate aluminium during estuarine mixing.

Further investigations of aluminium flocculation during the mixing of river- and sea-waters have been made by Eckert and Sholkovitz (1976). These authors studied the flocculation of aluminium from Scottish river-waters to which the major salts ($NaCl$, $MgCl_2$ and $CaCl_2$) of sea-water had been added in an attempt to elucidate the chemical and electrostatic controls on flocculation (Stumm and Morgan, 1962; Stumm and O'Melia, 1968). Eckert and Sholkovitz have emphasized four features of their results:

(i) the molarities of electrolytes required to achieve maximum flocculation decrease in the order $NaCl > MgCl_2 > CaCl_2$;

(ii) the maximum flocculation by the sea-water electrolytes increased in the same order, i.e. $NaCl < MgCl_2 < CaCl_2$;

(iii) the extent of aluminium flocculation observed for calcium is not reached for magnesium or sodium even at molarities three to eight times that of calcium;

(iv) maximum flocculation is reached for aluminium at sodium, magnesium and calcium molarities significantly above those found in sea-water.

The first of these points is consistent with the results of other workers, e.g. Wright and Schnitzer (1968), Orlov and Yeroshicheva (1967), Ong and Bisque (1968) and Khan (1969). The second demonstrates the dependence of aluminium flocculation on the charge and ionic size of the flocculating electrolytes in sea-water and this, together with the third observation, indicates that chemical as well as purely electrostatic interactions are involved in the flocculation. Finally, from the last feature, it is apparent that the proportion of aluminium flocculated with any of the individual salts in sea-water (at the molarities at which they occur in normal sea-water) is only a fraction of the total aluminium available.

41.4.4. THE BEHAVIOUR OF BORON DURING ESTUARINE MIXING

The data in Table 41.1 indicate that the average concentration of boron in river-water is almost 500 times less than that in sea-water. Levinson and Ludwick (1966) demonstrated that boron is adsorbed from solution in sea-water onto clay minerals and they suggested that this process may lead to the non-conservative behaviour of boron during estuarine mixing. This hypothesis has been investigated by Liss and Pointon (1973), who found a non-linear relationship between dissolved boron and salinity for a suite of water samples collected from the estuary of the Alde River, England, which they interpreted in terms of a ~ 25–30% loss of boron from solution. Boron removal to argillaceous sediments in estuaries is supported by the work of Boon and MacIntyre (1968) who found a significant correlation between the boron concentration in estuarine bottom-sediments and the salinity of the associated waters.

In contrast to the results of Liss and Pointon (1973) for boron behaviour in the estuary of the Alde River, Hosokawa *et al.* (1970) and Liss and Pointon (1973) found no significant removal from solution in the Chikugogawa and Beaulieu estuaries, respectively. The behaviour of boron during estuarine mixing is not well established, and the differing results for individual estuaries indicate that more research into this topic is required.

41.4.5. THE BEHAVIOUR OF THE HALOGENS DURING ESTUARINE MIXING

As mentioned above, chlorinity has frequently been used as an index of conservative behaviour for the investigation of other species during estuarine mixing processes. Thus, it is important that the halide ions which contribute to chlorinity are behaving perfectly conservatively themselves. The only halogen to have received any significant investigation in estuarine processes is fluorine. The fluoride concentration of sea-water is approximately an order of magnitude higher than that of river-waters (see Table 41.1). Most of the fluorine present in river-water is probably not derived from the weathering of crustal rocks, but is recycled marine fluoride from atmospheric precipitation; Carpenter (1969) has estimated that up to 80% of the fluoride present in river-water is recycled in this way. In the same paper, Carpenter also argued that the very similar fluorine concentrations in atmospheric dusts, river detritus and near-shore and deep-sea sediments imply that there is little transfer of fluorine from solution to particulate material in natural waters.

Although the results of Carpenter (1969) suggest that it is unlikely that fluorine behaves in a nonconservative fashion in estuaries, several workers have studied this topic. The first investigation was made by Windom (1971) who measured the dissolved fluoride concentrations in waters from selected estuaries in Georgia, U.S.A. Good agreement was found between the fluoride concentrations and the theoretical dilution curves, indicating conservative behaviour. Similar results have been obtained by Warner (1972) for Chesapeake Bay, and Hosokawa *et al.* (1970) for the Chikugogawa Estuary. Kullenburg and Sen Gupta (1973) have reported an average 17% loss of fluoride from solution for the Baltic Sea, although Liss (1976) has suggested that the complexity of the Baltic Sea system, and the numerous rivers running into it, together with the use of averaged data, make it difficult to accept their results.

In conclusion, it would seem very probable that fluoride behaves in a conservative manner similar to that of the other halides.

41.4.6. THE BEHAVIOUR OF PHOSPHORUS AND NITROGEN DURING ESTUARINE MIXING

The estuarine chemistry of the nutrients silicate, phosphate and nitrate has been dominated by intensive interest in the behaviour of silicon during the mixing of fresh water and sea-water (see Section 41.4.2). Recently, however, attention has been drawn to the behaviour of phosphate, nitrate and other inorganic species of nitrogen in estuaries. This interest stems partly from the importance of these substances in biological processes, including the control of productivity, and partly from concern about the effects of the input to estuaries of nutrient-rich effluents, e.g. domestic sewage.

Phosphorus occurs in the Earth's surface environment almost exclusively as *ortho*-phosphate, and *ortho*-phosphate ions are found in solution in river-water as a result of the weathering of the phosphate minerals of continental rocks. The supply of phosphate to estuaries is not limited to that derived from natural weathering, however, as some comes from anthropogenic sources. Examples of man's contribution to the phosphate in natural waters are fertilizers, detergents and both industrial and domestic effluents. In addition to the simple inorganic phosphate ions, river-waters and sea-water may contain a variety of dissolved organic phosphates derived from biological activity as well as particulate phosphates of both organic and inorganic origin. The marine chemistry of phosphorus has been reviewed in Chapter 11.

Interest in the estuarine chemistry of phosphorus has stemmed from two considerations: (i) the important nutrient role of phosphate in estuarine and coastal waters, and its influence on primary and secondary production; and (ii) the ubiquitous nature of phosphate pollution arising from the disposal of sewage effluents into rivers or directly into estuaries. The first investigation of the behaviour of dissolved (inorganic) phosphate in estuaries was carried out by Stefánsson and Richards (1963), who found a marked constancy of phosphate concentrations over a wide salinity range in the Columbia River estuary, the mean concentration being $37 \, \mu gP \, l^{-1}$. It was suggested that this lack of variation may have been due to the similarity of the phosphate concentrations in the river-water and sea-water, and also to a buffering effect which acts to maintain a constant value of $\sim 37 \, \mu gP \, l^{-1}$. The existence of a phosphate buffering mechanism has also been proposed by Butler and Tibbits (1972) for the Tamar Estuary, England. These workers collected samples of estuarine waters on two occasions, the first after a period of little rainfall, and the second following heavy rainfall in the Tamar catchment. Although other constituents, e.g. nitrogen compounds, showed a significant increase in their concentrations in the estuarine waters on the second survey, little difference was observed for either inorganic or organic dissolved phosphorus under the two extreme run-off conditions.

Aston and Hewitt (1977) have studied the dispersion of dissolved phosphate and particulate phosphate from sewage discharge sites adjacent to a semi-enclosed tidal area, with particular reference to phosphate distribution in intertidal sediments. Analysis of water and suspended matter samples collected on the flood and ebb tidal cycles showed that the coastal effluent outfalls were important sources of particulate phosphate in the estuary, with a net flux of particulate phosphate into the semi-enclosed area. Aston and Hewitt found that the deposition of particulate phosphate in tidal channels is greatly influenced by channel geometry, shallow, broad channels concentrating phosphate at their heads in contrast to narrow, steep-sided and deeper channels where phosphate deposition was concentrated at the mouth. These features of phosphate deposition in intertidal areas may be explained in terms of the particle size distributions of suspended matter and their relationship to phosphate concentrations in the suspended particles.

The interrelations of the phosphorus and biological cycles in estuaries have been studied by various workers. Much of this research has employed radiophosphorus tracer experiments and field studies to examine the routes of biological phosphorus cycling in estuaries, e.g. the work of Pomeroy and his co-workers. Pomeroy (1960) studied the residence times of phosphorus in estuaries and related the rates of uptake and turnover of dissolved *ortho*-phosphate by organisms in different water masses. Pomeroy *et al.* (1963) have investigated the amounts and forms of phosphorus excreted by zooplankton, and Pomeroy *et al.* (1965, 1966, 1969) have employed ^{32}P tracer in investigations of the pathways and rates of redistribution of phosphorus in salt marsh ecosystems. These studies on phosphorus cycling in estuaries have been criticized by Correll *et al.* (1975) on the grounds that in most instances detailed kinetic analyses of phosphorus fluxes were not carried out, and consequently their conclusions as to rates of phosphorus uptake, turnover times and transport vary with respect to time. These authors cite the studies of Rigler (1956, 1964) and Lean (1973) who have emphasized the importance of making very detailed time series measurements if rates of phosphorus uptake and transport are to be quantitatively evaluated for aquatic ecosystems.

Correll *et al.* (1975) have extended earlier studies of the cycling and biological fluxes of phosphorus in estuaries by an experimental approach using ^{32}P-labelled *ortho*-phosphate in Chesapeake Bay. Their investigations included detailed studies on phosphorus cycling in estuarine plankton, the major pathways of which are summarized in Fig. 41.7. These workers have concluded that *ortho*-phosphate is taken up from solution in estuaries by bacteria, mainly located on the surfaces of suspended sediments and detritus, and that phytoplankton also take up some phosphate from solution under light conditions. The bacteria and phytoplankton are consumed by filter-feeders, especially ciliate protozoans, which then release both dissolved *ortho*-phosphate and organic forms of phosphorus.

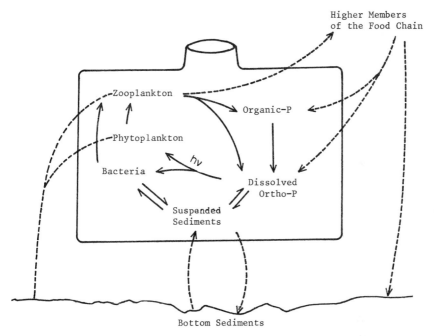

FIG. 41.7. A simple model of the pathways of phosphorus cycling in estuaries (after Correll *et al.*, 1975). Phosphate uptake by phytoplankton which requires light (*hν*) is indicated.

In general, river-waters contain more nitrogen, present to a great extent as dissolved nitrate, than does sea-water. Very few investigations have been made of the conservative/non-conservative behaviour of nitrate or other nitrogen compounds during estuarine mixing. Stefánsson and Richards (1963) found an almost linear relationship between dissolved nitrate and salinity in the Columbia River estuary, with some deviation from the theoretical dilution curve at very low salinities. This deviation may be interpreted as non-conservative behaviour, but the results are not conclusive. Butler and Tibbits (1972) reported an inverse relationship between salinity and total dissolved combined nitrogen for the Tamar Estuary. These authors noted that estuarine levels of dissolved nitrogen compounds are surprisingly low (~ 10–$\sim 80\ \mu g\,l^{-1}$ total nitrogen for the Tamar) considering the significant sources of dissolved nitrogen available, e.g. biological organics, detritus and terrestial run-off. No explanation of this observation was offered, but Butler and Tibbits (1972) have commented on the wide variations observed for the N/P ratio in the estuary, the range being 9/1–110/1 on a seasonal basis compared to a typical value of 15/1 for sea-water. It has been suggested that the temporal fluctuations in the N/P ratio of estuarine waters may be a reflection of variation in the addition of nitrite and nitrate resulting from fluctuations in rainfall.

Sewage effluents discharged directly or indirectly into estuaries and run-off from agricultural soils are both potential sources of phosphate and nitrate for estuaries. The influence of effluent and fertilizer sources on the nutrient budgets of estuaries will obviously vary geographically and be subject to an array of environmental controls. Two recent investigations on the Hudson River estuary, U.S.A., and the Firth of Clyde, Scotland, are useful case studies of nutrient pollution in estuaries.

Simpson et al. (1975) have examined the influence of the disposal of sewage from New York City on the nutrient budget of the Hudson River estuary, which receives a total discharge of sewage of somewhat more than two billion gallons (10^{10}l) per day. Discharge occurs at various locations in the estuary, but the mixing of effluent with receiving waters is very rapid and the effects of individual outfalls cannot be observed for significant distances.

The spatial distribution of phosphate is dominated in the Hudson River estuary by the effluent discharged, and the observed distributions suggest effective mixing, with phosphate concentration gradients closely following the vertical salinity gradients (see Fig. 41.8). Simpson et al. (1975) have concluded that the phosphate distribution is dominated by the balance between sewage inputs and water transport within and out of the system, and that biological activity has little influence on phosphate distribution in the estuary. This is not the case for some other estuarine systems, both polluted and unpolluted, and serves to emphasize the individuality of estuaries (see above). The cause of the low biological removal of phosphate derived from sewage in the Hudson River estuary is possibly the limited photosynthetic activity resulting from poor light penetration into the heavily silt-laden waters.

The Firth of Clyde is a complex estuarine environment which receives daily the domestic effluents from a population of ∼2·5 million people. The nutrient budget of this polluted estuary has been studied by Mackay and Leatherland (1976) as an example of an estuary under major urban influence. A typical salinity distribution for this firth is shown in Fig. 41.9; this illustrates the pronounced salt wedge which is the dominant feature of the estuary and which plays an important role in its chemistry. The salt wedge also effectively decreases the mean estuarine residence time of fresh water,

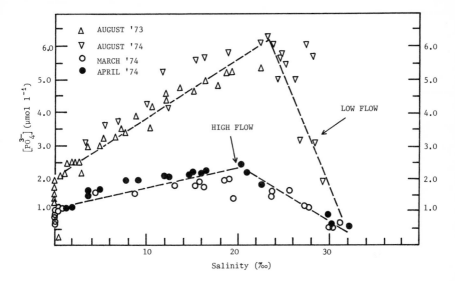

FIG. 41.8. Phosphate (molybdate-reactive) as a function of salinity in the Hudson Estuary (after Simpson *et al.*, 1975). Results for high-flow and low-flow conditions are illustrated.

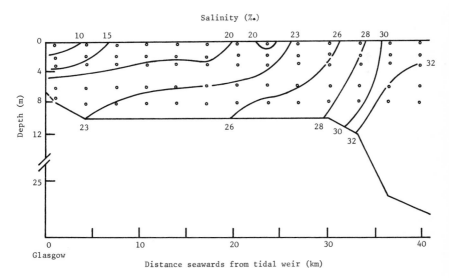

FIG. 41.9. Longitudinal profile of salinity between Glasgow and Gourock/Kilcreggan in the Clyde Estuary, Scotland, under a relatively low fresh water flow condition (from Mackay and Leatherland, 1976).

especially in summer when salinities of 20‰, or greater, are common in the bottom waters over much of the estuary. During periods of very high run-off, all traces of sea-water are completely flushed out of the uppermost layer over 15–20 km of the estuary (total length 40 km). In this respect, the Firth of Clyde is very different from most other British estuaries, many of which are vertically mixed or only partially stratified.

Consideration of the major nutrient ions, e.g. phosphate and nitrate, presents a problem in the Firth of Clyde because of the complex nature of the various natural effluent inputs to it. Mackay and Leatherland (1976) have found that on rare occasions phosphate has been removed during mixing in the Clyde Estuary, and have emphasized the danger of drawing conclusions about conservative behaviour on the basis of data from single surveys. Results from many surveys made during near-average conditions show that a linear relationship between phosphate and salinity exists, and that the few isolated points well above the theoretical dilution curve can always be traced to local sewage inputs. Mackay and Leatherland (1976) have suggested that, under normal conditions, phosphate does not have the opportunity to reach sorption equilibrium with the suspended solids and that, on an annual basis, there is no net phosphate removal from the estuarine waters to bottom sediments or biological material.

Evidence for the occasional removal of phosphate from solution in the Clyde Estuary has been provided by Mackay and Leatherland (1976) on the basis of phosphate–silicate relationships. These authors contended that if there is significant biological removal of phosphate in the estuary, this would be accompanied by nitrate removal. Figure 41.10 shows the plots of concentrations of phosphate versus nitrate, and phosphate versus silicate for a selected survey, and demonstrates the non-linear relationship between phosphate and nitrate which has led to the suggestion that phosphate removal does occasionally occur by non-biological processes. Furthermore, Mackay and Leatherland (1976) have argued that since silicate appears to have a conservative nature in the Clyde Estuary, the non-linear phosphate–silicate line in Fig. 41.10 is the result of phosphate removal.

Nitrate frequently behaves in a non-conservative manner in the Firth of Clyde, and Mackay and Leatherland (1976) have given a detailed account of its estuarine behaviour which includes the effect of seasonal changes. Each summer, during periods of dissolved oxygen depletion, the substantial removal of dissolved nitrate is a feature of the Clyde estuary. They have illustrated the non-conservative behaviour by considering nitrate–silicate relationships and dismiss salinity as a mixing index because of the problems arising from the input of low-nitrate fresh water from feeder rivers. An example of denitrification (see Chapter 11) is shown in Fig. 41.11. The two points which lie well below the dilution curve correspond to samples of

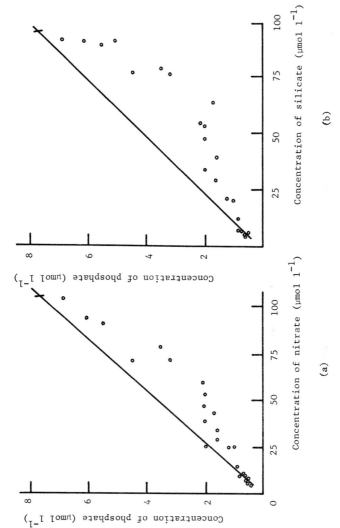

FIG. 41.10. Relationships between the concentration of phosphate and (a) nitrate and (b) silicate in the Clyde Estuary, Scotland (after Mackay and Leatherland, 1976).

bottom water from the two uppermost stations in the estuary at the tip of a salt wedge. These samples, within which oxygen had been depleted to $\sim 1\%$ of the saturation value, exhibit extensive denitrification. The linearity of the dilution line does, however, suggest that denitrification has a negligible effect on the output of nitrate from the estuary as a whole, and that it is restricted to the almost stagnant bottom waters.

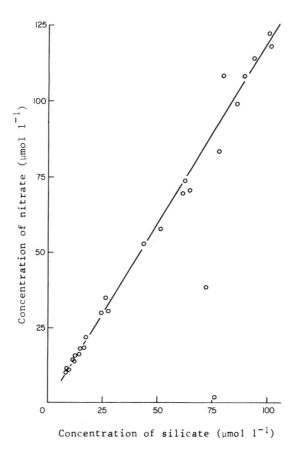

FIG. 41.11. Relationship between the concentrations of nitrate and silicate, showing the influence of nitrification in the Clyde Estuary, Scotland (from Mackay and Leatherland, 1976).

Mackay and Leatherland (1976) have suggested that estuarine denitrification may be complete in the upper estuary of the Clyde after long, dry periods. Under these conditions, both river-nitrate and sea-water-derived nitrate are lost by denitrification to molecular nitrogen. Losses from sea-water are relatively small compared to those from river-water, simply as a consequence of the much lower concentrations of nitrate in sea-water.

The annual cycles of nitrate and chlorophyll a in the Inner Firth of Clyde system are shown in Fig. 41.12. The effect of the spring phytoplankton bloom on nitrate concentration levels is clearly demonstrated. This seasonal

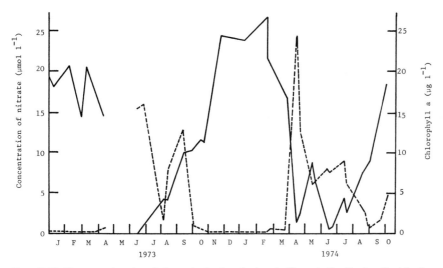

FIG. 41.12. The annual cycles of nitrate (unbroken line) and chlorophyll a (broken line) in the Inner Firth, Clyde Estuary, Scotland (from Mackay and Leatherland, 1976).

fluctuation is limited to those parts of the estuary in which the level of suspended solids is sufficiently low to allow photosynthesis to proceed, and to those in which the concentrations of toxic pollutants are not appreciable. For example, Mackay and Leatherland (1976) have reported that no significant seasonal changes of nitrate concentrations occur in the large sector of the Clyde Estuary subject to pollution.

41.5. PHYSICAL CHEMISTRY AND MODELLING OF ESTUARIES

There have been many recent advances in our understanding of the physical chemistry of natural waters, especially sea-water, and comprehensive reviews of this topic are available (see Millero, 1975a and Chapter 2). The chemical models of sea-water and other natural waters are based on ion–water and ion–ion interactions, and deal almost exclusively with the major cations and anions present. Despite these recent advances in our understanding of the physical chemistry of sea-water, only Millero and his co-workers have attempted to extrapolate the available chemical models to estuarine waters. Millero's model of estuarine physical chemistry is closely followed below; however the original sources cited should be consulted for a more comprehensive theoretical treatment.

Millero (1975b) has essentially applied the current models of sea-water and river-water to two aspects of the physical chemistry of estuaries. These are firstly, the investigation of physico-chemical properties of the major ions in estuarine waters in order to predict the densities of waters formed by the mixing of sea-water with river-water and secondly, the prediction of the speciation of some metal ions in estuarine waters using an ion-pairing model for ionic interactions.

41.5.1. THE DENSITY OF ESTUARINE WATERS FROM PHYSICO-CHEMICAL DATA

The densities of river-waters and the brackish waters formed as a result of mixing with sea-water in estuaries may be determined by the application of multicomponent electrolyte solution theory:

$$\Phi = \sum E_i \phi_i \qquad (1)$$

where Φ is any apparent equivalent property (such as volume, compressibility, enthalpy or heat capacity), E_i is the equivalent ionic fraction of a species i, i.e. $E_i = e_i/e_T$, where e_i is the equivalent molality of species i, and $e_T (= \sum e_i)$ and ϕ_i is the apparent equivalent property of species i at the ionic strength of the solution. The apparent equivalent property of the solution is related to the measured property (P) of the natural water by:

$$\Phi = \frac{(P - P^0)}{e_T} \qquad (2)$$

where P^0 is the physical property of pure water and e_T is the total equivalents of the dissolved salts in the natural water.

By dividing the apparent property into an infinite dilution term (Φ^0) and one or more concentration terms it is possible to simplify Young's rule at various concentrations:

$$\Phi = \Phi^0 + SI_v^{\frac{1}{2}} + bI_v + \ldots \qquad (3)$$

where b is an empirical constant, I_v is the molal ionic strength and:

$$\Phi^0 = \sum E_M E_X \phi_{MX} \quad \text{and} \quad S = \sum E_M E_X S_{MX} \qquad (4)$$

where E_M is the equivalent fraction of cation M and E_X is the equivalent fraction of anion X. Since Φ^0 and S are independent of specific ion–ion interactions, Young's rule need only be applied to b (and higher order terms if necessary):

$$b = \sum_{MX} E_M E_X b(MX) + E_v b(B) \tag{5}$$

where subscript v indicates equivalent volume.

Since the apparent equivalent property of a natural water is related to the physical properties by Equation (2), it is possible to determine the physical properties of the water from the estimated value of Φ:

$$P = P^0 + \Phi e_T. \tag{6}$$

Now $e_T = kCl_v$ and $I_v = k'Cl_v$ (where the volume chlorinity $Cl_v = Cl(\text{‰})$ $\times d$, in which d is the density) and a combination of Equations (3) and (6) gives:

$$P = P^0 + ACl_v + BCl_v^{3/2} + CCl_v^2 \tag{7}$$

where $A = k\Phi^0$, $B = k(k')^{\frac{1}{2}}S$, $C = kk'b$.

Equation (7) predicts the volume chlorinity dependence of the particular property of natural waters, e.g. sea-water diluted with pure water. Since Φ^0 is related to ion–water interactions and S and b are related to ion–ion interactions, any physical property of the water can be expressed as:

$$P = P^0 + \sum \text{ion–water interactions} + \sum \text{ion–ion interactions.} \tag{8}$$

Now, the ion–ion term can be divided into a theoretical Debye–Hückel limiting law term and a term due to deviations from the law:

$$\sum \text{ion–ion interactions} = \text{Debye–Hückel term} + \sum \text{deviations.} \tag{9}$$

The application of these concepts by Millero (1975b) and Millero and Lawson (1975) to natural waters has included the investigation of density as related to physico-chemical data. The apparent equivalent volume of sea-water is related to the density (d) by:

$$\Phi_v = \frac{1000(d^0 - d)}{e_T d^0} + \frac{M}{d^0} \tag{10}$$

where d^0 is the density of pure water, e_T is the normality (equivalents litre^{-1}) and M is the mean equivalent weight of sea-salt. By combining Equation (10) with Equation (3) one obtains:

$$d = d^0 + A_v Cl_v + B_v Cl_v^{3/2} + C_v Cl_v^2. \tag{11}$$

In order to determine the density of estuarine waters, Millero (1975b) has applied Young's rule to "World" river-water and sea-water. Sample calculations of Φ_v^0, b_v, and M_T for "World" river-water are shown in Table 41.4,

TABLE 41.4

Values of Φ_v^0, M_T *and* b_v *for river-salts at* 25°C[a]

Solute	Φ_v^0	$E_i\phi_v^0$	M_i	E_iM_i	$b_v(i)$	$E_ib_v(i)$
Ca^{2+}	$-8\cdot93$	$-4\cdot077$	$20\cdot0400$	$9\cdot150$	$0\cdot242$	$0\cdot110$
Mg^{2+}	$-10\cdot59$	$-2\cdot175$	$12\cdot1525$	$2\cdot496$	$-0\cdot197$	$-0\cdot040$
Na^+	$-1\cdot21$	$-0\cdot202$	$22\cdot9898$	$3\cdot839$	$1\cdot078$	$0\cdot180$
K^+	$9\cdot03$	$0\cdot325$	$39\cdot1020$	$1\cdot408$	$1\cdot129$	$0\cdot041$
HCO_3^-	$24\cdot29$	$14\cdot171$	$61\cdot0172$	$35\cdot597$	$2\cdot122$	$1\cdot238$
SO_4^{2-}	$6\cdot99$	$0\cdot993$	$48\cdot0288$	$6\cdot820$	$0\cdot134$	$0\cdot019$
Cl^-	$17\cdot83$	$2\cdot391$	$35\cdot4530$	$4\cdot754$	$-1\cdot030$	$-0\cdot138$
NO_3^-	$26\cdot20$	$0\cdot257$	$62\cdot0049$	$0\cdot608$	$-1\cdot000$	$-0\cdot010$
$Si(OH)_4$	$60\cdot0$[b]	$7\cdot974$	$96\cdot1156$	$12\cdot774$	—	—
	$\Phi_v^0 = 19\cdot657$		$M_T = 77\cdot446$		$b_v = 1\cdot400$	

[a] Data from Millero (1975b). See text for definitions of Φ_v^0, M_T and b_v.
[b] Data from Brewer and Bradshaw (1975).

and Table 41.5 gives values of these parameters for various rivers and sea-water.

By using Equation (11) and

$$\Phi_v = \Phi_v^0 + S_vI_v^{\frac{1}{2}} + b_vI \qquad (12)$$

Millero (1975a) has devised an equation for the density of river-water:

$$d = d^0 + 10^{-3}(M_T - d^0\Phi_v^0)e_T - 10^{-3}S_vd^0e_TI_v^{\frac{1}{2}} - 10^{-3}b_vd^0e_TI_v. \qquad (13)$$

Since e_T and I_v are small, Millero (1975b) neglects these terms and estimates the densities from the infinite dilution apparent molal volumes, $\Phi_v^0 = \sum E_i\phi_v^0$.

TABLE 41.5

Values of Φ_v^0, S_v, b_v, M_T *for river- and sea-salts at* 25°C[a]

Source	Φ_v^0	S_v	b_v	M_T
North America	$14\cdot934$	$2\cdot612$	$1\cdot417$	$74\cdot672$
South America	$25\cdot168$	$2\cdot055$	$1\cdot339$	$80\cdot734$
Europe	$14\cdot512$	$2\cdot739$	$1\cdot523$	$76\cdot407$
Asia	$18\cdot723$	$2\cdot432$	$1\cdot572$	$77\cdot151$
Africa	$22\cdot226$	$2\cdot104$	$1\cdot006$	$75\cdot856$
Australia	$23\cdot008$	$2\cdot276$	$1\cdot273$	$78\cdot540$
World	$19\cdot657$	$2\cdot371$	$1\cdot400$	$77\cdot446$
Sea-water	$13\cdot896$	$2\cdot150$	$-0\cdot101$	$58\cdot034$

[a] Data from Millero (1975b). See text for definitions of Φ_v^0, S_v, b_v and M_T.

Calculations of the relative densities of various river-waters at 25°C are given in Table 41.6, together with the densities of sea-water solutions diluted with pure water calculated at the same total concentrations of dissolved solids. The calculated densities of the river-waters are in good agreement with those determined from sea-water at the same total dissolved solids concentrations, with a mean variation of $\pm 3 \cdot 4 \times 10^{-6} \, \text{g cm}^{-3}$. Thus, the

TABLE 41.6

Relative densities of various river-waters at 25°C $((d - d^0) \times 10^6)^a$

Source	g_T (ppm)[b]	River	Sea-water	Δ (ppm)[c]
North America	147·8	118·3	111·8	6·5
South America	75·1	51·7	57·0	−5·3
Europe	185·4	150·3	149·6	0·7
Asia	148·8	112·8	112·5	−0·3
Africa	133·8	94·7	101·2	−6·5
Australia	61·4	43·5	46·8	−3·3
World	126·8	94·9	96·0	−1·1
			mean	$\pm 3 \cdot 4$

[a] Data from Millero (1975b).
[b] g_T (ppm) represents the total dissolved solid concentration, related to the salinity by $g_T = 1 \cdot 004847 \, S(\%_0)$.
[c] Δ indicates the agreement between the calculated densities of the various river-waters and those determined from sea-water at the same g_T value.

densities (and other physical properties) of river-waters are equal to those of sea-water diluted to the same total concentation of dissolved solids.

Extending Young's rule to estuaries themselves:

$$\Phi_v \text{(estuary)} = E_R \phi_v(R) + E_S \phi_v(S) \qquad (14)$$

where E_R and E_S are the equivalent fractions of river- and sea-salts, and $\phi_v(R)$ and $\phi_v(S)$ are the apparent equivalent volumes at the ionic strength of the estuarine water. Millero (1975b) has given values of Φ_v (estuary) derived using Equation (14) for various weight fractions (X_{SW}) of sea-water, and these are shown in Table 41.7. The table also gives values of the mean equivalent weight of the estuarine salt:

$$M_T = E_R M_R + E_S M_S \qquad (15)$$

and the total normality of the estuarine water

$$N_T = E_R N_R + E_S N_S.$$

Using these data, Millero (1975b) has devised values for the relative densities of the estuarine waters calculated from the equation:

$$d - d^0 = (M_T - d^0\Phi_v) N_T. \tag{16}$$

These densities are shown in Table 41.8 together with the densities of sea-water diluted to the same total dissolved solid concentrations as the estuarine waters.

TABLE 41.7

Apparent equivalent volume, mean molecular weight and total normality of estuary salts at various weight fractions of sea-salt $(X_{SW})^a$

X_{SW}	E_S	Φ_v (est)	M_T	N_T
0·00	0·0000	19·768	77·446	0·00164
0·02	0·8827	14·866	63·310	0·01370
0·04	0·9392	14·644	59·215	0·02576
0·06	0·9594	14·608	58·822	0·03785
0·08	0·9698	14·617	58·620	0·04995
0·10	0·9762	14·641	58·496	0·06207
0·20	0·9893	14·806	58·242	0·12280
0·30	0·9937	14·962	58·156	0·18388
0·40	0·9960	15·098	58·112	0·24525
0·50	0·9973	15·217	58·086	0·30694
0·60	0·9982	15·323	58·069	0·36895
0·70	0·9988	15·418	58·057	0·43127
0·80	0·9993	15·508	58·047	0·49393
0·90	0·9997	15·590	58·040	0·55689
1·00	1·0000	15·671	58·034	0·62019

[a] Data from Millero (1975b). See text for definitions of X_{SW}, E_S, Φ_v (est), M_T and N_T.

The agreement between the calculated densities and those of diluted sea-waters is excellent, and direct measurements made by Millero and Lawson (1975) of the densities of mixtures of artificial river-water and standard sea-water agreed with those calculated to $\pm 3 \times 10^{-6}$ g cm^{-3} over a salinity range of 0–35‰. By knowing the river-water input of dissolved salts (g_T river) to an estuary, the total dissolved solids in an estuarine water (g_T estuary) can be found from:

$$g_T(\text{estuary}) = g_T(\text{river}) + bS(‰) \tag{17}$$

where $S(‰)$ is the conductimetric salinity and b is a constant.

Millero *et al.* (1975) have used the equations given above to examine the densities of Baltic sea-waters and have found very good agreement between their calculated values and the density measurements of Cox *et al.* (1970) and Kremling (1972) for Baltic waters if a river input of g_T (river) = 0·120 g kg^{-1}

was assumed. This value for river input of dissolved salts is in excellent agreement with the data of Kremling (1972) for the composition of river-waters draining into the Baltic. Millero (1975b) has pointed out the interesting feature that the earlier measurements of the density of Baltic sea-waters given by Knudsen (1901) give a value of g_T (river) $= 0.073$ g kg^{-1} which is in agreement with compositional data by Lyman and Fleming (1940), suggesting that the river input or the evaporation to precipitation ratio for the Baltic Sea has increased over the last 70 years or so.

TABLE 41.8

Densities of estuarine waters at various fractions of sea-salt $((d - d^0) \times 10^3)^a$

X_S	g_T	Estuary	Sea-water	Δ (ppm)
0·00	0·127	0·095	0·096	−1
0·02	0·827	0·622	0·622	0
0·04	1·529	1·149	1·148	1
0·06	2·229	1·675	1·672	3
0·08	2·931	2·200	2·197	3
0·10	3·631	2·725	2·720	5
0·20	7·136	5·339	5·335	4
0·30	10·640	7·951	7·947	4
0·40	14·144	10·560	10·558	2
0·50	17·649	13·172	13·172	0
0·60	21·153	15·788	15·789	−1
0·70	24·657	18·408	18·410	−2
0·80	28·161	21·034	21·035	−1
0·90	31·666	23·665	23·666	−1
1·00	35·170	26·301	26·301	0

mean ± 1.9

^{a} Data from Millero (1975b). g_T and Δ are as defined for Table 41.6.

41.5.2. METAL-ION SPECIATION IN ESTUARIES

Millero (1975b) has suggested that, because the total ionic strength of river-water is very low, it is permissible to estimate the activity coefficients of free ions and ion pairs by the use of an equation derived from the Debye–Hückel theory:

$$\log \gamma_F = -(0.509) Z^2 \left[\frac{I^{\frac{1}{2}}}{(1 + I^{\frac{1}{2}})} - 0.3 I \right]. \qquad (18)$$

Where γ_F is the mean activity coefficient of the free species, Z is the electrostatic charge on the ion and I is the ionic strength.

The equation gives values of γ_F of 0·951 and 0·817 for free monovalent and for divalent ions or ion pairs respectively, at $I = 2.095 \times 10^{-3} \, mol \, kg^{-1}$. Solving this equation for "World" river-water, Millero (1975b) has produced speciation data for the major cations and anions and these are summarized in Table 41.9. In both river-water and sea-water, the major cations are predominantly in the free form. The anions, with the exception of CO_3^{2-}, are also mainly in the form of free ions. It would be reasonable to assume that during the mixing of fresh-water and sea-water in estuaries the major cations and anions are predominantly present as their free ions.

TABLE 41.9

Speciation of the major cations and anions in river-water at 25°C[a]

Cation	% M	% MSO$_4$	% MHCO$_3$	% MCO$_3$	% MOH
H$^+$	99·98	0·02	—	—	—
Na$^+$	99·83	0·05	0·12	—	—
Mg^{2+}	97·54	1·15	1·21	0·08	0·01
Ca^{2+}	96·89	1·45	1·32	0·33	0·01
K$^+$	99·92	0·08	—	—	—

Anion	% X	% NaX	% MgX	% CaX	% KX
Cl$^-$	100·00	—	—	—	—
SO$_4^{2-}$	93·55	0·11	1·66	4·64	0·04
HCO$_3^-$	99·23	0·04	0·21	0·52	—
CO$_3^{2-}$	31·03	0·03	6·50	62·44	—
NO$_3^-$	99·93	0·01	—	0·06	—
OH$^-$	94·64	0·01	4·83	0·52	—
F$^-$	98·79	0·01	0·88	0·32	—

[a] Data from Millero (1975b).

The speciation of the heavy metals, e.g. mercury, copper, lead, zinc and cadmium, in estuarine waters is of considerable interest in terms of the biological availability and toxicity of these metals. Theoretical prediction of the speciation of heavy metals from the Debye–Hückel theory can be made using Equation (18) and Millero (1975b) has calculated the total activity coefficients of copper, zinc, cadmium and lead in river-water in an attempt to compare speciation in riverine, estuarine and marine waters. He has pointed out the fact that organic ligands occur in natural waters at various (often unknown) concentrations, and their ability to form complexes with the heavy metals will make any theoretical calculations of metal speciation nothing better than a first approximation. Millero concluded that the preliminary data on heavy metal ion speciation in river- and sea-waters suggested that large changes in speciation occur during estuarine mixing, and that further work on the influence of the changing state of ionic solutes during estuarine mixing on chemical processes such as solubility and ion-exchange is required.

41.6. TRACE ELEMENTS

The trace element chemistry of both natural waters and sediments has been an area of considerable growth in the past two decades, and the estuarine chemistry of trace elements is an important feature of this advance. Early interest arose mainly from the fact that estuaries are an important stage in the supply of trace elements from continental weathering to the oceans. As previously indicated, the river–estuary transport system is quantitatively important in the supply of both dissolved and particulate matter to the oceans, so that a knowledge of the trace element composition of estuarine waters and particulates is essential to the understanding of the marine geochemical budget.

41.6.1. TRACE ELEMENT SPECIATION AND SOLID–SOLUTION EXCHANGE

The chemistry of the minor, or trace, elements in sea-water and their speciation has been reviewed in Chapters 7 and 3 respectively, and in view of this the following discussion will be specifically limited to estuaries.

41.6.1.1. *Trace element speciation*

Several major difficulties arise when any attempt is made to predict the chemical speciation of trace elements in estuarine waters. Burton (1976) has divided these problems into four categories.

(i) It is not reasonable to assume that the trace elements in estuarine waters are present at equilibrium. Biological cycling of trace elements produces unstable oxidation states of the elements and organic complexes. The net results of these biological processes is the production of chemical species which do not conform with those which are predicted by the concepts of physical chemistry alone. Models of trace element speciation, such as those of Sillén (1961), Goldberg (1965) and Whitfield (1975), which use thermodynamic principles and which assume equilibrium conditions, do not take into account the biological influences which may control speciation. Thus, the use of theoretical models for the prediction of speciation has been criticized by Burton (1976) on the grounds that estuaries are environments of generally high biological activity and, in consequence, they may be further from equilibrium than many other parts of the hydrosphere. Equilibrium models may be criticized further inasmuch as they invariably assume a closed thermodynamic system, whereas, in reality, estuaries are open systems within which the reactants and products are in states of flux in and out of the system. The residence times of trace elements are probably considerably less than those required for the attainmant of equilibrium conditions within most estuaries.

(ii) Complexation of trace elements in estuaries by organic compounds of marine and continental origins must be taken into account when speciation is considered. The nature of the dissolved and particulate organic matter in estuaries is not well defined as yet, and the concentrations of those organic compounds which have been identified in estuaries are rather variable (see Section 41.7). The amount and nature of the organic matter varies as a result of changes in river run-off conditions and biological activity within the river and estuary. This paucity of data, combined with speculations about the changes in composition of organic materials during estuarine mixing (see Section 41.7), does not allow a realistic prediction to be made of the extent of the organic complexation of trace elements in estuaries. Despite these general difficulties, some attempts have recently been made to predict the organic complexation of trace elements in natural waters. Theoretical models for speciation exist for fresh water and sea-water, but to date only one model of organic–metal speciation has been specifically attempted for brackish waters (Dickson et al., 1978). Reviews of fresh water and sea-water organic complexation models have been given by Stumm and Morgan (1970), Singer (1973) and Ahrland (1975) (see also Chapter 3).

(iii) Burton (1976) cited the uncertainty of the value of the redox potential of natural waters as another reason why the speciation of trace elements in estuarine waters cannot easily be predicted. Recent studies by Breck (1972) and Liss et al. (1973) have suggested values of around 8·5 for the pE of oxygenated sea-water. This contrasts with the value of 12·5 derived from the assumption that the pE of normal sea-water should be controlled by the oxygen–water redox couple. Since the oxidation states of the transition elements are most often directly dependent on the pE of the medium, any discrepancy in the pE value taken for estuarine or other natural waters will lead to serious errors in the prediction of trace element speciation.

(iv) Finally, Burton (1976) suggested that the modelling of trace element speciation in estuaries will be affected by errors in the existing data on metal–ligand stability constants. These errors arise mainly from the extrapolation of activity coefficients for ideal, dilute solutions to the higher ionic strengths found in marine and estuarine waters. These errors have led Millero (1975b) to conclude that any models of trace element speciation in estuaries can only be viewed as order of magnitude estimates (see Section 41.5.2).

To illustrate the discrepancies between predicted (i.e. thermodynamically modelled) and observed trace element speciations in estuarine waters, Burton (1976) has considered copper as an example and some of his important conclusions may be summarized as follows.

(a) The thermodynamic model of inorganic copper speciation presented by Zirino and Yamamoto (1972) for normal sea-water cannot be extrapolated to estuarine conditions. For normal sea-water, their data suggest that copper is present mainly in the form of $Cu(OH)_2^0$, with $CuCO_3^0$ as the next most abundant species. Decreasing the pH to that of brackish waters would then lead to more of the copper being present as $CuCO_3^\circ$ at the expense of $Cu(OH)_2^\circ$. Burton (1976), however, has pointed out that in a real estuarine situation the pH decrease in going from sea-water to brackish water is accompanied by changes in the anionic composition of the water. Hence, a simple extrapolation on the basis of pH alone is inadequate, and a new thermodynamic model for estuarine conditions is required to account for the variations in the anion concentrations due to mixing, and also for the change in ionic strength (see Section 41.5.2).

(b) Estuarine waters, in common with other natural waters, contain a variety of organic constituents which are capable of forming stable complexes with copper ions. Burton (1976) cited the studies of Slowey *et al.* (1967), Matharu (1975) and Stiff (1971), all of which indicate that a considerable fraction of the total dissolved copper in estuarine and river-waters may be extracted into suitable organic solvents, e.g. chloroform and hexanol. This suggests that some of the copper is present as organic complexes. Furthermore, several workers have carried out experiments which were designed to extract copper from estuarine and other natural waters with organic chelating reagents. These have indicated that a proportion of the total copper is present as stable organic complexes even after photo-oxidation of the waters (Corcoran and Alexander, 1964; Alexander and Corcoran, 1967; Williams, 1969; Slowey and Hood, 1971; Foster and Morris, 1971; Matharu, 1975).

(c) The presence of both inorganic and organic associations of copper in estuarine waters will certainly give rise to erroneous assumptions in the analytical determination of "dissolved copper" by the various analytical techniques which only determine specific forms of the metal, e.g. anodic stripping voltammetry, and the solvent extraction pre-concentration procedures used with atomic absorption spectrophotometry.

The speciation of copper in estuarine and associated waters has been studied in more detail than that of other trace elements; however, it is reasonable to suppose that the preceding remarks will also apply to the latter. Examples of the occurrence of organic complexes (especially humic acid complexes) of other metals in estuarine and coastal waters include those described for iron (Head, 1971; Butler and Tibbits, 1972; Theis and Singer, 1974), mercury (Fitzgerald and Lyons, 1973; Andren and Harriss, 1975; Millward and Burton, 1975), lead (Matson, 1968), cadmium (Gardiner, 1974) and copper and other heavy metals (Dickson *et al.*, 1978).

41.6.1.2. Trace element solid–solution exchange

It is well established that the adsorption of trace elements from solution onto mineral and organic solids takes place in natural waters. The first detailed study of trace element adsorption from sea-water was that by Krauskopf (1956), and many other workers have since investigated trace element adsorption and desorption in the marine environment (for a review of this topic see Chapter 4).

Trace element adsorption results from several mechanisms including electrostatic attraction, changes in the hydration state of the adsorbate, covalent bonding, Van de Waal's bonding and hydrogen bonding (see Chapter 4). Mineral and other solid particles suspended in natural waters carry an electrical surface charge, so that attraction and surface adsorption of charged species from solution is a widespread mechanism by which solid–solution exchange can occur. This form of interaction is rapid and readily reversible (see Chapter 4), and may be postulated as an important process by which cationic and anionic trace element species are adsorbed by, or desorbed from, suitably charged particles.

In an estuarine situation, the suspended solids capable of providing adsorption sites are subjected to important physico-chemical changes which result from the mixing of river- and sea-waters. These changes, which affect, for example pH, pE and ionic strength, influence the adsorption–desorption behaviour of the suspended matter and modify the colloidal nature of much of it (see Section 41.8.1). Thus, estuaries represent a situation in which the solid–solution exchange of trace elements might conceivably lead to the enrichment or depletion of the trace element content of either suspended particles or the "dissolved" constituents of estuarine waters. Exchange reactions are obviously major factors in the control of the forms in which elements are transported from the continents to the oceans and marine sediments.

The solid–solution exchange processes in estuaries can be categorized, each category loosely corresponding to an "adsorption" or "desorption" effect. Some of the estuarine exchange processes which have been investigated under field or laboratory conditions will be discussed below.

41.6.1.2.1. *Exchange of Trace Elements Between Pre-Existing Solids and Solution.* The ability of pre-existing solids to adsorb trace elements from solution under both freshwater and marine conditions has frequently been cited (e.g. Krauskopf, 1956; Turekian, 1965; Chester, 1965; Aston and Duursma, 1973). On the other hand, the desorption of trace elements from river-borne particulate matter during laboratory experiments intended to simulate estuarine mixing has been reported by Kharkar *et al.* (1968) and Rickard

(1971). Field evidence for trace element desorption in estuaries has been given by Turekian (1971) and Evans and Cutshall (1973).

There is somewhat conflicting evidence on trace element behaviour during solid–solution exchange in estuaries and this has led to considerable controversy. Much of the disparity between the results of different researchers may be accounted for in one or more of the following ways.

(i) The destruction or alteration of pre-existing solids (e.g. organic matter, clay minerals) can lead to the release of incorporated trace elements. This is not strictly desorption, but is better regarded as loss of trace elements to solution resulting from the decomposition and/or a compositional change of the solids.

(ii) Differences in (a) the rate of estuarine mixing, (b) the nature of the river- and sea-water end-members, and (c) the type of solids supplied from marine and up-river sources often result in the solid–solution exchange processes differing from one estuary to another.

(iii) Differences in the grain size distributions of solids taken from various locations in a river–estuary system may lead to an apparent uptake or loss of trace elements. This grain size distribution effect is a result of the fact that a given weight of small-size particles presents a greater surface area for adsorption/desorption processes than does the same weight of large-size particles of the same material, i.e. the smaller particles have a higher specific surface area. This effect has been illustrated by several workers (e.g. Turekian, 1965; Cranston and Buckley, 1972; de Groot et al., 1971). Figure 41.13 shows the relationship between the trace element composition and the grain size distribution ($\% < 16$ μm) of sediment samples from the River Ems. It is clear that a comparison of suspended solids of different grain size distributions could lead to an apparent change in the trace element content of the solids from a river–estuary system. Conclusions on the adsorption or desorption of trace elements during passage through an estuary must take particular care to eliminate grain size and other sedimentological factors.

The problems of the interpretation of solid–solution exchange processes for trace elements in estuaries are illustrated rather well by the work of de Groot (1966), de Groot et al. (1971) and Müller and Förstner (1975). The transport of heavy metals into the North Sea by the suspended solids in the highly polluted Rhine River has been studied by de Groot (1966) and de Groot et al. (1971). The results of this work are summarized in Fig. 41.14. If it is assumed that at Biesbosch, a freshwater area (chlorinity, $0\%_0$), the total amount of a given heavy metal in the sedimentary material is 100%, a steady decrease in the percentage concentration is observed towards the Haringvliet (chlorinity, $2\%_0$), with a further decrease towards the Dutch Wadden Sea (chlorinity, $16\%_0$). At the Wadden Sea, cadmium, mercury, copper, zinc, lead and chromium show a loss of $70–95\%$, whereas copper, nickel, and iron decrease by 45–60% of the Biesbosch values. Manganese, samarium, scandium and lanthanum concentrations do not change significantly. De Groot and his co-workers (de Groot, 1966; de Groot *et al.*, 1968, 1971) have explained the decreases as being the result of desorption. They have also envisaged that intensive decomposition of the organic matter in the sedimentary solids occurs,

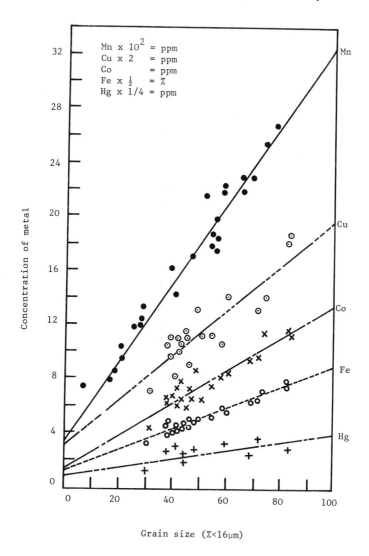

Fig. 41.13. The relationship between the trace element composition and grain size distribution (% <16 μm) for sediment samples from the River Ems (after de Groot *et al.*, 1971).

and that the decomposition products of this organic matter then form soluble organo-metallic complexes with the metals released from the sediments. This is an example of the type of process outlined in (i) above as a possible "desorption" mechanism.

Müller and Förstner (1975) have studied the heavy metal content of sedimentary solids in the highly polluted Elbe Estuary which, like the Rhine,

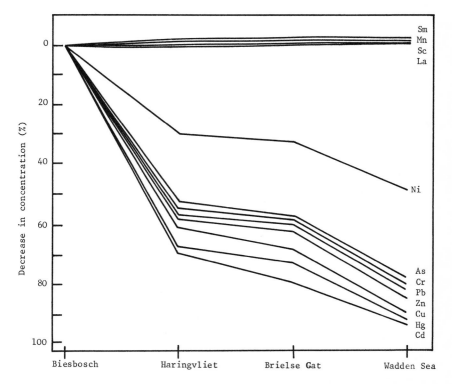

FIG. 41.14. The heavy metal concentrations of sediments from various locations on the Rhine Estuary (after de Groot *et al.*, 1971). The decreases in heavy metal concentrations are expressed as percentages of the concentration in fluvial Rhine sediment (at Biesbosch).

drains into the North Sea. Their investigations clearly indicate that a mixing process between suspended material from the highly polluted Elbe River and sedimentary material from the North Sea, which has a relatively low heavy metal concentration, is the most important mechanism leading to the gradual decrease of metal concentrations in the Elbe Estuary sediments toward the North Sea. They have suggested that the seaward decreases of metals in the Rhine Estuary may be explained in a similar manner, since upstream transport of marine sediments from the North Sea into the Rhine is well established (van Veen, 1936). These authors also reported that the percentages of the metal released from Rhine River sediments when treated with sea-water for prolonged times were not great; e.g. losses over 75 hours at 20°C lay in the ranges: copper 0·9–1·0%, lead 0·3–2·4%, cadmium 0·5–3·0%, cobalt 1·2–2·7% and zinc 0·9–1·7%.

Overall, the work of Müller and Förstner (1975) provides a good example of the problems of the interpretation of solid–solution exchange in estuaries and points to the need for attention to physical and sedimentological processes when element exchange processes are investigated. The present confusion surrounding the question of whether or not estuaries are, in general, environments in which trace element exchange between solution and pre-existing sedimentary material occurs will only be resolved when careful attention is paid to the three points stated above.

Despite the complications mentioned here, the higher salinity of estuarine water compared to that of river-water might be expected to give rise to desorption from pre-existing solids for those trace elements which can take part in ion-exchange reactions. Simple cationic trace element species held on the ion-exchange sites of clay minerals can take part in exchange reactions with the major ions present in solution; e.g.

$$T^{n+} - \text{clay} + n\text{Na}^+_{aq} \rightleftarrows n\text{Na}^+ - \text{clay} + T^{n+}_{aq}$$

(where T^{n+} is a cationic trace element species).

Since the concentrations of dissolved major ions, e.g. Ca^{2+}, Mg^{2+}, Na^+ and K^+, increase in the transition from river-water to sea-water in estuaries, a law of mass action effect will give rise to reactions of the above type in a forward direction.

Aston and Duursma (1973) have studied the effect of salinity changes on the adsorption–desorption behaviour of caesium, zinc, cobalt and ruthenium with marine sediments. Figure 41.15 shows the equilibrium absorption distribution coefficients for Cs (labelled as ^{137}Cs) between sea-water and suspended sediments as a function of the illite content of the sediments. The distribution coefficient (defined as $K_{ds} = (M)_s/(M)_w$, where $(M)_s$ and $(M)_w$ are the concentrations of caesium in sediment and water respectively) is directly proportional to the illite content of the sediment. Cheng and Hamaguchi (1968) have proposed that Cs^+ adsorption on illite takes place by exchange with K^+ ions. Although caesium uptake is proportional to the illite content of the sediment (and thus to the exchangeable K^+ content), this relationship was not observed for zinc, cobalt or ruthenium, which suggests that these latter trace elements may not be adsorbed by simple ion-exchange reactions. Aston and Duursma (1973) have further demonstrated an ion-exchange-type uptake for caesium by examining the partition of the element between solid–solution phases as a function of salinity. Figure 41.16 illustrates the effect of salinity on the uptake of ^{137}Cs, and, for comparison, ^{65}Zn, on an argillaceous near-shore sediment. The amount of caesium per unit weight of sediment decreases with the increase of salinity, as predicted above. When the concentration of $(Cs^+ + Na^+)$ per unit weight of sediment is expressed as a function of salinity, an increase in this quantity is found as salinity increases. The value tends to a maximum as the normal salinity of sea-water is reached, with the

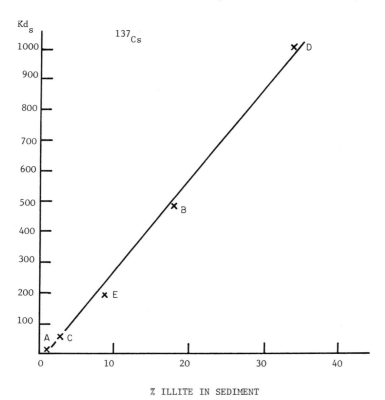

% ILLITE IN SEDIMENT

FIG. 41.15. Distribution coefficient (K_{ds}) values at equilibrium for ^{137}Cs adsorption on suspended sediment as a function of the illite content of the sediment (after Aston and Duursma, 1973). Sediment (A) Atlantic 24°48′N., 77°32′W.; (B) Mediterranean 43°37′N., 67°23′W.; (C) Pacific 50°02′S., 127°31′W.; (D) Indian 18°60′N., 72°57′E.; (E) Atlantic 04°35′N., 06°45′W.

maximum number of exchange sites occupied by Na^+ ions. In contrast, adsorbed zinc does not appear to be in competition with major cations for exchange sites on the sediment, and the mechanism of zinc uptake cannot be explained by ion-exchange processes.

In conclusion, the uptake or release of trace elements by pre-existing sedimentary solids in estuaries is probably related to ion-exchange processes for some elements, e.g. caesium. This should give rise to desorption as major cation concentrations increase during estuarine mixing. Other elements do not exhibit simple ion-exchange behaviour, and their solid–solution exchange in estuaries is poorly understood in terms of their reactions with pre-existing sedimentary solids.

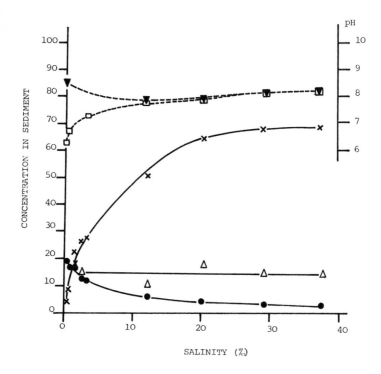

FIG. 41.16. Salinity effects on the uptake and distribution of ^{137}Cs and ^{65}Zn by sediment B (see legend to Fig. 41.15). Continuous lines: $[Cs + Na] \times 10^{-2} g\,g^{-1}$ (x); $[Cs] \times 10^{-7} g\,g^{-1}$ (●); $[Zn] \times 10^{-7} g\,g^{-1}$ (△). Broken lines show the pH of the waters without suspended sediment (□), and with suspended sediment (▼). (After Aston and Duursma, 1973.)

41.6.1.2.2. *Precipitation of dissolved matter to give new solid phases.* The hydrogenous precipitation of, for example, hydrated oxides of iron and manganese from sea-water has important consequences for trace element solid–solution exchange (see Chapters 28 and 34). These hydrogenous oxides are very effective in the removal of trace elements from solution in natural waters by lattice substitution and surface adsorption processes. The formation of these oxides is dependent on the hydrolysis of dissolved iron and manganese ions, and for this reason it is a function of the pH of the solution .Estuaries are a transition area between river-water pH values, typically 6·0–7·0, and that of sea-water, ~8·1. Thus, it is within estuaries that formation of new solid phases, e.g. oxides or hydroxides, might be expected to occur.

Iron has received more attention than other transition metals with regard to the possible formation from it of new solid phases during estuarine mixing. This is due in part to the recognised importance of iron oxides in controlling the uptake of other trace elements from solution, but is also due to the fact that the hydrolysis of iron(III) in river-water is one of the most likely reactions of this type to occur when the pH is raised by mixing with sea-water.

Coonley et al. (1971) have reported the precipitation of iron from solution in the Mullica River estuary, in which acidic river-water with a high dissolved iron concentration is mixed with sea-water. Similar losses of dissolved iron during estuarine mixing have been described by Windom et al. (1971) for the Satilla Estuary, and by Holliday and Liss (1976) for the Beaulieu Estuary, England. For the Southampton Water estuary, which receives high-pH water from a chalk catchment, Head (1971) reported dissolved iron in the river-water input at concentrations similar to that of sea-water, no loss from solution occurring during estuarine mixing.

In an attempt to determine the influence of sedimentary particles in suspension on the loss of dissolved iron(III) from solution, Aston and Chester (1973) studied ^{59}Fe behaviour as a function of salinity. Their results showed that suspended sediment particles greatly enhance the removal of iron from solution, and that the removal is also more rapid and complete with increasing salinity. Two forms of iron(III) solution were used for the study. The first was a fresh, or unaged, solution of ferric chloride, whereas the other was iron-(III) which had been equilibrated with sea-water. Figure 41.17 illustrates that there is loss of iron(III) from solution in the presence of suspended sediment, and demonstrates that the iron removal is particularly important for the so-called unaged iron, i.e. that equivalent to the inorganic iron entering an estuary in acidic river-water. The effect for "aged" iron which had been equilibrated with sea-water (Chow, 1965) was less dramatic.

The effects of the suspended particles on iron precipitation are shown in Fig. 41.18; this demonstrates removal of dissolved iron(III) to the extent of 75–90% in the first five hours under experimental conditions. Electrophoretic measurements which were used to gauge the surface charge of the sediment particles showed that the initial negative charge was markedly neutralized as the iron precipitation proceeded. Aston and Chester (1973) suggested that this loss of negative charge on the suspended particles is the result of interaction with cationic iron hydrolysis products and the formation of an iron oxide–hydroxide coating on the particles. The hydrolysis reaction may be represented by a Bronstëd-type reaction:

$$Fe(H_2O)_6^{3+} \xrightarrow{OH^-} Fe(OH)(H_2O)_5^{2+} + H_3O^+ \xrightarrow{OH^-} Fe(OH)_2(H_2O)_4^+$$
$$+ H_3O^+ \xrightarrow{OH^-} Fe(OH)_3(H_2O)_3^0 + H_3O^+$$
$$\text{(solid)}$$

The occurrence of iron oxide coatings on suspended sediment particles from

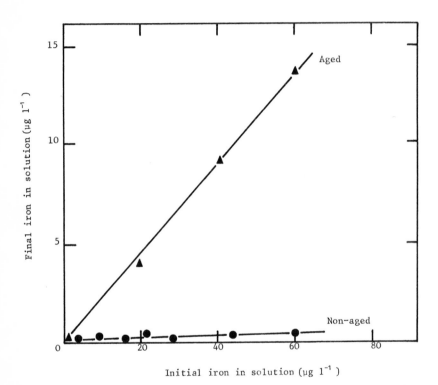

FIG. 41.17. The precipitation of iron from solution in sea-water (S = 34·2‰, pH = 7·9) in the presence of suspended sediment for "aged" and "unaged" iron solutions (after Aston and Chester, 1973).

the Amazon and Yukon River estuaries has been reported by Gibbs (1973). The hydrolysis history of inorganic iron in river-water would appear to be important to its association with the negatively charged particles which act as nuclei. Burton (1976) has pointed out that the history of the iron present in rivers (e.g. whether it is derived from weathering or anthropogenic sources) may affect its estuarine behaviour.

Windom *et al.* (1971) have provided data on the behaviour of dissolved

iron, together with manganese, zinc, copper and nickel, in the Satilla Estuary.
These workers found rapid removal of dissolved iron when the river-water
encountered saline waters, some loss of dissolved manganese, but little effect
on the concentrations of dissolved copper and nickel. The results of later
studies by Holliday and Liss (1976) on the Beaulieu Estuary agreed with those
of Windom *et al.* (1971); the iron in solution exhibiting rapid losses during

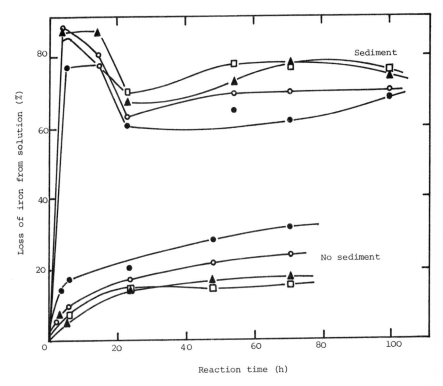

FIG. 41.18. The precipitation of iron in waters of various salinities (after Aston and Chester, 1973).
(●) 34·2‰, pH 7·90; (○) 23·7‰, pH 7·90; (▲) 17·1‰, pH 7·85; (□) 8·6‰, pH 7·15.

mixing, whereas other dissolved trace elements showed only negligible loss.
In contrast, Gibbs (1973) has shown that the iron oxide coatings which cover
the suspended sediment grains of the Yukon and Amazon Estuaries are
important in the transport of manganese, nickel, cobalt, chromium and
copper. This incorporation of trace elements may be attributed to co-
precipitation during iron oxide formation (Gibbs, 1973).

The results of Windom *et al.* (1971) and Holliday and Liss (1976), which indicate a lack of reactivity for several trace elements, are surprising, espe-cially when iron is being formed. Liss (1976) has commented that even in the absence of co-precipitation with iron, three other processes might be expected to lead to the non-conservative behaviour of trace elements.

(i) Manganese hydrolysis to form hydrated manganese oxides in a manner analogous to iron oxide formation. Fukai and Huynh-Ngoc (1968) have, however, shown experimentally that manganese precipitation requires conditions of higher pH than those normally encountered in estuaries.

(ii) Losses of trace elements from clay surface adsorption sites should lead to a non-conservative behaviour (see e.g. Kharkar et al., 1968; Gibbs, 1973; but see also Section 41.6.1.2.1).

(iii) Biological activity might be expected to affect the dissolved and particulate forms of trace elements in estuaries (see Section 41.6.2).

This discussion has so far been limited to the reactant approach, as defined by Sholkovitz (1976); however, it is worthwhile considering the experimental results obtained by Sholkovitz using a product approach (see Section 41.4.1). The only trace elements considered by this worker were iron and manganese, for which the results may be summarized as follows. The river-waters from Scotland used in Sholkovitz's laboratory studies were rich in colloidal humic material derived from peat drainage. After mixing with sea-water, the river-waters showed an almost complete loss of iron and manganese from solution. Sholkovitz (1976) concluded that organic matter, especially flocculated humic substances, plays an important role in controlling the concentrations of inorganic trace elements in river-waters and their non-conservative behaviour behaviour during estuarine mixing.

Several criticisms can be raised about the suggestion that humic materials are dominant in controlling the non-conservative behaviour of iron, manga-nese and other trace elements in estuaries. The major problems arise from the field observations of, for example, Windom et al. (1971) and Holliday and Liss (1976), who have found no evidence for the non-conservative behaviour of certain trace elements, and the observations of Spencer and Sachs (1970) which indicate that the trace element composition of estuarine and coastal suspended matter is controlled by alumino-silicates and not organic floccu-lants. No doubt, in a situation where river-water of excessive humic content enters sea-water, flocculation may result in inorganic trace element removal. For instance, Price and Calvert (1973) have shown that there is a co-variance between non-silicate iron and particulate phosphorus in suspended matter from Loch Etive, Scotland, a drainage area similar to that examined by Sholkovitz (1976). The latter author suggested that this co-variance may be attributed to iron–humic–phosphorus flocculants.

When the composition of sediments is considered, there is no evidence of iron accumulation, even in a confined area such as Great Bay (New Jersey) where river-borne iron is flocculating (Coonley et al., 1971). Sholkovitz (1976) has concluded that humic flocculation will only significantly alter the chemical composition of suspended matter in the confined waters of estuaries if a pycnocline inhibits the deposition of the flocculants so that they are rapidly dispersed by currents and turbulence. In the light of these conclusions, humic flocculation may be regarded as being quantitatively trivial in the trace element chemistry of estuarine waters.

41.6.2. BIOLOGICAL AND TIME-DEPENDENT CONTROLS OF TRACE ELEMENTS

The partitioning of trace elements between solid and solution phases in estuaries is modified by biological activity and by the influence of tidal and seasonal variations in the supply of the individual elements. Aquatic organisms are well known for their ability to take up trace elements from solution and particulates (Bowen, 1966; Riley and Chester, 1971). Biological productivity in estuaries is controlled by factors such as temperature, light penetration, nutrient availability, etc., many of which are time dependent. The rhythmic nature of tides and the variations in river-water supply, which are a result of changes, both rapid and seasonal, in catchment conditions, give rise to temporal trends in the factors controlling biological productivity and to changes in trace element supply itself. The characteristics of individual estuaries may lead to a dominance of either biological or hydrological influences on the temporal variations in dissolved and particulate trace element concentrations. For example, Williams and Chan (1966) have shown that in the Fraser River estuary, the seasonal cycle of particulate iron concentrations has an inverse trend to that of the dissolved iron. Seasonal variations in river run-off give rise to changes in the particulate iron input to the estuary, but in addition a definite biological cycle of the element can be identified and has been ascribed by Williams and Chan (1966) to the assimilation of particulate iron by phytoplankton and to the subsequent excretion of iron in a dissolved form. In contrast, Knauer and Martin (1973) concluded that seasonal changes in phytoplankton production have little or no effect on the dissolved concentrations of copper, manganese, lead and zinc in the waters of Monterey Bay, California. In this coastal water mass, hydrographical fluctuations appear to exert a dominant control over seasonal changes in dissolved metal concentrations.

Carpenter et al. (1975) found that a significant biological influence is exerted on zinc in northern Chesapeake Bay and the Susquehanna River estuary. Biogenic uptake in the water column and subsequent transport to the bottom sediments is counteracted by the return of zinc to solution during organic decomposition processes at the sediment surface, the latter having turnover times of about a week. The zinc is regenerated into solution in the form of organic complexes, and the occurrence of complexed zinc correlates well with the seasonal variations in phytoplankton standing-crops and productivity (Carpenter et al., 1975). This is an example of a marked biological influence on the trace element zinc in estuarine waters.

The influence of vegetation on the flux of trace metals through a salt marsh estuary (Windom, 1975) centres on the ability of the plants inhabiting this environment (e.g. *Spartina* spp.) to transfer the metals from deposited sediment to the water mass. In salt marshes in the south-eastern U.S.A., the regeneration of metals into the estuarine waters by the decay of *Spartina* is most significant for mercury, and Windom (1975) proposed that such a mechanism is important in the supply of organically bound mercury to solution. In the absence of this regeneration by decaying vegetation, the mercury would presumably be removed to the sediments.

In addition to the influence of phytoplankton and other vegetation, the effects of fish on the estuarine chemistry of trace elements must be taken into account. Cross et al. (1975) have considered the role of juvenile fish in the cycling of trace elements in coastal plain estuaries, and have concluded that in those estuaries where there are substantial fish populations these exert a significant effect on trace metal cycling. In the Newport River estuary, with a population of the order of ten million fish, 0.3% of the zinc, 0.7% of the iron and 0.06% of the manganese budgets of the water mass are ingested daily by filter-feeding fish. Because the assimilation efficiences of fish for metals are in general low, unassimilated metals that are egested from these fish must be quantitatively important in the cycling of metals in this estuary. In a shallow, coastal plain estuary, this may result in the trace metals being rapidly moved to the surface sediments in forms in which they would be subject to ingestion by other organisms. Cross et al. (1975) emphasized that the paucity of data on faecal deposition, feeding patterns and growth rates of estuarine fish does not allow a quantified model of trace element cycling in estuaries to be produced at present. There is no doubt that a better knowledge of estuarine ecology is needed if our understanding of the chemistry of trace elements and other constituents in estuaries is to be put on a firm footing.

Apart from biological influences, which tend to produce seasonal changes in the estuarine chemistry of trace elements, there are those temporal changes which result from tidal and seasonal modifications of estuarine mixing and river run-off. Boyden et al. (1978) have found that the dissolved and particulate trace element concentrations at locations in two estuaries (SW. England) show considerable variations during tidal and seasonal cycles. One estuary, the Restronguet Creek, receives drainage from a catchment which has considerable heavy metal pollution from past and present mining activities. In contrast, the other, the Helford Estuary, is almost free from pollution. The results of this two-year study of tidal and seasonal variations in the estuaries by Boyden *et al.* may be summarized as follows.

(i) During tidal cycles in both estuaries, the maximum concentrations of cadmium, copper, cobalt, iron, manganese, nickel, lead and zinc, in both soluble and particulate forms, occurred at low water for the whole water column. The extent of the differences in concentrations found at low and high water varied from element to element. For example, iron, lead and copper tended to occur in the particulate fraction of low water samples to an extent significantly greater than that for other elements.

(ii) Except for nickel and lead, there were higher trace element concentrations in both the low and high tide surface and bottom waters in winter than in summer. This observation probably reflects the increased weathering and transport of trace elements in the catchments during winter.

(iii) Unlike the estuarine waters, the surface sediments of the estuaries studied by these authors showed very little variation in their trace element contents on either a tidal or a seasonal basis.

41.7. ORGANICS AND BIOCHEMISTRY

Estuaries are environments which receive relatively high inputs of both inorganic and organic matter. They are also ecosystems of relatively high biological productivity, and it is important to consider the organic constituents of estuaries for two reasons. First, to understand the organic cycles of estuaries in their own right, and second, to appreciate the influence of organic matter on the inorganic constituents. The latter point arises both from the ability of organic substances to form complexes with inorganic ions, and from the influence of organic matter on the pH and Eh which is exerted through carbon dioxide and oxygen variations.

For convenience, organic substances in estuaries and other natural waters may be classified into their dissolved and particulate fractions. The distinction between these fractions is normally made by the adoption of an arbitrary filter pore size, e.g. ~ 0.45 μm (see Section 41.4.1). Organic matter unable to pass through the filter will include living and dead biological material and is termed "particulate organic carbon" (POC). That which passes through the filter is composed of organic compounds either in true solution or present as colloidal material; this latter fraction is generally termed 'dissolved organic carbon" (DOC).

All natural organic matter, whether dissolved or particulate, is derived from biological activity, and the organic material present in estuarine waters and sediments may have three sources: sea-water, river-water or *in situ* formation. In many estuaries, a small, but obviously important, proportion of the organic matter has an anthropogenic origin (e.g. industrial organic wastes and domestic sewage).

41.7.1. DISSOLVED ORGANIC CARBON SOURCES AND REMOVAL PROCESSES

The biological and chemical aspects of dissolved organic material in sea-water have been reviewed in Chapter 12 and much of the discussion in that chapter on the supply, composition and removal of dissolved organic matter is applicable to estuaries. In view of this, attention will be given here to those features of the chemistry of DOC which are peculiar to estuaries.

The DOC in estuaries results from three inputs. Sea-water entering estuaries typically contains ~ 1 mgCl^{-1} of DOC (McAllister et al., 1964), although some coastal waters may have up to 20 mgCl^{-1} (Strickland, 1965). River-water contains variable DOC concentrations because of differences in the catchment area types, e.g. forested, peaty moorland and agricultural. Beck et al. (1974) have suggested a mean DOC concentration of 10 mgCl^{-1} for river-water, which would imply that it will supply DOC concentrations to estuaries about ten times those supplied, on average, by sea-water. Rather smaller concentrations of DOC (0.3–2 mgCl^{-1}) have been reported for rivers draining forested catchments (Hobbie and Likens, 1973), whereas the DOC content of river-water derived from swamps may reach values up to ~ 100 mgCl^{-1} (Beck et al., 1974).

Estuarine waters tend to have DOC concentrations intermediate between those of river-and sea-waters, and Riley and Chester (1971) have suggested that a value of 5 mgCl^{-1} is a reasonable average under non-polluted conditions. Observed DOC concentrations reported in the recent literature lie in the range 1–5 mgCl^{-1} for British and North American estuaries (Morris and Foster, 1971; Maurer and Parker, 1972; Head, 1976). In addition to the DOC supplied to estuaries by river- and sea-waters, the in situ formation of dissolved organic compounds must be taken into account. Head (1976) has suggested that $\sim 10\%$ of the primary production in estuaries will be lost as exudations of DOC. Unlike the open ocean, in which most of the standing-crop of phytoplankton is grazed by zooplankton, in coastal and estuarine waters consumption of the standing-crop by the benthos is greater than that by zooplankton. The higher productivity of coastal and estuarine waters leads to a signficant deposition of organic detritus. In Departure Bay, British Columbia, the proportion of the primary production deposited was found by Stephens et al. (1967) to be $\sim 50\%$, which suggests that benthic feeding and subsequent exudation are important factors in controlling the DOC con-concentration in estuarine waters.

In addition to that produced by phytoplankton, some estuarine DOC results from in situ production by macrophytes. Mann (1972, 1973) has reviewed the productivity of estuarine and coastal macrophytes and has compared this with the primary productivity of coastal and open-ocean environments (see Fig. 41.19). He has pointed out that in coastal environments the fixation of carbon per unit area by macrophytes is frequently an order of magnitude greater than that by phytoplankton. Head (1976) has concluded that this high degree of productivity by macrophytes in coastal and estuarine situations will lead to the formation of considerably more DOC than is derived from phytoplankton. This deduction is supported by the studies of

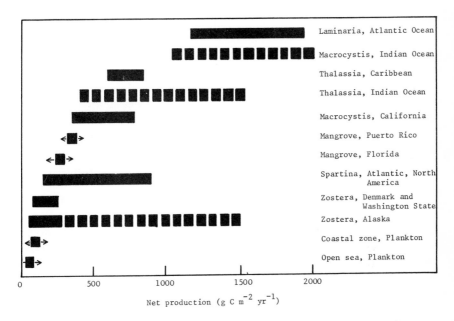

FIG. 41.19. The range of net annual primary production of the major marine macrophyte communities compared to that of phytoplankton (after Mann, 1972). Broken lines indicate extrapolation from short-term data. The arrows provide an estimate of the variability of the measurements.

Sieburth and Jensen (1970) who have shown that ~40% of the net carbon fixed by the macroalga *Fucus* spp. is lost by exudation as DOC. Similar conclusions were drawn by Khailov and Burlakova (1969) for macroalgal exudation in the Black and Barents Seas; these workers also suggested that there is further formation of DOC after the death of the algae.

The biological removal of DOC by incorporation into the aquatic food chain in the estuarine environment has been examined by Stephens and

Schinske (1961) and by Stephens (1967). Invertebrates have been shown to take up amino acids directly from solution in laboratory experiments. Parsons and Seki (1970) have suggested that filter-feeding organisms may be capable of taking up DOC constituents, e.g. amino acids, onto their mucus. Head (1976) has pointed out that heterotrophic bacteria in estuaries will remove DOC by mineralization, and by incorporation into bacterial cell tissue.

41.7.1.1. *The conversion of dissolved organic carbon to particulate matter*

It is generally accepted that DOC is removed from coastal and estuarine waters by mechanisms additional to biological utilization. The aggregation of DOC to form particulate organic matter was demonstrated experimentally by Sutcliffe *et al.* (1963), and particulate matter similar to that formed in these laboratory experiments has been reported to occur in Long Island Sound (Riley, 1963). The possible need for bacteria or other organisms as a prerequisite for aggregation of DOC is not clear. Barber (1966) concluded that non-sterile conditions are necessary for aggregation, but Batoosingh *et al.* (1969) have reported DOC aggregation in sea-water sterilized by filtration through 0·2 μm filters. Sieburth (1965) has suggested that bacteria liberate ammonium ions and raise the pH of waters so that hydrolysed inorganic nuclei for DOC aggregation are formed. Riley (1970) has concluded that although nucleation is very probably needed for DOC aggregation, there is no real evidence for the involvement of bacteria in the process. Particulate matter formed by the aggregation of DOC has been found to contain a high proportion (i.e. $\sim 50\%$) of inorganic matter (Baylor and Sutcliffe, 1963), and this may be interpreted as an indication of inorganic nucleation in aggregation.

The mechanism of DOC aggregation is not clear, and even the nature and chemical composition of the aggregates has yet to be investigated in detail. Kane (1967) and Gordon (1970) have examined the nature of the organic aggregates in ocean waters, but no data is available for estuarine aggregates. It was pointed out in Chapter 12 that the concentrations of the individual compounds such as lipids, amino acids, carbohydrates, vitamins and fatty acids which make up DOC are unlikely to exceed their solubilities. Adsorption onto pre-existing organic detritus and inorganic nuclei, e.g. clay particles, cannot be excluded. In estuaries, it is possible that certain organic substances present in solution in river-water will flocculate when they encounter saline waters. Shapiro (1964), Ong and Bisque (1968) and Bondarenko (1971) have shown that humic substances are flocculated by electrolyte solutions, and the humic material of terrestrial origin present in river-water has been shown experimentally to flocculate in sea-water (Swanson and Palacas, 1965; Sieburth and Jensen, 1968). Humic materials are a major constituent of the DOC of river-waters and consist of hydrophilic colloids (Swanson and Palacas, 1965; Kalle, 1966; Khailov, 1968; Lamar, 1968; Schnitzer and Khan,

1972; Beck *et al*, 1974) of fulvic and humic acids (Black and Christman, 1963; Lamar, 1968; Schnitzer and Khan, 1972; Beck *et al*., 1974). The general mechanisms of humic material flocculation have been described by various workers, e.g. Black (1960), Stumm and O'Melia (1968) and Stumm and Morgan (1970), but it is only recently that the flocculation of dissolved humics in estuarine mixing processes has been investigated. In a new approach to the problem (see also Section 41.4.1), Sholkovitz (1976) has found experimental evidence for the flocculation of humics during the mixing of river-water and sea-water. Figures 41.20A and B illustrate the removal of DOC and humic material from four Scottish river-waters as a function of salinity. The most important feature of the salinity–flocculation relationships is that the maximum amount of removal is always reached between salinities of 15‰ and 20‰. It appears that the salinity of maximum removal is fixed irrespective of the large variations in the concentrations of the organic and inorganic constituents in these four river-waters. The greatly different removal gradients for the four rivers indicate that the flocculation of humates and total DOC is non-stoichiometric with respect to salinity (see also Stumm and O'Melia, 1968).

Eckert and Sholkovitz (1976) have found that the flocculation of humics during estuarine mixing is controlled, to the greatest extent, by the destabilizing effect of calcium ions on the hydrophilic humic colloids (see also Sholkovitz *et al*., 1978). This observation is consistent with the conclusion of Ong and Bisque (1968) that the large ionic radius Ca^{2+} ion should be effective in destabilizing hydrophilic colloids if the mechanism proceeds by the Shulze–Hardy rule of coagulation. The extent of DOC removal by flocculation in the simulated estuarine mixing processes is of the order of 3–11 % for the four river-waters used by Sholkovitz (1976), but the results of this study do not enable the relative flocculation of humic and non-humic substances to be firmly established.

Sea-water and river-water may introduce artifical dissolved organic substances into estuaries as a result of pollution. Examples of these anthropogenic contributions to the DOC of estuaries are soluble petroleum constituents from maritime oil spills and pesticides from the run-off from agricultural land. It has been estimated that the river supply of pesticides, such as the chlorinated hydrocarbon DDT, via estuaries to the oceans is very small compared to the atmospheric flux of these substances to the marine environment (National Academy of Sciences, 1971). Risebrough *et al.* (1968) have estimated the input of chlorinated hydrocarbon pesticides by the San Joaquin and Mississippi Rivers to be 1900 and 10 000 kg year^{-1} respectively. Polychlorinated biphenyls (PCBs) are undoubtedly transported through estuaries which receive waste from industrial processes. The importance of river transport of these pollutants has been discussed by Nisbet and Sarofim (1972). The subject of marine pollution by various dissolved organic substances, including their transport to the oceans, has been reviewed in Chapter 17.

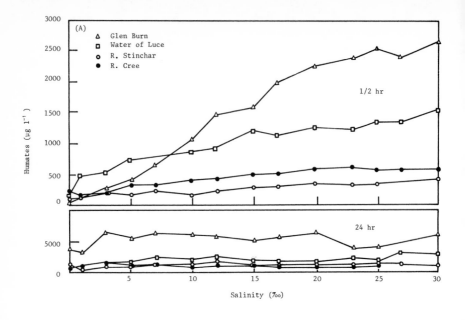

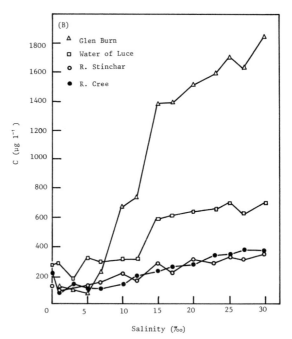

FIG. 41.20. (A) Humate flocculation as a function of salinity. Flocculant concentrations are shown for $\frac{1}{2}$ hour and 24 hour experimental runs. (B) Dissolved organic carbon flocculation as a function of salinity for a $\frac{1}{2}$ hour experimental run. (After Sholkovitz, 1976.)

41.7.2. PARTICULATE ORGANIC CARBON SOURCES AND REMOVAL PROCESSES

The formation of particulate organic carbon (POC) from DOC by aggregation and flocculation processes in estuaries has been discussed in the previous section. Other forms of POC present in estuaries are: (i) *in situ* living organisms, e.g. phytoplankton and macrophytes; (ii) the dead remains of organisms and their solid decay products; (iii) living and dead organic matter transported into estuaries by rivers and the sea; and (iv) POC of anthropogenic origin, e.g. sewage sludges.

The occurrence of particulate organic carbon in the oceans has been reviewed by Riley (1970) and in Chapter 13. In a similar manner to DOC, much of the chemistry of POC in estuaries can be described by analogy to the open ocean, and the reader is directed to the above reviews. Attention will be paid here to some aspects of POC with particular reference to estuaries.

Particulate organic carbon concentrations in most natural waters are much less than those of DOC, but in estuaries the concentrations of these two forms of organic matter are usually similar. Head (1976) has suggested that this is the result of two estuarine features. First, the production of POC from DOC in river-water during estuarine mixing (see Section 41.7.1.1), and second, the high biological productivity of estuaries which leads to the generation of large amounts of DOC and POC, the former being subsequently converted to the latter by aggregation and biological uptake.

Estuarine phytoplankton production is, on average, in the range 100–500 $gC\,m^{-2}\,year^{-1}$ (Riley and Conover, 1956; Gilmartin, 1964; Qasim *et al.*, 1969; Platt, 1971; Wood *et al.*, 1973), compared with an average of $\sim 100\,gC\,m^{-2}\,year^{-1}$ in coastal waters (Ryther, 1963; Ryther, 1969). The amount of POC produced by primary productivity in an estuary varies seasonally, so that the average values quoted above are not necessarily representative of the extent of production at any particular instant. The production of macrophytes in coastal and estuarine environments is of the order of 500–1000 $gC\,m^{-2}\,year^{-1}$ (Mann, 1972), so that they are potentially more important sources of POC than are phytoplankton. Furthermore, Mann (1972) has estimated that, in general, $\sim 90\%$ of macrophyte production results in POC as detritus and DOC, with only $\sim 10\%$ entering the grazing food chain.

The contribution of POC from land drainage via rivers is the dominant source of estuarine POC in some instances. Qasim and Sankaranarayanan (1972) have found, for example, that in a tropical estuary the contribution of *in situ* POC was only 1% of the total input from the river. In the Amazon system, the loads of river-borne POC delivered to the estuary are in the range 2000–10 000 $\mu gC\,l^{-1}$ under maximum run-off conditions and these dominate the POC content of the estuarine and coastal environments (Williams, 1968). Marine and terrestrial POC have different $^{13}C/^{12}C$ ratios (Williams and

Gordon, 1970), and it may well be possible to use this property to estimate the relative contribution of these two sources to estuarine POC.

Anthropogenic additions of POC to estuaries consist almost entirely of municipal sewage. The POC content of sewage effluents is composed of carbohydrates, fats, proteins, tars, textile fibres and faecal products (Føyn, 1971). The sewage disposed of directly and indirectly into estuaries may be untreated or may have been subjected to the activated sludge purification process which is capable of removing more than 90% of the organic and bacterial content. The quantity of POC entering an estuary as sewage is obviously a function of the population of the area concerned.

Removal of POC from estuaries takes place by three principal processes: (i) the physical transport of suspended solids from estuaries into coastal and open-ocean areas; (ii) the settling out of POC to estuarine sediments; and (iii) the biochemical decomposition of POC in suspension. The physical removal of POC from estuaries depends upon the characteristic circulations, tidal movements and flushing times of individual estuaries (see Section 41.3). Sholkovitz (1976) has concluded that flocculated POC will only significantly alter the chemical composition of the particulates in the confined waters of those estuaries in which a pycnocline inhibits the deposition of flocculants. He has further suggested that currents and turbulence will rapidly disperse the flocculated POC into coastal waters. Thus, contributions of POC to estuarine sediments are not composed of flocculated organic matter, but usually consist of the remains of dead organisms. The occurrence of POC in estuarine sediments will then be a function of the rate of supply of organic detritus to the sediment, and of the extent of its dilution by inorganic detrital solids. Aston and Hewitt (1977) have shown that channel geometry in tidal areas exerts an important influence on the distribution of organic carbon in sediments. Shallow, broad channels tend to concentrate POC at their heads, whereas steep-sided, deeper channels concentrate POC at their mouths.

The biochemical oxidation of POC by micro-organisms in natural waters has been modelled by Richards (1965a), and recently reviewed in an estuarine context by Head (1976). The enhanced concentrations of organic matter in estuaries compared to open-ocean areas does not normally lead to the complete depletion of dissolved oxygen. Even in polluted estuaries, oxygen is supplied from the atmosphere and adjacent water bodies at a sufficient rate to replenish that consumed in POC oxidation. In estuaries having restricted circulations, e.g. fjords, a complete removal of oxygen can occur as a result of POC oxidation (Richards, 1965b; Head, 1976, see also Chapter 15). The relative importance of the three major processes of POC removal in estuaries will vary from one estuary to another, and no general conclusions can be drawn about this important stage of the organic carbon cycle.

41.8. SOME ASPECTS OF ESTUARINE SEDIMENTARY PROCESSES

Estuaries are one stage in the transport of sediments from the continents to the ocean basins (Section 41.1), but the deposition of sediments in estuaries is not simply the trapping of a proportion of these solid continental weathering products. Estuaries are complex sedimentary environments, and estuarine sediments may originate from a number of sources, both upstream and marine. The proportions of sediments from different sources at any particular location in an individual estuary will depend on a variety of hydrological and geological factors, e.g. circulation patterns, tidal movements, weathering conditions and source rocks. Gorsline (1967) has concluded that most estuaries are dominated by sediment derived from seaward sources, and that when stream supply is dominant the estuary will yield to a delta. Aston and Chester (1976) have pointed out that the two extreme sources of sediments in estuaries, i.e. landward and seaward, together with intermediate sources such as river mouths and estuarine slopes, impose severe limitations on the geochemical interpretation of estuarine sedimentary processes.

The most important sedimentary and chemical interactions in estuaries may be subdivided into two aspects: (i) the modification of sedimentary detritus during its transport to the oceans; and (ii) the modification of estuarine sediments after deposition, i.e. diagenetic changes.

41.8.1. MODIFICATIONS OF SEDIMENTARY MATERIAL IN ESTUARIES

During transport from the continents to the oceans via estuaries, sedimentary detritus is subjected to changes in salinity, pH and other physico-chemical parameters. Some of these changes, and their effect on the chemical composition of detritus, have already been discussed in Section 41.6, with particular attention to trace elements. Some of the other ways in which sedimentary detritus becomes modified in estuaries will be considered below (see also Chapter 27 for a discussion of the halmyrolysis of lithogenous material during estuarine mixing).

On passing into an estuary those suspended sediment particles which are commonly $< 2\,\mu m$ in diameter and which are mainly composed of clay minerals, will experience a change in their surface charge characteristics. As a consequence of lattice discontinuities, clay mineral particles carry a net negative charge which is balanced by a double layer of hydrated cations. The stability of this surface charge is dependent on the ionic strength of the water, and the increasing ionic strength experienced in passing from river to brackish to sea-water causes surface charge disturbances. The theories of clay coagulation and flocculation are complex and beyond the scope of detailed discussion here (see Chapter 4). Van Olphen (1963) has reviewed the surface chemistry and colloidal behaviour of clays, and Stumm and Morgan (1970) have discussed coagulation theory in relation to natural waters. In river-waters, the stability ratios* of clays, such as montmorillonite, are of the order of 10^{100} or higher, suggesting extreme and long-term colloidal stability. In sea-water, stability ratios have been calculated to approach unity, indicating the colloidal instability of clays in saline waters.

Various workers have studied the flocculation of clays in mixtures of fresh water and sea-water. Whitehouse and McCarter (1958) carried out early investigations of clay flocculation in estuarine waters with respect to the selective transport of different clay minerals. These authors found that mixing at river–sea boundaries with high current velocities can give rise to selective transport. For example, Whitehouse and McCarter (1958) showed that in low-salinity waters illites and kaolinites flocculate faster than do montmorillonites. In consequence, conditions would appear more favourable for the deposition of illites and kaolinites in estuaries than for montomoril-lonites which remain in colloidal suspension and pass into coastal and deep-sea waters. Further studies on the differential settling tendencies of clay minerals by Whitehouse et al. (1960) showed that the flocculated assem-blages of clays formed in estuaries settle out as a solid-rich fluid phase. On deposition, water is expelled and the flocculated structure is destroyed by compaction. This process is reversible, and under the influence of fresh-water currents the flocculates may be dispersed and redistributed in suspen-sion (Dyer, 1972). More recent investigations of sediment coagulation and deposition in estuaries have been reported by Hahn and Stumm (1970) and Edzwald et al. (1974).

Hahn and Stumm (1970) have described three models of coagulation and sedimentation of colloidal matter in natural waters. One of these models is particularly relevant to the estuarine situation because it considers the superimposition of horizontal transport and vertical settling on a coagulating colloid. These authors have suggested that if the flow of water, and the rate of sedimentation in an estuary are known, then the variation of turbidity of

---

* The stability ratio (W) for a colloidal system is an expression of the tendency of the substance to remain stable in colloidal suspension, and not to precipitate by flocculation. The stability ratio is related to the potential energy barrier (V_{max}) by which a colloidal dispersion is characte-rized. Thus:

$$W = \frac{K^{-1}}{2a}\, e^{V_{max}/kT}$$

where K is the double-layer thickness, a is the particle radius and kT is the Boltzmann constant × absolute temperature. The equation shows that the stability ratio (W) increases with an increasing ratio of the double-layer thickness to the particle radius, and depends exponentially on the energy barrier of the colloidal system. For a comprehensive discussion of stability ratios see Stumm and Morgan (1970).

the brackish water and the composition of the underlying sediments may be predicted. Figure 41.21 gives an example of the result of such a calculation for the northern Gulf of Mexico. The relative amounts of illite and montmorillonite in the sediments of a deltaic region change along a longitudinal section of the delta because of differences in the stabilities of these colloidal clays. Illite coagulates at lower ionic strengths than does montmorillonite,

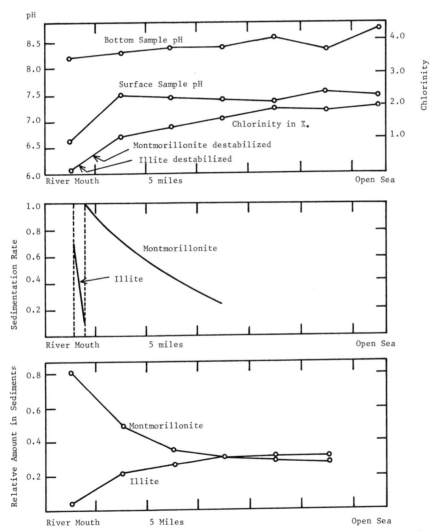

FIG. 41.21. Variation in sediment clay mineral composition as in function of distance from the point of fresh water input, northern Gulf of Mexico, and (centre) the predicted rate of sedimentation of clays as a function of distance (after Hahn and Stumm, 1970).

giving rise to the phenomenon of differential flocculation and settling which was first observed by Whitehouse and Jeffrey (1953), and then studied by Whitehouse and McCarter (1958). These observations do not take any account of possible subsequent halmyrolytic changes in the clay mineral composition in the Gulf of Mexico sediments.

Edzwald et al. (1974) have used both experimental and field approaches to study the coagulation of colloids under estuarine conditions. Their laboratory experiments may be divided into two types, in one of which they used buffered sodium chloride solution and in the other synthetic estuarine water. The results of these laboratory studies are summarized in Table 41.10.

TABLE 41.10

Stability values for clay minerals in buffered sodium chloride solutions[a]

Ionic strength	Montmorillonite	Stability value (α) Kaolinite	Illite
0·05 M	0·075 ± 0·009	0·0245 ± 0·007	0·0128 ± 0·003
0·09 M	0·089 ± 0·006	0·0308 ± 0·004	0·0275 ± 0·005
0·30 M	0·125 ± 0·005	0·0724 ± 0·007	0·0455 ± 0·005

[a] Data from Edzvald et al. (1974).

The stability values, α^*, for the clay minerals indicate that destabilization increases with increasing ionic strength as predicted by colloid theory. When the α-values obtained by destabilization with NaCl solutions are compared with those from the synthetic estuarine water experiments, it is evident that destabilization is lower for the sodium chloride solutions. This is in agreement with the theoretical prediction that the divalent cations present

---

* The term α, the stability value of a clay colloid, has been used by Edzwald et al. (1974) in the context of Smoluchowski's model of particle aggregation where the collisions are due to fluid motion, i.e. orthokinetic flocculation. Smoluchowski (1917) expressed the number of aggregated particles (n) as a function of time (t) by the equation:

$$\frac{dn}{dt} = \frac{-4\alpha\Phi Gn}{\pi}$$

where Φ is the volume of colloidal particles per unit volume of suspension, and G is the root-mean-square velocity of the particles. A completely destabilized suspension has a stability value of 1, whereas stable suspensions are characterized by values of $\ll 1$.

in the estuarine waters should be more effective in colloid destabilization than the monovalent Na^+ ion. Edzwald et al. (1974) concluded that the colloidal form of illite is more stable than that of kaolinite, which in its turn is more stable than that of montmorillonite. Edzwald et al. (1974) extended their laboratory studies to a field situation, using sediments from the well-mixed Pamlico River estuary, U.S.A. Figures 41.22A and B show the salinity of the estuarine waters as a function of distance downstream, and the clay mineral composition of the sediments sampled over a downstream section of the estuary. Kaolinite is the dominant clay mineral in the upper end of the estuary where the salinity is lowest and its concentration in the sediment decreases towards the mouth of the estuary. Illite occurs in minor amounts at the upper end and increases towards the mouth, whereas montmorillonite is present in minor amounts along the entire length of the estuary. The authors have suggested that the distribution of kaolinite and illite in the sediments of the Pamlico Estuary can be explained in terms of coagulation and the experimentally determined stability values.

Burton (1976) has summarized recent laboratory and field studies on estuarine coagulation, and has concluded that four mechanisms of coagulation should be recognized. These are as follows.

(i) Neutralization of the negative charge on clay colloids by specific adsorption of positively charged species. Aston and Chester (1973) have shown that the negative surface charge on estuarine sediment particles is effectively reduced, and under some extreme conditions, reversed, to a positive charge by the uptake of cationic iron hydrolysis products from brackish waters. Other observations on the reduction of the negative surface charge of particles during the transition from fresh to saline waters have been made by Pradvić (1970), Neihof and Loeb (1972) and Martin et al. (1971). Martin et al. have suggested that the changes in negative charge may result as a consequence of both salinity increases and the uptake of organic matter from sea-water.

(ii) Compression of the electrical double-layer on particles takes place as a result of the increasing ionic strength of estuarine waters. This process is a function of the increased charge on the counter-ions as well as of the increased concentration of electrolyte in the bulk solution.

(iii) The formation of interparticle bridging bonds by the adsorbed materials, e.g. hydrolysis products and organic substances.

(iv) Enmeshment of clay and hydroxide particles.

The processes of coagulation in estuaries are closely linked to those which control both the soluble and particulate distribution of minor elements and the distribution of clay minerals in sediments. In this respect, coagulation is a process of considerable importance in estuarine chemistry and it offers an area in which there is great potential for future research.

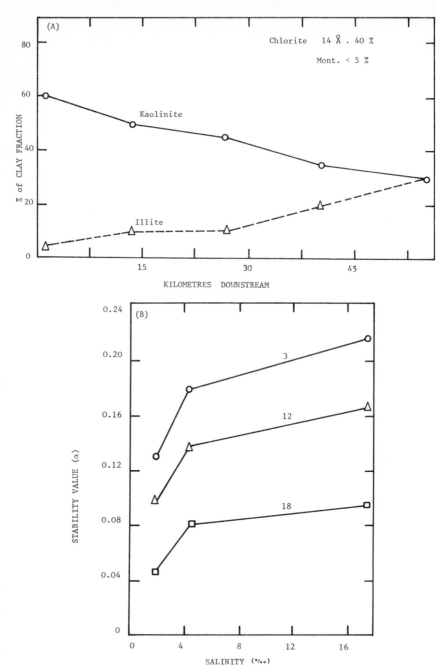

FIG. 41.22. (A) Clay mineral composition of Pamlico River estuary sediments from railroad bridge reference point, Washington D.C. (B) The stability value (α) as a function of salinity for the clay size fraction of the Pamlico sediments. (After Edzwald *et al.*, 1974)

41.8.2. MODIFICATIONS OF SEDIMENTARY DEPOSITS IN ESTUARIES

In common with other modern sediments, those deposited in estuaries are subjected to modifications of both their solid materials and pore waters. These changes of sediment composition arising from burial are termed diagenesis (see Chapter 30). Discussion of diagenetic processes will be limited here to those investigations which have been specifically direct towards estuarine sediments.

A fundamental change undergone by estuarine and other sediments upon burial is that from aerobic to anaerobic conditions in the pore waters. In estuarine sediments this change in redox potential is frequently abrupt and occurs at depths of some centimetres below the sediment–water interface. The reversal of redox conditions from oxidizing to reducing is a consequence of the bacterial utilization of dissolved oxygen in the pore waters and the biochemical reduction of dissolved oxygenated ions (e.g. sulphate and nitrate) in pore waters by certain bacteria. The organic debris upon which bacteria subsist is normally abundant in estuarine sediments, and this leads to intense bacterial activity. A result of this activity is the relatively shallow depths to which aerobic conditions prevail in estuarine sediments. At depths of a few centimetres, the rate of bacterial utilization of dissolved oxygen in pore waters is sufficient to balance the supply of dissolved oxygen by diffusion from the overlying water.

An important consequence of the bacterial processes described above is the great abundance of sulphide-rich sediments in estuaries. The formation of such sediments has been reviewed in Chapter 30 and will not be discussed further here. It is pertinent, however, to examine the role of such anoxic deposits in estuaries as a control on the distribution of some geochemically important elements. In order to establish the normal background levels of trace elements in estuarine sediments, Aston and Chester (1976) have summarized the composition of near-shore sediments in general to enable comparison to be made between aerobic and anaerobic estuarine sediments. Table 41.11 shows the compositions of sediments from various locations. Certain elements, e.g. Ni, Co, Cr, V, Ba, Sr, Pb, Zn and Y have similar concentrations in both aerobic and anaerobic estuarine sediments, but Mn, Cu, Sc, Zr and Mo tend to be enriched in estuarine anaerobic sediments relative to both aerobic sediments and near-shore sediments in general.

The enrichment of trace elements in estuarine anaerobic sediments has been investigated in some detail for two elements only, molybdenum and uranium. Gross (1967) has reported that the anaerobic sediments of the Saanich Inlet, British Columbia, have an average molybdenum concentration of 26 ppm, whereas the element could not be detected in the aerobic sediments from this estuary. Bertine (1973) has shown that the concentration range of molybdenum in estuarine anoxic sediments is similar to that observed for deep-sea sediments; the faster sedimentation rates of the former imply a faster accumulation of molybdenum in these than in the deep-sea sediments. The mechanism of molybdenum enrichment in anaerobic sediments is not understood, but may involve both inorganic and organic adsorption and complexation reactions.

TABLE 41.11

Average trace element composition of near-shore sediments and comparison of oxic and anoxic sediments[a]

Element	Average near-shore sediment[b]	Oxic near-shore sediments		Anoxic near-shore sediments	
		Gulf of Paria[c]	Saanich Inlet[d]	Baltic Sea[e]	Saanich Inlet[d]
Ni	55	31	33	43	26
Co	13	12	9	22	8
Cr	100	93	86	90	35
V	130	146	110	130	37
Ba	750	394	—	750	—
Sr	250	210	—	130	—
Pb	20	22	20	25	trace
Zn	95	—	88	110	80
Y	—	—	26	—	25
Mn	850	2000	370	4030	400
Cu	48	17	38	78	45
Sc	—	—	17	—	12
Zr	—	—	140	—	760
Mo	—	—	n.d.	35	26

[a] After Aston and Chester (1976). All values are in $\mu g\,g^{-1}$; n.d. signifies not detected.
[b] Wedepohl (1960). [c] Hirst (1962). [d] Gross (1967). [e] Manheim (1961).

Uranium, in a similar manner to molybdenum, is enriched in estuarine anaerobic sediments, and further illustrates the importance of such modified deposits in the geochemical budget of certain elements. Veeh (1967) has shown that anoxic sediments may contain up to 39 ppm of uranium compared to an average concentration of 3 ppm for other marine sediments. Taking an average river-water uranium concentration of $0.04\,\mu g\,l^{-1}$, Veeh (1967) has concluded that estuarine anaerobic sediments would have to comprise only $\sim 0.4\%$ of the total oceanic area in order for removal to these deposits to balance the total input of dissolved uranium via river-waters. Turekian and Chan (1971) have suggested that a more realistic average concentration of dissolved uranium in river-waters would be $\sim 0.3\,\mu g\,l^{-1}$, and that on this assumption these modified deposits would have to comprise several percent of oceanic sediments to exert a significant influence on the geochemical balance of uranium (see also Chapter 33).

In view of the fact that estuarine and near-shore anaerobic sediments are capable of taking up and retaining certain elements, e.g. Mo, U, Mn, Cu, Sc and Zr, it is possible that these deposits may act as a major source of metals if disturbed. On resuspension by tidal currents and storms, previously buried sediments are transferred from an anoxic environment to normal oxic waters. This transition will result in the oxidation of reduced sediments, e.g. sulphides, and may involve the release of enriched trace elements into estuarine waters.

REFERENCES

Ahrland, S. (1975). *In* "The Nature of Sea Water" (E. D. Goldberg, ed.). Dahlen Konferenzen, Berlin.

Alexander, G. B. (1953). *J. Am. Chem. Soc.* **75**, 5655.

Alexander, J. E. and Corcoran, E. F. (1967). *Limnol. Oceanogr.* **12**, 236.

Andren, A. W. and Harriss, R. C. (1975). *Geochim. Cosmochim. Acta,* **39**, 1253.

Aston, S. R. (1978). *In* "Practical Estuarine Chemistry" (J. Phillips, ed.). Cambridge University Press, Cambridge.

Aston, S. R. and Chester, R. (1973). *Estuar. Coastal Mar. Sci.* **1**, 225.

Aston, S. R. and Chester, R. (1976). *In* "Estuarine Chemistry" (J. D. Burton and P. S. Liss, eds). Academic Press, London and New York.

Aston, S. R. and Duursma, E. K. (1973). *Neth. J. Sea Res.* **6**, 225.

Aston, S. R. and Hewitt, C. N. (1977). *Estuar. Coastal Mar. Sci.* **5**, 243.

Barber, R. T. (1966). *Nature, Lond.* **211**, 257.

Batoosingh, E., Riley, G. A. and Keshwar, B. (1969). *Deep-Sea Res.* **16**, 213.

Baylor, E. R. and Sutcliffe, W. H. (1963). *Limnol. Oceanogr.* **8**, 369.

Beck, K. C., Reuter, J. H. and Perdue, E. M. (1974). *Geochim. Cosmochim. Acta,* **38**, 341.

Bertine, K. K. (1973). *Mar. Chem.* **1**. 43.

Bewers, J. M., Macawley, I. D, and Sundby, B. (1974). *Can. J. Earth Sci.* **11**, 939.

Bien, G. S., Contois, D. E. and Thomas W. H. (1958). *Geochim. Cosmochim. Acta,* **14**, 35.

Black, A. P. (1960). *J. Am. Wat. Wks Ass.* **52**, 492.

Black, A. P. and Christman, R. F. (1963). *J. Am. Wat. Wks Ass.* **55**, 897.

Bondarenko, G. P. (1972). *Geochem. Int.* **9**, 702.

Boon, J. D. and MacIntyre, W. B. (1968). *Chesapeake Sci.* **9**, 21.

Bowen, H. J. M. (1966). "Trace Elements in Biochemistry". Academic Press, London and New York, 241 pp.

Boyden, C. R., Aston, S. R. and Thornton, I. (1978). *Estuar. Coastal Mar. Sci.* (in press).

Boyle, E., Collier, R., Dengler, A. T., Edmond, J. M., Ng, A. C. and Stallard, R. F. (1974). *Geochim. Cosmochim. Acta,* **38**, 1719.

Breck, W. G. (1972). *J. Mar. Res.* **30**, 121.

Brewer, P. G. and Bradshaw, A. (1975). *J. Mar. Res.* **33**, 157.

Burton, J. D. (1970). *J. Cons. Perm. Int. Explor. Mer,* **33**, 141.

Burton, J. D. (1976). *In* "Estuarine Chemistry" (J. D. Burton and P. S. Liss, eds). Academic Press, London and New York.

Burton, J. D. and Liss, P. S. (1973). *Geochim. Cosmochim. Acta,* **37**, 1761.

Burton, J. D., Leatherland, T. M. and Liss, P. S. (1970a). *Limnol. Oceanogr.* **15**, 473.

Burton, J. D., Liss, P. S. and Vemigopalan, K. K. (1970b). *J. Cons. Perm. Int. Explor. Mer,* **33**, 134.

Butler, E. I. and Tibbits, S. (1972). *J. Mar. Biol. Ass. U.K.* **52**, 681.

Cameron, W. M. and Pritchard, D. W. (1963). *In* "The Sea" (M. N. Hill, ed.), Vol. 2. John Wiley and Sons, New York.

Carpenter, J. H., Bradford, W. L. and Grant, V. (1975). *In* "Estuarine Research" (L. E. Cronin, ed.), Vol. 1. Academic Press, New York and London.

Carpenter, R. (1969). *Geochim. Cosmochim. Acta*, **33**, 1153.

Cheng, H. S. and Hamaguchi, H. (1968). *Hlth Phys.* **14**, 353.

Chester, R. (1965). *Nature, Lond.* **288**, 664.

Chow, T. J. (1965). *In* "Proceedings of the 1st International Congress on Fouling and Marine Corrosion" Cannes, 53pp.

Coonley, L. S., Jr., Baker, E. B. and Holland, H. D. (1971). *Chem. Geol.* **7**, 51.

Corcoran, E. F. and Alexander, J. E. (1964). *Bull. Mar. Sci. Gulf Caribb.* **14**, 594.

Corner, E. D. S., Head, R. N., Kilvington, C. C. and Marshall, S. M. (1974). *J. Mar. Biol. Ass. U.K.* **54**, 319.

Correll, D. L., Faust, M. A. and Severn, D. J. (1975). *In* "Estuarine Research" (L. E. Cronin, ed.), Vol. 1. Academic Press, New York and London.

Cox, R. A., McCartney, M. J. and Culkin, F. (1970). *Deep-Sea Res.* **17**, 679.

Cranston, R. E. and Buckley, D. E. (1972). *Environ. Sci. Tech.* **6**, 274.

Cross, F. A., Willis, J. N., Hardy, L. H., Jones, N. Y. and Lewis, J. M. (1975). *In* "Estuarine Research" (E. L. Cronin, ed.), Vol. 1. Academic Press, New York and London.

de Groot, A. J. (1966). *In* "Comm. II and IV, Internat. Soc. Soil Sci. Trans." Aberdeen.

de Groot, A. J., Zschuppe, K. H., de Bruin, M., Houtman, J. P. W. and Singgih, P. A. (1968). *In* "International Conference on Modern Trends in Activation Analysis", Gaithersberg.

de Groot, A. J., de Goeij, J. J. M. and Zegers, C. (1971). *Geologie Mijnb.* **50**, 393.

Dickson, A. G., Mantoura, R. F. C. and Riley, J. P. (1978). *Estuar. Coastal Mar. Sci.* **6**, 387.

Dyer, K. R. (1972). *In* "Estuarine Environment" (R. S. K. Barnes and J. Green, eds). Applied Science, London.

Dyer, K. R. (1973). "Estuaries: A Physical Introduction". John Wiley and Sons, London, 140 pp.

Eckert, J.M. and Sholkovitz, E. R. (1976). *Geochim. Cosmochim. Acta*, **40**, 847.

Edzwald, J. K., Upchurch, J. B. and O'Melia, C. R. (1974). *Environ. Sci. Tech.* **8**, 58.

Evans, D. W. and Cutshall, N. H. (1973). *In* "Radioactive Contamination of the Marine Environment", pp. 125–140. International Atomic Energy Agency, Vienna.

Fanning, K. A. and Pilson, M. E. Q. (1973). *Geochim. Cosmochim. Acta*, **37**, 2405.

Fa-Si, L., Yu-Duan, W., Long-Fa, W. and Ze-Hsia, C. (1964). *Oceanologia Limnol. Sin.* **6**, 311.

Fitzgerald, W. F. and Lyons, W. B. (1973). *Nature, Lond.* **242**, 452.

Foster, P. and Morris, A. W. (1971). *Deep-Sea Res.* **18**, 231.

Føyn, E. (1971). *In* "Impingement of Man on the Oceans" (D. W. Hood, ed.). Wiley–Interscience, New York.

Fukai, R. and Huynh-Ngoc, L. (1968). "Studies on the Chemical Behaviour of Radionuclides in Sea-water I", Pub. No. 22. International Atomic Energy Agency, Vienna.

Gardiner, J. (1974). *Wat. Res.* **8**, 23.

Garrels, R. M. and Mackenzie, F. T. (1971). "Evolution of the Sedimentary Rocks". W. W. Norton, New York, 317 pp.

Gibbs, R. J. (1973). *Science, N.Y.* **180**, 71.

Gilmartin, M. (1964). *J. Fish. Res. Bd Can.* **21**, 505.

Goldberg, E. D. (1965). *In* "Chemical Oceanography" (J. P. Riley and G. Skirrow, eds), Vol. 1, 1st Edn. Academic Press, London and New York.

Gordon, D. C. (1970). *Deep-Sea Res.* **17**, 233.

Gorsline, D. C. (1967). *In* "Estuaries" (G. H. Lauff, ed.). Amer. Assoc. Adv. Sci., Washington, Pub. No. **83**.

Grim, R. E. and Johns, W. D. (1953). *In* "Clays and Clay Minerals" (A. Swineford and N. V. Plummer, eds). National Research Council Pub. No. **327**, 81.

Gross, M. G. (1967). *In* "Estuaries" (G. H. Lauff, ed.). Amer. Assoc. Adv. Sci., Washington, Pub. No. **83**.

Hahn, H. H. and Stumm, W. (1970). *Am. J. Sci.* **268**, 354.

Hair, M. E. and Bassett, C. R. (1973). *Estuar. Coastal Mar. Sci.* **1**, 107.

Harder, H. (1965). *Geochim. Cosmochim. Acta*, **29**, 429.

Head, P. C. (1970). *Mar. Poll. Bull.* **1**, 138.

Head, P. C. (1971). *J. Mar. Biol. Ass. U.K.* **51**, 891.

Head, P. C. (1976). *In* "Estuarine Chemistry" (J. D. Burton and P. S. Liss, eds). Academic Press, London and New York.

Hedgpeth, J. W. (1967). *In* "Estuaries" (G. H. Lauff, ed.). Amer. Assoc. Adv. Sci., Washington, Pub. No. **83**.

Hirst, D. M. (1962). *Geochim. Cosmochim. Acta*, **26**, 1147.

Hobbie, J. E. and Likens, G. E. (1973). *Limnol. Oceanogr.* **18**, 734.

Holliday, L. M. and Liss, P. S. (1976). *Estuar. Coastal Mar. Sci.* **4**, 349.

Hosokawa, I., Ohshima, F. and Kondo, N. (1970). *J. Oceanogr. Soc. Japan*, **26**, 1.

Hughes, P. (1958). *Geophys. J. R. Astron. Soc.* **1**, 271.

Hydes, D. J. (1974). Ph.D. Thesis, University of East Anglia, 168 pp.

Iler, R. K. (1955). "The Colloid Chemistry of Silica and Silicates". Cornell, New York, 324 pp.

Kalle, K. (1969). *Oceanogr. Mar. Biol. Ann. Rev.* **4**, 91.

Kane, J. W. (1967). *Limnol. Oceanogr.* **12**, 287.

Ketchum, B. H. (1951). *J. Mar. Res.* **10**, 18.

Ketchum, B. H. (1955). *Sewage Ind. Wastes*, **27**, 1288.

Khailov, K. M. and Burlakova, Z. P. (1969). *Limnol. Oceanogr.* **14**, 521.

Khan, S. U. (1969). *Proc. Soil Sci. Soc. Am.* **33**, 851.

Kharkar, D. P., Turekian, K. K. and Bertine, K. K. (1968). *Geochim. Cosmochim. Acta*, **32**, 285.

Khailov, K. M. (1968). *Geochem. Int.* **5**, 497.

Knauer, G. A. and Martin, J. H. (1973). *Limnol. Oceanogr.* **18**, 597.

Knudsen, M. (1901). "Hydrographische Tabellen". G.E.C., Gad., Copenhagen.

Kobayaski, J. (1967). *In* "Chemical Environment in the Aquatic Habitat" (H. L. Golterman and R. S. Clymo, eds). North-Holland, Amsterdam.

Krauskopf, K. B. (1956). *Geochim. Cosmochim. Acta*, **9**, 1.

Kremling, K. (1972). *Deep-Sea Res.* **19**, 377.

Kullenberg, B. and Sen Gupta, R. (1973). *Geochim. Cosmochim. Acta*, **37**, 1327.

Lamar, W. L. (1968). *Prof. Pap. U.S. Geol. Surv.* **600-D**, D24.

Lean, D. R. S. (1973). *Science, N.Y.* **179**, 678.

Levinson, A. A. and Ludwick, J. C. (1966). *Geochim. Cosmochim. Acta*, **30**, 855.

Liss, P. S. (1976). *In* "Estuarine Chemistry" (J. D. Burton and P. S. Liss, eds). Academic Press, London and New York.

Liss, P. S. and Pointon, M. J. (1973). *Geochim. Cosmochim. Acta*, **37**, 1493.

Liss, P. S. and Spencer, C. P. (1970). *Geochim. Cosmochim. Acta*, **34**, 1073.

Liss, P. S., Herring, J. R. and Goldberg, E. D. (1973). *Nature, Phys. Sci.* **242**, 108.

Livingstone, D. A. (1963). *Prof. Pap. U.S. Geol. Surv.* **440-G**, 64 pp.

Lyman, J. and Fleming, R. H. (1940). *J. Mar. Res.* **3**, 134.

Mackay, D. W. and Leatherland, T. M. (1976). *In* "Estuarine Chemistry" (J. D. Burton and P. S. Liss, eds). Academic Press, London and New York.

Mackenzie, F. T., Garrels, R. M., Bricker, O. P. and Bickely, F. (1967). *Science, N.Y.* **155**, 1404.

Maéda, H. (1952). *Publs Seto Mar. Biol. Lab.* **2**, 249.

Maéda, H. (1953). *J. Shimonoseki Coll. Fish.* **3**, 167.

Maéda, H. and Takesue, K. (1961). *Rec. Oceanogr. Wks Japan*, **6**, 112.

Maéda, H. and Tsukamoto, M. (1959). *J. Shimonoseki Coll. Fish.* **8**, 121.

Makimoto, H., Maéda, H. and Era, S. (1955). *Rec. Oceanogr. Wks Japan*, **2**, 106.

Manheim, F. T. (1961). *Geochim. Cosmochim. Acta*, **25**, 52.

Mann, K. H. (1972). *Memorie Ist. Ital. Idrobiol.* **29**, 353.

Mann, K. H. (1973). *Science, N.Y.* **182**, 975.

Martin, J.-M., Jednačak, J. and Pravdić, V. (1971). *Thalassia Jugosl.* **7**, 619.

Matharu, H. S. (1975). Ph.D. Thesis, University of Southampton, 186 pp.

Matson, W. R. (1968). Ph.D. Thesis, Massachusetts Institute of Technology, 258 pp.

Maurer, L. G. and Parker, P. L. (1972). *Contr. Mar. Sci.* **16**, 109.

McAllister, C. D., Shah, N. and Strickland, J. D. H. (1964). *J. Fish. Res. Bd Can.* **21**, 159.

Millero, F. J. (1975a). *In* "The Sea" (E. D. Goldberg, ed.), Vol. 5. John Wiley and Sons, New York.

Millero, F. J. (1975b). *In* "Marine Chemistry in the Coastal Environment" (T. M. Church, ed.). Am. Chem. Soc. Symp. Ser. No. **18**.

Millero, F. J. and Lawson, D. R. (1975). Cited in Millero (1975b).

Millero, F. J., Gonzalez, A. and Ward, G. K. (1975). Cited in Millero (1975b).

Millward, G. E. and Burton, J. D. (1975). *Mar. Sci. Comm.* **1**, 15.

Morris, A. W. and Foster, P. F. (1971). *Limnol. Oceanogr.* **16**, 987.

Müller, G. and Förstner, U. (1975). *Environ. Geol.* **1**, 33.

National Academy of Sciences, Washington, D.C. (1971). "Chlorinated Hydrocarbons in the Marine Environment".

Neihof, R. A. and Loeb, G. I. (1972). *Limnol. Oceanogr.* **17**, 7.

Nisbet, I. C. T. and Sarofim, A. F. (1972). *Environ. Hlth Persp.* **1**, 21.

Officer, C. B. (1976). "Physical Oceanography of Estuaries". Wiley–Interscience, New York, 465 pp.

van Olphen, H. (1963). "An Introduction to Clay Colloid Chemistry". Wiley–Interscience, New York.

Ong, H. L. and Bisque, R. E. (1968). *Soil Sci.* **106**, 220.

Orlov, D. S. and Yeroshicheva, N. L. (1967). *Dokl. Soil Sci.* **28**, 1799.

Park, P. K., Catalfomo, M., Webster, G. R. and Reid, D. H. (1970). *Limnol. Oceanogr.* **15**, 70.

Parsons, T. R. and Seki, H. (1970). *In* "Symposium on Organic Matter in Natural Waters" (D. W. Hood, ed.), pp. 1–27. Institute of Marine Science, University of Alaska.

Pickard, G. L. (1975). "Descriptive Physical Oceanography", 2nd Edn. Pergamon Press, Oxford, 214 pp.

Platt, T. (1971). *J. Cons. Perm. Int. Explor. Mer*, **33**, 324.

Pomeroy, L. R. (1960). *Science, N.Y.* **131**, 1731.

Pomeroy, L. R., Mathews, H. M. and Min, H. S. (1963). *Limnol. Oceanogr.* **8**, 50.

Pomeroy, L. R., Smith, E. E. and Grant, G. M. (1965). *Limnol. Oceanogr.* **10**, 167.

Pomeroy, L. R., Odum, E. P., Johannes, R. E. and Roffman, B. (1966). *In* "Disposal of Radioactive Wastes into Seas, Oceans and Surface Waters". International Atomic Energy Agency, Vienna.

Pomeroy, L. R., Johannes, R. E., Odum, E. P. and Roffman, B. (1969). *In* "The 2nd National Symposium on Radioecology" (D. J. Nelson and F. C. Evans, eds). Ann Arbor Science Publishers, Ann Arbor, U.S.A.

Pravdić, V. (1970). *Limnol. Oceanogr.* **15**, 230.

Price, N. B. and Calvert, S. E. (1973). *Mar. Chem.* **1**, 169.

Pritchard, D. W. (1952). *Adv. Geophys.* **1**, 243.

Pritchard, D. W. (1957). *Trans. Am. Geophys. Un.* **38**, 581.

Qasim, S. Z. and Sankaranarayanan, V. N. (1972). *Mar. Biol.* **15**. 193.

Qasim, S. Z., Wellershaus, S., Bhattathiri, P. M. A. and Abidi, S. A. H. (1969). *Proc. Indian Acad. Sci. B*, **69**, 51.

Richards, F. A. (1965a). *In* "Advances in Water Pollution Research" (E. A. Pearson, ed.), Vol. 3. Pergamon Press, Oxford.

Richards, F. A. (1965b). *In* "Chemical Oceanography" (J. P. Riley and G. Skirrow, eds), Vol. 1, 1st Edn. Academic Press, London and New York.

Rickard, D. T. (1971). *Stock. Contr. Geol.* **23**, 1.

Rigler, F. H. (1956). *Ecology*, **37**, 550.

Rigler, F. H. (1964). *Limnol. Oceanogr.* **9**, 511.

Riley, G. A. (1963). *Limnol. Oceanogr.* **8**, 372.

Riley, G. A. (1970). *Adv. Mar. Biol.* **8**, 1.

Riley, G. A. and Conover, S. A. M. (1956). *Bull. Bingham Oceanogr. Coll.* **15**, 47.

Riley, J. P. and Chester, R. (1971). "Introduction to Marine Chemistry". Academic Press, London and New York, 465 pp.

Risebrough, R. W., Huggett, R. J., Griffin, J. J. and Goldberg, E. D. (1968). *Science, N.Y.* **159**, 1233.

Ryther, J. H. (1963). *In* "The Sea" (M. H. Hill, ed.), Vol. 2. Wiley–Interscience, New York.

Ryther, J. H. (1969). *Science, N.Y.* **166**, 72.

Sackett, W. and Arrhenius, G. (1962). *Geochim. Cosmochim. Acta*, **26**, 955.

Schink, D. R. (1967). *Geochim. Cosmochim. Acta*, **31**, 987.

Schnitzer, M. and Khan, S. U. (1972). "Humic Substances in the Environment". Marcel Dekker, New York, 327 pp.

Shapiro, J. (1964). *J. Am. Wat. Wks Ass.* **56**, 1062.

Sholkovitz, E. R. (1976). *Geochim. Cosmochim. Acta*, **40**, 831.

Sholkovitz, E. R., Boyle, E. A. and Price, N. B. (1978). *Earth Planet. Sci. Lett.* **40**, 130

Sieburth, J. McN. (1965). *J. Gen. Microbiol.* **41**, XX.

Sieburth, J. McN. (1969). *J. Expl Mar. Biol. Ecol.* **3**, 290.

Sieburth, J. McN. and Jensen, A. (1968). *J. Expl Mar. Biol. Ecol.* **2**, 174.

Sieburth, J. McN. and Jensen, A. (1969). *J. Expl Mar. Biol. Ecol.* **3**, 275.

Sieburth, J. McN. and Jensen, A. (1970). *In* "Symposium on Organic Matter in Natural Waters" (D. W. Hood, ed.), pp. 203–223. Institute of Marine Science, University of Alaska.

Siever, R. (1971). *In* "Handbook of Geochemistry" (K. H. Wedepohl, ed.). Springer-Verlag, Berlin.

Siever, R. and Woodford, N. (1973). *Geochim. Cosmochim. Acta*, **37**, 1851.

Sillén, L. G. (1961). *In* "Oceanography" (M. Sears, ed.). Amer. Assoc. Adv. Sci., Washington, Pub. No. **67**.

Simon, W. G. (1953). Landesamt, No. **11**, Hamburg, 153 pp.

Simpson, H. J., Hammond, D. E., Deck, B. L. and Williams, S. C. (1975). *In* "Marine Chemistry in the Coastal Environment" (T. M. Church, ed.). American Chemical Society, Washington.

Singer, P. C. (1973). "Trace Metals and Metal–Organic Interactions in Natural Waters". Ann Arbor Science Publishers, Ann Arbor, U.S.A., 380 pp.

Slowey, J. F. and Hood, D. W. (1971). *Geochim. Cosmochim. Acta*, **35**, 121.

Slowey, J. F., Jeffrey, L. M. and Hood, D. W. (1967). *Nature, Lond.* **214**, 377.
Smoluchowski, M. (1917). *Z. Phys. Chem.* **92**, 129.
Spencer, D. W. and Sachs, P. L. (1970). *Mar. Geol.* **9**, 117.
Stefánsson, U. and Richards, F. A. (1963). *Limnol. Oceanogr.* **8**, 394.
Stephens, C. F. and Oppenheiner, C. H. (1972). *Contr. Mar. Sci.* **16**, 99.
Stephens, G. C. (1967). *In* "Estuaries" (G. H. Lauff, ed.). Amer. Assoc. Adv. Sci., Washington, Pub. No. **83**.
Stephens, G. C. and Schinske, R. A. (1961). *Limnol. Oceanogr.* **6**, 175.
Stephens, K., Sheldon, R. W. and Parsons, T. R. (1967). *Ecology*, **48**, 852.
Stiff, M. J. (1971). *Wat. Res.* **5**, 171.
Stommel, H. and Farmer, H. G. (1952). "On the Nature of Estuarine Circulation", Rept. No. 52, Woods Hole Oceanographic Institution, Woods Hole, Massachusetts.
Strickland, J. D. H. (1965). *In* "Chemical Oceanography" (J. P. Riley and G. Skirrow, eds), Vol. 1, 1st Edn. Academic Press, London and New York.
Stumm, W. and Morgan, J. J. (1962). *J. Am. Wat. Wks Ass.* **54**, 971.
Stumm, W. and Morgan, J. J. (1970). "Aquatic Chemistry". Wiley–Interscience, New York, 583 pp.
Stumm, W. and O'Melia, C. R. (1968). *J. Am. Wat. Wks Ass.* **60**, 514.
Sutcliffe, W. H., Baylor, E. R. and Menzel, D. W. (1963). *Deep-Sea Res.* **9**, 120.
Swanson, V. E. and Palacas, J. G. (1965). *Bull. U.S. Geol. Surv.* **1214-B**, B1.
Theis, T. L. and Singer, P. C. (1974). *Environ. Sci. Tech.* **8**, 569.
Turekian, K. K. (1965). *In* "Chemical Oceanography" (J. P. Riley and G. Skirrow, eds), Vol. 1, 1st Edn. Academic Press, London and New York.
Turekian, K. K. (1971). *In* "Impingement of Man on the Oceans" (D. W. Hood, ed.). Wiley–Interscience, New York.
Turekian, K. K. and Chan, L. H. (1971). *In* "Activation Analysis in Geochemistry and Cosmochemistry" (A. O. Brunfelt and E. Steinnes, eds). Universitetsforlaget, Oslo.
van Olphen, H. (1963). "An Introduction to Clay Colloid Chemistry". Wiley–Interscience, New York.
van Veen, J. (1936). Cited in Simon (1953).
Veeh, H. H. (1967). *Earth Planet. Sci. Lett.* **3**, 145.
Warner, T. B. (1972). *J. Geophys. Res.* **77**, 2728.
Wedepohl, K. H. (1960). *Geochim. Cosmochim. Acta*, **18**, 200.
Whitehouse, V. G. and Jeffrey, L. M. (1953). *Texas A. and M. Res. Found. Proj.* 34, Tech. Rept. No. 1.
Whitehouse, V. G. and McCarter, R. S. (1958). *In* "Proceedings of the 5th National Conference on Clays and Clay Minerals", 81.
Whitehouse, V. G., Jeffrey, L. M. and Debrecht, J. D. (1960). *In* "Proceedings of the 7th National Conference on Clays and Clay Minerals", 1.
Whitfield, M. (1975). *Geochim. Cosmochim. Acta*, **39**, 1545.
Williams, P. M. (1968). *Nature, Lond.* **218**, 937.
Williams, P. M. (1969). *Limnol. Oceanogr.* **14**, 156.
Williams, P. M. and Chan, K. S. (1966). *J. Fish. Res. Bd Can.* **23**, 575.
Williams, P. M. and Gordon, L. I. (1970). *Deep-Sea Res.* **17**, 19.
Windom, H. L. (1971). *Limnol. Oceanogr.* **16**, 806.
Windom, H. L. (1975). *In* "Estuarine Research" (L. E. Cronin, ed.), Vol. 1. Academic Press, New York and London.
Windom, H. L., Beck, K. C. and Smith, R. (1971). *S.East Geol.* **12**, 169.
Wollast, R. (1973). Internal Report, Inst. de Chemie Industrielle, University Libre de Bruxelles.
Wood, B. J. B., Tett, P. B. and Edwards, A. (1973). *J. Ecol.* **61**, 569.
Wright, J. R. and Schnitzer, M. (1963). *Proc. Soil Sci. Soc. Am.* **27**, 171.
Zirino, A. and Yamamoto, S. (1972). *Limnol. Oceanogr.* **17**, 661.

Chapter 42

Coastal Lagoons

Centro de Ciencias del Mar y Limnología,
Universidad Nacional Autónoma de México,
México 20, D. F.

42.1. INTRODUCTION

Many large expanses of relatively protected, shallow marine waters exist along the coastal margins of the world's land masses, and in some areas they are associated with coastal lagoons. Such lagoons are important commercially because they support extensive fisheries and provide protected harbours, and scientifically as they provide examples of a wide variety of marine ecosystems. In the present review, the published information concerning the lagoonal environment is presented with the intention of illustrating how the chemical and biochemical processes operating in any lagoon depend on the interactions between certain basic factors.

Unfortunately, it is difficult to define even a simple coastal lagoon. For example, Pritchard (1967) prefers to consider coastal lagoons as estuaries. In contrast, Lankford (1977) defines a coastal lagoon as:

> a coastal zone depression below mean higher high water, having permanent or ephemeral communication with the sea, but protected from the sea by some type of barrier,

a definition which could of course embrace many estuaries. Nevertheless, this definition is perhaps more broadly based than that given by Emery and Stevenson (1958a) which states that coastal lagoons are:

> bodies of water, separated in most cases from the ocean by offshore bars or islands, of marine origin and are usually parallel to the coastline.

Such a description, though not definitive, provides a clear picture of what a coastal lagoon is generally considered to be. However, it must be recognized that no clear distinction can be drawn between lagoons, estuaries and bays, all of which exist in the coastal marine environment, but the characteristics of which range from "salt marsh" at one extreme to "open bay" at the other.

42.2. Formation of the Lagoonal Environment

Coastal lagoons are all recent and transient geological features. The fundamental process leading to their formation is a relative change in sea-level, brought about by either coastal emergence or submergence (Zenkovitch, 1969). In the case of an emergent coastline, shallowing of the water overlying an offshore submerged beach may lead to the formation of a bar which isolates inshore water and thus forms a lagoon. For a submergent coastline, one of two major mechanisms may operate (Fig. 42.1); (a) a gradual sea-level rise across a gently sloping coastal plain may initiate the formation of an offshore bar which sometimes migrates towards the new coast if the sea-level continues to rise (Zenkovitch, 1969), or (b) with a more steeply-sloping coast, the rising sea-level may fill coastal depressions, a process which, under certain circumstances, may be followed by the formation of a bar which encloses the depressions and so forms lagoons (Lankford, 1977). The actual formation of an offshore bar (or barrier) depends on sufficient building material (sand, shells or pebbles) being transported and deposited in the zone of wave action and also requires that the slope of the area of deposition is sufficiently gentle for the deposited material to remain in place (Zenkovitch, 1969).

The barriers of many present-day lagoons were formed as a result of the Holocene sea-level rise which occurred 6000–7000 years ago (Curray *et al.*, 1969; Emery and Stevenson, 1958a; McIntire and Ho, 1969; Phleger, 1969;

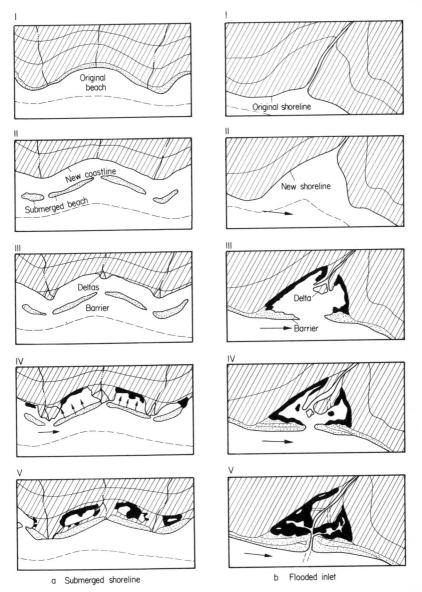

FIG. 42.1. Two typical sequences in the formation and maturity of lagoons following coastal submergence. Sequence (a) results from submergence of a gently sloping coastline (based on work by Zenkovitch (1969) and Phleger (1969)). Sequence (b) results from flooding of a river inlet and associated low-lying land (after Lankford, 1977). The initial rise in sea-level is represented by stages I and II, bar formation occurs in stage III and the lagoon matures through stages IV and V. Dark areas represent mangrove swamps or other intertidal formations and stippled areas represent sand barrier and beach formations.

Zenkovitch, 1969). Since the rate of rise in sea-level during the late Holocene was only about 15–20 cm (100 year)$^{-1}$, the formation of such bars was often only a gradual process (Phleger, 1969). In some instances the supply of bar-building material was spasmodic and therefore led to the formation of large numbers of roughly parallel beach ridges, sometimes sandwiching long parallel lagoons (McIntire and Ho, 1969; Curray *et al.*, 1969). The initial length of the offshore bar was controlled by the supply of building material by longshore drift, and by the morphology of the coast (Zenkovitch, 1969), the number of inlet channels breaking the bar being related to the amount of water exchanged between the newly formed lagoon and the open ocean (Phleger, 1969).

In geological terms, a coastal lagoon is a highly unstable environment. From the time of its formation it is subjected to processes of sedimentation and erosion which often finally reduce it to marshland which can then be further eroded by the sea (Emery and Stevenson, 1958a). The present generation of lagoons is not the first one. For example, foraminiferal evidence shows many generations of "fossil lagoons" buried under the Texas mainland (Walton and Smith, 1969), and the presence of an offshore submerged bar shows the existence of a former lagoon in what is now the estuary of the River Plate, Argentina (Urien, 1977). It is the processes of lagoonal decay that have led to the great variety of conditions found at the present time and which greatly affect the physical and chemical conditions of each system.

Except in very arid areas, the bulk of the sediments in coastal lagoons are contributed by rivers (Emery and Stevenson, 1958b). However, sedimentary material may also be introduced from the seaward side of lagoons by tidal movements or by onshore winds transporting dune sand (Phleger, 1969). Wind-transported sand is important in desert regions such as the Arabian Desert where the growth of a wide sandy coastal plain is gradually reducing the area of the fringing lagoons (Evans and Bush, 1969).

The manner in which a lagoon "matures" depends upon the balance between the sedimentation and the erosion within it. The initial abrupt fall in the current velocities of river and tidal inputs into a lagoon generally causes deposition of the larger-size fractions of the suspended material which they transport. For this reason, deltas are often observed at river discharges of lagoons or on both sides of a bar inlet channel (Phleger, 1969). The extension of river deltas is generally rapid, for example, 0·5 km (100 year)$^{-1}$ (Shepard and Moore, 1960), and may occasionally be more spectacular and lead to the rapid division of a lagoon (Bouma and Bryant, 1969). The division of lagoons may also result from the growth of mangroves in tropical coastal areas, or from a high rate of sedimentation occurring at tidal null points when a lagoon has two or more bar inlet channels (Phleger, 1969). The processes of division are physically, chemically and ecologically important as they frequently lead to the isolation of lagoons from the direct influences of both the rivers and the sea, or, as often happens, they leave the lagoons connected to the river estuary or the sea by a long narrow channel (*estero*).

Even the lagoon barrier itself may be subjected to continual changes in structure. The bar inlet generally migrates in the direction of the prevailing longshore drift (Phleger, 1969), an effect which further isolates the enclosed lagoon. This migration has been measured in the Guerrero Negro Lagoon, Mexico, as 200 m in 3 years (Phleger, 1969) and in the Mugu Lagoon, California, U.S.A., as 100 m in 100 years with an annual cyclic migration of 200 m (Warme, 1969). Annual variations, such as this cyclic migration, are particularly pronounced in tropical and arctic regions in which the river flow is highly seasonal. In such regions, the changes in the balance between the scouring action of the currents flowing seawards across the inlet (tidal currents and river run-off) and the depositional action of longshore drift and waves, both of which transport sediments (reinforcing the bar), may, in extreme cases, produce lagoons that seasonally open and close to the sea (Faas, 1974; Lankford, 1977; Lawson, 1966; Warme, 1969).

Since lagoons are generally very shallow, their sediments are readily exposed to resuspension by current and wave action. This effect causes a re-sorting of the sediments, the coarsest fractions often being found at the bar inlet where the tidal currents are strongest (Emery and Stevenson, 1958b; Phleger, 1969). In contrast, the finest fractions are deposited in the areas furthest away from the tidal inlets (Postma, 1965; Phleger, 1969), and in basins where the water depth is greater than 5 m (Emery and Stevenson, 1958b).

The combination of each of the above processes leads to a wide range of lagoonal morphologies in which sedimentation rates vary widely. A study by Shepard and Moore (1960) of the lagoon systems in Texas showed an average shoaling rate of ~ 38 cm $(100 \text{ year})^{-1}$, ranging from 7 cm $(100 \text{ year})^{-1}$ in isolated bays to 107 cm $(100 \text{ year})^{-1}$ near river discharges. It is sedimentation rate differences such as these which, combined with the effects due to large seasonal climatic variations observed in some areas, are major causes of the instability of lagoon systems. This geological instability, together with climatic stresses, will be shown to directly influence their hydrography and chemistry.

42.3. PHYSICAL PROCESSES

In any study of coastal lagoons it is impossible to describe chemical processes without a previous consideration of the physical processes acting on the system. For example, the residence time of a substance within a lagoon may be related to its flushing time, the salinity and temperature may control certain biogeochemical cycles, and changes in lagoonal water level may result in mineral deposition or solution on resubmerged mudflats.

The hydrographical regime of a lagoon is largely determined by the relationship between the fresh and saline water inputs into the system and the rate of evaporation. In the simplest models, lagoons are often shown as basins having a two-layer steady-state flow across their barrier inlets (Emery and Stevenson, 1958a; Groen, 1969). Although this model will be shown to be grossly oversimplified, it will serve as an introduction to explain the basic salinity patterns observed in coastal lagoons.

In this "two-layer" lagoonal model, an "estuarine" circulation pattern develops when input (run-off) exceeds evaporation, that is to say there is a surface outflow of water of low salinity and a sub-surface inflow of sea-water (Fig. 42.2). The exact nature of this pattern may be considerably modified by the local tidal regime, but it generally results in a net outflow and gives rise to a lagoon having a mean salinity somewhat lower than that of the sea.

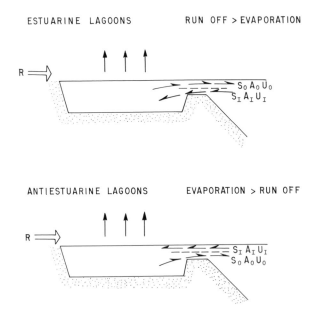

FIG. 42.2. Two-layer lagoons showing the basis of steady-state models. S_I, A_I and U_I respectively represent the salinity, cross-sectional area and velocity of the inflowing sea-water at the barrier channel. S_O, A_O and U_O similarly represent conditions for the outflowing lagoon water. At steady-state $S_O = S_I$, $A_O = A_I$ and $U_O = U_I$.

In those instances in which evaporation exceeds the freshwater input, an "anti-estuarine" circulation develops with a surface inflow of sea-water and a sub-surface outflow of more saline water. In lagoons having this type of circulation, the salinity is generally higher than that of the sea, and the lagoon may be classified as hypersaline.

Unfortunately, although these two basic types of lagoon exist, very few lagoons can be considered to be steady-state environments. It is important to realize that because there is such a multiplicity of lagoonal types it is not possible to make a rigid hydrographical or chemical classification of such systems. Perhaps more realistic approaches to classification are those based on mode of formation and geological structure (Lankford, 1977), or on the multidisciplinary approach (Por, 1971). The latter worker based his classification on the types of biological communities present, the salinity regime and the degree of isolation of the lagoon. In this classification, 13 types of lagoon were distinguished but, although this approach is useful, it introduces a high degree of complexity to the description of these systems. The actual hydrographical and chemical states of each lagoon merely reflect its state of maturity and the physical stresses imposed upon it. In addition to run-off and evaporation, these stresses include tides, wind, the longshore drift of sediments and, in polar environments, ice formation (Fig. 42.3).

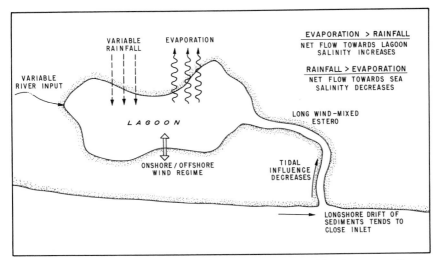

FIG. 42.3. The variable conditions of a non-steady-state lagoon with a long *estero* (barrier channel).

The tidal influence on a lagoon is largely controlled by the size of the lagoon, the tidal range and the morphology of the barrier inlet. A small inlet may cause strong damping of the tide, and a shallow lagoon may considerably modify the harmonic pattern of the incident tide, especially if the travel time for the incident tidal wave is large compared with its period (Groen, 1969). Thus, in a typical lagoon which has developed a very long *estero* (Fig. 42.3), the tidal wave might be completely damped out before reaching the lagoon. This effect prevents tidal flushing. In those lagoons with deep channels and a large tidal range e.g. the Ojo de Liebre Lagoon in Baja California, Mexico (Phleger and Ewing, 1962), the tide may be observed in all areas. In contrast, in lagoons receiving a large run-off and having a small tidal range, the inflowing tide may be completely suppressed, leading to very low salinity conditions (Phleger, 1969). In addition to diurnal tides, there may be other tidal effects such as those arising from the cyclic annual variations which result from very long period tides. On the Mexican Pacific coast this may result in the summer mean tides being 25 cm higher than those in winter (Instituto de Geofísica, 1976); this causes a considerable change in the volume of a shallow lagoon. Storm surges may also considerably affect water levels in lagoons and cause floods or damage (Lankford, 1977). The huge tides resulting from Hurricane Carla in Texas, U.S.A., resulted in considerable structural damage and produced physical and ecological changes in the coastal lagoons lying in their path (Oppenheimer, 1963).

Winds, too, can exert a considerable influence on the pattern of circulation within a lagoon. Apart from mixing the water column, a steady wind stress may tend to pile up sufficient water at one side of a lagoon to cover (or expose) mudflats (Oppenheimer and Ward, 1963), or to promote a surface water circulation (Groen, 1969).

The combination of the effects briefly mentioned above, together with those due to latitude, are best illustrated by means of a few examples. The most extreme variations of hydrography occur in tropical and polar lagoons, which are environments in which steady-state conditions do not appear to be achieved.

Arctic lagoon systems have been reviewed by Faas (1974). The Arctic coast, a typical example of which is shown in Fig. 42.4, is characterized by permafrost, low precipitation (< 25 cm) and a large seasonal temperature range (− 35°C to 15°C). In small shallow lagoons, such as Esatkuat (Fig. 42.4), low salinity conditions prevail in summer due to run-off of melted ice, even though the lagoon is connected to the sea. In the winter freeze-up the barrier inlet closes because there is insufficient scouring by the outward flowing tide and an ice layer 2 m thick is formed. Since the lagoon is generally less than 3·5 m deep and totally enclosed, hypersaline conditions develop in the water below the ice layer. The annual cycle is completed by the summer thaw.

Tropical coastal lagoons have very large seasonal salinity variations and these are dependent upon factors such as the run-off which they receive and their morphology. A comparison of two adjacent Mexican Pacific lagoons has been made by Mee (1977) (Fig. 42.5). The Laguna Chautengo has a surface area of 36 km² and a small direct opening to the sea. It is quite shallow (0·5–1·5 m) and has seasonal river inputs. The Laguna Apozahualco is very small (2 km²) and shallow (0·5 m), and has no direct river input, run-off being that directly associated with the rainfall. The annual salinity cycles of these lagoons illustrate the effect of both depth and river run-off. The inlets

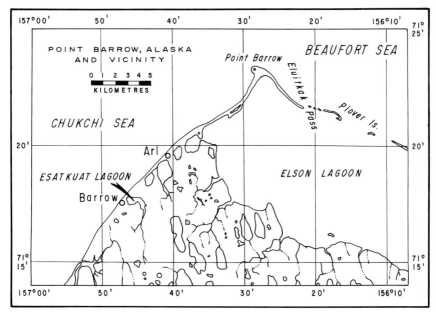

FIG. 42.4. A typical Arctic coastline; the coastline near Point Barrow in northern Alaska (after Faas, 1974).

of both lagoons close at the end of the rainy season, since without net transport of river-water to the sea the tidal exchange across them is, by itself, insufficient to overcome the effects of longshore drift which tends to seal them off. During the dry season, evaporation dominates, its effect being particularly pronounced in the shallower Laguna Apozahualco in which salinities may reach 140‰; as a consequence, a local salt collection industry has developed. The start of the rainy season produces very low salinities in both systems and extensive flooding. The resultant opening of the barrier inlets facilitates a large discharge of water (and its associated dissolved and suspended material) from the lagoon. This is followed by a period in which sea-water enters the lagoon, the resultant salinity being determined by the river run-off. In

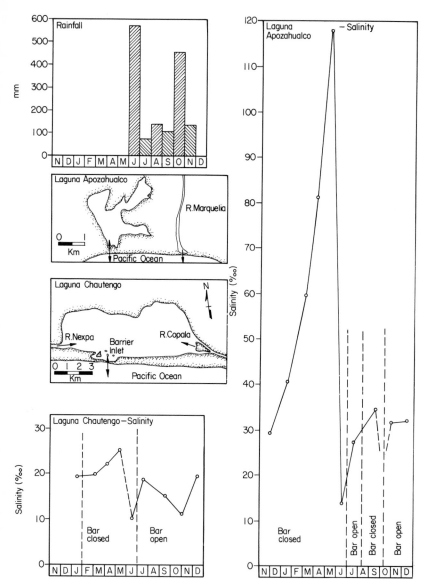

FIG. 42.5. A comparison of seasonal salinity variations in two Mexican Pacific coastal lagoons (data from Mee, 1977).

the case of Apozahualco, the bar may open and close several times, indicating that the system has a very low stability.

The type of system described above may be considerably modified in those lagoons which have a very long *estero*. Such a system was studied and modelled by Ocampo and Emilsson (1975) who considered a system consisting of 70 km of lagoons and channels in which the tide is completely damped out before reaching the lagoons (Figs 42.6, 42.7). During the dry season, evaporation from the lagoons causes an intrusion of sea-water to move along the *estero* at a velocity of up to 1 km (day)$^{-1}$, with the result that the salt content of the system increases steadily. When the rainy season commences, the sea-water front is rapidly driven out of the system and the salinity of the entire lagoon drops to 10‰. The lagoon system is thus flushed out only once a year.

Perhaps the most extreme lagoonal system which has been described is the Solar Lake in the Sinai Desert (Cohen *et al.*, 1976a). This lagoon is really a small pond (140 × 50 m) totally isolated by a gravel bar through which seepage provides the only contact with the sea. The lagoon is generally extremely hypersaline as evaporation predominates because the climate is arid. A short period of irregular rainfall causes the salinity to fall and water seeps from the lagoon to the sea. During the remainder of the year the seepage is reversed and the lagoon salinity increases. The residence time of the water in the lagoon has been calculated to be 5·5 months. In the winter months the lagoon is highly stratified and has a reverse temperature gradient of up to 36°C, the underlying water having a maximum temperature of 56·6°C and a salinity of up to 180‰. Stratification is a common feature in lagoons having poor mixing and exchange with the sea, and has important chemical consequences such as the development of anoxic waters.

The examples presented here are all for lagoons for which steady-state models are not applicable (since the total salt content of each system is subject to seasonal variations). Even in the apparently stable lagoonal conditions found in temperate latitudes, the salt content may vary from year to year. This is well illustrated by the work of Emery (1969) on Oyster Pond, a small brackish coastal lagoon in Massachusetts, U.S.A., which is almost completely isolated from the sea. During the period from 1963 to 1965 there was a steady increase in the salinity of the lagoon which Emery attributed to the sharp decrease in rainfall during the period of study. Interestingly, this rather isolated system also exhibited stratification and anoxia, illustrating that these are results of isolation rather than latitude.

The predominant physical stresses on the ecology of a lagoon appear to change with both isolation and latitude. In tropical and arid latitudes the predominant stress results from both the large temperature changes and the extremes in salinity. In temperate latitudes the predominant seasonal stresses are those of changing temperature, whereas, in polar latitudes the stresses

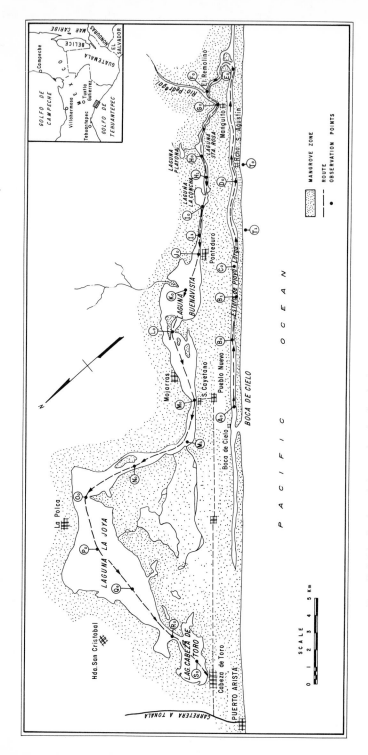

FIG. 42.6. Distribution of observational points in the La Joya–Buenavista Lagoon system, Mexico.

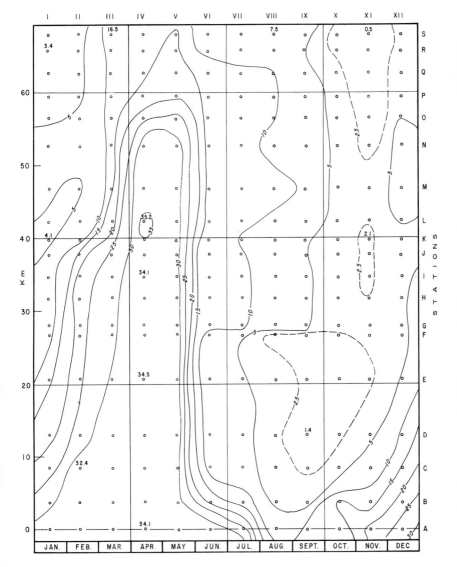

FIG. 42.7. Isopleth showing the spatial and temporal variation of salinity (mean for the water column) along the La Joya–Buenavista Lagoon system, Mexico. The data are based on weekly averages during a three-year period. (By kind permission of R. Ocampo and I. Emilsson.)

are those of salinity and temperature. These stresses, together with flushing, the isolation of the lagoon and the presence of stratification, provide the physical framework upon which the chemical processes of the lagoon act.

42.4. The Chemistry of Coastal Lagoons

42.4.1. general aspects

The previous discussion has emphasized the diversity of lagoonal types. However, lagoons have sufficient common properties to enable their chemistry to be described within a common framework. All lagoons are relatively protected and shallow, and their saline properties are strongly influenced by continental climatic processes. For this reason their chemistry is generally controlled by continental and marine interactions.

Unfortunately, the present state of knowledge of chemical processes in coastal lagoons is strongly biased towards those involving biologically important substances such as nutrients, micronutrients and certain contaminants, and very little is known of purely geochemical processes. It might be anticipated that the geochemistry of a lagoon having an estuarine circulation would be rather similar to that of an estuary, but there is little published evidence to support this hypothesis.

In hypersaline lagoons, significant chemical compositional changes may result from the precipitation of major and minor ions during the process of evaporation. The formation of salt minerals in epicontinental basins has been extensively reviewed by Borchert (1965) and a summary of the principal phases of salt deposition is given in Table 42.1. It can be seen from this table that a very considerable reduction in the volume of a lagoon is necessary to initiate the precipitation of halite (NaCl); such precipitation is very uncommon except in certain natural and artificial "salinas" basins in areas of high evaporation which become temporally isolated from the sea permitting almost total evaporation of their water content. Such evaporation has been observed by Copeland (1967) for the Laguna Madre, Tamaulipas, Mexico, in which salinities as high as 300‰ were recorded; these were sufficient to cause the precipitation of all but the phase IV minerals of Table 42.1. Such a large chemical change has a direct biological impact due to the osmotic stress produced by increased ionic strength, the decrease in alkalinity (resulting from the precipitation of carbonates and borates) and the decrease in oxygen solubility (which arises from the increasing salinity of the lagoon water). The more subtle, but nevertheless important, changes in the interaction of dissolved ions, and their activities, do not appear to have been studied in such hypersaline environments, and provide a wide field for further chemical investigations.

TABLE 42.1

Precipitation of mineral salts during the evaporation of sea-water in an isolated basin[a]

Density[b]	% Of original volume[b]	Principal salt mineral(s) deposited	% Of total dissolved salts	Minor salt deposition
1·026	100	—	—	—
1·050	53·3	I Calcium carbonate + dolomite	1	Fe_2O_3
—	—	—	—	Borate salts
1·126	19·0	II Gypsum	3	—
1·214	9·5	III Halite	69	Some $MgSO_4$, $MgCl_2$ and NaBr
—	—	IV Na, K, Mg, sulphates; K, Mg, chlorides	27	—

[a] After Borchert (1965) and Copeland (1967).
[b] The figures in these columns represent approximate values at which precipitation first occurs.

The present review is mainly concerned with the cycles of nutrients, and attempts to demonstrate how lagoons and similar shallow water environments maintain a primary productivity which is higher than that of the adjacent open ocean. A lagoon is not a completely closed system with respect to the dissolved and suspended material within it and it receives inputs of dissolved and particulate nutrients which may or may not be balanced by losses from the system. Supply of nutrient materials, whether from continental, marine or aerial reservoirs, is important in initiating the chemical cycles associated with productivity in lagoons and must receive special attention before such cycles can be described. Losses of such materials arise from both the flushing of dissolved and suspended particulate material to the sea, and the incorporation of the latter into the sediments. The relative importance of each of these processes largely depends on the hydrography of the lagoon.

42.4.2. SUPPLY OF NUTRIENTS

The supply of dissolved and particulate materials to a coastal lagoon depends upon the external reservoir of each of the materials and the availability of a transport mechanism to carry them into the lagoon. It seems reasonable to assume that by far the most important sources of dissolved and particulate nutrients, and probably also of dissolved organic substances and trace metals, are continental reservoirs. Mobilization from these reservoirs may take place by weathering, biological processes and human activities (such as agriculture, industry and domestic activities), and transport occurs during run-off. Thus, for lagoons having a river input, the river can often be identified as the principal source of nutrient material (Emery and Stevenson, 1958a). Indeed, river discharges may in some cases be shown to completely support and regulate the phytoplankton bloom within a lagoon (Howmiller and Weiner, 1968). Conversely, a lack of river input, or a depleted continental nutrient reservoir, may cause a lagoon to have a low primary productivity (Tampi, 1969). Consequently, the nutrient inputs into tropical lagoons in areas which have highly seasonal rainfall vary considerably over a year. In the Laguna Chautengo, Mexico, the first rains of the rainy season cause considerable flash-flooding and run-off from the low-lying land bordering the lagoon. Such run-off temporarily provides a more important source of nutrients than does the river discharge into the lagoon (Fig. 42.8). During the period of less changeable climatic conditions following the initial rains, the river again assumes the dominant role as a nutrient source. Okuda (1969) compared two morphologically similar tropical lagoons, receiving seasonal and continuous river run-off respectively, and showed the importance of a steady input of river-water to the maintenance of a stable lagoonal ecosystem.

An important part of the run-off into many lagoons results from direct rainfall and, according to Collier and Hedgpeth (1950), some Texas lagoons may receive many times their own volume of rain-water in any one year. Whilst the effect of such an input is often to "dilute" the lagoon by flushing, the rain may carry a quantity of nutrients which is sufficient to cause eutrophication (Reimold and Daiber, 1967).

Many lagoons, however, do not have a significant continental run-off and these may range from so-called "neutral" lagoons having a similar salinity to that of the sea (Glooschenko and Harriss, 1974), to classical hypersaline lagoons which are constantly open to the sea. Despite the absence of continental run-off, such lagoons generally contain a total concentration of dissolved and combined nutrients which is relatively higher than that in the adjacent sea (Copeland and Nixon, 1974; Nichols, 1966; Postma, 1969). The elevated dissolved nutrient concentrations may be caused by the long-term retention of irregularly introduced nutrients (e.g. those resulting from sporadic run-off), and the combined nutrient levels arise from biota introduced with sea-water. This effect is probably a result of the long flushing time of hypersaline systems combined with the rapid incorporation and cycling of nutrient material within the lagoon biosphere. A study of phosphorus in a small open hypersaline lagoon by Nichols (1966) revealed such a concentration effect. The total concentration of this element (largely in the form of suspended particulate matter) increased with distance inwards from the lagoon mouth.

Increasingly important sources of nutrient materials for many lagoons are

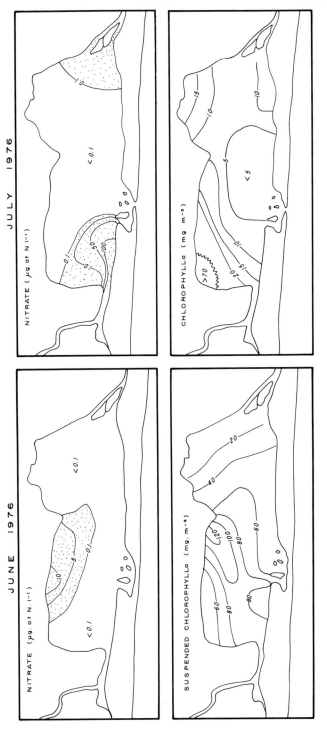

FIG. 42.8. Spatial distributions of concentrations of dissolved nitrate and suspended particulate chlorophyll *a* in the Laguna Chautengo, Mexico, during surveys in June and July 1976 (from Mee, 1977).

domestic, agricultural and industrial wastes. For example, Copeland and Wohlschlag (1968) observed that eutrophication resulted from industrial and domestic contamination of the coastal lagoons of Texas, U.S.A. Sewage discharges and agricultural run-off have also been shown to support large algal blooms in estuarine environments (Barlow *et al.*, 1963), and even a minor source, such as power-plant cooling water (Bader and Roessler, 1972), can have considerable local effect. In addition to sewage, intense human activity can lead to the introduction of toxic pollutants, which, because of the long flushing times of coastal lagoons and biological re-cycling, may have long residence times in the lagoonal environment. In this context it should be noted that high concentrations of pollutants have been observed in estuarine invertebrates (Butler, 1966).

42.4.3. NUTRIENT LIMITATION AND MOBILIZATION

From the various chemical studies which have been conducted in coastal lagoons there is considerable evidence to suggest that primary production is chemically limited by available inorganic nitrogen nutrients. Thus, the fact that all of the published studies indicate that excess nutrient phosphorus is present, but only limited quantities of inorganic nitrogen nutrients (even when nitrogen fixation occurs) shows that the overall lagoonal systems are very rarely phosphorus-limited. The work of Ryther and Dunstan (1971) on the lagoons bordering Long Island, U.S.A., and on coastal and open-ocean environments, led them to conclude that most natural and domestic discharges contain a nitrogen/phosphorus ratio much lower than the 15/1 which is generally recognized as approximating to the demands of primary producers. This leads to an effective excess of phosphorus nutrients in coastal marine environments. In areas well away from the influence of continental discharges an inorganic phosphorus excess is also observed; this probably occurs because it is regenerated more quickly from organic detritus than are the consequently limiting inorganic nitrogen nutrients (Ryther and Dunstan, 1971).

It is worthwhile to examine the effects of a continental nutrient supply on a typical coastal lagoon. There is usually an intense phytoplankton bloom in the vicinity of a nutrient discharge, and the dissolved inorganic nitrogen nutrients become rapidly exhausted (Fig. 42.8). However, the standing-crop is widely distributed throughout most lagoons (apart from those areas rapidly flushed by the sea), even though there is no obvious direct source of supply for the limiting nutrient. This observation may be explained by bacterial oxidation of dissolved organic nitrogen, by regeneration from detritus (including nutrient recycling), or simply by the redistribution of phytoplankton by surface water movements. Forunately, it is possible to distinguish in practice which of these three processes is operative.

In some tropical lagoons, the barrier inlet closes during the dry season and there is no continental run-off; such lagoons therefore provide natural "experimental tanks". The Laguna Chautengo (Mee,1977) is an example of such a system. In Table 42.2 the results of some seasonal measurements of standing-crop of phytoplankton are listed, and it can be seen that throughout the dry season the crop remained high despite the lack of an external nutrient

TABLE 42.2

Seasonal variations in the standing-crop during 1976 in the Laguna Chautengo, Mexico (data from Mee, 1977)

	Hydrographic conditions			Total chlorophyll *a* (tons)
Survey date	Season	Barrier inlet	River flow	
Jan. 16	Dry	Open	Minimal	0·24
Mar. 2	Dry	Closed	Negligible	0·24
Apr. 4	Dry	Closed	Negligible	0·38
May 10	Dry	Closed	Negligible	0·36
June 16	Rainy	Closed	Considerable run-off	3·10
July 18	Rainy	Open	Flowing	0·37
Sept. 13	Rainy	Open	Flowing	0·34
Nov. 1	Rainy	Open	Flowing	0·15
Dec. 9	Dry	Open	Minimal	0·12

supply. This strongly suggests that the regeneration of inorganic nitrogen was of a sufficient magnitude to maintain the entire primary production of the lagoon during this period. The lagoon system described by Ocampo and Emilsson (1975) provides a further example of high primary production in the absence of continental run-off. This is demonstrated by the presence of large areas of day-time oxygen supersaturation during the dry season (see Fig. 42.9) and particularly high oxygen concentrations are found during February and March, well before the onset of the summer rains. Nutrient regeneration would, therefore, seem to be of primary importance in establishing high productivity within the lagoonal environment. This is not unreasonable when it is considered that most lagoons are very shallow and completely mixed. These two factors provide: (i) a surface (the sediment–water interface) upon which detrital material may collect and at which considerable nutrient regeneration may occur, and (ii) a process by which regenerated nutrients can be rapidly transported into areas of maximum utilization.

Processes occurring across, and at, the sediment–water interface have been reviewed by Hayes (1964). This interface provides an important site for the

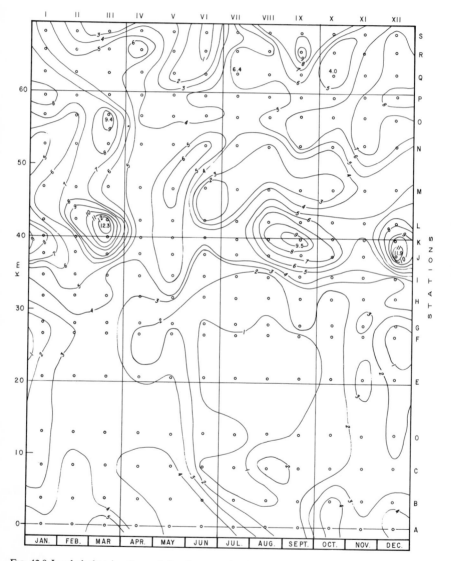

FIG. 42.9. Isopleth showing the spatial and temporal variation of dissolved oxygen concentrations (day-time mean for the water column) along the La Joya–Buenavista Lagoon system, Mexico, based on weekly averages during a three year period. Station positions are shown in **Fig.** 42.6. (By kind permission of R. Ocampo and I. Emilsson.)

bacterially-mediated oxidation of organic detritus, it supports an active benthic fauna and flora and it may play an important role in chemical adsorption and desorption phenomena, particularly for phosphate-phosphorus. Quantitative *in situ* studies of the chemical processes have generally used enclosed, or semi-enclosed, bottom chambers (Bruce and Hood, 1959; Hale, 1975). The study by Hale (1975) is particularly interesting as the chambers used were provided with communication with the outside water column (via simple baffles) in order to maintain neutral physical conditions within each chamber. This study demonstrated that ammonia was generally only released from the interface to the water column ($+4\cdot3$ to $+276\cdot1$ µmol m^{-2} hour^{-1}). In contrast, nitrate could be either released or taken up ($-1\cdot2$ to $+6\cdot1$ µmol m^{-2} hour^{-1}), and a distinct two-way flux was also shown by phosphate ($-9\cdot4$ to $+41\cdot6$ µmol m^{-2} hour^{-1}).

The fact that ammonia is released from the sediment–water interface is important evidence for the regeneration of nutrients in shallow water environments. Both the *in situ* observations by Hale (1975) and the laboratory observations by Okuda (1960) show that there is a practically linear relationship between the rate of ammonia release and temperature (within the experimental range of 5–30°C), and this reflects a bacterial control of the regeneration process. Hence, temperature may represent one controlling factor in the supply of the limiting nutrient to the water column in coastal lagoons.

Recent studies have examined the possibility of modelling the inorganic nitrogen flux in the surface layers of shallow water sediments. The model devised by Vanderborght *et al.* (1977) considers two distinct sediment layers. The upper layer is oxygenated and the rate of ammonification within it is related to the rate of oxygen consumption. In practice, oxygenation of this layer is often facilitated by the presence of benthic burrowing animals (Grundmanis and Murray, 1977); nitrification is also assumed to take place in the upper layer. The second layer in the model consists of anoxic sediments in which the ammonification is assumed to be stoichiometrically related to the sulphate utilization, and in which denitrification occurs. The stoichiometric NH_4^+–SO_4^{2-} relationship is one of the drawbacks of the model in view of the fact that Berner (1977) has shown that (i) certain organic compounds are broken down preferentially and (ii) there is variable differential diffusion of the regenerated species, which together with adsorption effects make stoichiometric modelling impossible. Further, the authors have admitted that it is unsatisfactory to derive kinetic constants by adjusting the theoretical profiles of dissolved species in interstitial waters to the experimental ones. Nevertheless, despite these drawbacks, such models provide a useful basis for further investigations.

The role of phosphorus exchange across the sediment–water interface of coastal lagoons has been the subject of studies by Bruce and Hood (1959),

Pomeroy et al. (1965) and Hale (1975), all of whom showed that inorganic phosphate could be readily adsorbed and desorbed at the sediment–water interface. In situ studies by Hale (1975) have shown that this process is temperature dependent, with increasing release taking place at temperatures above 4–10°C, and increasing uptake at lower temperatures. In addition, Pomeroy et al. (1965) have found that phosphate exchange is impeded, although not prevented, by the poisoning of sediment samples with formalin. These two observations suggest that the phosphate exchange process is, at least partially, bacterially mediated, although some chemical adsorption and desorption also appear to take place. The in situ and laboratory $^{32}PO_4^{3-}$ experiments by Pomeroy et al. (1965) led them to suggest that the rapid exchange between the phosphate reservoir in sediments (and suspended sediments) and that of the water column acts as a buffer to maintain relatively stable phosphate concentrations in the water column. These concentrations would thus depend upon the exchange capacity of the sediments, the exchange rate between the water and sediments, the flushing time of the system and the rapidity of vertical mixing. For a lagoonal system which is well mixed and supports a large load of suspended sediment, the exchange process should be extremely rapid, thus providing a continuous supply of phosphorus for the requirements of photosynthetic organisms.

Phosphorus is not the only element which has been shown to be readily exchanged between dissolved and sedimentary reservoirs. Tracer studies of ^{60}Co, ^{59}Fe, ^{54}Mn and ^{65}Zn have been made in the coastal lagoons of Texas, U.S.A. by Parker et al. (1963). The tracers which had been inoculated into the water column exhibited a rapid initial loss to the sediments with a residence half-time of only a few hours. The rapidity of this exchange suggests that the supply of micronutrients and phosphate is unlikely to be a limiting factor in the primary production of any shallow lagoonal environment.

Mobilization of phosphorus is not limited to the above exchange processes. For example, winds blowing across drying mudflats may promote capillary action, drawing phosphorus compounds released by bacterial regeneration to the surface (Oppenheimer and Ward, 1963). Continuation of the drying process leaves a thin crust of phosphate-rich salts on the mudflat and these may provide an important phosphorus source when the mudflat is resubmerged. The activities of filter-feeding benthic organisms provide a further mechanism for the rapid remobilization of suspended particulate nutrients. This is well illustrated by the excellent study made by Kuenzler (1961) of the phosphorus turnover of a population of mussels (Modiolus demissus) in a salt marsh in Georgia, U.S.A. This mussel population was shown to be able to turnover all the particulate phosphorus in the water column (~ 450 µg-at.P m^{-2}) in a period of only 2·6 days, releasing some 96% of the phosphorus as faeces and pseudofaeces to be subsequently regenerated bacterially.

No quantitative studies appear to have been made of the nutrient role of organic detritus in lagoons which, in many respects, may be regarded as sediment traps for detritus from the bordering communities. The large export of organic detritus from salt marsh and mangrove areas associated with coastal lagoons (Heald, 1971; Cooper, 1974) may well considerably augment the fertilization of these water bodies, following nutrient regeneration.

Although the preceding discussion has concentrated on the importance of sediment–water interactions in nutrient mobilization, the regeneration of nutrients within the water column cannot be discounted, particularly in very turbid lagoons in which there is constant resuspension of particulate material. A study by Postma (1965) of the Guerrero Negro Lagoon, Baja California, Mexico, shows that the resuspended particulate material consists of suspended sand, material larger than 35 μm (probably organic detritus and clay–organic material conglomerates) and phytoplankton. Nichols (1966) showed that such suspended matter may be a large reservoir of particulate nutrient material. Thus, it is extremely difficult to separate the effects of respiration and death of suspended phytoplankton from those of regeneration from suspended detritus and resuspended sediments. No studies have been reported of the relative importance of each of these processes of nutrient regeneration; any speculation here would therefore be unprofitable.

The most important outcome of the various processes of nutrient regeneration in a coastal lagoon is that the resultant products are immediately available for utilization in primary production by planktonic and benthic plants. A considerable proportion of the cycle of all nutrients is spent in living tissues or detritus, and the period during which they reside in the dissolved reservoir is therefore relatively short. This, in turn, means that they are only influenced by water movements during a limited period of their residence within the lagoon. The important consequence of this is that the residence time of dissolved nutrients in a lagoon may be somewhat longer than the flushing time of the water. Furthermore, in a mixed water column away from direct nutrient sources the rate of primary production must be directly related to the rate of regeneration of the limiting nutrient. These two factors may contribute considerably to the relatively high productivity of the coastal lagoonal environment.

42.4.4. THE EFFECT OF STRATIFICATION

The discussion of nutrient mobilization in coastal lagoons has, until now, been limited to well-mixed systems. Although coastal lagoons are generally very shallow, isolation from wind or tidal mixing, coupled with an input of water of markedly different density from that of the lagoon (e.g. river- or sea-water), may result in stratification of the water column. Where renewal of the bottom-water layer is impeded by partial isolation from the sea (e.g. by a shallow barrier inlet), anoxic conditions frequently develop. This susceptibility to anoxia often results from the high oxygen demand of detritus-rich bottom sediments.

An interesting example of this effect occurs in the Laguna Mitla in Mexico (Mee, 1977). This lagoon, which is of moderate size (36 km^2) with a maximum depth of 7 m, has become almost completely isolated from the sea and receives both rainfall and local run-off only seasonally. This isolation, which prevents nutrient loss to the sea, together with the effects of nutrient regeneration in the water column and at the sediment–water interface, has led to the development of an enormous standing-crop of microalgae, the maximum biomass of which corresponded to 0·73 g chlorophyll a m^{-2}. During the tropical dry season (January–June) this standing-crop is maintained by a delicate balance between nutrient regeneration and primary production. Figure 42.10 illustrates the distribution of various chemical and physical parameters in the water column at one station during a 24-hour period. The high productivity and oxygen demand of the system leads to a tremendous diurnal variation in oxygen concentration with anoxic conditions and production of ammonia occurring at the bottom of the water column during the night. This demonstrates the high rate of nutrient regeneration in the system and the ready conversion of the regeneration mechanism from the aerobic to the anaerobic form.

Following the start of the 1976 rainy season, the volume of the lagoon increased and the large freshwater input led to the development of thermohaline stratification (Fig. 42.11); this had the effect of inhibiting advection of nutrients from the major zone of regeneration in the lower water layer (possibly the sediment–water interface at which detrital material collects) to the 1-m-thick productive euphotic zone. This in turn reduced the productive capacity of the upper water layer and resulted in a loss of about two-thirds of the total standing-crop. Interestingly, the increase in ammonia and phosphate in the anoxic water layer roughly corresponded to the nutrient content of the lost standing-crop. The effect of additional run-off into the lagoon was thus to temporarily decrease its primary production and not to increase it, an observation which illustrates the importance of a mixed water column in the maintenance of high productivity in a coastal lagoon.

Solar Lake, Sinai, provides an interesting exception to this generalization (Cohen et al., 1976b). In this hypersaline system, productivity actually increases following the stratification resulting from summer rains. However, about 91 % of the primary production has been shown to result from photosynthetic sulphur bacteria within the anoxic zone which was well illuminated and thus provided conditions conducive to the development of a large population of these organisms.

Another example illustrating the effect of heavy fertilization on an open lagoon is provided by the Lago Lungo in Italy (Carrada and Rigillo Troncone, 1975). This lagoon consists of a small (~ 0.7 km^2) basin, having a maximum depth of 7 m which communicates with the sea via a shallow canal. Typical nutrient, salinity, oxygen and hydrogen sulphide profiles for the deepest part

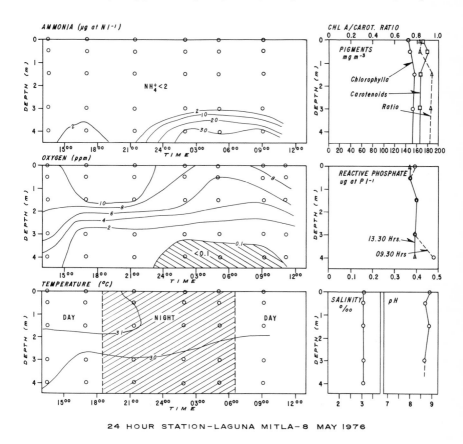

24 HOUR STATION—LAGUNA MITLA—8 MAY 1976

FIG. 42.10. Variations of some physical and chemical parameters during a 24 hour study in May 1976 at one observational point in the Laguna Mitla, Mexico (from Mee, 1977).

of the lagoon are shown in Fig. 42.12, from which it can be seen that the upper 1·5 m of the water column consists of well-oxygenated nitrate-rich brackish water. The high nitrate concentration and high productivity (shown by oxygen supersaturation) of the water probably arise from the agricultural run-off and lake discharge which the lagoon receives. These surface waters provide an interesting example of a phosphate-depleted environment which reflects

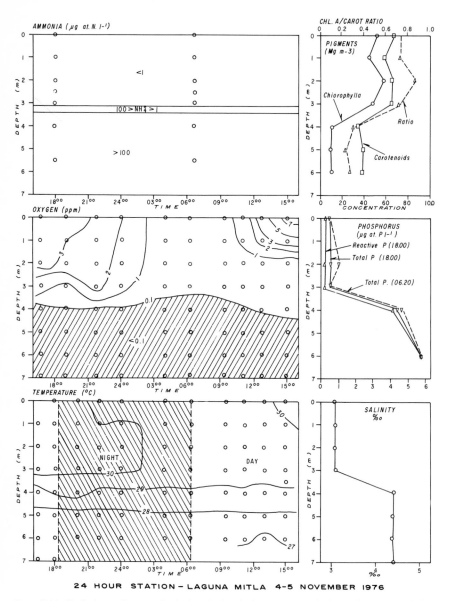

FIG. 42.11. Variations of some physical and chemical parameters during a 24-hour study in November, 1976 at one observational point in the Laguna Mitla, Mexico (from Mee, 1977).

the discharge of lake water into the lagoon (almost freshwater conditions) and the absence of contact with lagoon sediment phosphates. The remainder of the water column is virtually anoxic, with high levels of hydrogen sulphide at depths below 3 m and large concentrations of ammonia and phosphate, all of which are characteristic products of anaerobic respiration (Richards, 1965). At the bottom of the lagoon, conditions in the water are slightly oxic

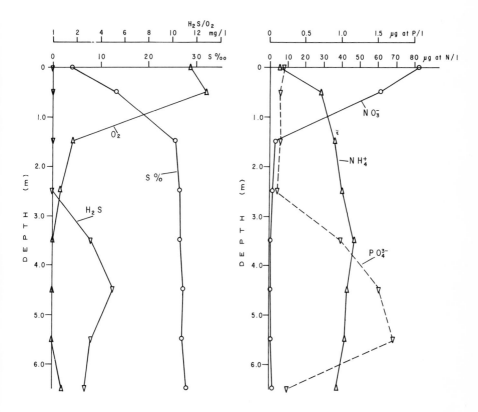

FIG. 42.12. Profiles of some chemical parameters at a central observational point in the Lago Lungo Lagoon, Italy. Original data from Carrada and Rigillo Troncone (1975).

and hydrogen sulphide, ammonia and phosphate concentrations decrease. This may well result from slight seepage of sea-water into the lagoon from the barrier inlet. The anoxic conditions in the Lago Lungo thus appear to result from a combination of the heavy fertilization of the surface waters with a tidal renewal of the sub-surface waters which is insufficient to overcome the high oxygen demand of the sinking detritus. This oxygen demand is reflected by the presence of a 2-m-thick "red water" layer containing hetero-trophic bacteria which lies at the chemocline between the two water layers (Carrada and Rigillo Troncone, 1975).

42.4.5. UTILIZATION OF NUTRIENTS AND THE PRODUCTION OF ORGANIC MATTER

42.4.5.1. *Introduction*

Following the description of the manner in which inorganic nutrients are introduced and mobilized in the coastal lagoon environment, it is now neces-sary to consider how they are reincorporated into organic matter. This conver-sion is very important for two distinct reasons. (i) It provides the first step in a food chain leading to the production of animal protein and (ii) it initiates the second half of the full lagoonal nutrient cycle which is completed by the production of organic detritus. This system of cycling permits the continuous re-use of nutrient substances within the lagoonal environment in order to maintain a steady rate of primary and secondary production.

In coastal lagoons, the plants responsible for the primary production of organic matter may be planktonic or benthic. Benthic plant communities have a special importance within lagoons as they themselves maintain a system of chemical recycling which results in a net exportation of organic matter to the rest of the area. In the present section, the importance of each of these cycles to the lagoon as a whole will be considered from a chemical view-point.

42.4.5.2. *Primary producers within the lagoonal system*

Primary productivity is a useful parameter in chemical investigations of coastal lagoons as it reflects both the rate of fixation of inorganic carbon and the rate of utilization of nutrient substances within the environment. These in turn provide information on the importance of nutrient recycling in any particular lagoon. For example, a lagoon maintaining high primary pro-ductivity, despite a poor nutrient supply, must support a relatively efficient system of nutrient recycling.

It is of interest to compare primary production both within individual lagoons and between specific lagoonal plant communities. Unfortunately, productivity measurements in coastal lagoons have been made by means of such a variety of methods and sampling programmes that the comparative value of the available data is extremely limited. Several authors have based their productivity measurements on the ^{14}C method of Steemann–Nielsen (1952). However, there are wide variations in the incubation times used, which range from 1 hour (Mee, 1977) to 4 hours (Tundisi *et al.*, 1973). Simi-larly, the extent to which data have been extrapolated is highly variable. For example, despite evidence of very large diurnal fluctuations in lagoon pro-ductivity, some authors (e.g. Cohen *et al.*, 1976b; Tundisi *et al.*, 1973) have chosen to base their measurements on a depth profile during only one short period of the day.

Gross productivity estimates have been made from the large diurnal changes in oxygen concentration in the water column. The method developed by Odum (Odum and Hoskin, 1958; Odum and Wilson, 1962; Odum et al., 1963), specifically for shallow, mixed water bodies such as coastal lagoons, involves interpretation of 24-hour dissolved oxygen measurements at defined stations. The diffusive exchange of oxygen between the water column and the atmosphere is estimated from a "diffusion constant" calculated from two points on the night-time respiration curve. Diffusion estimates thereby obtained are only very approximate, since diffusion is assumed to vary linearly with the oxygen saturation of the water column, and no account is taken of wind variations and water surface conditions. The method also assumes that respiration is constant throughout the measurement period. However, in situations in which the oxygen demand is very high, the amount of oxygen available during the night may be insufficient to permit aerobic respiration, thus producing a diurnal variation in respiration (Odum and Wilson, 1962). Under such conditions, the assignment of a respiration value by the Odum method, constitutes little more than an educated guess. Despite these limitations, however, this method is extremely useful for application to coastal lagoons as it provides a productivity estimate for the entire water column (including benthic communities) which is difficult to achieve using ^{14}C techniques.

Oxygen measurements in light and dark bottles have also been used in coastal lagoons (see e.g. Emery, 1969; Nichols, 1966). The interpretation of oxygen variations in closed bottles is not particularly precise and, apart from isolating the productivity of planktonic communities, does not present any advantages over the Odum method. Similarly, bell jar measurements of oxygen variations have been used to elucidate the relative importance of benthic production (Pomeroy, 1960).

The results of representative studies of lagoonal productivity are shown in Table 42.3. Gross productivity estimates are expressed in terms of g carbon m^{-2} day^{-1} and not as g oxygen as in many of the original studies. This conversion, although it introduces a rather arbitrary conversion factor, does permit comparison of the various data. The productivities of communities of organisms associated with individual lagoons are shown in Tables 42.4 and 42.5 for comparative purposes.

It is of considerable interest to compare the primary productivity data for coastal lagoons (Table 42.3) and those given by Cushing (1975) for the average ^{14}C productivity of both open-ocean areas (0.1 gC m^{-2} day^{-1}) and upwelling areas (1 gC m^{-2} day^{-1}). From the table it can be seen that coastal lagoons; shallow though they are, have a productivity which equals, and in

TABLE 42.3

Comparative values for primary productivity in some coastal lagoons

Latitude	Description	Season, etc.	Productivity (gC m^{-2} day^{-1})		Source
			^{14}C method	Gross (Odum method)[b]	
41°33′N	Oyster Pond, Massachusetts, U.S.A. Small isolated brackish lagoon, ice cover in winter.	Summer[a]	—	0·535-4·60	Emery (1969)
45°30′N.	Venice Lagoon, Italy. Mediterranean climate, large, tidally flushed lagoon.	Winter[a]	0·010	—	Emery (1969)
		High tide	0·12-0·99	—	Vatova (1961, 1963)
29°20′N.	Solar Lake, Sinai. Small isolated hypersaline desert lagoon.	Low tide	0·10-0·25	—	Vatova (1961, 1963)
		Stratified (March)	8·02 (max)[c]	—	Cohen et al. (1976b)
28°N.	Estero Tastiota, Mexico. Small turbid desert lagoon, tidally flushed.	Mixed (July)	0·136 (max)	—	Cohen et al. (1976b)
		Winter	—	0·5-7·3	Nichols (1966)
26°-28°N.	Laguna Madre system, Texas, U.S.A. Large hypersaline lagoon in low rainfall coast, indirect sea communication.	Summer	—	1·2-9·4	Nichols (1966)
			—	1·0-28·0	Odum and Wilson (1962), Odum (1967) (cited in Copeland and Nixon, 1974)
17°00′N.	Laguna Mitla, Mexico. 36 km^2, completely isolated brackish tropical lagoon, seasonal rainfall.	Stratified[a]	1·1	16·0	Mee (1977)
16°35′N.	Laguna Chautengo, Mexico. 36 km^2, seasonally isolated and flushed lagoon, seasonal rainfall.	Mixed[a]	2·8	25·5	Mee (1977)
		Rainy season[a] (bar closed)	1·4	—	Mee (1977)
		(bar open)	0·32	4·5	Mee (1977)
		Dry season[a] (bar closed)	—	10·4	Mee (1977)

TABLE 42.3 (*cont.*)

| 25°00'S. | Cananeia Lagoon, Brazil. Tidally flushed tropical mangrove lagoon with summer rains. | Summer | 0·80 | — | Tundisi *et al.* (1973) |
| | | Winter | 0·10 | — | Tundisi *et al.* (1973) |

[a] Not representative of a full spatial/seasonal survey.
[b] Converted from the original data (in gO_2) using $gC = 0.77 \times gO_2$ (factor derived from formulae given by Richards (1965)).
[c] Principally photosynthetic sulphate-reducing bacteria.

TABLE 42.4

Comparative data on the primary productivity of specific communities in coastal lagoons

Productivity (gC m^{-2} day^{-1})[a]

Community	Description	Net		Gross		Source
		Method	Range	Method	Range	
Marine grasses	Biscayne Bay, Florida, U.S.A.; open lagoon	—	—	Harvesting[b]	0·75–3·4	Wood et al. (1967)
	Similar Florida lagoons	—	—	Odum	0·6–6·3	Odum (1974a)
	Open Texan lagoons, U.S.A.	—	—	Odum	1·2–8·8	Odum (1974a)
Algal mats	Laguna Madre, Texas, U.S.A.	O$_2$ variations	0·4–0·6	—	—	Sollins (1969) (cited in Birke, 1974).
				Odum	0·8–9·2	Odum (1974a)
	Solar Lake, Sinai	^{14}C under laboratory conditions	1·5–12·0	—	—	Cohen et al. (1976b)
Benthic microflora	Boca Ciega Bay, Florida, U.S.A.	—	—	O$_2$/bell jar[c]	1·2 ± 0·2	Pomeroy (1960)

[a] Data originally expressed in gO$_2$ has been converted to gC using gC = 0·77 × gO$_2$ (factor derived from formulae given by Richards (1965).
[b] Calculated from dry weight by using gC = 0·34 × dry weight (Mee, unpublished).
[c] Estimated from original 1 hour data by multiplication by 12.

TABLE 42.5

Comparative data on the primary productivity of communities associated with coastal lagoons

Community and latitude	Description	Productivity (gC m^{-2} day^{-1})		Source
		Net	Gross	
Mangroves	Various mean values reported for mixed mangrove communities in Florida, U.S.A.	4·8–7·5	10·3–13·9	Carter (1973) (cited in Lugo and Snedaker, 1974)
26°–28°N.	Red mangrove forests Florida, U.S.A.	0–4·4	1·4–6·3	Lugo and Snedaker (1974)
Salt marshes	*Spartina* spp.[b]			
40°N.	New Jersey, U.S.A.	0·36[a]	—	Good (1965)
38°N.	Delaware, U.S.A.	0·40[a]	—	Morgan (1971)
35°N.	North Carolina, U.S.A.	0·70[a]	—	Williams (1965)
31°N.	Georgia, U.S.A.	1·80[a]	—	Teal (1962)
35°N.	*Juncus roemarianus* North Carolina, U.S.A.	1·10[a]	—	Foster (1968) (cited in Marshall, 1974)

[a] Calculated from dry weight by using gC = 0·40 × dry weight (factor for pure carbohydrate).
[b] Estimated from original 1 year data divided by 365 to give daily average.

many cases exceeds, that of an entire open-ocean water column. The productivity of lagoons is highly variable, and insufficient data exist to determine whether there is any relationship between primary productivity and latitude. However, it is worth noting that the lowest recorded lagoonal productivity is that of the temperate Oyster Pond while it was covered with ice (Emery, 1969). In contrast, the highest productivities are observed in relatively isolated tropical and sub-tropical lagoons such as the Laguna Madre in Texas and the Laguna Mitla in Mexico. The high productivity of those lagoons which are not subject to rapid flushing by the sea may result from a combination of a low nutrient loss to the sea and an efficient recycling of the nutrients. However, it is difficult to draw valid conclusions for lagoonal environments in general until more comparative data are available.

Specific communities within lagoons (e.g. marine grasses, algal mats and benthic microflora) may be responsible for a relatively high primary production, and in some areas may, in fact, be dominant. The relative importance of the productivity of phytoplankton, benthic microflora and marine grasses are illustrated by the study carried out by Pomeroy (1960), who made bell jar and light and dark bottle oxygen productivity measurements in water of different depths in Boca Ciega Bay, an open lagoon in Florida, U.S.A. (see Fig. 42.13). He found that in water less than 1 m deep, benthic microflora and the marine grass, *Thalassia testudinum*, dominate the primary production, whereas in deeper water areas phytoplankton communities predominate. Maximum productivity was observed in water 5 m deep, a depth at which only phytoplankton productivity was observed.

The chemistry of each of the benthic plant communities in the lagoonal environment is somewhat analagous to the chemistry of the lagoon as a whole. Each community exists by recycling a large proportion of the nutrients it requires for primary production. The input of nutrients which it receives must therefore balance the net loss of plant detritus to the sediments or to the rest of the lagoon in order to maintain a steady plant biomass. A small nutrient supply and a large plant biomass therefore indicate highly efficient recycling within the community. Similarly, the contribution of nutrients from the community to the lagoon can never exceed the original input. Marine grasses provide an example of such a system. Communities of marine grasses are found in shallow, clear water lagoons ranging from those of tropical latitudes (Odum, 1974a) to those of almost tundral environments (McRoy and Allen, 1974). The grass beds appear to be initially fertilized by both dissolved and suspended nutrients transported by the currents flowing through them. The deposition of particulate material may take place when the currents are slowed down by the grass beds. Furthermore, the rate at which water currents transport dissolved substances to the growing plants may strongly influence the rate of organic production (Conover, 1968).

In some areas, large colonies of epiphytes may cover the leaves of the plants and account for up to 25–33% of the total metabolism of the community (Jones 1968). A recent study by Capone and Taylor (1977) has shown that of such epiphytes, the epiphytic heterocystous blue-green algae are capable of fixing up to 38% of the nitrogen required daily for leaf production by *Thalassia testudinum*, a common marine grass. This nitrogen fixation thus represents an important nutrient source for the community.

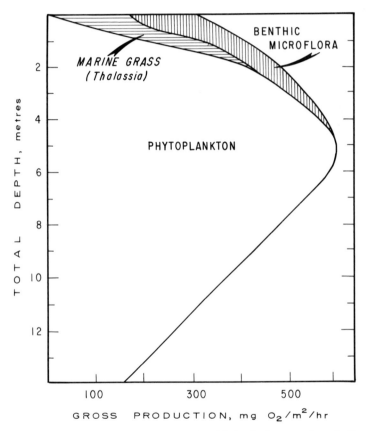

FIG. 42.13. Variation of gross primary productivity with depth of water column in Boca Ciega Bay, an open coastal lagoon in Florida, U.S.A. (after Pomeroy, 1960).

Marine grasses produce large quantities of detrital material. Few secondary organisms graze these plants, and as much as 2% of the plant leaves may fall each day (Wood *et al.*, 1969), a proportion of which are lost from the community to other areas of the lagoon. Regeneration of nutrients from dead plant material retained within the community probably provides the principal source of nutrients for its continued growth. Such regeneration is rapid, a typical half-time for plant decomposition being about one month (*T. testudinum*) (Wood et al., 1969). Indeed, the oxygen demand of grass bed sediments may be sufficient to promote the development of sulphate-reducing bacteria (Wood *et al.*, 1969). The marine grass community thus provides a good example of nutrient recycling. Such recycling has the effect of producing large quantities of organic carbon from a relatively poor nutrient supply. The efficiency of recycling might thus be assessed by comparing the net external nutrient supply to the community with the carbon productivity. Unfortunately, no such assessment has yet been made.

If conditions in a shallow coastal lagoon become too extreme (high salinities, intense sunlight, large diurnal temperature ranges) for the growth of large planktonic or sea-grass communities, mats of blue-green algae often develop, these organisms constituting a highly efficient and almost completely closed ecosystem. Birke (1974) has described three major zones within the upper few centimetres of such a mat. The first is the layer responsible for photosynthesis and consists of closely packed algal filaments. The second is a largely heterotrophic yellowish zone, and the third is a highly anaerobic zone, rich in organic matter and hydrogen sulphide, gradually merging downwards into the sediment. The upper (photosynthetic) zone can be further sub-divided into three layers. The uppermost of these layers is dark brown in colour and acts as a light and temperature shield; although it is only 1–2 mm thick, it is capable of reducing the intensity of the incident light by 95%. Photosynthesis commences in the second layer, and reaches a maximum in the third layer. It is not known by what mechanism external nutrients are supplied to the algal mats, nor whether nitrogen fixation occurs in them. However, the mats are generally distributed in hypersaline lagoons and areas which do not receive a large input of nitrogen nutrients from the land. Their recycling efficiency must therefore be very high in order to maintain a high productivity, and detrital nutrient losses from the system must consequently be small.

In his review of the chemistry of algal mat systems, Birke (1974) has described how a diurnally variable redox potential of up to 0.5 V may develop between the top and the bottom of the mat. The strongly reducing environment at the bottom of the mat probably causes rapid regeneration of nutrients which migrate upwards to fertilize the photosynthetic layers. The sulphur cycle of the mat also shows an intense redox cycle. A quantitative study of this cycle in the Solar Lake, Sinai, has been made by Jørgensen and Cohen (1977) using an *in situ* $^{35}SO_4^{2-}$ tracer technique (Fig. 42.14). This showed that the algal mat was capable of reducing up to 6.7 µmol SO_4^{2-} cm^{-3} day^{-1}. The presence in the reducing layer of grains of aragonite, which are probably produced during the decomposition of the mat by ammonifying bacteria, provides still further evidence of the active recycling of nutrients which maintains the high organic production of the mat system (Dalrymple, 1965).

$$NH_3 + HCO_3^- \longrightarrow CO_3^{2-} + NH_4^+$$
$$CO_3^{2-} + Ca^{2+} \longrightarrow CaCO_3.$$

The system is astonishingly compact, concentrating into a thickness of few

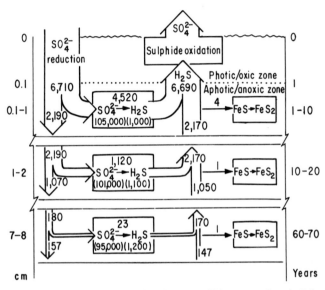

FIG. 42.14. Diagram showing the sulphur flux in a mat of blue-green algae in Solar Lake, Sinai (after Jørgensen and Cohen, 1977). Numbers in vertical arrows illustrate rates of transfer in nmol cm^{-2} day^{-1}. Numbers on horizontal arrows and in boxes illustrate the rates of transfer in nmol cm^{-3} day^{-1}. The numbers in parentheses indicate reservoir sizes in nmol cm^{-3}.

centimetres a chlorophyll a content comparable to that of some forests (Birke, 1974) and having a primary productivity well above the average for entire oceanic water column (Table 42.4). It has even been suggested (Birke, 1974) that the organic matter deposited from the mat may be of a type suitable for oil petrogenesis.

42.4.5.3. *Intertidal plant communities associated with the lagoonal environment*

The margins of the coastal lagoon environment are not very well defined since they are subjected to frequent changes by tidal movements, seasonal and wind-driven flooding, and rapid sedimentation. The marginal plant

communities, tidal marsh vegetation and, in low latitudes, mangroves, contribute a considerable proportion of their productivity to the lagoon as detritus.
Their chemistry is similar to that of the basic recycling system described for
other lagoonal plant communities and this provides further evidence for the
overall system of nutrient recycling within the lagoon.

Comparative data for the productivity of mangroves and salt marsh plants
are shown in Table 42.5, from which it may be seen that mangrove communities have a productivity much higher than that generally observed for
coastal lagoons. Salt marsh productivity, however, is similar to that of
lagoons and is very strongly influenced by latitude because the growing period
of salt marsh plants is much shorter at high than at low latitudes (Cooper,
1974).

Tidal marshes are typical marginal features of temperate lagoon systems.
They are usually associated with intertidal creeks and mudflats, and are thus
influenced by both continental run-off and tidal flushing. The initial source
of nutrients for marsh communities is usually particulate and dissolved
materials transported by continental run-off. Much of the organic detritus
produced by decaying marsh plants is flushed into the adjacent estuary or
lagoon by tidal action. In this way, up to 45% of the original net primary
production may be exported from the marsh (Cooper, 1974). The remaining
detritus is deposited within the marsh and, following regeneration, re-enters
the production cycle of the marsh organisms. Thus, almost all of the loss of
nitrogen from the marsh is in the form of organic nitrogen with the detritus,
and practically no inorganic nitrogen is exported (Gooch, in Daiber, 1974).
Fertilization experiments by Broome et al. (1975) have revealed that growth
of Spartina alterniflora, a common intertidal grass, is nitrogen-limited.

Salt marshes occur in temperate regions and this factor causes their
chemistry to exhibit great seasonal variability. Measurements made by
Aurand (in Daiber, 1974) on the water overlying a Delaware marsh in the
U.S.A. showed that the nitrate concentration varies between almost zero in
the summer growing season and ~ 30 μg-at.l^{-1} in the winter (when primary
production is very limited). Phosphate concentrations in the interstitial
water in the marsh show the opposite effect, with higher concentrations in
summer than winter (Gooch, in Daiber, 1974). As a possible explanation for
this paradox, Gooch has suggested that much of the phosphorus may be
bound seasonally as iron phosphate in the sediments. The high productivity
and respiration of marsh grasses in summer results in the presence of anaerobic, sulphate-reducing bacteria and consequently, a highly reducing environment. In such an environment, iron is reduced forming soluble $Fe(II)$ (and
insoluble FeS), thus releasing phosphate to the water. When biological
activity diminishes in winter, the surface sediment layer becomes oxic and
$Fe(II)$ is oxidized to $Fe(III)$. In those locations in which the dissolved con-

centration of iron is higher than that of phosphate, a mixed ferric hydroxo-phosphate may be precipitated (Stumm and Morgan, 1970), and this process may account for the lower dissolved phosphate concentration found during the winter.

At lower latitudes, seasonal changes become decreasingly significant and mangrove forests may develop. These tolerate a wide range of salinity conditions and grow in shallow water along most of the coastal margins and the banks of estuaries and lagoons between latitudes 25°S. and 25°N. They are amongst the most productive of all brackish water coastal ecosystems (Kuenzler, 1974; Table 42.5). Lugo and Snedaker (1974) in their excellent review of mangrove ecology have described the energy and nutrient cycles of these systems. As with all the plant communities previously described, the high productivity of mangroves is maintained by nutrient recycling; the net exportation of detrital nutrients from the community is balanced by inputs from upland areas transported by continuous or occasional run-off or by tidal action. A considerable part of the debris (leaves and twigs) falling from the mangrove trees is exported from the community to adjacent lagoonal or estuarine waters. Thus, Heald (1971) has estimated that up to half of the debris produced annually in some Florida mangrove forests is exported in this manner; this proportion represents a daily average of 1·2 g dry weight of debris per square metre of mangrove communities. A large mangrove forest fringing a coastal lagoon may therefore be a tremendous source of detritus to the lagoonal sediments.

Each of the communities discussed above converts inorganic nutrients to organic detrital material which is subsequently exported to the lagoon. This export is important beause it often includes relatively large detrital material, such as leaves and twigs, which may require periods of up to weeks or months for complete bacterial decomposition. This gradual and continuous regeneration may provide a well-distributed continuous nutrient supply which sustains primary production within the lagoon, thus masking the effects of irregular run-off and flushing, as well as stabilizing the system as a whole.

42.4.6. OVERALL CHEMICAL CYCLES

In order to study the chemical fluxes in lagoons, several parameters must be identified. These include the elemental reservoirs and the relative importance of the various transport paths between them. Unfortunately, sufficient information of this type is not yet available to permit the construction of even the outlines of the cycles of the common elements in most lagoons. The nitrogen cycle, which limits production, provides an example of this because the occurrence and importance of nitrogen fixation have yet to be fully investigated. Furthermore, even the extent to which such basic processes as nitrification and denitrification take place is still not known. However, for phosphorus it is possible to show the lagoonal cycle in at least a schematic form (Fig. 42.15b). This model is based on that presented by Riley and Chester (1971) for the open ocean (Fig. 42.15a) and has been modified to take into account the various processes described above. The principal modifications are necessitated by the need to allow for both the presence of benthic plants and animals, and the interactions between the sediments and the water column. The latter include the interchange of phosphorus between the suspended particulate and sediment reservoirs (by resuspension and settling), and the adsorption and desorption of inorganic phosphate to, and from, the sediments. A further important difference between the open ocean and lagoonal cycles arises from the large input of dissolved and particulate organic phosphorus to lagoons which results either from the conversion of inorganic phosphate to organic, and subsequently detrital, phosphorus by plant communities in, for example, mangroves or salt marshes, or from agricultural and domestic wastes.

The principal difference between the phosphorus cycles of lagoons and the open ocean is the greater number of pathways for the transport of this element in the former environment. For example, in coastal lagoons the sedimentary phosphorus reservoir may be mobilized by mixing, chemical desorption, bacterial action, benthic plants and by the feeding of benthic animals and nekton. Although this suggests that there is likely to be a greater nutrient flux in the lagoonal environment than in the open ocean, this cannot be verified until more data are available on reservoir sizes and on the transfer between individual reservoirs. The acquisition of such data is one of the principal challenges confronting the marine chemist investigating chemical cycles in lagoons. Several questions must be answered before this challenge can be met, and some of these can be illustrated by considering the limiting nutrient, inorganic nitrogen. In any lagoon in which the nutrient supply is dominated by nutrient recycling, nitrogen exists in three major reservoirs (Fig. 42.16). These are the dissolved reservoir, the "living tissue" reservoir (generally dominated by primary producers such as phytoplankton) and the detrital reservoir, which is located principally at the sediment surface. In practice, it is at present only possible to make direct measurements of the dissolved nitrogen reservoir because detritus and living tissue become mixed by effects such as sediment resuspension and benthic production, thereby making distinction between them impossible. However, dissolved inorganic nitrogen measurements usually prove futile as they reveal only insignificant concentrations of the element, even though the organic production of a lagoon is known to be high. The small size of the dissolved inorganic nitrogen reservoir is indicative of a process which is analogous to an organic chemical reaction having several intermediate steps, for example:

$$A \xrightarrow{k_1} B \xrightarrow{k_2} C \xrightarrow{k_3} E \ldots.$$

If the rates of reaction k_1 or k_3 are much slower than k_2, the overall reaction rate will be controlled by k_1 and k_3 and the intermediate B will be present in comparatively small quantities. In the present case, A, B and C may be considered to be the detrital, dissolved and living tissue reservoirs respectively, and the rate of nitrogen flux in the cycle is controlled by the rate of nutrient regeneration and/or the rate of conversion of living tissue to detritus. This cycle is obviously grossly oversimplified and ignores many intermediate stages, such as bacterial conversions, but nevertheless it indicates that the only way to obtain useful information about the system is to measure rates of transfer between reservoirs. Measurements of this type are extremely difficult to make, requiring either the use of isotopic tracers or the measurement of the way in which the rate of accumulation in one reservoir changes in response to the alteration of one of the intermediate stages, for example by stopping primary production and measuring variations in the dissolved nutrient reservoir. Neither of these sophisticated techniques have yet been applied to coastal lagoons; however, they may prove to be the only practical means of obtaining sufficient data to successfully model chemical processes in them. This is of practical importance as such models are essential for the responsible management of lagoonal resources.

42.5. HUMAN INFLUENCES ON LAGOONS

Coastal lagoons have been estimated to occupy as much as 13 % of the world's coastlines (Lankford, 1976). They are extensively used as harbours (e.g. Alexandria, Egypt; Lagos, Nigeria; Miami, U.S.A.; Venice, Italy), as recreational areas, for their mineral resources (salt, sand, gravel, fossil hydrocarbons) and, most importantly, for their fisheries. Each of these human demands on the lagoonal system presents additional stresses which may affect its chemistry and biology.

Two general types of stress are imposed on the chemistry of lagoons by human activities. The first arises from physical changes in the lagoon and its surrounding area and the second arises from chemical contamination. Examples of some of these stresses and their consequences are shown in diagrammatical form in Fig. 42.17.

The majority of physical stresses on the lagoonal environment arise from the failure of planners to realize how sensitive such environments are. For example, the desirability of lagoon-side residences provokes private planners to construct "dredge-and-fill" areas where sediments dredged from the lagoon

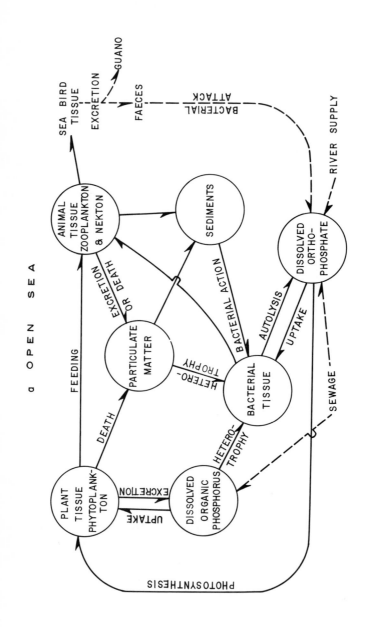

a OPEN SEA

b SHALLOW COASTAL LAGOON

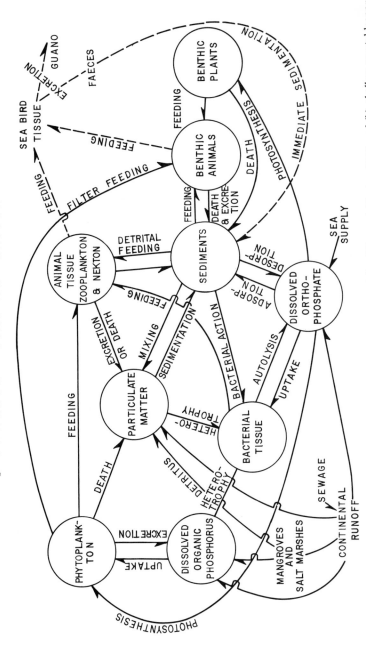

FIG. 42.15. The principal features of the phosphorus cycles in (a), the open ocean (after Riley and Chester, 1971) and (b), shallow coastal lagoons. Reservoirs within the environment are encircled and relatively minor processes are shown with broken lines. For simplicity, mangrove and salt marsh communities are shown as being external features and processes leading to net loss of phosphorus are not shown.

area are used to construct land areas within it. This has the effect of reducing the size of the lagoon, thus inhibiting circulation and lowering productivity (Smith, 1966). Similarly, Smith has noted that the dredging of navigation channels may produce large spoil heaps which effectively reduce the area of the lagoon and may cover areas of highly productive marine grasses, adversely affecting primary production (Odum, 1963). Changes in the physical circulation may also be caused by inadequately wide bridges, by the filling-in of flood channels and by the diversion of river inputs.

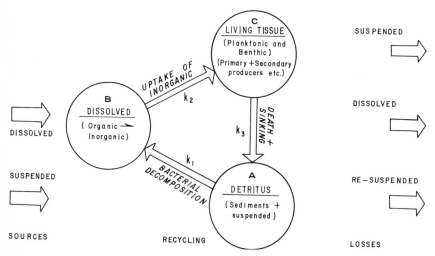

FIG. 42.16. The three basic reservoirs of nitrogen in the lagoon environment. The diagram is highly simplified and, for the purposes of flux calculations, it is probably more realistic to separate the dissolved organic and inorganic reservoirs.

From the chemical point of view, perhaps the most important physical changes imposed on lagoons are changes in land use in the area. In the author's personal experience important mangrove margins may be cut for no other reason than to improve the appearance of shoreside property. In Delaware, U.S.A., 45 000 acres of saltmarsh were lost between 1954 and 1964 to development that was 99% non-agricultural (factories, roads, spoil areas, etc.); this represents a tremendous loss of detritus-producing land (Schmidt, 1966). The area of human influence may extend well beyond the immediate area of the lagoon. For example, upland drainage, and a consequent loss of nutrients, may cause the death of mangrove communities within a period of about ten years (Lugo and Snedaker, 1974). The planning of coastal lagoon management must therefore take into account projected changes over a considerable land area.

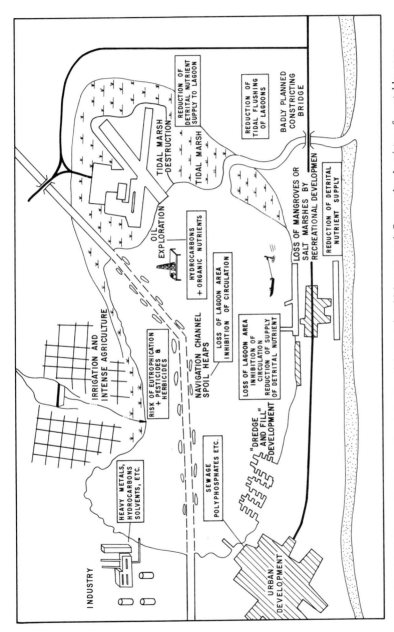

INDUSTRY

HEAVY METALS,
HYDROCARBONS
SOLVENTS, ETC.

IRRIGATION AND
INTENSE AGRICULTURE

RISK OF EUTROPHICATION
+ PESTICIDES &
HERBICIDES

NAVIGATION CHANNEL
SPOIL HEAPS

LOSS OF LAGOON AREA
INHIBITION OF CIRCULATION

SEWAGE
POLYPHOSPHATES ETC.

OIL
EXPLORATION

HYDROCARBONS
+ ORGANIC NUTRIENTS

LOSS OF LAGOON AREA
INHIBITION OF
CIRCULATION
REDUCTION OF SUPPLY
OF DETRITAL NUTRIENT

"DREDGE
AND FILL"
DEVELOPMENT

URBAN
DEVELOPMENT

TIDAL MARSH
DESTRUCTION

TIDAL MARSH

REDUCTION OF
DETRITAL NUTRIENT
SUPPLY TO LAGOON

REDUCTION OF
TIDAL FLUSHING
OF LAGOONS

BADLY PLANNED
CONSTRICTING
BRIDGE

LOSS OF MANGROVES OR
SALT MARSHES BY
RECREATIONAL DEVELOPMEN

REDUCTION OF DETRITAL
NUTRIENT SUPPLY

FIG. 42.17. Composite diagram showing some of the ways in which man can influence the chemistry of coastal lagoons.

Lagoons are also very sensitive to chemical contamination, partly because, as a result of chemical cycling, residence times of substances within them are long, and this leads to the long-term retention of contaminants. This will be particularly acute for contamination by heavy metals, which are often present in industrial wastes, and also for fossil hydrocarbons. A recent study of sediments, molluscs and sea-grasses from several coastal lagoons has revealed that the highest concentrations of fossil hydrocarbons occur near petrochemical complexes and oil refineries (Botello, 1978), thus suggesting that these hydrocarbons are being released to the coastal lagoon environment. Insecticides, herbicides and PCBs may also present a considerable problem in some lagoonal environments, although their presence in them, has not been extensively investigated.

Perhaps the most important immediate chemical problem in coastal lagoons arises from eutrophication induced by nutrient contamination. The source of contaminant nutrients may be agricultural, domestic and/or industrial waste-waters (Copeland and Wohlschlag, 1968). The results of such contamination, which is clearly demonstrated by the discharge of the Houston Ship Canal into the open lagoons of Texas, U.S.A., (Odum *et al.,* 1963), are a lowering of the dissolved oxygen concentration in the water and even the production of toxic hydrogen sulphide.

These various examples of the effect of human activities upon lagoonal environments demonstrate the need for careful management in order to optimize the possible uses of such areas. For example, the cost of a sewage treatment plant in Biscayne Bay, Florida, U.S.A., was far outweighed by the recreational benefits derived from the resultant clean water environment (Sampedro, 1972). Unfortunately, the resolution of such management problems is rarely so simple. For example, the geological ideal of reducing sedimentation in lagoons may be opposed to that of the fish-farmer who wants to optimize the transport of detritus into them. A full multidisciplinary investigation of such environments is the only way in which to achieve constructive planning. Perhaps the only solution to the problem of lagoonal management lies in legislation which defines and separates the various areas of human activity in each lagoon (Odum, 1974b). This type of legislation can only be effective if there is a better understanding of how lagoons function.

42.6. Conclusions

Coastal lagoons are areas of transition between the oceanic and terrestrial environments. They are thus strongly influenced by continental processes such as run-off and the effects of human activities. This influence, combined with the rapid recycling of nutrients observed in such shallow water regions, results in a high production of organic matter.

The study of the coastal lagoonal environment, important for optimum management, remains at an early stage of development. Almost no investigations have been made of the geochemistry of coastal lagoons, and the few chemical studies which have been published are generally related to nutrients and primary production. For the marine chemist, the problems of studying rapid nutrient and mineral cycling presents one of the most interesting challenges for the future.

ACKNOWLEDGEMENT

The author wishes to thank Dr Enrique F. Mandelli for critical reading of the manuscript.

REFERENCES

Bader, R. G. and Roessler, M. A., sr. rptrs. (1972). "Progress Report to U.S. Atomic Energy Commission", (AT (40-1)-3801-4). University of Miami, U.S.A.
Barlow, J. P., Lorenzen, C. J. and Myren, R. T. (1963). *Limnol. Oceanogr.* **8**, 251.
Berner, R. A. (1977). *Limnol. Oceanogr.* **22**, 781.
Birke, L. (1974). *In* "Coastal Ecological Systems of the United States" (H. T. Odum, B. J. Copeland and E. A. McMahan, eds), pp. 331–345. Conservation Foundation, Washington D.C.
Borchert, H. (1965). *In* "Chemical Oceanography" (J. P. Riley and G. Skirrow, eds), Vol. 2, pp. 205–276. Academic Press, London and New York.
Botello, A. V. (1978). *Anal. Cent. Cienc. Mar Limnol.* (in press).
Bouma, A. H. and Bryant, W. R. (1969). *In* "Lagunas Costeras, un Simposio". Mem. Simp. Intern. Lagunas Costeras. UNAM–UNESCO, Mexico D.F.
Broome, S. W., Woodhouse, W. W., Jr. and Seneca E. D. (1975). *Proc. Soil Sci. Soc. Am.* **39**, 301.
Bruce, H. E. and Hood, D. W. (1959). *Univ. Tex. Inst. Mar. Sci. Publs*, **6**, 133.
Butler, P. A. (1966). *Am. Fish. Soc. Spec. Publ.* **3**, 110.
Capone, D. G. and Taylor, B. F. (1977). *Mar. Biol.* **40**, 19.
Carrada, G. C. and Rigillo Troncone, M. (1975). *Rapp. P.-v. Réun. Commn Int. Explor. Scient. Mer Méditerr.* **23**, 81.
Cohen, Y., Krumbein, W. E., Goldberg, M. and Shilo, M. (1976a). *Limnol. Oceanogr.* **22**, 597.
Cohen, Y., Krumbein, W. E. and Shilo, M. (1976b). *Limnol. Oceanogr.* **22**, 609.
Collier, A. and Hedgpeth, T. W. (1950). *Univ. Tex. Inst. Mar. Sci. Publs*, **1**, 121.
Conover, J. T. (1968). *Botanica Marina*, **11**, 1.
Cooper, A. W. (1974). *In* "Coastal Ecological Systems of the United States" (H. T. Odum, B. J. Copeland and E. A. McMahan, eds). Conservation Foundation, Washington D.C.

Copeland, B. J. (1967). *Univ. Tex. Contr. Mar. Sci.* **12**, 207.

Copeland, B. J. and Nixon, S. W. (1974). *In* "Coastal Ecological Systems of the United States" (H. T. Odum, B. J. Copeland and E. A. McMahan, eds), pp. 312–329. Conservation Foundation, Washington D.C.

Copeland, B. J. and Wohlschlag, D. E. (1968). *In* "Advances in Water Quality Improvement", (E. F. Gloyna and W. W. Eckenfelder, eds). Water Resources Symp. 1. University of Texas Press, Austin, Texas.

Curray, J. R., Emnel, F. J. and Crompton, P. J. S. (1969). *In* "Lagunas Costeras, un Simposio". Mem. Simp. Intern. Lagunas Costeras, pp. 66–100. UNAM–UNESCO, Mexico, D.F.

Cushing, D. H. (1975). "Marine Ecology and Fisheries". Cambridge University Press, England, 278 pp.

Daiber, F. C. (1974). *In* "Coastal Ecological Systems of the United States" (H. T. Odum, B. J. Copeland and E. A. McMahan, eds), pp. 99–149. Conservation Foundation, Washington D.C.

Dalrymple, D. W. (1965). *Univ. Tex. Inst. Mar. Sci. Publs*, **10**, 187.

Emery, K. O. (1969). "A Coastal Pond Studied by Oceanographic Methods". Elsevier, New York, 77 pp.

Emery, K. O. and Stevenson, R. E. (1958a). *In* "Treatise on Marine Ecology and Paleoecology" (J. W. Hedgpeth, ed.), Vol. 1, pp. 673–693. *Geol. Soc. Amer. Mem.* **67**.

Emery, K. O. and Stevenson, R. E. (1958b). *In* "Treatise on Marine Ecology and Paleoecology" (J. W. Hedgpeth, ed.), Vol. 1, pp. 729–734. *Geol. Soc. Amer. Mem.* **67**.

Evans, G. and Bush, P. (1969). *In* "Lagunas Costeras, un Simposio". Mem. Simp. Intern. Lagunas Costeras, pp. 155–170. UNAM–UNESCO, Mexico, D. F.

Faas, R. W. (1974). *In* "Coastal Ecological Systems of the United States" (H. T. Odum, B. J. Copeland and E. A. McMahan, eds). Conservation Foundation, Washington D.C.

Glooschenko, W. A. and Harriss, R. C. (1974). *In* "Coastal Ecological Systems of the United States" (H. T. Odum, B. J. Copeland and E. A. McMahan, eds), pp. 488–498. Conservation Foundation, Washington D.C.

Good, R. E. (1965). *New Jers. Acad. Sci. Bull.* **10**, 1.

Groen, P. (1969). *In* "Lagunas Costeras, un Simposio". Mem. Simp. Intern. Lagunas Costeras, pp. 275–280. UNAM–UNESCO, Mexico D.F.

Grundmanis, V. and Murray, J. W. (1977). *Limnol. Oceanogr.* **22**, 804.

Hale, S. S. (1975). *In* "Mineral Cycling in Southeastern Ecosystems" (F. G. Howell, J. B. Gentry and M. H. Smith, eds), pp. 291–308. ERDA Symp. Series, 1975, (CONF-740513).

Hayes, F. R. (1964). *Oceanogr. Mar. Biol. Ann. Rev.* **2**, 121.

Heald, E. J. (1971). *Sea Grant Tech. Bull.* **6**. University of Miami, U.S.A., 110 pp.

Howmiller, R. P. and Weiner, A. (1968). *Ecology*, **49**, 1184.

Instituto de Geofísica (1976). *An. Inst. Geofís. Univ. Mex.* **21**. Parte B, Apendice 1.

Jones, J. A. (1968). "Primary Productivity by the Tropical Turtle Grass *Thalassia testudinum* Konig and its Epiphytes". Ph.D. Thesis, University of Miami, U.S.A., 196 pp.

Jørgenson, B. B. and Cohen, Y. (1977). *Limnol. Oceanogr.* **22**, 644.

Kuenzler, E. J. (1961). *Limnol. Oceanogr.* **6**, 400.

Kuenzler, E. J. (1974). *In* "Coastal Ecological Systems of the United States" (H. T. Odum, B. J. Copeland and E. A. McMahan, eds), pp. 346–371. Conservation Foundation, Washington D.C.

Lankford, R. R. (1976). Paper presented in Gen. Symp. G4, Joint Oceanographic Assembly, Edinburgh, Sept. 1976.

Lankford, R. R. (1977). *In* "Estuarine Processes" (R. Wiley, ed.), Vol. 2, p. 182. Academic Press, New York and London.

Lawson, G. W. (1966). *Oceanogr. Mar. Biol. Ann. Rev.* **4**, 405.

Lugo, A. E. and Snedaker, S. C. (1974). *Ann. Rev. Ecol. Syst.* **5**, 39.

Marshall, H. L. (1974). *In* "Coastal Ecological Systems of the United States" (H. T. Odum, B. J. Copeland and E. A. McMahan, eds), pp. 150–170. Conservation Foundation, Washington D.C.

McRoy, C. P. and Allen, M. B. (1974). *In* "Coastal Ecological Systems of the United States" (H. T. Odum, B. J. Copeland and E. A. McMahan, eds), pp. 17–36. Conservation Foundation, Washington D.C.

McIntire, W. G. and Ho, C. (1969). *In* "Lagunas Costeras, un Simposio". Mem. Simp. Intern. Lagunas Costeras, pp. 49–62. UNAM–UNESCO. Mexico D.F.

Mee, L. D. (1977). "The Chemistry and Hydrography of Some Coastal Lagoons, Pacific Coast of Mexico". Ph.D. Thesis, University of Liverpool, 117 pp.

Morgan, M. H. (1971). M.Sc. Thesis, University of Delaware, U.S.A., 34 pp.

Nichols, M. M. (1966). *Univ. Tex. Inst. Mar. Sci. Publs,* **11**, 159.

Ocampo, R. E. and Emilsson, I. (1975). *An. Inst. Geofís. Univ. Mex.* **20**, 21.

Odum, H. T. (1963). *Univ. Tex. Inst. Mar. Sci. Publs,* **9**, 48.

Odum, H. T. (1974a). *In* "Coastal Ecological Systems of the United States" (H. T. Odum, B. J. Copeland and E. A. McMahan, eds), pp. 442–487. Conservation Foundation, Washington D.C.

Odum, H. T. (1974b). *In* "Coastal Ecological Systems of the United States" (H. T. Odum, B. J. Copeland and E. A. McMahan, eds), pp. 141–151. Conservation Foundation, Washington D.C.

Odum, H. T. and Hoskin, C. M. (1958). *Univ. Tex. Inst. Mar. Sci. Publs,* **5**, 16.

Odum, H. T. and Wilson, R. F. (1962). *Univ. Tex. Inst. Mar. Sci. Publs,* **8**, 23.

Odum, H. T., Cuzon du Rest, R. P., Beyers, R. J. and Allbaugh, C. (1963). *Univ. Tex. Inst. Mar. Sci. Publs,* **9**, 404.

Okuda, T. (1960). *Trab. Inst. Oceanogr. Univ. Recife, Brazil,* **2**, 7.

Okuda, T. (1969). *In* "Lagunas Costeras, un Simposio". Mem. Simp. Intern. Lagunas Costeras, pp. 291–300. UNAM–UNESCO, Mexico D.F.

Oppenheimer, C. H. (1963). *Bull. Mar. Sci. Gulf Caribb.* **13**, 59.

Oppenheimer, C. H. and Ward, R. A. (1963). *In* "Symposium on Marine Microbiology" (C. H. Oppenheimer and C. C. Thomas, eds), pp. 664–673. Springfield, Illinois.

Parker, P. L., Gibbs, A. and Lawler, R. A. (1963). *Univ. Tex. Inst. Mar. Sci. Publs,* **9**, 28.

Phleger, F. B. (1969). *In* "Lagunas Costeras, un Simposio". Mem. Simp. Intern. Lagunas Costeras, pp. 5–26. UNAM–UNESCO, Mexico D.F.

Phleger, F. B. and Ewing, G. C. (1962). *Bull. Geol. Soc. Am.* **73**, 145.

Pomeroy, L. R. (1960). *Bull. Mar. Sci. Gulf Caribb.* **10**, 1.

Pomeroy, L. R., Smith, E. E. and Grant, C. M. (1965). *Limnol. Oceanogr.* **10**, 167.

Por, F. D. (1971). *Mar. Biol.* **14**, 111.

Postma, H. (1965). *Neth. J. Sea Res.* **2**, 566.

Postma, H. (1969). *In* "Lagunas Costeras, un Simposio". Mem. Simp. Intern. Lagunas Costeras, pp. 421–430. UNAM–UNESCO, Mexico, D.F.

Pritchard, D. W. (1967). *In* "Estuaries" (G. H. Lauff, ed.), Amer. Assoc. Adv. Sci., Washington, Pub. No. **83**, pp. 3–5.

Reimold, R. J. and Daiber, F. C. (1967). *Chesapeake Sci.* **8**, 132.

Richards, F. A. (1965). *In* "Chemical Oceanography" (J. P. Riley and G. Skirrow, eds), Vol. 1, pp. 611–644. Academic Press, London and New York.

Riley, J. P. and Chester, R. (1971). "Introduction to Marine Chemistry". Academic Press, London and New York, 465 pp.

Ryther, J. H. and Dunstan, W. M. (1971). *Science, N.Y.* **171**, 3975.

Sampedro, R. M. (1972). *Sea Grant Tech. Bull.* **24**. University of Miami, U.S.A.

Schmidt, R. A. (1966). *Am. Fish. Soc. Spec. Publ.* **3**, 102.

Shepard, F. P. and Moore, D. G. (1960). *In* "Recent Sediments, Northwest Gulf of Mexico" pp. 117–152. Am. Ass. Pet. Geol., Tulsa, U.S.A.

Smith, S. H. (1966). *Am. Fish. Soc. Spec. Publ.* **3**, 93.

Steemann-Nielsen, E. (1952). *J. Cons. Perm. Int. Explor. Mer.* **18**, 117.

Stumm, W. and Morgan, J. J. (1970). "Aquatic Chemistry". Wiley–Interscience, New York, 583 pp.

Tampi, P. P. S. (1969). *In* "Lagunas Costeras, un Simposio". Mem. Simp. Intern. Lagunas Costeras, pp. 479–484. UNAM–UNESCO, Mexico D.F.

Teal, J. M. (1962). *Ecology,* **43**, 614.

Tundisi, J., Tundisi, T. M. and Kutner, M. B. (1973). *Int. Rev. Ges. Hydrobiol.* **58**, 925.

Urien, C. M. (1977). Paper delivered at "Simposio Latinoamericano sobre Lagunas Costeras", OAS–UNAM, Mexico D.F., Nov. 1977.

Vanderborght, J. P., Wollast, R. and Billen, G. (1977). *Limnol. Oceanogr.* **22**, 794.

Vatova, A. (1961). *J. Cons. Perm. Int. Explor. Mer,* **27**, 148.

Vatova, A. (1963). *Rapp. P.-v. Réun. Commn Int. Explor. Scient. Mer Méditerr.* **17**, 753.

Walton, W. R. and Smith, W. T. (1969). *In* "Lagunas Costeras, un Simposio". Mem. Symp. Intern. Lagunas Costeras, pp. 237–248. UNAM–UNESCO, Mexico D.F.

Warme, J. E. (1969). *In* "Lagunas Costeras, un Simposio". Mem. Simp. Intern. Lagunas Costeras, pp. 137–154. UNAM–UNESCO, Mexico D.F.

Williams, R. B. (1965). Paper delivered at AERS Meeting, Hampton, Virginia, Nov. 12th, 1965.

Wood, E. J. F., Odum, W. E. and Zieman, J. C. (1969). *In* "Lagunas Costeras, un Simposio". Mem. Simp. Intern. Lagunas Costeras, pp. 495–502. UNAM–UNESCO, Mexico D.F.

Zenkovitch, V. P. (1969). *In* "Lagunas Costeras, un Simposio". Mem. Simp. Intern. Lagunas Costeras, pp. 27–38. UNAM–UNESCO, Mexico D.F.

Subject Index

A

Abietane, 260, 262
Abyssal clays (see Clays)
Abyssal plains, 3, 57
Acantharia, 156
Acartia clausii, 159
Accretion, 49, 55
Acetobacter xylinium, 257
β-N-Acetyl-D-glucosamine, 271
Acidic rocks, 34
Acoustic profiles, 108
Actinium-227, 165
Active ridges (see Ridge, active)
Active zones (seismic), 17
Activity coefficients, 401
Activity ratios, 316 ff
Aden, Gulf of, 11, 13, 244, 288, 289
Adenine, 267, 268
Adiantone, 259, 260
Adsorption of trace metals, 406 ff
Aeolian dust, 190
Aeolian transport (see Wind transport)
Aerial photography, 85
Aerosols, atmospheric (see Atmospheric aerosols)
Aerosol chemistry of marine atmosphere, 173–231
Aerosol generation chamber, 208
Africa, 24
  continental margins of, 22
  drift of, 21, 22
  separation of, 26
  South, 24
  West, 24
African plate, 19, 20, 26, 65
Air–sea transfer, 177, 182, 183, 195
Alanine, 267
Alaska, Gulf of, 27

Alcohols, in aerosols, 195
  branched, in sediments, 254
  in leaf waxes, 286
  origin of, 274, 278
Alde Estuary, 378, 386
Adehydes, in leaf waxes, 286
Aleutian arc, 27
  Islands, 26, 27
  Trench, 45
Algae, blue-green, 245, 246, 255, 257, 267, 475, 476, 477
  brown, 246
  red, 246
Algal mat, 245, 472, 474, 476
n-Alkanes, in leaf waxes, 286
  origin of, 274, 278
  oxidation of, 254
  in sediments, 243, 246, 247, 248, 250, 253
n-Alkanoic acids, 247 ff
  origin of, 274, 278
Alkenes, 245–246
  origins of, 278
n-Alkanols (see Alcohols)
Alkylbenzenes, 245
Alkylcyclohexanes, 245
Alkylpolycyclanes, 245
Alloisoleucine, 267
Alpha-Mendeleyev Ridge, 24
Aluminium-26, 314, 315, 351, 353–356
Aluminium, behaviour in estuaries, 382–385
  coagulation of, 383, 384
  in marine aerosols, 189, 192
  organic matter, interaction with, 384
  particulate, 131, 140, 151, 153, 154, 160
  polymerization of, 383